Reservoir Characterization II

Academic Press Rapid Manuscript Reproduction

Reservoir Characterization II

Edited by

Larry W. Lake

Department of Petroleum Engineering
The University of Texas at Austin
Austin, Texas

Herbert B. Carroll, Jr.

National Institute for Petroleum and Energy Research (NIPER)
Bartlesville, Oklahoma

Thomas C. Wesson

United States Department of Energy
Bartlesville, Oklahoma

ACADEMIC PRESS, INC.
Harcourt Brace Jovanovich, Publishers
San Diego   New York   Boston
London   Sydney   Tokyo   Toronto

Academic Press, Inc.
San Diego, California 92101

United Kingdom Edition published by
Academic Press Limited
24–28 Oval Road, London NW1 7DX

Library of Congress Cataloging-in-Publication Data

Reservoir characterization II / edited by Larry W. Lake, Herbert B.
  Carroll, Jr., Thomas C. Wesson.
      p. cm.
    Proceedings of the Second International Reservoir Characterization
  Conference, held in Dallas, Tex., June 1989.
    Includes index.
    ISBN   0-12-434066-0
    1. Oil fields--Congresses.   2. Gas reservoirs--Congresses.
  3. Petroleum--Geology--Congresses.   4. Gas, Natural--Geology-
  -Congresses.   5. Rocks--Permeability--Congresses.   I. Lake, Larry
  W.   II. Carroll, Herbert B.   III. Wesson, Thomas C.
  IV. International Reservoir Characterization Conference (2nd : 1989
  : Dallas, Tex.)   V. Title: Reservoir characterization two.
  VI. Title: Reservoir characterization 2.
  TN870.5.R426   1991
  622'.3382--dc20
                                                        90-25653
                                                            CIP

PRINTED IN THE UNITED STATES OF AMERICA
91 92 93 94   9 8 7 6 5 4 3 2 1

Contents

Session 1: Microscopic

Session 2: Mesoscopic

Session 3: Macroscopic

Session 4: Workshop/Discussions

Session 5: Megascopic

Session 6: Posters

Contributors

Numbers in parentheses indicate the pages on which the authors' contributions begin.

P. M. Alder (662), *Laboratoire D'Aérothermique du C.N.R.S., 92 190 Meudon, France*

James R. Ammer (644), *United States Department of Energy, Morgantown Energy Technology Center, Morgantown, West Virginia 26507*

Helga Baardsen (102), *Production Laboratories, Statoil, N-4001 Stavanger, Norway*

Stefan Bachu (226), *Alberta Research Council, Edmonton, Alberta T6H 5X2, Canada*

Ibrahim Bahralolom (77), *New Mexico Petroleum Recovery Research Center, New Mexico Institute of Mining and Technology, Socorro, New Mexico 87801*

S. H. Begg (613), *BP Research, Sunbury Research Centre, Sunbury TW16 7LN, England*

R. A. Behrens (402), *Chevron Oil Field Research Company, La Habra, California 90631*

Tor Bjornstad (656), *Institute for Energy Technology, N-2007 Kjeller, Norway*

Allen C. Brummert (644), *United States Department of Energy, Morgantown Energy Technology Center, Morgantown, West Virginia 26507*

Herbert B. Carroll, Jr. (478), *National Institute for Petroleum and Energy Research (NIPER), Bartlesville, Oklahoma 74005*

M. M. Chang (524), *National Institute for Petroleum and Energy Research (NIPER), Bartlesville, Oklahoma 74005*

Mark A. Chapin (677), *Petroleum Engineering Department, Colorado School of Mines, Golden, Colorado 80401*

Marcel Chin-A-Lien (698), *South American Earth Sciences Department, INTEVEP, Caracas 1070A, Venezuela*

Douglas E. Craig (289), *Production Geosciences Department, Mobil Exploration and Producing U.S., Inc., Midland, Texas 79702*

Timothy A. Cross (493), *Department of Geology and Geological Engineering, Colorado School of Mines, Golden, Colorado 80401*

David L. Cuthiell (226), *Alberta Research Council, Edmonton, Alberta T6H 5X2, Canada*

Magnar Dale (652), *Rogaland Research Institute, Rogaland Regional College, Ullandhaug, 4004 Stavanger, Norway*

M. de Buyl (557), *Western Geophysical, Houston, Texas 77252*

P. M. Doyen (557), *Western Geophysical, Houston, Texas 77252*

Oyvind Dugstad (656), *Institute for Energy Technology, N-2007 Kjeller, Norway*

Kelly A. Edwards (497), *Chevron Oil Field Research Company, La Habra, California 90631*

Alan S. Emanuel (497), *Chevron Oil Field Research Company, La Habra, California 90631*

Milton B. Enderlin (277), *Halliburton Logging Services, Fort Worth, Texas 76101*

R. Eschard (686), *Institut Francais Du Petrole, 92 506 Rueil-Malmaison, France*

Graham E. Fogg (355), *Department of Land, Air, and Water Resources, University of California, Davis, California 95616*

Joseph G. Gallagher, Jr. (382), *Phillips Petroleum Company, Bartlesville, Oklahoma 74004*

W. Wendell Givens (27), *Mobil Research and Development Corporation, Dallas, Texas 75244*

J. Jaime Gómez-Hernández (251), *Departmento de Ingenieria Hidráulica y Medio Ambiente, Universidad Politécnica de Valencia, 46071 Valencia, Spain*

D. Guerillot (662), *Institut Francais Du Petrole, 92 506 Rueil-Malmaison, France*

T. M. Guidish (557), *Western Geophysical, Houston, Texas 77252*

Terence Hamilton-Smith (659), *K&A Energy Consultants, Inc., Tulsa, Oklahoma 74112*

John S. Hancock (644), *United States Department of Energy, Morgantown Energy Technology Center, Morgantown, West Virginia 26507*

Diana K. T. Hansen (277), *Halliburton Logging Services, Fort Worth, Texas 76101*

C. K. Harris (2), *Koninklijke Shell, Exploratie en Produktie Laboratorium, 2280 AB Rijswijk ZH, The Netherlands*

John Heller (77), *New Mexico Petroleum Recovery Research Center, New Mexico Institute of Mining and Technology, Socorro, New Mexico 87801*

A. Henriette (662), *Institut Francais Du Petrole, 92 506 Rueil-Malmaison, France*

T. A. Hewett (402), *Chevron Oil Field Research Company, La Habra, California 90631*

M. M. Honarpour (524), *National Institute for Petroleum and Energy Research (NIPER), Bartlesville, Oklahoma 74005*

Arvid Hove (102), *Production Laboratories, Statoil, N-4001 Stavanger, Norway*

Brian R. Hoyt (277), *Halliburton Logging Services, Fort Worth, Texas 76101*

Andrew Hurst (166), *Unocal UK, Sunbury-on-Thames, TW16 7LU, England*

S. R. Jackson (524), *National Institute for Petroleum and Energy Research (NIPER), Bartlesville, Oklahoma 74005*

Susan Jackson (689), *IIT Research Institute, National Institute for Petroleum and Energy Research (NIPER), Bartlesville, Oklahoma 74005*

Torgrim Jacobsen (315), *Department of Sedimentology and Stratigraphy, Continental Shelf and Petroleum Technology, Research Institute A/S, Jarlesletta, N-7002 Trondheim, Norway*

C. G. Jacquin (662), *Institut Francais Du Petrole, 92 506 Rueil-Malmaison, France*

Jerry L. Jensen (313), *Department of Petroleum Engineering, Heriot-Watt University, Edinburgh, Scotland EH14 4A5*

Zaixing Jiang (680), *Department of Petroleum Exploration, University of Petroleum, China, Dongying, Shandong, People's Republic of China*

Andrè G. Journel (251), *Stanford Center for Reservoir Forecasting, Department of Applied Earth Sciences, Stanford University, Stanford, California 94035*

John W. Kramers (226), *Alberta Research Council, Edmonton, Alberta T6H 5X2, Canada*

Larry W. Lake (478, 704), *Department of Petroleum Engineering, The University of Texas at Austin, Austin, Texas 78712*

Douglas A. Lawson (442), *ARCO Oil and Gas Company, Plano, Texas 75075*

Jørgen Leknes (102), *Production Laboratories, Statoil, N-4001 Stavanger, Norway*

Jan R. Lien (656), *University of Bergen, Bergen, Norway*

Menghui Liu (680), *Department of Petroleum Exploration, University of Petroleum, China, Dongying, Shandong, People's Republic of China*

Patrick H. Lowry (659), *K&A Energy Consultants, Inc., Tulsa, Oklahoma 74112*

Philip Lowry (665), *Institute for Energy Technology, N-2007 Kjeller, Norway*

F. Jerry Lucia (355), *Bureau of Economic Geology, The University of Texas at Austin, Austin, Texas 78712*

David F. Mayer (677), *Petroleum Engineering Department, Colorado School of Mines, Golden, Colorado 80401*

Rune Mjos (701), *Rogland Research Institute, Ullandhaug, 4004 Stavanger, Norway*

A. H. Muggeridge (197), *BP Research, Sunbury Research Centre, Sunbury-on-Thames, Middlesex, TW16 7LN, England*

Victor Nilsen (102), *Production Laboratories, Statoil, N-4001 Stavanger, Norway*

Manmath N. Panda (704), *Department of Petroleum Engineering, The University of Texas at Austin, Austin, Texas 78712*

Mary Passaretti (659), *K&A Energy Consultants, Inc., Tulsa, Oklahoma 74112*

Björn N. P. Paulsson (460), *Chevron Oil Field Research Company, La Habra, California 90631*

D. Pavone (683), *Institut Francais Du Petrole, 92 506 Rueil-Malmaison, France*

Dalian V. Payne (497), *Chevron Oil Field Research Company, La Habra, California 90631*

R. Michael Peterson (659), *Terra Vac, Belle Meade, New Jersey 08502*

Susan E. Pool (644), *United States Department of Energy, Morgantown Energy Technology Center, Morgantown, West Virginia 26507*

Mark E. Portman (644), *United States Department of Energy, Morgantown Energy Technology Center, Morgantown, West Virginia 26507*

Arne Raheim (665), *Institute for Energy Technology, N-2007 Kjeller, Norway*

Hans Rendall (315), *Department of Sedimentology and Stratigraphy, Continental Shelf and Petroleum Technology, Research Institute A/S, Jarlesletta, N-7002 Trondheim, Norway*

James D. Robertson (340), *ARCO Oil and Gas Company, Dallas, Texas 75221*

Eduardo Rodriguez (698), *South American Earth Sciences Department, INTEVEP, Caracas 1070A, Venezuela*

Kjell Johan Rosvoll (166), *Department of Mineral Resources Engineering, Imperial College, London SW7 2BP, England*

Yoram Rubin (251), *Department of Civil Engineering, University of California, Berkeley, California 94720*

J. L. Rudkiewicz (686), *Institut Francais Du Petrole, 92 506 Rueil-Malmaison, France*

R. A. Schatzinger (524), *National Institute for Petroleum and Energy Research (NIPER), Bartlesville, Oklahoma 74005*

R. K. Senger (355), *Bureau of Economic Geology, The University of Texas at Austin, Austin, Texas 78712*

Bijon Sharma (689), *IIT Research Institute, National Institute for Petroleum and Energy Research (NIPER), Bartlesville, Oklahoma 74005*

E. H. Smith (52), *Petroleum Reservoir Technology Division, United Kingdom Atomic Energy Authority, Dorset DT2 8DH, England*

Michael P. Stephens (695), *M-1 Drilling Fluids Company, Houston, Texas 77242*

Hans-Henrik Stølum (579), *Norwegian Petroleum Directorate, Stavanger, Norway*

M. Szpakiewicz (524), *National Institute for Petroleum and Energy Research (NIPER), Bartlesville, Oklahoma 74005*

Mojtaba Taheri (698), *South American Earth Sciences Department, INTEVEP, Caracas 1070A, Venezuela*

Lawrence W. Teufel (565), *Geomechanics Division, Sandia National Laboratories, Albuquerque, New Mexico 87185*

L. Tomutsa (524), *National Institute for Petroleum and Energy Research (NIPER), Bartlesville, Oklahoma 74005*

Olav Walderhaug (701), *Rogland Research Institute, Ullandhaug, 4004 Stavanger, Norway*

Thomas C. Wesson (478), *United States Department of Energy, Bartlesville, Oklahoma 74005*

J. K. Williams (613), *BP Research, Sunbury Research Centre, Sunbury TW16 7LN, England*

Paul F. Worthington (123), *BP Research, Sunbury Research Centre, Sunbury-on-Thames, Middlesex, TW16 7LN, England*

Gordon R. Young (704), *Department of Petroleum Engineering, The University of Texas at Austin, Austin, Texas 78712*

Li-Ping Yuan (226), *Alberta Research Council, Edmonton, Alberta T6H 5X2, Canada*

Chenglin Zhao (680), *Department of Petroleum Exploration, University of Petroleum, China, Dongying, Shandong, People's Republic of China*

Preface

We are pleased to present the Proceedings of the Second International Reservoir Characterization Technical Conference. Together with my co-chairmen Herb Carroll, formerly of the National Institute for Petroleum and Energy Research (NIPER), and Tom Wesson of the U.S. Department of Energy, we acknowledge the support, financial and otherwise, that NIPER and the Department of Energy have provided.

The first conference on reservoir characterization was also held in Dallas in the summer of 1985. The result of that conference was a volume of presentations that has become a standard reference on reservoir characterization. We hope that this present volume will make a worthwile contribution, perhaps even more so than the 1985 volume, since each article has been given a technical review by at least two peer reviewers. We thank the reviewers for their insightful comments and the authors for their patience.

Many people have said that the 1985 conference was the beginning of reservoir characterization. While there has been a surge of interest recently, reservoir characterization as an area of study existed before 1985 and it will be around for a long time. Actually, 1985 was not even the beginning of this conference; it grew out of a DOE workshop a few years earlier where we first learned that reservoir characterization meant many different things to many people which it still does.

We did not alter the basic formula of either the conference or this proceedings volume. However there were a few changes. First, thanks to Matt Honarpour in particular, we organized the conference around the scales of heterogeneity rather than along disciplines. We felt that this organization would be more likely to mix the participants and provide more cross-pollination. Second, we added tutorial sessions before the regular meeting. These provided common background to the anticipated diversity of the participants without taking up conference time. Our thanks to the tutorial leaders, Tom Hewett, Tom Burchfield, and Jack Caldwell, for providing this service. We retained the popular workshop sessions and also included their results in this volume. Poster sessions were added to the conference and were excellently managed by Jack Caldwell. Poster sessions are fairly rare in oil-recovery meetings, but they are effective in transferring information and we were pleased with the way they turned out. Finally, our thanks to Duane Babcock and ResTech, Inc. for sponsoring the social hour.

Our support staff was once again excellent. I.H. Silberberg of The University of Texas at Austin did a superb job editing all the manuscripts for language and grammar consistency. We are especially indebted to Bill Linville of NIPER, Virginia Foreman, formerly of NIPER, and Heidi Epp and the late Marjorie J. Lucas of The University of Texas at Austin for yeoman service. Many of the authors got to know Heidi quite well through her terror faxes about the ever-fading deadlines.

While it is true that there are still many unresolved questions in reservoir characterization, a comparison of the current volume with the 1985 proceedings will reveal that there has been much progress. We understand reservoir processes, on all scales, much better; we are able to implement and interpret geostatistical procedures and tie them back to outcrops and forward to applications; we are able to factor in seismic analysis in truly astounding detail; and, above all, we now have a much better appreciation for the diversity of our co-workers in this challenging and stimulating area.

No conference goes on without the help of an able organizing committee. We recognize the diligent work and support of this committee:

Edith Allison
U.S. Department of Energy

H. Duane Babcock
ResTech, Inc.

Thomas E. Burchfield
NIPER

Larry Byrd
K & A Energy Consultants

Jack Caldwell
Schlumberger Well Services

Aaron Cheng
NIPER

Rob Finley
University of Texas Bureau of Economic Geology

Mike Fowler
Oxy USA Inc.

John Heller
New Mexico Institute of Mining and Technology

Thomas A. Hewett
Chevron Oil Field Research Company

Matt Honarpour
NIPER

Gary Hoover
Phillips Petroleum Company

Doug Lawson
Stanford University

C. C. (Bill) Linville
NIPER

Dave C. Martin
New Mexico Institute of Mining and Technology

E. B. Nuckols
U.S. Department of Energy

Mike Peterson
K & A Energy Consultants

Mike Stephens
M-I Drilling Fluids

Min K. Tham
NIPER

Lynn Watney
Kansas Geological Survey

Session 1
Microscopic

The scale of several hundred grain or pore diameters, usually analyzed as networks or through microscopes

CHARACTERISATION OF SURFACE ROUGHNESS IN POROUS MEDIA

C.K. Harris

Koninklijke Shell
Exploratie en Produktie Laboratorium
2280 AB Rijswijk ZH,
The Netherlands

ABSTRACT

The fact that the boundary between matrix and pore space in a reservoir rock is non-smooth at the pore scale can have a significant effect on hydrocarbon recovery processes. This is particularly true in carbonate reservoirs, for which the ratio of roughness amplitude to pore-wall separation is often of order unity. In the present contribution a method of characterising pore-wall roughness is presented. The method involves the use of a model of a rough pore wall having the property that features pertaining to a profile of the model seen on a random section through it, can be related to the bulk model parameters characterising the roughness. Such a consideration is an important one when formulating models of the pore space in general, because it means that the wealth of detailed quantitative information concerning the pore geometry that can now be obtained by image analysis of the pore-space profile can be fully utilised. The model parameters can be chosen to yield a fractal pore wall and this important special case is treated in some detail.

I.   INTRODUCTION

   The complicated structure of the pore space in many
carbonates presents a formidable challenge to the theorist
wishing to characterise and model it at the pore scale.   Such
a characterisation is a prerequisite for understanding the
influence of pore geometry on various aspects of hydrocarbon
recovery in carbonate reservoirs.   A recurring feature of many
types of carbonate reservoir is the presence of significant
surface roughness at length scales comparable with the pore
scale and smaller.   An example is the oomoldic limestone
pore seen in section in Fig. 1.

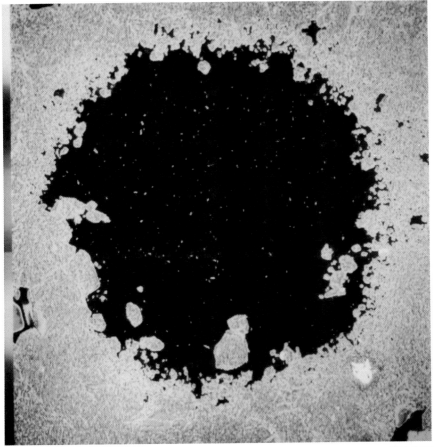

FT 118305

   Fig. 1.   Back-scattered electron image of oomoldic pore
profile.

In the present contribution, the problem of quantifying the morphology of a rough pore wall is addressed. The approach used is to construct an explicit model of pore-wall roughness, built up in such a way that properties both of the surface itself and of random thin sections through it can easily be related to the parameters of the model. This model is described in the following section. In view of the apparent ubiquity of fractals in nature in general [1] and their applicability to the surfaces of rocks in particular [2-5], the construction procedure has been designed to include the case of fractal surface behaviour; this is explained in Section III.

The use of the model in assessing the influence of surface roughness on the behaviour of reservoir fluids is considered next, in Section IV. The cases of bulk and wetting-phase flow and the adsorption of surfactants are briefly discussed.

The fractal theme is taken up again in Section V where sectioning of the model is discussed. The principal result proved there is that, when the surface of the model is fractal, the total perimeter of the profile of the pore seen in section (including that of the islands of matrix that appear separated from the outer boundary in the section plane) and the outer pore-profile boundary both exhibit fractal behaviour but have different fractal dimensions. Only the former is simply related to the surface fractal dimension. Finally Section VI contains some concluding remarks and proposals for future work.

II.   CONSTRUCTION OF THE MODEL

The surface roughness model introduced here is built up by starting with a closed surface S_2, smooth on a length scale b_2, and modifying it by the successive addition of features of decreasing length scale down to b_1. We shall denote the resulting irregular surface by S_1 and the surface at an intermediate stage, when features down to a length scale b have been added, by S. Let the corresponding surface areas be given by Σ_2, Σ_1 and Σ respectively. Randomness is incorporated into the model by stipulating that a feature whose size is in the range $b \rightarrow b - \delta b$, where δb is an infinitesimally small quantity, is equally likely to be added to any point on the surface with a probability $\sigma(b)\delta b$ per unit area. $\sigma(b)$ is a positive function that we are free to choose; in the next section we make a special choice that yields a fractal pore wall.

If a feature of size $b \to b - \delta b$ is added to S, an additional surface area $\Delta\Sigma$ will be created. We assume that $\Delta\Sigma$ is a random process with some probability distribution, which would be the case if the features added to the surface have a distribution of shapes. Let S_* denote the pore surface at length scale $b^* = b - \delta b$ and Σ^* its surface area, and define the quantities

$$\Sigma(b) \equiv \langle\Sigma\rangle \tag{1}$$

$$\Sigma(b^*) = \Sigma(b - \delta b) \equiv \langle\Sigma^*\rangle \tag{2}$$

where the angular brackets in Eqs. (1) and (2) denote averages over all possible realisations of S and S^* respectively, starting from S_2. The average increase in surface area that results on forming S^* from S is given by $\overline{\Delta\Sigma}\langle\Sigma\rangle\sigma(b)\delta b$, where $\overline{\Delta\Sigma}$ is the expectation of $\Delta\Sigma$ (i.e. $\Delta\Sigma$ averaged over the feature shape distribution), and this leads to the following result:

$$\Sigma(b - \delta b) - \Sigma(b) = \Sigma(b)\overline{\Delta\Sigma}\sigma(b)\delta b \tag{3}$$

Eq. (3) may be integrated, subject to the boundary condition that $\Sigma(b_2) = \Sigma_2$, to obtain the result

$$\Sigma(b) = \Sigma_2 \exp\{ \int_b^{b_2} db^* \sigma(b^*)\overline{\Delta\Sigma}\} \tag{4}$$

An important special case of the model considered so far is that for which it is constructed in such a way that the distribution of feature shape is independent of b. $\overline{\Delta\Sigma}$ is then given by

$$\overline{\Delta\Sigma} = b^2\overline{a} \tag{5}$$

where $\overline{a}$ is a dimensionless constant. Substituting (5) into (4), we obtain the result

$$\Sigma(b) = \Sigma_2 \exp\{\overline{a} \int_b^{b_2} db^* (b^*)^2 \sigma(b^*)\} \tag{6}$$

In this case the manner in which $\Sigma(b)$ changes with length scale in the range $b_1 < b < b_2$ is entirely determined by the function $\sigma(b)$ and a single parameter $\overline{a}$. A concrete example of such a model results from considering the features to be formed by embedding spheres in the pore wall, as shown in Fig. 2. A feature of size b is added to S by choosing a point Q at

random on the pore surface, with a probability which is
uniformly distributed over the available area, drawing the
inward normal to the surface at that point, and placing the
centre O of a sphere with radius r on the normal so that it
overlaps the surface. The feature then consists of that
portion of the embedded sphere which projects into the pore
space. The distance PQ, where P is the point at which the
normal intersects the feature, is drawn from a probability
distribution that is independent of b when expressed in units
of b. The sphere radius r is chosen so that the maximum
diameter of the feature is always b. Provided that $\sigma(b)$ is
small, which means that the probabilities that Q lies on a
portion of the pore surface which is significantly curved on a
scale of b, and of multiple overlap are both small, Eq. (5)
will hold with $\bar{a}$ given by

$$\bar{a} = \pi \int_0^1 ds\, f(s) s^2 \qquad\qquad (7)$$

where f(s) is the probability density function of the scaled
distance s = PQ/b. (Since this scaled distance controls the
shape of the feature, we have denoted it by s.) Before
treating sectioning of the model, we obtain the form for $\sigma(b)$
required for the pore surface to be a fractal.

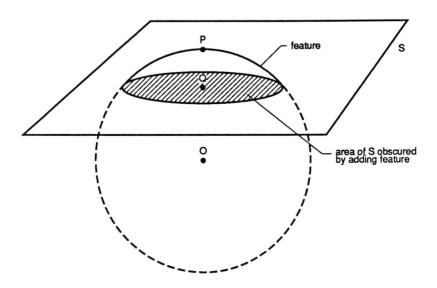

Fig. 2.   Addition of embedded-sphere feature to S.   The
area of S obscured by adding the feature is shown shaded.

III. FRACTAL BEHAVIOUR

     Fractals are irregular objects whose irregularity varies
with scale in a particular way. The term "object" here can
refer to an actual physical entity or set of entities, or it
can be a function describing how a physical quantity varies
with time or position. For example, the boundary of the pore
space in a porous medium [2-5], the fluid-fluid interface
during displacement processes in porous media [6,7] and the
fracture network in fractured rock [8] are objects of the
former type that have been argued to be fractals. On the
other hand fractal functions [9] have been used to model
terrain (height v. position, [10]), river flow cycles (annual
discharge v. time, [11]) and porosity logs (porosity v. depth,
[12]). In the present contribution, we shall mainly be
concerned with fractals of the first type. Consider for
example the oomoldic pore profile shown in Fig. 1. This
profile is of a well-defined shape but with a very irregular
boundary. One way in which the degree of irregularity
manifests itself is the way that the measured length of the
boundary increases as the length of the smallest feature that
can be resolved decreases. For example, in a digitised image
of Fig. 1, the smallest length scale b is the pixel size and
the length $P(b)$ of the pore-profile perimeter measured at this
scale could be taken to be the number of pixels $N(b)$
containing part of the perimeter, multiplied by b. The length
b can be changed either by making new images of the pore at
different magnifications or by modifying the original image by
applying so-called opening operations that successively smooth
the boundary [13]. Suppose that the length of the pore-
profile perimeter at length scale b_2 is $P(b_2)$. If this
perimeter is a fractal, its length measured at another length
scale b is given by

$$P(b) = P(b_2)(b/b_2)^{-(D_p - 1)} \tag{8}$$

with $D_p > 1$. The constant D_p is called the fractal dimension.
     Objects with a fractal irregularity obeying the law (1)
and its extension to surfaces in three dimensions are
widespread in nature [1]. The generalisation of (8) to
surfaces in three dimensions is straightforward. It is

$$\Sigma(b) = \Sigma(b_2)(b/b_2)^{-(D_s - 2)} \tag{9}$$

with $D_s > 2$. Here $\Sigma(b)$ is the surface area measured when the
length scale of the measuring unit is b. For example, $\Sigma(b)$
could be taken to be the number of three-dimensional pixels,

or "voxels", of side b which intersect the surface, multiplied
by b. D_S is the fractal dimension of the surface. If the
logarithm of P(b) or $\Sigma(b)$ is plotted against the logarithm of
b, the relation (8) or (9) yields a straight line from whose
slope the fractal dimension may be obtained. In practice, a
linear regime is found only for a finite range of b. It is
convenient to take b_2 in Eqs. (8) or (9) to be the upper limit
of this range. There will also be a lower limit below which
(8) or (9) ceases to be valid and which will be denoted by b_1.
The power-law behaviour in Eqs. (8) and (9) follows from the
property that a portion of a fractal object is similar to the
whole in some way. This characteristic also serves as a
definition of a fractal [14]. The scale dependence of the
outline of the pore profile shown in Fig. 1 was found to
exhibit the fractal behaviour of Eq. (8) with b_2 about 5μm and
b_1 less than 0.3μm [15]. Fractal behaviour of the pore-
profile outlines was also measured for other pore profiles
seen on the section and for sections through four other
oomoldic limestone samples. The resulting fractal dimensions
were found to correlate well with a petrophysical parameter,
the cementation exponent m [15].

    To obtain an power-law areal dependence on the length
scale b of the smallest features resolved for the pore surface
model introduced in the preceding section we choose

$$\sigma(b) = \sigma_0/b^3 \tag{10}$$

where σ_0 is a dimensionless constant. Substituting Eq. (10)
into Eq. (6), integrating, and comparing with Eq. (9) then
yields the following result for the surface fractal dimension:

$$D_S = 2 + \overline{a}\sigma_0 \tag{11}$$

    We now show that, while the surface area $\Sigma(b)$ of the pore
diverges as b → 0, its volume V(b) tends to a finite constant
in this limit, provided that $D_S < 3$. V(b) evolves according
to the equation

$$V(b - \delta b) - V(b) = -\Sigma(b)\overline{\Delta V}\sigma(b)\delta b \tag{12}$$

where $\overline{\Delta V}$ is the shape-averaged volume of a feature of size b.
For the case of a fractal pore wall, we have

$$\overline{\Delta V} = \overline{a}_V b^3 \tag{13}$$

where $\bar{a}_v$ is a dimensionless constant. Substituting Eqs. (9), (10) and (13) into Eq. (12), and integrating, yields the following expression for $V(b)$:

$$V(b) = V_2 - \sigma_0 \bar{a}_v \Sigma_2 b_2 \{1 - (b/b_2)^{(3 - D_s)}\}/(3 - D_s) \qquad (14)$$

The right-hand side of Eq. (14) clearly tends to a finite limit as $b \to 0$.

IV. APPLICATION TO PORE-SPACE FLUID DYNAMICS

Before discussing the effect of surface roughness on one- and two-phase flow in the pore space, we describe the relationship between the present model and roughness descriptors used in the literature.

A. Comparison with Conventional Roughness Descriptors

We start by emphasising that the model of a rough pore wall developed here can treat a more general class of rough surfaces than those that are fractally rough. For example, we are free to choose both the form of the function $\sigma(b)$ and the distribution of feature shapes. A further generalisation would be the lifting of the restriction that the feature shape distribution is independent of scale.

When we try to relate the feature density and shape distributions in the present model to conventional roughness descriptors, we run into the problem that the picture of roughness used to formulate the latter is of a height variation above a plane which is realisation of a random function (see, e.g., Ref. 16). The roughness descriptors then pertain to this function. The most important ones are the form of the probability distribution function for the height at a single point and the structure function, which describes the correlation between the heights at a pair of points as a function of their absolute and relative positions. The structure function is given by

$$S(\underline{r}) = \frac{1}{2}<\{h(\underline{r}_1) - h(\underline{r}_1 + \underline{r})\}^2> \qquad (15)$$

where the points are at $\underline{r}_1$ and $\underline{r}_1 + \underline{r}$ and the angular brackets in Eq. (15) denote an average over all realisations of the random function.

It is usual to assume that increments of the random process are stationary, (i.e. invariant under translation), so that $S(\underline{r})$ depends only on the relative position $\underline{r}$, and ergodic, which means that the ensemble average in Eq. (15) can be replaced by a single realisation of the random process, averaged over all positions, i.e.,

$$S(\underline{r}) = \lim_{A \to \infty} (2A)^{-1} \int dx_1 dx_2 \{h(\underline{r}_1) - h(\underline{r}_1 + \underline{r})\}^2 \qquad (16)$$

If Eq. (16) holds, an estimate of $S(\underline{r})$ is

$$S_{est}(\underline{r}) = (2N)^{-1} \sum_{i=1}^{N} \{h(\underline{r}_i) - h(\underline{r}_i + \underline{r})\}^2 \qquad (17)$$

Thus $S(\underline{r})$ can be estimated by taking N pairs of points, with the same separation and orientation, from a single realisation of the random process, which is of course all that one has in practice.

The random function picture of rough surfaces has been successfully applied in modelling ocean waves and topography (for details and references see Refs. 1, 10 and 16). Gravity plays an important role during the formation of these surfaces and ensures that their height above a reference plane that is perpendicular to the local gravitational field is well described by a single-valued function of position - i.e. that overhangs are rare. This situation is illustrated schematically in Fig. 3a. In contrast, a glance at SEM micrographs of the pore wall in carbonates, or at a thin-section image such as Fig. 1 shows that such a picture is inappropriate in this case. This situation is illustrated by the one-dimensional surface shown in Fig. 3b.

In what follows we shall denote the types of roughness depicted in Figs. 3a and 3b type I and type II roughness respectively. The present approach has been designed to model type II rough surfaces, an example of which is the pore wall in many reservoir rocks, and thus there is no direct relation between the quantities describing it and those applied in the literature to characterise type I rough surfaces. In spite of this, it is possible to envisage various ways of applying random function methods to type II rough surfaces. For example, a pair of functions $h_{max}(\underline{r})$ and $h_{min}(\underline{r})$ pertaining to the maximum and minimum heights respectively of the surface above a smooth reference surface, as a function of position $\underline{r}$ on the reference surface could be defined. In this case the average height difference divided by the standard deviation of the mean height would yield a measure of the degree of importance of overhangs, and $h_{max}(\underline{r})$ is probably useful in

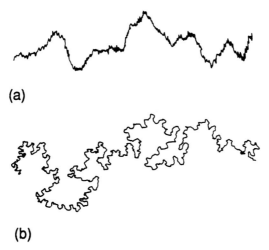

(a)

(b)

Fig. 3. Schematic illustration of rough surfaces without overhangs (a) and with overhangs (b).

modelling fluid flow over the surface as discussed in the next subsection.

B. Single and Two-Phase Flow in Irregular Geometries

1. Single-phase flow

There are two different approaches which could be applied in modelling single-phase flow in geometries where type II roughness is present on the boundaries. In the first approach, the flow field is successively perturbed by the addition of features of size b_2 down to b_1, and in the second, the type II roughness is replaced by a type I roughness with height variations given by $h_{max}(\underline{r})$, defined above. It is assumed that the effects of making this replacement will be small sufficiently far from the boundaries. The advantage of the second approach is that results in the literature pertaining to type I roughness can be used and developed. For example, Phan-Thien [17] has shown that the mean contribution ΔR to the flow resistance per unit length of a circular pipe, due to corrugations in an axial plane (but not in cross-section), with a structure function $S(z)$, is given by

$$\Delta R/R_0 = a^{-2} \int_{-\infty}^{\infty} dz g(z) S(z) + O(\epsilon^3) \qquad (18)$$

The quantities appearing in Eq. (18) are the mean pipe radius a, the flow resistance per unit length R_0 for a straight pipe of radius a, and the ratio ϵ of the maximum height of the corrugations (measured with respect to the mean pipe radius) to a. g(z) is a known function [17]. Phan-Thien's method can be generalised to other geometries.

2. Two-phase flow

Surface roughness plays an important role in determining the processes that occur when wetting and non-wetting phases flow together in a porous medium. To give an example, when a droplet of non-wetting fluid that is unstable to snap-off is passing through a pore throat, the rate of supply of wetting fluid to the bridge region is dependent in part on the degree of surface roughness. Thus the surface roughness is a factor determining whether a droplet of a given length travelling at a given speed can pass through a given pore throat before snapping off. Translated into the language of oil recovery, this means that the surface roughness influences the processes by which oil is trapped during a water flood and has an effect on the distribution of residual oil. Thus the surface roughness is a geometrical feature to be taken into account when formulating EOR strategies.

It should be possible to relate the flow of wetting fluid along a rough surface to a few simple parameters describing the roughness. Indeed, results in this direction have already been obtained by Lenormand and Zarcone [18], who used grooves of a characteristic depth, width and separation to model the roughness, and by Katz and Trugman [19], who modelled a random surface using a triangular net with vertices at random heights above a plane, creating a network of grooves and ridges between adjacent triangular faces. Once expressions for the wetting fluid flow rate have been obtained, they can be used to draw qualitative conclusions concerning the likelihood of the snap-off process and the distribution of residual oil. These conclusions can subsequently be quantified by using the expressions in pore-level laws which are incorporated into pore-network simulators.

It is also worth remarking that the presence of roughness increases the surface area available for adsorption of polymers or surfactants, perhaps by hundreds of times. This is another reason why surface roughness should be taken into account when formulating an EOR strategy.

V.   SECTIONING THE MODEL

We now consider taking random plane sections through the pore surface model. The quantities we shall concentrate on are the scale-dependences of the lengths of the total pore-profile perimeter, the pore-profile outline or external pore-profile perimeter and the perimeters of individual islands of matrix within the pore-profile outline that are cut off by the section plane. The scale dependence of the total number of islands contained within the external perimeter of the pore profile is also discussed.

A.   The Total Pore-Profile Perimeter

Let $B(S,P)$ denote the length of the total perimeter of the profile of S seen on a random section P through a connected region R containing S. An expression for the expected length $B(b)$, defined as

$$B(b) \equiv <<B(S,P)>_P> \qquad (19)$$

is easily derived using a well-known result from the stereological literature. The inner set of angular brackets on the right-hand side of Eq. (19) denotes an average over all section plane positions and orientations for which the section plane intersects R and the outer set an average over all realisations of S. The result just referred to states that

$$<B(S,P)>_P \propto \Sigma \qquad (20)$$

with a constant of proportionality depending only on R [20]. Averaging Eq. (20) over all realisations of S, applying Eqs. (1) and (19) and dividing by a corresponding result for S_2 yields

$$B(b)/B(b_2) = \Sigma(b)/\Sigma_2 \qquad (21)$$

Substituting Eq. (4) into Eq. (21) then leads to the following equation for $B(b)$:

$$B(b) = B(b_2)\exp\{ \int_b^{b_2} db^* \sigma(b^*)\overline{\Delta\Sigma}\} \qquad (22)$$

For the special case of a fractal pore surface we obtain

$$B(b) = B(b_2)(b/b_2)^{-(D_p - 1)} \qquad (23)$$

with

$$D_P = D_S - 1 \qquad (24)$$

The total pore-profile perimeter fractal dimension is then simply the pore surface fractal dimension less unity, a result well-known in the literature [Ref. 1, p. 365].

Before turning to the far less trivial case of the pore-profile outline, we remark that B(b) is not the same as P(b), the perimeter of S_1 measured at scale b. This is because a section through a feature of size b or larger can result in a profile which is smaller than b. The perimeter of this would be included in B(b) but not in P(b). Hence P(b) is less than B(b). This difference between B(b) and P(b) is similar to the differences in the results obtained for the length of a perimeter measured at resolution b using different techniques: the leading power-law dependence on the scale b is unaffected. This argument can be confirmed by explicit calculations and is also assumed to hold for $B_{ext}(b)$ and $P_{ext}(b)$, the analogous pair of quantities for the external pore-profile boundary.

B. The Pore-Profile Outline

1. General formalism

Let $B_{ext}(S,P)$ denote the length of the external boundary of the profile of S, seen on a random section P through a connected region R containing S. The expected length $B_{ext}(b)$ is then defined analogously to B(b) as

$$B_{ext}(b) \equiv <<B_{ext}(S,P)>_P> \qquad (25)$$

To obtain an expression for $B_{ext}(b)$ it is convenient to work with a more general quantity $B_{ext}(b,[g])$ given by

$$B_{ext}(b,[g]) = <<B_{ext}(S,P,[g])>_P> \qquad (26)$$

where

$$B_{ext}(S,P,[g]) = \int_{C_{ext}(S,P)} d\lambda g(\theta) \qquad (27)$$

$g(\theta)$ is an arbitrary function of the angle θ subtended by the normals to S and P at a point on the external pore-profile boundary $C_{ext}(S,P)$. This point is at a distance λ measured along the curve from a fixed point on it and the integration on the right-hand side of Eq. (27) is around the boundary. Clearly,

$$B_{ext}(b) = B_{ext}(b,[1])$$ (28)

We now work towards a differential equation for $B_{ext}(b,[g])$ by considering the difference between $B_{ext}(b - \delta b,[g])$ and $B_{ext}(b,[g])$. This difference will be non-zero if (a) a feature is added to S and (b) this feature intersects the section plane in such a way as to modify the external pore-profile perimeter. The probability of (a) above is just $\Sigma\sigma(b)\delta b$ and we define the quantity

$$\Delta B_{ext}(S,P,P_F,s,[g]) \equiv \text{contribution to } B_{ext}(S,P,[g]) \text{ made by adding a feature of size } b,$$
$$\text{shape } s, \text{ and position and orientation } P_F \text{ to S.} \quad (29)$$

Then

$$B(b - \delta b,[g]) - B(b,[g])$$
$$= \sigma(b)\delta b<\Sigma<<\overline{\Delta B_{ext}(S,P,P_F,s,[g])}>_{P_F}>_P> \quad (30)$$

The bar on the right-hand side of Eq. (30) denotes an average over the feature shape distribution and the innermost set of angular brackets is an average over the position and orientation of the feature.
$\Delta B_{ext}(S,P,P_F,s,[g])$ is zero if the area of S obscured by adding the feature does not intersect $C_{ext}(S,P)$. If the obscured area does intersect $C_{ext}(S,P)$ as illustrated in Fig. 4, $\Delta B_{ext}(S,P,P_F,s,[g])$ is given by

$$\Delta B_{ext}(S,P,P_F,s,[g]) = \int_0^{C_{new}} d\lambda_{new} g(\theta_{new})$$
$$- \int_0^{C_{old}} d\lambda_{obs} g(\theta_{obs}) \quad (31)$$

Referring to Fig. 4, C_{new} is the length of the portion of the external perimeter loop created by adding the feature and C_{obs} is the portion of the old external perimeter loop $C_{ext}(S,P)$ obscured by adding the feature. θ_{new} is the angle subtended by the normal to P and the normal to the feature surface at a point a distance λ_{new} measured from the point A along the portion of the perimeter loop created by the feature, while θ_{obs} is the angle subtended by the normals to S and P at a

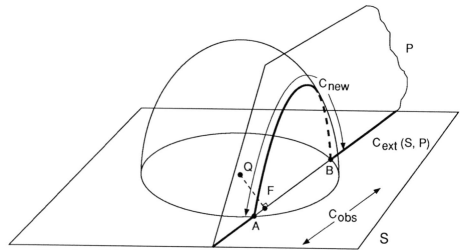

Fig. 4. Addition of feature contributing to the external
perimeter loop $C_{ext}(S,P)$.

point a distance λ_{obs} along the obscured portion of the loop
measured from A.

In adding the feature to S, a point Q is chosen at random
on S and this is then taken to be the centroid of the area
obscured by the feature. We assume that $\sigma(b)$ is small enough
for the probability that Q lies in a region significantly
curved on a lengthscale b to be negligible. Making this
assumption, so that the obscured area can be taken to be
plane, and introducing the dimensionless quantities

$$c_{new} = C_{new}/b \qquad (32a)$$

$$c_{obs} = C_{old}/b \qquad (32b)$$

$$d\eta = d\lambda_{new}/C_{new} \qquad (32c)$$

Eq. (31) becomes

$$\Delta B_{ext}(S,P,P_F,s,[g]) = bc_{new} \int_0^1 d\eta g(\theta_{new}) - bc_{obs}g(\theta) \qquad (33)$$

where θ is the angle subtended by the normals to S and P at F,
the foot of the perpendicular from Q onto the portion of
$C_{ext}(S,P)$ obscured by adding the feature. Denoting the
average of $\Delta B_{ext}(S,P,P_F,s,[g])$ over all positions and
orientations of the added feature by $\Delta B_{ext}(S,P,s,[g])$, we have

$$\Delta B_{ext}(S,P,s,[g]) = b^2 \int_{C_{ext}(S,P)} d\lambda \Phi(s,\theta,[g])/\Sigma \qquad (34)$$

where

$$\Phi(s,\theta,[g]) = (2\pi b)^{-1} \int_0^{2\pi} d\phi \int_{x_{min}(\phi)}^{x_{max}(\phi)} dxc_{new} \int_0^1 d\eta g(\theta_{new})$$

$$- a_{obs}(s)g(\theta) \qquad (35)$$

x here is the displacement of F from Q and $x_{min}(\phi)$, $x_{max}(\phi)$ are the minimum and maximum values of x, respectively, for which a feature at orientation ϕ intersects the loop $C_{ext}(S,P)$. ϕ is the angle between QF and a line fixed with respect to the obscured area and $a_{obs}(s)$ is the value of the obscured area in units of b^2. Substituting Eq. (34) into Eq. (30), dividing by δb and taking the limit $\delta b \to 0$ yields the differential equation

$$dB_{ext}(b,[g])/db = -b^2\sigma(b)\beta(b,[g]) \qquad (36)$$

with

$$\beta(b,[g]) = <<\int_{C_{ext}(S,P)} d\lambda\overline{\Phi(s,\theta,[g])}>_p> \qquad (37)$$

The next step is to rearrange the right-hand side of Eq. (37) to eliminate the awkward integration over λ. We write

$$\overline{\Phi(s,\theta,[g])} = \int_0^{\pi} d\theta''\delta(\theta - \theta'')\overline{\Phi(s,\theta'',[g])} \qquad (38)$$

where $\delta(\theta - \theta'')$ denotes the generalised or Dirac delta function. Substituting Eq. (38) into Eq. (37) and interchanging the order of integration over θ'' with that over λ and the pair of averages leads to the result

$$\beta(b,[g]) = \int_0^{\pi} d\theta''\overline{\Phi(s,\theta'',[g])}B_{ext}(b,\theta'') \qquad (39)$$

where

$$B_{ext}(b,\theta'') = <<\int_{C_{ext}(S,P)} d\lambda\delta(\theta - \theta'')>_p> \qquad (40)$$

Comparing Eq. (40) with Eqs. (26) and (27) we see that $B_{ext}(b,\theta'')$ is a special case of $B_{ext}(b,[g])$ with $g(\theta)$ given by

$\delta(\theta - \theta")$. Substituting Eq. (39) into Eq. (36) yields the following differential equation for $B_{ext}(b,[g])$:

$$dB_{ext}(b,[g])/db = -b^2\sigma(b) \int_0^\pi d\theta"\overline{\Phi(s,\theta",[g])}B_{ext}(b,\theta")$$

(41)

Eq. (41) can be solved in principle for any $B_{ext}(b,[g])$. The first step is to choose $g(\theta) = \delta(\theta - \theta')$ in Eq. (41) to obtain

$$dB_{ext}(b,\theta')/db = -b^2\sigma(b) \int_0^\pi d\theta"\overline{\Phi(s,\theta",\theta')}B_{ext}(b,\theta")$$

(42)

where we have used $\Phi(s,\theta",\theta')$ to denote $\Phi(s,\theta",[g])$ with $g(\theta")$ = $\delta(\theta" - \theta')$. Eq. (42) is then solved subject to the initial function $B_{ext}(b_2,\theta")$ and the solution substituted into Eq. (41) which is then solved for $B_{ext}(b,[g])$, subject to the initial functional $B_{ext}(b_2,[g])$. An explicit solution can be obtained when the pore surface is fractal, which requires scale-independent $\overline{\Phi(s,\theta",\theta')}$ and $\sigma(b)$ given by Eq. (10). We proceed to consider this special case in some detail.

2. Fractal pore surface

We now obtain an explicit expression for $B_{ext}(b,[g])$ when the pore surface is fractal. Setting $g(\theta) = \delta(\theta' - \theta")$ in Eq. (35), substituting the latter into Eq. (42), rearranging and making use of Eq. (10) yields the equation

$$-(b/\sigma_0)dB_{ext}(b,\theta')/db + \overline{a}_{obs}B_{ext}(b,\theta')$$

$$- \int_0^\pi d\theta"\overline{\Phi}_{new}(s,\theta",\theta')B_{ext}(b,\theta") \quad (43)$$

where

$$\Phi_{new}(s,\theta",\theta') = (2\pi b)^{-1} \int_0^{2\pi} d\phi \int_{x_{min}(\phi)}^{x_{max}(\phi)} dxc_{new} \int_0^1 d\eta\delta(\theta_{new} - \theta')$$

(44)

We introduce the new variables

$$t = -\ln(b/b_2)$$

$$v' = \cos(\theta')$$

$$v'' = \cos(\theta'') \tag{45}$$

Further, let

$$u(v',t) = B_{ext}(b,\theta')/\sin^2(\theta'),$$

$$L(s,v',v'') = \sin(\theta'')\Phi_{new}(s,\theta'',\theta')/\sin^2(\theta') \tag{46}$$

In terms of the new quantities introduced in Eqs. (45) and (46), Eq. (43) becomes:

$$(1/\sigma_0)\frac{d}{dt}u(v',t) + \bar{a}_{obs}u(v',t) = \int_{-1}^{1} dv''\overline{L(s,v',v'')}u(v'',t) \tag{47}$$

Laplace transforming Eq. (47) yields a Fredholm integral equation of the second kind in the Laplace-transformed quantity $\tilde{u}(v',p)$:

$$\lambda\tilde{u}(v',p) = u(v',0) + \int_{-1}^{1} dv''L(s,v',v'')\tilde{u}(v'',p) \tag{48}$$

with

$$\lambda = \bar{a}_{obs} + p/\sigma_0 \tag{49}$$

The kernel $\overline{L(s,v',v'')}$ is symmetric for embedded-sphere features and is square-integrable for non-pathological shape distributions. These results follow by evaluating the integrals in Eq. (44) and using Eq. (46) to obtain an explicit expression for $L(s,v',v'')$ in this special case. $L(s,v',v'')$ is conjectured to be symmetric for general feature shapes but no proof is available at the time of writing. Assuming that $\overline{L(s,v',v'')}$ is symmetric, square-integrable and well-behaved, the solution of Eq. (48) is [21]

$$\tilde{u}(v',p) = \sum_{k=0}^{\infty} a_k u_k(v')/\{\sigma_0(\lambda - \lambda_k)\} \tag{50}$$

$u_k(v')$ and λ_k; $k = 0, 1, \ldots$ are the eigenvectors and eigenvalues of the kernel. The λ_k are real, discrete and of finite multiplicity and a_k is given by

$$a_k = \int_{-1}^{1} dv' u_k(v')u(v',0) \tag{51}$$

Substituting Eq. (49) into Eq. (50), inverting the Laplace tranform and making use of Eqs. (45) and (46) leads to the following expression for $B_{ext}(b,\theta')$:

$$B_{ext}(b,\theta') = \sin^2(\theta') \sum_{k=0}^{\infty} a_k u_k(v')(b/b_2)^{-\mu_k \sigma_0} \qquad (52)$$

with

$$\mu_k = \lambda_k - \bar{a}_{obs} \qquad (53)$$

Finally, Eq. (52) is substituted into Eq. (41) and the latter integrated to obtain an expression for $B_{ext}(b,[g])$:

$$B_{ext}(b,[g]) = B_{ext}(b_2,[g])$$
$$+ \sum_{k=0}^{\infty} \{(b/b_2)^{-\mu_k \sigma_0} - 1\} a_k c_k[g]/\mu_k \qquad (54)$$

where

$$c_k[g] = \int_0^{\pi} d\theta'' \overline{\Phi(s,\theta'',[g])} \sin^2(\theta'') u_k(v'') \qquad (55)$$

Let us label the exponents μ_k ($k = 0, 1, \ldots$) in order of decreasing magnitude. μ_0 is easily shown to be positive by proving that the lower bound for λ_0 given by the familiar Raleigh quotient is larger than $\bar{a}_{obs}$. The fractal dependence of the length of the external pore-profile perimeter $B_{ext}(b)$ on the scale b then follows by setting g = 1 in Eq. (54) and making use of Eq. (28). This leads to the result

$$B_{ext}(b) = (b/b_2)^{-\mu_0 \sigma_0} \{a_0 c_0[1]/\mu_0$$
$$+ O[(b/b_2)^{\mu_0 \sigma_0}, (b/b_2)^{(\mu_0 - \mu_1)\sigma_0}]\} \qquad (56)$$

For $b_1 \leq b \ll b_2$, and provided that $a_0 c_0[1]/\mu_0$ is greater than zero, the correction terms in square brackets on the right-hand side of Eq. (56) are small and $B_{ext}(b)$ exhibits fractal behaviour with a fractal dimension

$$D_P^{ext} = 1 + \mu_0 \sigma_0 \qquad (57)$$

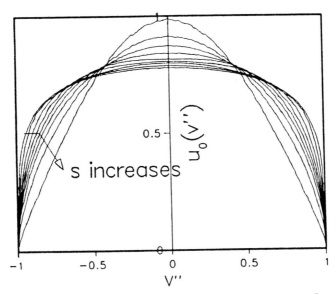

Fig. 5. Eigenfunctions of largest eigenvalue of embedded-sphere kernel for various shapes s.

In the case of embedded-sphere features with monodisperse shape distributions, the largest eigenvalue λ_0 of the kernel $L(s,v',v'')$ and its eigenfunction $u_0(v')$ have been obtained numerically for a number of values of the shape s. The eigenfunctions are plotted in Fig. 5 and the eigenvalues were used to compute the ratio $(D_p^{ext} - 1)/(D_p - 1)$. This ratio, which is a measure of how irregular the external pore-profile perimeter is compared to the total pore-profile perimeter, is plotted as a function of shape s in Fig. 6. The influence of s on the feature geometry is illustrated in the inset diagrams, which show the appearance of the features in a plane passing through OQP in the notation of Fig. 2.

The calculations carried out in this subsection demonstrate two things. First, the fractal surface construction presented here yields a power-law scale-dependence for the external pore-profile perimeter seen in section and thus can account for the fractal behaviour observed in Ref. 5, page 584, and in Ref. 15. Second, it demonstrates that the resulting fractal dimension D_p^{ext} depends in a complicated way on the details of the surface. This is

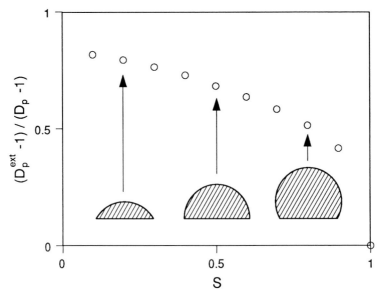

Fig. 6. Ratio of excess external to excess total fractal
dimensions of pore-profile perimeters in the embedded sphere
model versus shape (inset diagrams).

in sharp contrast to the total pore-profile perimeter fractal
dimension D_p, which is directly related via Eq. (24) to the
surface fractal dimension D_s, a fundamental surface property.

C. Island Perimeters and their Number v. Area Relation

We conclude the present section by discussing the scale
dependence of the perimeters of individual islands of matrix
cut off by the section plane and the number v. area relation
for these islands. It is convenient to define the quantity

$B(S,P,b')$ ≡ Total perimeter of all islands formed at a
                   range of length scales $b' \rightarrow b' + \delta b'$ during
                   the generation of S, seen on a
                   section plane P through S.   (58)

together with the averaged quantity

$$B(b,b') \equiv \lim_{\delta b' \rightarrow 0} <<B(S,P,b')>_P>/\delta b'$$ (59)

A theory for the evolution of $B(b,b')$ precisely analogous to that for $B_{ext}(b)$ expounded in the preceding subsection may be developed by starting with the quantity

$$B(S,P,b',[g]) = \int_{C(S,P,b')} d\lambda g(\theta) \qquad (60)$$

instead of $B_{ext}(S,P,[g])$. The integration on the right-hand side of Eq. (60) is over the boundary $C(S,P,b')$ of all perimeter loops on the section plane P formed in a range of length scales $b' \to b' + \delta b'$ during the generation of S. (Since we take $\delta b'$ infinitesimally small this means at most one loop.)

For the case of a fractal pore surface, $B(b,b')$ exhibits the following scaling form:

$$B(b,b') = B(b',b')\tilde{f}(b/b') \qquad (61)$$

When $b_1 \leq b \ll b'$, $\tilde{f}(b/b')$ exhibits a power-law dependence on b/b' of the same form as the dependence of $B_{ext}(b)$ on b/b_2 given in Eq. (56). That is,

$$\tilde{f}(b/b') \propto (b/b')^{-\mu_0\sigma_0} + \text{small corrections.} \qquad (62)$$

Thus the island perimeters have the same fractal dimension as that of the pore-profile outline, D_p^{ext}.

To obtain an expression for the number of islands $N(A)dA$ with areas in the range $A \to A + dA$ seen on a section through a fractal pore surface S_1, we make use of the following result, which is explained below:

$$N(A)dA \propto \int_{b_1}^{b_2} P(A|b)dA\{\Sigma(b)\sigma(b)bdb\} \qquad (63)$$

The integration on the right-hand side of Eq. (63) is over b and the quantity $P(A|b)dA$ is defined by

$$P(A|b)dA \equiv \text{Probability (an island first seen at a range of length scales } b \to b - \delta b \text{ has an area } A \to A + dA \text{ on forming } S_1)$$

$$(64)$$

The term in braces on the right-hand side of Eq. (63) arises because $\Sigma(b)\sigma(b)\delta b$ is the probability of adding a feature at a range of length scales $b \to b - \delta b$ while the probability that the section plane intersects it to form an island is proportional to b. The constant of proportion in Eq. (63)

depends on the region R through which the random section is
taken (cf. Eq. (20)) and on the feature shape distribution.
Eliminating $\Sigma(b)$ in favour of $B(b)$ using Eq. (20) yields the
result

$$\overline{N(A)}dA = \overline{a}_A dA \int_{b_1}^{b_2} P(A|b)B(b)\sigma(b)bdb \tag{65}$$

where $\overline{a}_A$ is a constant depending only on the feature shape
distribution.

We now consider the special case $b_1 = 0$, remarking that
the pore-space profile area and that of the islands is finite
in this case because of the finiteness of $V(0)$. In this case,
the only length scales relevant to the dimensionless quantity
$P(A|b)dA$ are b and $\sqrt(A)$. Hence this quantity must have the
scaling form

$$P(A|b)dA = b^{-2}\tilde{P}(A/b^2)dA \tag{66}$$

The scaling function $\tilde{P}(x)$ is expected to fall off rapidly
as $x \to \infty$. To obtain its behaviour for small x we note that,
for the section plane to cut off an island with an area
smaller than $A_0 \ll b^2$, it must lie a small distance of order
A_0/b from a plane tangential to the feature surface. Then

$$\int_0^{A_0} P(A|b)dA \sim A_0/b^2 \tag{67}$$

Substituting Eq. (66) into Eq. (67) then leads to the
conclusion that $\tilde{P}(x)$ tends to a finite constant $\tilde{P}(0)$ as $x \to 0$.

Substituting Eqs. (10) and (23) into Eq. (65), setting b_1
= 0 and applying Eq. (66) leads to the result that

$$\overline{N(A)}dA = \{dA/b_2^2\}\{b_2^{-1}B(b_2)\sigma_0\overline{a}_A/2\}a^{-(1 + D_p/2)}I(a) \tag{68}$$

where a is the reduced area A/b_2^2 and $I(a)$ is given by

$$I(a) = \int_a^\infty dx\tilde{P}(x)x^{D_p/2} \tag{69}$$

Assuming that a is small, and making use of the fact that $\tilde{P}(x)$
tends to a constant for small x, we deduce that

$$\overline{N(A)}dA \propto dA/b_2^2 a^{-(1 + D_p/2)}\{1 + O[a^{(1 + D_p/2)}]\} \tag{70}$$

When b_1 is non-zero, but much less than $\sqrt{(A)}$, a small correction that is a power of b_1^2/A must be included inside the braces on the right-hand side of Eq. (70). The leading-order dependence of $N(A)dA$ is the same as obtained by Mandelbrot for his "Brown" islands [22].

VI. CONCLUDING REMARKS

The principal novel aspects of the surface roughness model introduced in this paper are the facts that it addresses type II roughness and has been constructed so that it is feasible to extract the model parameters from measurements on thin-section images. For the case of a fractal pore surface, it has been shown that the fractal dimension of the pore-profile outline is not simply related to the surface fractal dimension D_s. To estimate D_s, it is recommended either to measure the scale dependence of the total pore-profile perimeter or to use the island area distribution law given in Eq. (70).

A problem for the future is the general one of extracting $\sigma(b)$ and $\overline{\Delta\Sigma}$ starting from thin section data. A combination of analytic work (especially for the embedded sphere case) and analysis of sections through simulated surfaces is expected to yield useful results. It would also be worthwhile to relate the present surface model to the small-angle scattering experiments reported in Refs. 2 and 3.

ACKNOWLEDGEMENTS

I should like to thank Fritz Rambow, whose enthusiasm for fractals motivated the present work, and the management of Shell Internationale Research Maatschappij for granting permission to publish this paper.

REFERENCES

1. Mandelbrot, B.B. (1982). "The Fractal Geometry of Nature". Freeman, San Francisco.
2. Wong, P., Howard, J., and Lim, J.-S. (1986). Phys. Rev. Lett. 57, 637.

3. Mildner, D.F.R., Rezvani, R., Hall, P.L., and Borst, R.L., (1986). Appl. Phys. Lett. 48, 1314.
4. Krohn, C.E., and Thompson, A.H. (1985). Phys. Rev. B 33, 6366.
5. Jacquin, C.G., and Adler, P.M. (1982). Transport in Porous Media 2 571.
6. Lenormand, R., and Zarcone, C. (1985). Phys. Rev. Lett. 54 2226.
7. Feder, J. (1988). "Fractals". pp. 49-61 and Refs. therein. Plenum Press, New York.
8. Barton, C.C., and Larson, E. (1985) in "Proceedings of the International Symposium on Fundamentals of Rock Joints" (O. Stephansson, ed.), p. 77. Centek Publishers, Lulea.
9. Mandelbrot, B.B., and van Ness, J.W. (1968). SIAM Journal 10, 422.
10. Voss, R.F. (1985) in "Fundamental Algorithms in Computer Graphics" (R.A. Earnshaw, ed.), p. 805. Springer-Verlag, Berlin.
11. Mandelbrot, B.B., and Wallis, J.R. (1969). Water Resour. Res. 5, 321.
12. Hewett, T.A. (1986). SPE paper 15386, presented at the 61st Annual Technical Conference of the Society of Petroleum Engineers.
13. Serra, J. (1982). "Image Analysis and Mathematical Morphology". Academic Press, London.
14. Mandelbrot, B.B. (1987). Private communication reported in Ref. 7, p. 11.
15. Rambow, F. (1989). To be published.
16. Thomas, T.R. (1982). "Rough Surfaces". Longman, New York.
17. Phan-Thien, N. (1981). Phys. Fluids 24, 579.
18. Lenormand, R., and Zarcone, C. (1984). SPE paper 13264 presented at the 59th Annual Technical Conference of the Society of Petroleum Engineers.
19. Katz, A.J., and Trugman, S.A. (1988). J. Colloid and Interface Sci. 123, 8.
20. Weibel, E.R. (1980). "Stereological Methods. Vol. 2, Theoretical Foundations". Academic Press, New York.
21. Jerri, A.D. (1985). "Introduction to Integral Equations with Applications". pp. 137-146. Marcel Dekker, New York.
22. Mandelbrot, B.B. (1975). Proc. Natl. Acad. Sci. USA 72, 3825.

A DUAL-POROSITY, SURFACE, AND MATRIX ELECTRICAL CONDUCTION MODEL FOR LOW-CONTRAST RESISTIVITY SANDSTONE FORMATIONS

W. Wendell Givens

Mobil Research and Development Corporation
Dallas, Texas

I. ABSTRACT

Based upon the Archie and other often used electrical conduction models, low-contrast resistivity (LCR) formations have high calculated water saturations. However, many LCR formations produce hydrocarbons with a very low watercut. Scanning-electron-microscope micrographs, capillary pressure curves, and proton magnetic resonance data for an extensive class of LCR sandstones suggest the following: a free-fluid pore space formed by sand grains and a capillary bound-water pore space formed by authigenic clay minerals lining the sand grains and in the pore throats. Log-log plots of resistivity index versus partial water saturation for most LCR rocks are nonlinear. A new dual-porosity, surface, and matrix (DPSM) electrical conduction model of fluid-filled rocks predicts these nonlinearities.

The DPSM model treats bulk electrical conductivity as three parallel conductance paths: (1) a free-fluid macropore space, (2) a capillary bound-water micropore space, and (3) the matrix (surface conductance and conductance of non-clay minerals). Archie equations are applied to each pore space and can have different values of m and n. Routine laboratory macroscopic core measurements provide all DPSM model parameters needed for log analysis.

RESERVOIR CHARACTERIZATION II

27

II. INTRODUCTION

Based upon the Archie (Archie, 1942) and the often used electrical conduction models, low contrast resistivity (LCR) formations have high calculated water saturations. However, many LCR formations produce hydrocarbons with a very low watercut. The extent and abundance of potentially productive LCR formations are now being recognized. Not all LCR formations produce hydrocarbons. The problems are to identify productive LCR formations and to evaluate their hydrocarbon potential quantitatively. Evaluation of the hydrocarbon potential of a formation is heavily based upon its electrical conductivity. Therefore, to create a well-suited conductivity model, one must know the rock and fluid properties that contribute to electrical conductivity in reservoir rocks.

Several long-held and widely accepted concepts and premises have probably slowed the development of interpretation models for the more complex reservoir rocks. This is especially true for clay-bearing sandstones. One such premise is that the passage of fluids and electric current follow identically the same paths through a rock and therefore that capillary bound water does not conduct electric current (Pirson, 1958). A general acceptance of this premise may have led to an initial overacceptance of surface conduction (Waxman and Smits, 1968) as the only real contributor to electrical conductivity by clay minerals in sandstones. Our studies suggest the conductance of capillary-bound water is much more prevalent and usually contributes more to rock conductivity than surface conductance. Another premise is that the empirical Archie formation factor and resistivity index equations, $F=1/\phi^m$ and $I=1/S_w^n$, apply universally to all rocks (Archie, 1942). Some suggested characteristics of Archie and non-Archie rocks (Herrick, 1988) are given in Table I. The Archie concepts of a formation factor ($F=R_o/R_w$) and a resistivity index ($I=R_t/R_o$) are satisfying. These concepts and the empirical Archie porosity and water saturation equations can be used effectively as building blocks in models of non-Archie rocks.

A conductive rock matrix model (CRMM) was an initial attempt at developing an electrical analog of a non-Archie rock (Givens, 1986, 1987). The CRMM treats the bulk electrical conductivity of LCR rocks as two parallel conductance paths. One path is a conductive pore network containing fluids that are free to move and the second path is the remainder of the rock (the matrix). The matrix may be conductive due to one or more of the following: surface conduction, capillary-bound water, or conductive materials, such as pyrite.

The proposed dual-porosity, surface, and matrix (DPSM) electrical conduction model separates capillary-bound water from the matrix by treating a rock's bulk electrical conductivity as three parallel conductance paths. The three paths are: (1) a free-fluid pore space, subject to preferential displacement of water by hydrocarbons and which can have a water saturation S_w between 1 and 0, (2) capillary-bound water in the remainder of the interconnected pore space ($S_w=1$), and (3) the matrix (surface conductance and conductive minerals).

TABLE I. Factors Affecting Electrical Rock Type

Rock type	Geometric factors	Non-geometric factors
	Unimodal Pore System	
Archie rocks	Intergranular porosity	Water wet No surface conduction No conductive minerals
	Polymodal Pore Systems	
Non-Archie rocks	Moldic/vuggy porosity (oomoldic carbonates) Discontinuous micro-porosity (structural shale) (microporous chert) Continuous microporosity (authigenic clay coatings)	Oil wet Surface conduction (clay minerals) Conductive minerals

(Herrick, 1988)

III. PHYSICAL PROPERTIES OF A LARGE CLASS OF LOW-CONTRAST RESISTIVITY (LCR) SANDSTONES

Scanning electron microscope (SEM) micrographs of samples from LCR sandstone reservoirs in a large number of different geographical regions shows a common characteristic - a bimodal distribution of large and small pore sizes. Figure 1 shows a micrograph of a typical LCR sandstone. The small pores are formed primarily by authigenic clay minerals. They line the surface of the primary pore regime formed by the sand grains, and the large and small pores are interconnected. Mineral analysis shows that all the common clay minerals can contribute to the network of smaller pores.

Oil-brine capillary pressure data suggest a bimodal distribution of pore sizes, a sequential desaturation of the pores in the order of large to small pores, and a range of irreducible water saturation S_{wir} from about 10-70 percent. Pore sizes (surface to volume ratios) determined by proton magnetic resonance (PMR) also display a bimodal distribution of pore sizes in most LCR sandstones.

 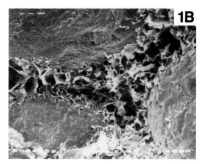

FIGURE 1. SEM micrographs 1A (magnification 100X) AND 1B (magnification 400X).

Resistivity data obtained concurrently with capillary pressure data show no meaningful correlation with either the amount or type of clay minerals. Log-log plots of resistivity index versus partial water saturation for most LCR sandstones are nonlinear.

Presumed in the Archie resistivity index equation $I=S_w^{-n}$ is that all pores desaturate equally. Even pores in the most ideal rocks desaturate preferentially. Thus, an electrical conduction model must include the conductance of that fraction of the interconnected pore space that desaturates preferentially and separately the unperturbed electrical conductance due to irreducible capillary-bound water.

IV. CONCEPT OF A DUAL-POROSITY, SURFACE, AND MATRIX (DPSM) ELECTRICAL CONDUCTION MODEL

The SEM micrograph shown in Figure 1 is typical of the LCR sandstone studied. The common characteristic is a bimodal distribution of large and small pores. The small pores line the surface of the large pores and are formed primarily by clay minerals. The small pores appear to be open to the larger pores and therefore can contain the same type of water as the larger pores. Mineral analyses show that all the common clay minerals can contribute to the network of small pores. Since the small pore regime is on the surface of the larger pores, the large and small pore regimes should act electrically as parallel conductance paths.

A schematic of this type of bimodal pore network is shown in Figures 2(A), (C), and (D). For definitions of symbols, see the nomenclature. The total interconnected porosity ϕ consists of two parts, a free-fluid porosity ϕ_f and an irreducible water porosity ϕ_i, i.e., $\phi = \phi_f + \phi_i$. Water saturation S_{wf} can range from 0 to 1 in the ϕ_f pore space and the water saturation S_{wi}

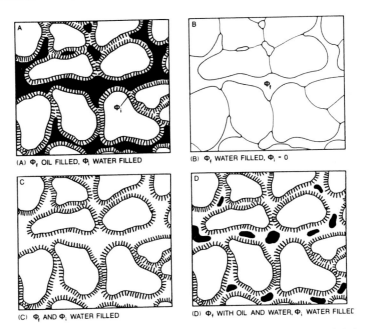

(A) Φ_f OIL FILLED, Φ_i WATER FILLED   (B) Φ_f WATER FILLED, Φ_i = 0

(C) Φ_f AND Φ_i WATER FILLED   (D) Φ_f WITH OIL AND WATER, Φ_i WATER FILLED

FIGURE 2. Schematic of typical LCR sandstone with clay fabric lining the major pores.

equals 1 in the ϕ_i pore space. Figure 2(A) shows the ϕ_f pore space totally filled with oil ($S_{wf} = 0$) and the ϕ_i pore space filled with water ($S_{wi} = 1$). A measurement of the resistivity for this condition gives R_{io} (R_0 for the ϕ_i pore space), and the ratio R_{io}/R_w is the formation factor for the ϕ_i pore space. We assume the Archie equation $F = 1/\phi^m$ is valid for the ϕ_i pore space and has a porosity exponent m_i. This would be the same F and m measured if the total porosity was ϕ_i, and the ϕ_f pore space was a nonconductor. To a more or less degree, depending on the type of clay minerals forming the ϕ_i pore space, two types of conduction can take place in this pore regime, conduction in capillary-bound bulk water and surface conduction at the bulk-water clay-surface interface. These two conductance paths appear to be in parallel and are so treated in DPSM. In terms of electrical conductance, the porosity exponent m is a "quantitative" measure of how effectively the water-filled pore space conducts current. Earlier work (Givens, 1987), showed that matrix conductance in parallel with the pore space makes the water-filled pore space appear to be a more effective current conductor than it actually is. This is reflected in a value of m that is less than would be measured if matrix conductance was not present.

Figure 2(B) illustrates the same rock where the total porosity has been reduced to ϕ_f by enlarging the rock grains to the size of the original grains plus the ϕ_i pore space. The total porosity is now ϕ_f and is water-filled. A measurement of the resistivity for this condition gives R_{fo} (R_0 for the ϕ_f pore space), and the ratio R_{fo}/R_w is the formation factor for the ϕ_f pore space. Again, we assume the Archie formation factor equation $F = 1/\phi^m$ is valid for the ϕ_f pore space and has a porosity exponent m_f. The electrical effectiveness of the water-filled ϕ_f pore space and the water-filled ϕ_i pore space need not be the same. Therefore, m_f and m_i are allowed to be different in the DPSM.

Thus, the rock illustrated consists of two pore spaces that can have different electrical properties and are electrically in parallel. A measurement of the resistivity for this condition gives R_o for the total ϕ pore space, and the ratio R_o/R_w is the measured formation factor for the rock. Without knowing any better, we assume the formation factor $F = 1/\phi^m$. The question is, how is m (shown later to be m_a) related to m_f and m_i?

The problem to be solved by an interpretion model of LCR rocks is to determine the oil saturation when the free-fluid pore space is partially saturated with oil, as illustrated by Figure 2(D).

V. MATHEMATICAL DEVELOPMENT OF DPSM

Based upon the concepts developed in the preceding section, DPSM treats the bulk electrical conductivity of a LCR rock in terms of three parallel conductance paths: (1) a free-fluid pore space that is subject to preferential desaturation and can have a water saturation S_{wf} from 0 to 1, (2) capillary-bound (irreducible water) in the remainder of the interconnected pore space ($S_{wi} = 1$), and (3) the matrix including surface conduction. The model assumes the Archie formation factor equation is valid for both the free-fluid and capillary-bound water pore regimes and the porosity exponents m_f and m_i can be different. The model also assumes the Archie resistivity index equation is valid for the free-fluid pore space and can be characterized by a saturation exponent n_f. It may appear that DPSM is a dual-porosity model, and in a sense it is; but as will be apparent in the development, DPSM is nothing like the dual-porosity model of Raiga-Clemenceau (Raiga-Clemenceau et al., 1984).

We assume that the total or bulk electrical conductance can be represented by three resistances in parallel. The electrical conductances of three resistances in parallel is equal to the sum of the separate conductances, i.e.,

$$\frac{1}{r} = \frac{1}{r_{fx}} + \frac{1}{r_{io}} + \frac{1}{r_r} \ , \tag{1}$$

where r_{fx} is the resistance of the free-fluid pore network, r_{io} is the resistance of the capillary-bound water network, and r_r is the resistance of the rock matrix that includes surface conductance. In my previous work on the development of the CRMM (Givens, 1987), it was shown that the geometric factors were the same for the whole rock and separately for the individual pore networks and the rock framework when expressed in terms of the appropriate bulk resistivity. Therefore, in terms of bulk resistivities, Equation (1) for a fully water saturated rock ($S_w = 1$) becomes

$$\frac{1}{R_o} = \frac{1}{R_{fo}} + \frac{1}{R_{io}} + \frac{1}{R_r} \ . \tag{2}$$

For a partially saturated rock ($S_{wf} < 1$), Equation (1) becomes

$$\frac{1}{R_t} = \frac{1}{R_{ft}} + \frac{1}{R_{io}} + \frac{1}{R_r} \ . \tag{3}$$

A. Porosity Mixing Rules

DPSM treats a rock as having a total interconnected porosity ϕ consisting of two parts, a free-fluid porosity ϕ_f and a capillary-bound water porosity ϕ_i. An obvious mixing rule is

$$\phi = \phi_f + \phi_i \ . \tag{4}$$

DPSM also attributes to both pore spaces a formation factor that can have different porosity exponents m_f and m_i and a formation factor for the total pore space ϕ with a porosity exponent m_a. The question, asked in a preceding section, is how is m_a related to m_f and m_i? First, assume the matrix conductance $1/R_r = 0$. Then Equation (2) becomes

$$\frac{1}{R_o} = \frac{1}{R_{fo}} + \frac{1}{R_{io}} \ . \tag{5}$$

Multiplying both sides of Equation (5) by R_w, it becomes

$$\frac{1}{F} = \frac{1}{F_f} + \frac{1}{F_i} \ , \tag{6}$$

where F, F_f, and F_i are the formation factors for the total porosity, the free-fluid porosity, and the capillary-bound water porosity, respectively. Applying the Archie formation factor relationship $F = 1/\phi^m$, Equation (6) becomes

$$\phi^{m_a} = \phi_f^{m_f} + \phi_i^{m_i} \ , \tag{7}$$

which is a second and very interesting mixing rule. Backward theoretical calculations and analysis of experimental data with DPSM at the best confirm this relationship, and at the worst show an internal consistency in our calculations and in applying the DPSM to resistivity data. More will be said about this mixing rule in a later section. In addition to the two mixing rules, it is assumed that ϕ and ϕ_i are also related by the irreducible water saturation S_{wir}, i.e.,

$$S_{wir} = \left[\frac{\phi_i}{\phi} \right] .$$

(8)

If surface or any other form of matrix conductance exists, the last two terms in Equations (2) and (3) are both constant, since $S_{wi} = 1$ in the ϕ_i pore space, and can be replaced by an equivalent single term $1/R_{ir}$.

$$\frac{1}{R_o} = \frac{1}{R_{fo}} + \frac{1}{R_{ir}} ,$$

(9)

and Equation (5) becomes

$$\frac{1}{R_t} = \frac{1}{R_{ft}} + \frac{1}{R_{ir}} .$$

(10)

Multiplying both sides of Equation (12) by R_w and applying the Archie formation factor equation $F = 1/\phi^m$, Equation (12) becomes

$$\phi^{m_a} = \phi_f^{m_f} + \phi_i^{m_{ir}} ,$$

(11)

which is the more general form of Equation (7), because the effect of any matrix conductance is reflected in the porosity exponent m_{ir}. The porosity exponent m_{ir} is less than the porosity exponent m_i in Equation (7), because matrix conductance, including surface conductance, makes the water-filled ϕ_i pore space appear to be a more effective conductor of electric current than it is alone. If clay minerals line the primary pores, there is almost certain to be some surface conductance, the amount depending on the type of clay minerals present. Backward theoretical calculations that include matrix conductance satisfy this mixing rule.

B. DPSM and CRMM Equations

The Archie definitions of the formation factor and resistivity index are applied to Equations (9) and (10) to develop formation factor, resistivity versus water saturation, resistivity index versus water saturation, and water saturation versus porosity equations for the DPSM. The development of these equations is the same as in the development of the CRMM (Givens, 1987) and will not be repeated here. Corresponding DPSM and CRMM equations are shown for comparison. DPSM Equations (9) and (10) and the corresponding CRMM equations are as follows:

<u>DPSM</u>   <u>CRMM</u>

$$\frac{1}{R_o} = \frac{1}{R_{fo}} + \frac{1}{R_{ir}} \qquad (9) \qquad \frac{1}{R_o} = \frac{1}{R_{po}} + \frac{1}{R_r}$$

$$\frac{1}{R_t} = \frac{1}{R_{ft}} + \frac{1}{R_{ir}} \qquad (10) \qquad \frac{1}{R_t} = \frac{1}{R_{pt}} + \frac{1}{R_r}$$

These equations have the same form but are conceptually different for reasons discussed in the preceding section on DPSM concepts and later in this section. DPSM and CRMM equations derived from the above equations are as follows:

<u>DPSM</u>   <u>CRMM</u>

$$F = \frac{B}{\phi_f^{m_f}} \qquad (12) \qquad F = \frac{b}{\phi^m}$$

where

$$B = \frac{R_{ir}}{(R_{fo} + R_{ir})} , B \le 1 \qquad (13) \qquad b = \frac{R_r}{(R_{po} + R_r)} , b \le 1$$

$$\frac{1}{R_t} = \frac{S_{wf}^{n_f}}{R_{fo}} + \frac{1}{R_{ir}} \qquad (14) \qquad \frac{1}{R_t} = \frac{S_w^n}{R_{po}} + \frac{1}{R_r}$$

$$\frac{1}{I} = BS_{wf}^{n_f} + (1 - B) \qquad (15) \qquad \frac{1}{I} = bS_{wf}^n + (1 + b)$$

$$S_{wf} = \left[\frac{R_w}{R_t \phi_f^{m_f}} + \left(1 - \frac{1}{B} \right) \right]^{\frac{1}{n_f}} \quad (16) \quad S_w = \left[\frac{R_w}{R_t \phi^m} + \left(1 - \frac{1}{b} \right) \right]^{\frac{1}{n}}$$

The DPSM and CRMM models differ in a very fundamental way. For some rocks, this difference is not significant; however, for rocks having a large amount of capillary-bound water, the difference in calculated water saturations using DPSM Equation (18) or CRMM Equation (18) can be large.

The DPSM and CRMM differ in the treatment of conductance in capillary-bound water. The CRMM treats capillary-bound water conductance as a part of matrix conductance; the DPSM treats them as separate parallel conductance paths. The result of this distinction is the DPSM produces equations in terms of the free-fluid or effective porosity ϕ_f

and its associated Archie parameters S_{wf}, m_f, and n_f. It follows that in plots of the resistivity index versus water saturation the plot should be I versus S_{wf} and not I versus S_w, the average water saturation!

Implicit in the Archie resistivity index equation $I = 1/S_w^n$ is the assumption that all pores, regardless of size, desaturate equally. Capillary pressure curves indicate this is not the case. The common practice in making resistivity measurements is to displace brine in an initially fully saturated core by oil or gas. For a given volume of displaced brine, the resistivity index is measured and the partial water saturation is calculated in terms of the volume of brine displaced and the total pore volume. This water saturation is the average water saturation S_w for the total pore space. In the case of rocks characterized by a high S_{wir}, plotting I versus S_w can produce a variety of nonlinearities in a Log (I) versus Log (S_w) plot that are impossible to interpret in terms of the simple Archie equation.

C. Irreducible Water Saturation S_{wir}

Irreducible water saturation in the interconnected pore space of a rock is a valid concept but lacks a widely accepted definition. It is likely dependent on both the pore structure and the capillary pressure conditions in a reservoir. Capillary pressure curves give good estimates of S_{wir}.

In the section on porosity mixing rules for a bimodal pore system, it was assumed that

$$S_{wir} = \left[\frac{\phi_i}{\phi} \right] ,$$
(8)

which, in fact, is not an unreasonable definition of S_{wir}. It is easily shown that if this relation is valid, then

$$S_{wf} = \frac{\left(S_w - S_{wir} \right)}{\left(1 - S_{wir} \right)} .$$
(17)

Equation (17) is used both in theoretical calculations and in the analysis of experimental resistivity index data.

VI. THEORETICAL RESISTIVITY INDICES

Theoretical resistivity index curves based upon the DPSM are shown in Figures 3, 5, 7, and 9. The theoretical values of I are calculated using the DPSM with the following input parameters: ϕ, S_{wir}, R_w, m_i, m_f, n_f, and R_r. The independent variable is S_{wf}, and the dependent variable is I. The theoretical data are correctly plotted as I versus S_{wf}. For input parameters $S_{wir} = 0$ and $1/R_r = 0$, $S_{wf} = S_w$, $\phi_f = \phi$, $n_f = n$, and the logarithmic plot of I versus S_{wf} is a straight line with a slope of -n. For $0 < S_{wir} < 1$, logarithmic plots of I versus S_{wf} show varying degrees of curvature toward the S_{wf} axis. The same values of I are also plotted versus S_w, because this is the customary way of plotting resistivity index data. Depending on the values of the input parameters, these plots may show a wide range of curvature toward or away from the S_w axis, inflections, and even a straight line. Figures 3, 5, 7, and 9 are examples that show all these features.

Figures 4, 6, 8, and 10 show a comparison of apparent water saturation using the standard Archie and CRMM models to the DPSM. These comparisons are for the same DPSM properties used to generate the resistivity index curves shown in Figures 3, 5, 7, and 9, respectively. The Archie (1) analysis assumes m = 2 and n = 2. The Archie (2) analysis uses a value of n_a obtained by fitting a straight line to the I versus S_w curve in the S_w range 1 to 0.5. The value of m_a is calculated using the Archie equation $F = 1/\phi^{m_a}$. The CRMM parameters are obtained by fitting the CRMM model to the I versus S_w curve as detailed in the CRMM paper (Givens, 1987). The DPSM and CRMM water saturations are calculated using DPSM Equation (16) and CRMM Equation (16).

The comparisons in Figure 4 show the Archie (2) analysis and the CRMM analysis agree quite well with the DPSM over a wide range of S_w, especially at the higher values of S_w. The Archie (1) analysis underestimates S_w considerably except at S_{wir}. It should be noted the m_a and n_a used in Archie (2) and the CRMM parameters m and n are significantly different from 2, the values of DPSM parameters m_f and n_f used in generating the needed values of R_t. The comparisons in Figure 6 show good agreement between the DPSM and CRMM over most of the range of S_w, but rather large overestimates of S_w by both Archie (1) and Archie (2) in the S_w range 0.2 to 0.6. The comparisons in Figure 8 show good agreement of Archie (2) and CRMM with DPSM over most of the S_w range. Archie (1) shows a very large disagreement (higher S_w) over the entire range of S_w, including a large range of apparent $S_w > 1$. The comparisons in Figure 10 also show rather good agreement of Archie (2) and CRMM with DPSM over most of the S_w range. Archie (1) shows an

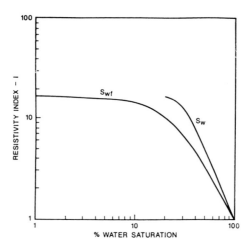

FIGURE 3. Theoretical DPSM resistivity index I versus S_{wf} and S_w. Input: $\phi = 0.2$, $S_{wir} = 0.2$, $R_w = 0.05$ Ω-m, $m_i = 2$, $m_f = 2$, $n_f = 2$, and $R_r = \infty$. Calculated rock parameters: $m_a = 2.24$ and $n_a = 2.39$.

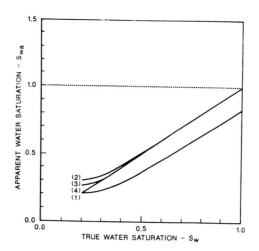

FIGURE 4. Comparison of apparent water saturation S_{wa} for Archie (1), Archie (2), and CRMM to DPSM (4) for the same rock properties as for Figure 3.

Archie (1)	Archie (2)	CRMM (3)	DPSM (4)
$n = 2$	$n_a = 2.39$	$n = 2.54$	$n_f = 2$
$m = 2$	$m_a = 2.24$	$m = 2.25$	$m_f = 2$
		$b = 0.973$	$B = 0.941$

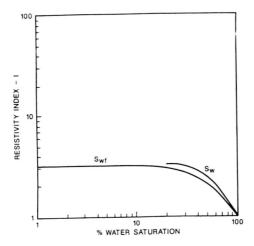

FIGURE 5. Theoretical DPSM resistivity index I versus S_{wf} and S_w.
Input: $\phi = 0.2$, $S_{wir} = 0.2$, $R_w = 0.05$ Ω-m, $m_i = 2$, $m_f = 2$, $n_f = 2$, and
$R_r = 5$ Ω-m. Calculated rock parameters: $m_a = 2.05$ and $n_a = 1.51$.

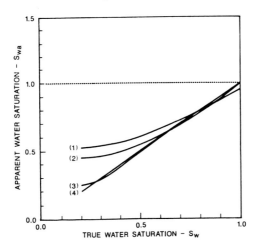

FIGURE 6. Comparison of apparent water saturation S_{wa} for Archie
(1), Archie (2), and CRMM (3) to DPSM (4) for the same rock properties
as for Figure 5.

Archie (1)	Archie (2)	CRMM (3)	DPSM (4)
$n = 2$	$n_a = 1.51$	$n = 2.54$	$n_f = 2$
$m = 2$	$m_a = 2.05$	$m = 2.25$	$m_f = 2$
		$b = 0.713$	$B = 0.688$

W. *Wendell Givens*

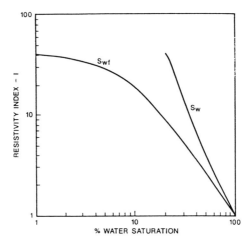

FIGURE 7. Theoretical DPSM resistivity index I versus S_{wf} and S_w. Input: $\phi = 0.2$, $S_{wir} = 0.2$, $R_w = 0.05$ Ω-m, $m_i = 2$, $m_f = 1.5$, $n_f = 1.5$, and $R_r = \infty$. Calculated rock parameters: $m_a = 1.64$ and $n_a = 1.69$.

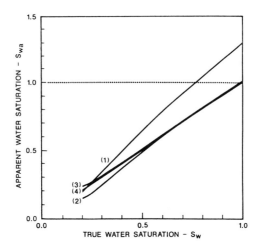

FIGURE 8. Comparison of apparent water saturation S_{wa} for Archie (1), Archie (2), and CRMM (3) to DPSM (4) for the same rock properties as for Figure 7.

Archie (1)	Archie (2)	CRMM (3)	DPSM (4)
$n = 2$	$n_a = 1.91$	$n = 1.81$	$n_f = 1.5$
$m = 2$	$m_a = 1.69$	$m = 1.66$	$m_f = 1.5$
		$b = 1.00$	$B = 0.976$

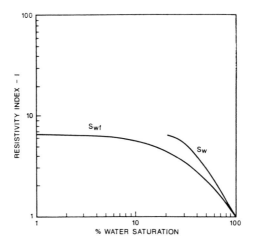

FIGURE 9. Theoretical DPSM resistivity index I versus S_{wf} and S_w. Input: $\phi = 0.2$, $S_{wir} = 0.2$, $R_w = 0.05$ Ω-m, $m_i = 2$, $m_f = 1.5$, $n_f = 1.5$, and $R_r = 5$ Ω-m. Calculated rock parameters: $m_a = 1.6$ and $n_a = 1.56$.

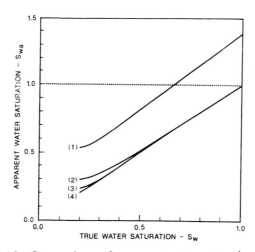

FIGURE 10. Comparison of apparent water saturation S_{wa} for Archie (1), Archie (2), and CRMM (3) to DPSM (4) for the same rock properties as for Figure 9.

Archie (1)	Archie (2)	CRMM (3)	DPSM (4)
$n = 2$	$n_a = 1.56$	$n = 1.81$	$n_f = 1.5$
$m = 2$	$m_a = 1.60$	$m = 1.66$	$m_f = 1.5$
		$b = 0.911$	$B = 0.847$

even larger disagreement (much higher S_w) over the entire range of S_w, including a larger range of $S_w > 1$, than is shown in Figure 8. Note that one or both of the Archie (2) values of m_a and n_a used in the comparisons in Figures 8 and 10 are significantly different from 2. These values are obtained from the generated resistivity data in exactly the same way values of m and n are usually obtained from experimental data. These comparisons suggest that better estimates of S_w result by using experimental values of m and n obtained in the customary way, even though the values may be uncomfortably different from 2.

VII. DPSM ANALYSIS OF CORE RESISTIVITY DATA

Figure 11 shows a plot of core resistivity index data with the measured values of I plotted against the corresponding average values of water saturation S_w. The solid curve is calculated using parameters determined by a fit of the DPSM to the I versus S_{wf} data. The values of S_{wf} corresponding to the measured values of average water saturation S_w are calculated using Equation (17). However, to make this transformation, a value of S_{wir} is needed.

A. Three Methods of Estimating S_{wir}

An estimate of S_{wir} can generally be obtained from a good quality capillary pressure curve.

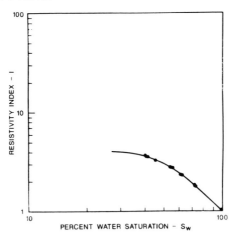

FIGURE 11. Measured resistivity index I versus average water saturation S_w for a LCR rock. The solid curve is a DPSM fit to the data.

PMR-determined pore size distributions synthetic capillary pressure curves generated from PMR data provide an easy visual way of estimating S_{wir}. The basis for estimating S_{wir} from a synthetic capillary pressure curve is discussed later.

The major premise of the DPSM is that capillary-bound water affects the bulk electrical properties of a fluid-filled rock. If this premise is true and the DPSM is an adequate model of a rock with an effective amount of capillary-bound water, then it may be possible to extract the amount of capillary water from resistivity index data. A third method then involves transforms of S_w to S_{wf} by iteration of S_{wir}. The DPSM is fit to R_t or resistivity index data plotted as R_t versus S_{wf} or I versus S_{wf}. S_{wf} is the transform of S_w for each value of S_{wir}. DPSM is fit to data exactly in the same way as the CRMM (Givens, 1987). Thus, a fit of DPSM gives values for the DPSM parameters R_{fo}, R_{ir} or B, and n_f. With the parameters R_{fo} and R_{ir} or B, the resistivity index can be corrected for the effect of capillary-bound water. If the parameter values correctly represent the electrical properties of the rock, a plot of the corrected data should be a straight line. The iteration of S_{wir} stops when the values of n_f from the DPSM fit and the slope (n_f) of a linear least-squares (LLS) fit to the corrected data differ by less than 0.005.

B. Oil-Brine and PMR Synthetic Capillary Pressure Curves

Figure 12 shows an oil-brine capillary pressure curve and a synthetic capillary pressure curve for the same core sample as the resistivity index data shown in Figure 11. The synthetic capillary pressure curve was generated from pore surface-to-volume ratios determined from PMR T1 measurements shown in Figure 13. Moving from higher to lower S_w, each vertical step in the curve corresponds to a transition from one size group of pores to the next smaller size group. Rather than giving the details of calculating the synthetic curve, the curve is interpreted as the largest pores make up about 15% of the total pore volume, the next smaller pore group (dual peaks) about 35%, the next smaller group about 20%, and so on. The size of the increase in capillary pressure at each transition depends on how different in size are the adjacent sized group of pores.

Estimating S_{wir} from either of these curves is subjective. The oil-brine capillary curve shows a gradual rather than a distinctly sharp increase in capillary pressure beginning at about 60% water saturation. The maximum capillary pressure that could be achieved was limited by the porous plate to a little more than 20 psi. Due to the shape of the curve, S_{wir} was estimated to be about 0.35 to 0.4. The synthetic curve shows a large transition in capillary pressure at a water saturation of about 0.3. Again, an estimate of S_{wir} from the synthetic curve is subjective, but a S_{wir} of 0.3 seems more likely than 0.5, the value of S_w at the next higher S_w transition.

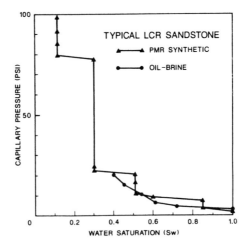

FIGURE 12. Experimental and PMR synthetic oil-brine capillary pressure curves for LCR rock cited in Figure 11.

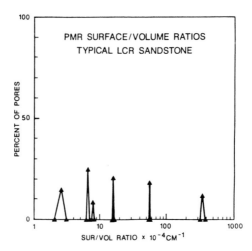

FIGURE 13. PMR surface/volume ratios for LCR rock cited in Figure 11.

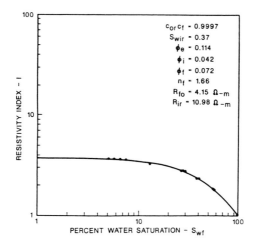

FIGURE 14. The I data shown in Figure 11 but with S_w transformed to S_{wf} using Equation 19 and $S_{wir} = 0.37$. The solid curve is a DPSM fit to the transformed data. All parameters shown are from the DPSM analysis except S_{wir} and ϕ.

C. S_{wir} From Resistivity Data

The S_{wir} iteration method described earlier was applied to the resistivity index data shown in Figure 11, and the results are shown in Figures 14-19. A value of $S_{wir} = 0.37$ was selected to begin the iteration process. The initial value is not too critical because the iteration process can proceed in either direction from this value. A reasonable value reduces the iteration process. For each S_{wir}, S_{wf} is calculated using Equation (19), and DPSM is fit to the transformed data. The transformed data for the initial value of $S_{wir} = 0.37$ and the DPSM fit (the solid curve) are shown in Figure 14. The DPSM fit is very good. The resistivity index, corrected for the conductance parallel to the ϕ_f pore space, is shown in Figure 15. The straight line is not a LLS fit in this case but is calculated using the value of n_f that came from the DPSM fit. Not only is the corrected data plot curved, but a LLS fit would give a slope larger than n_f. The convergence criterion for terminating the iteration process, the absolute value of the slope by a LLS fit differing from n_f from the DPSM fit by less than 0.005, is satisfied by an $S_{wir} = 0.27$, and the results are shown in Figures 16 and 17. The resistivity index is calculated using the convergent value of S_{wir} and values of the parameters obtained from the DPSM fit. I versus S_{wf} and I versus S_w curves are shown in Figure 18. The I versus S_w curve is also shown in Figure 11. A comparison of apparent water saturation S_{wa}, using the

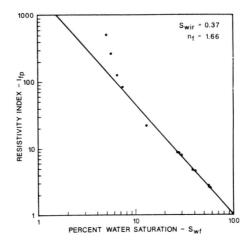

FIGURE 15. The data shown in Figure 14 are corrected for the effect of R_{ir}. If $S_{wir} = 0.37$ is the best value for this rock, the data should plot as a straight line. The solid straight line is calculated using the $n_f = 1.66$ from the DPSM fit shown in Figure 14.

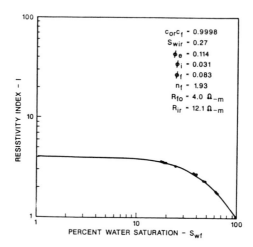

FIGURE 16. The I data shown in Figure 11 but with S_w transformed to S_{wf} using Equation 19 and $S_{wir} = 0.27$. The solid curve is a DPSM fit to the transformed data. All parameters shown are from the DPSM analysis except S_{wir} and ϕ.

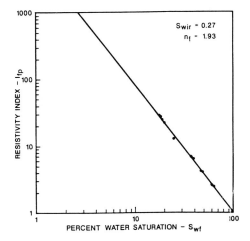

FIGURE 17. The data shown in Figure 16 are corrected for the effect of R_{ir}. $S_{wir} = 0.27$ was obtained by iteration and was the best value that satisfied the convergence criterion,

$$| \, n_f \, (\text{DPSM}) - n_f \, (\text{LLS - Corrected I}) \, | \leq 0.005.$$

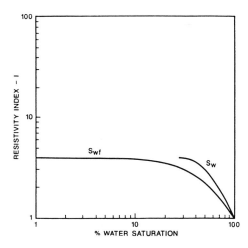

FIGURE 18. Calculated resistivity index I versus S_{wf} and S_w using parameters from the DPSM fit of the transformed data shown in Figure 16.

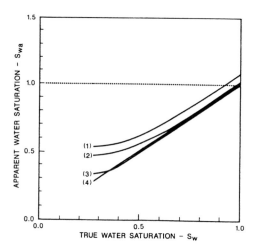

FIGURE 19. A comparison of apparent water saturation S_{wa} for Archie (1), Archie (2), and CRMM (3) to DPSM (4) using the parameters obtained by the DPSM fit to the transformed data shown in Figure 16.

Archie (1)	Archie (2)	CRMM (3)	DPSM (4)
$n = 2$	$n_a = 1.85$	$n = 2.67$	$n_f = 1.93$
$m = 2$	$m_a = 1.93$	$m = 2.04$	$m_f = 1.80$
		$b = 0.793$	$B = 0.752$

standard Archie model and the CRMM to the DPSM, is shown in Figure 19. These comparisons show that Archie (1) shows the largest difference; Archie (2) shows good agreement above $S_w > 0.7$, and CRMM and DPSM agree over almost the entire range of water saturation.

VIII. CONCLUSIONS

1. Many LCR sandstones are characterized by a bimodal pore distribution of large and small pores.
2. The small and large pore networks may be different electrically, e.g., each pore network may have a different value of m.
3. The distribution of the clay minerals, rather than the amount or type of minerals, determines the electrical effectiveness of capillary-bound bulk water.
4. To be electrically effective, the amount of capillary-bound water does not need to be large.

5. Surface conduction can be active but need not be dominant in the small pore regime.
6. Due to preferential desaturation of large pores and the existence of capillary-bound water, resistivity index data should be interpreted in terms of S_{wf} rather than S_w.
7. Resistivity index curves for rocks with a bimodal pore system show different degrees and types of nonlinearity.
8. Good resistivity index data are invaluable in analyzing LCR sandstones.
9. DPSM is not only applicable to LCR rocks, but also other types of rocks.

NOMENCLATURE

b	=	A parameter in the CRMM model that is the ratio $R_r/(R_{po} + R_r)$.
B	=	A parameter in the DPSM model that is the ratio $R_{ir}/(R_{fo} + R_{ir})$.
F	=	Formation factor that is the ratio R_o/R_w, in the DPSM model is also the ratio B/ϕ_f^{mf}, and in the CRMM model is the ratio b/ϕ^m.
F_f	=	Formation factor for the free-fluid pore space in the DPSM model. $F_f = R_{fo}/R_w = 1/\phi_f^{mf}$.
F_i	=	Formation factor for the capillary-bound water pore space in the DPSM model. $F_i = R_{io}/R_w = 1/\phi_i^{mi}$.
I	=	Resistivity index that is the ratio R_t/R_o.
I_{fp}	=	Resistivity index for the free-fluid pore space that is the ratio R_{ft}/R_{fo}.
m	=	Porosity exponent for an ideal Archie rock and the CRMM model.
m_a	=	The apparent porosity exponent when setting $F = 1/\phi^{m_a}$ in either the DPSM or CRMM models. For an ideal Archie rock, $m_a = m$.
m_f	=	Free-fluid porosity (ϕ_f) exponent in the DPSM model.
m_i	=	Capillary-bound water porosity (ϕ_i) exponent in the DPSM model.
m_{ir}	=	Quasi-porosity exponent of ϕ_i when there is matrix conductance.
n	=	Saturation exponent for an ideal Archie rock and the CRMM model.
n_a	=	Apparent porosity exponent for any other than an ideal Archie rock when the rock is treated as an ideal Archie rock.
n_f	=	Saturation exponent for the free-fluid pore space in the DPSM model.
ϕ	=	Total interconnected porosity in a rock.
ϕ_f	=	Free-fluid component of ϕ.
ϕ_i	=	Capillary-bound water component of ϕ.
r	=	Electrical resistance (ohms).
r_r	=	Electrical resistance (ohms) of a rock's matrix including surface conductance.

r_{fx} = Electrical resistance (ohms) of a rock's interconnected porosity filled with fluids that are free to move.

r_{io} = Electrical resistance (ohms) of a rock's interconnected porosity filled with capillary-bound water having a resistivity R_w.

R_o = Bulk electrical resistivity (ohm-meters) of a rock's totally interconnected porosity totally filled with water having a resistivity R_w.

R_r = Bulk electrical resistivity (ohm-meters) of a rock's matrix, including surface conductivity.

R_t = Bulk electrical resistivity (ohm-meters) of a rock's total interconnected porosity partially filled with water having a resistivity R_w.

R_{fo} = Bulk electrical resistivity (ohm-meters) of a rock's free-fluid porosity fully saturated with water having a resistivity R_w.

R_{ft} = Bulk electrical resistivity (ohm-meters) of a rock's free-fluid porosity partially saturated with water having a resistivity R_w.

R_{io} = Bulk electrical resistivity (ohm-meters) of a rock's interconnected porosity filled with capillary-bound water having a resistivity R_w.

R_{ir} = Combined bulk electrical resistivity (ohm-meters) of a rock's interconnected porosity filled with capillary-bound water having a resistivity R_w and a rock's bulk matrix resistivity including surface conductivity.

R_{po} = Same as R_{fo} except fluids are free to move in the total interconnected porosity. R_{po} appears in the CRMM model.

R_{pt} = Same as R_{ft} but with the same exception as for R_{po}. R_{pt} appears in the CRMM model.

R_w = Electrical resistivity of water (ohm-meters).

S_w = Fractional water saturation of a rock's total interconnected porosity ϕ.

S_{wa} = Apparent water saturation.

S_{wf} = Partial water saturation in free-fluid porosity ϕ_f.

S_{wi} = Water saturation always equal to 1 in the capillary-bound water porosity ϕ_i.

S_{wir} = Irreducible or capillary-bound water saturation equal to the ratio ϕ_i/ϕ.

ACKNOWLEGEMENTS

The authors wish to thank Mobil Research and Development Corporation for the continuing support of this work and for permission to publish this material. Special thanks to E. J. Schmidt for this contribution of PMR data and to L. D. Smallwood and Bill Rakocy for their careful preparation of core samples and in taking of data. Special thanks also to R. E. Maute, J. Zemanek, W. R. Mills, and other colleagues for their many helpful discussions.

REFERENCES

Archie, G. E. (1942). "The Electrical Resistivity Log as an Aid in Determining Some Reservoir Characteristics," Trans. AIME, Vol. 146, pp. 54-56

Givens, W. W. (1986). "Formation Factor, Resistivity Index, and Related Equations Based Upon a Conductive Rock Matrix Model (CRMM)," SPWLA 27th Annual Logging Symposium, Houston, Paper P.

Givens, W. W. (1987). "A Conductive Rock Matrix Model for the Analysis of Low-Contrast Resistivity Formation," The Log Analyst, Vol. 28, No. 2, pp. 138-151.

Herrick, D. C. (1989). "Conductivity Models, Pore Geometry, and Conduction Mechanisms, SPWLA 29th Annual Logging Symposium, San Antonio, Paper D.

Pirson, S. J. (1958). Oil Reservoir Engineering, p. 111, McGraw-Hill Books Inc., New York.

Raiga-Clemenceau, J., Fraisse, C., and Grosjean, Y. (1984). "The Dual-Porosity Model, a Newly Developed Interpretation Method for Shaly Sands, " SPWLA 25th Annual Logging Symposium, New Orleans, Paper F.

Waxman, J. H., and Smits, L. J. M. (1968). "Electrical Conductivity in Oil-Bearing Shaly Sands," Soc. Pet. Eng. J. Trans., Vol. 243, pp. 107-122.

THE INFLUENCE OF SMALL-SCALE HETEROGENEITY ON AVERAGE RELATIVE PERMEABILITY

E H Smith

Petroleum Reservoir Technology Division
AEE Winfrith, U.K.

I. INTRODUCTION

This paper assesses the effect of small-scale heterogeneity on the performance of a reservoir. For small heterogeneities with scale lengths less than 1 foot, gravity can be neglected and capillary forces dominate. Consequently any variations in the capillary pressure functions between nearby heterogeneities will result in an uneven saturation distribution. If the capillary pressure correlates with the permeability of the rock, the effective average relative permeability of the medium is likely to be significantly affected.

Recently, several authors have considered the behaviour of heterogeneous systems in which high permeability rock has a lower capillary pressure than low permeability rock. Lasseter et al. (1) considered a system containing a random distribution of the two types of rock, whilst Kortekaas (2) modelled the effects of cross-bedding. In both of these papers models were constructed to represent small regions of a reservoir. Simulations were then performed in which water was injected at one side of the model with fluids being produced at the opposite boundary. The evolving water saturation distributions were then analysed for the effects of the heterogeneity. The results of Lasseter et al. showed the importance of trapping in 2-dimensional systems whilst Kortekaas showed that flow parallel to the layers in a layered system gave better recovery than that obtained from flow perpendicular to the layers. Pseudo relative permeability functions were calculated using procedures based on the Kyte and Berry technique (3). This enabled the effects of the heterogeneity to be incorporated in larger-scale models. It was not however possible to obtain average relative permeability curves over the entire saturation range.

In this paper dynamic simulations are avoided and advantage is taken of the fact that, at the small scale in a reservoir away from the vicinity of the Buckley-Leverett shock front, the saturations change very slowly. Steady-state conditions can therefore be assumed, enabling the saturation distribution in a small region of a reservoir to be found. Using the saturation distribution the average relative permeability functions for such a region can be calculated. Pseudo relative permeability functions, for use in larger scale numerical simulation, can be obtained from these functions using a standard method such as the Kyte and Berry technique (3).

II. JUSTIFICATION FOR STEADY-STATE ASSUMPTION

For the steady-state assumption to be valid, the state of a region of the reservoir must be changing very slowly. Therefore in a two-phase system the conditions

$$\frac{dS_w}{dt} \approx 0 \tag{1}$$

and

$$C_t \frac{dP}{dt} \approx 0 \tag{2}$$

where

    t is time
    S_w is wetting phase saturation
    P is the pressure
    C_t is the total compressibility

should be satisfied. These two conditions ensure that the volumes of both phases in a small region of the reservoir are essentially constant. In reservoirs with good pressure support from either a large aquifer or water injection, the pressure is maintained relatively constant throughout reservoir life. The condition in equation 2 is therefore satisfied. However, the saturation will change from the

initial high oil saturation to a final oil saturation close
to the residual oil saturation. For good reservoirs this
saturation change is relatively large, in the range 30-70%.

If the mobility ratio is favourable, a Buckley-Leverett
shock front will be formed at which most of the change of
saturation will occur. Ahead of the shock front the saturation remains constant at the initial saturation. Behind the
shock front the saturation slowly rises from the shock front
value up to the saturation at which the displaced phase
becomes immobile. As this saturation change is normally
signficantly less than the saturation change at the shock
front and occurs over hundreds rather than tens of feet,
values of dS_w/dt are in the range 0.1-0.01% per day. The
approximation that dS_w/dt is approximately zero is therefore
good for small regions of the reservoir containing small-
scale heterogeneities.

At the front, the saturation will be changing rapidly and
the steady-state approximation will not be valid. However
shock wave theory shows that the conditions at a shock front
do not affect its height or velocity. It is therefore valid
to use the steady-state approximation to generate average
relative permability functions over the entire saturation
range, in order to calculate the shock front saturation. The
steady-state approximation therefore provides a valid
technique for assessing the effects of small-scale
heterogeneity, which avoids the limitations of the dynamic
approach described in the following section.

III. LIMITATIONS OF THE DYNAMIC APPROACH

The effects of small-scale heterogeneity are usually
estimated by the dynamic approach using readily available
reservoir simulators. For this approach a model is constructed of a small representative sample of rock that
contains small scale heterogeneities. As shown in Figure 1,
a displacing fluid, often water, is injected at one of the
faces of the model whilst the displaced fluid, initially oil,
is produced at the opposite face. Effective pseudo relative
permeabilities are then calculated in a central region, away
from possible end effects.

Due to the expense of the simulation, typically no more
than about 5,000 grid blocks are used. Consequently 2-dimen-
sional calculations are normally performed with about 100
grid blocks between the inflow and outflow faces. The
central region can therefore be no more than about 50 grid

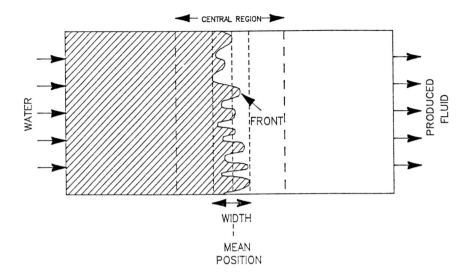

FIGURE 1. DYNAMIC CALCULATION OF THE EFFECTS OF HETEROGENITY.

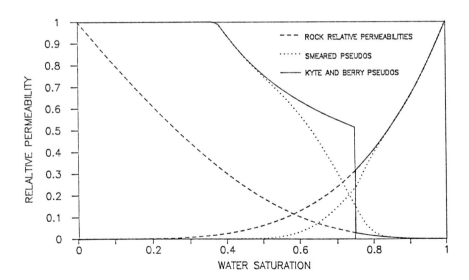

FIGURE 2. TYPICAL EFFECTS OF STOCHASTIC NOISE IN THE POSITION OF THE
SHOCK—FRONT ON PSEUDOS.

blocks across. As the smallest heterogeneities being
modelled must be at least as large as a grid block, the
length of the entire system must be less than 100 times the
length of the shortest heterogeneities. If the heterogeneities of interest range from 2-6 inches in size, the
total model length is less than 20 ft whilst the length of
the central region is less than 10 ft.

If the rock sample was homogeneous and the mobility ratio
favourable, Kyte and Berry (3) pseudos would be produced as
shown in Figure 2. The heterogeneities will however cause
distortion in the shock front over widths which are probably
several times the size of the largest heterogeneities.
Therefore for a system containing heterogeneities in the size
range 2-6 inches, the front might be distorted ±18 inches
from its mean position. The total width of the distorted
shock front could therefore be about 3 ft or 30% of the
thickness of the central 10 ft section of the model over
which pseudos are being generated. As shown in Figure 2 this
would cause a significant smearing of the pseudo relative
permeabilities that could easily mask any other effect of the
heterogeneity. In a larger system the stochastic distortions
of the front would not be significantly larger and would
therefore have a much smaller effect.

As the total model length is only 20 ft, compared with
reservoir well spacings of hundreds of feet, saturations will
be changing at least an order of magnitude faster for the
same flow rates, due to the steeper saturation gradients.
There may therefore be insufficient time for the model to
reach the same degree of equilibrium found in the reservoir.
The dynamic approach therefore has serious limitations.
Stochastic effects tend to dominate, models are often limited
to 2-dimensions, and as the injectors and producers are very
close it is impossible to match both the flow rate and the
rate at which saturations change.

IV. FEATURES OF A STEADY-STATE SYSTEM

In order to calculate the average relative permeability
of a heterogeneous medium that satisfies steady-state conditions, the saturation distribution needs to be found. In
general this distribution is dependent on the relative permeability as a function of position, which is itself dependent
on the saturation distribution. However under the limiting
conditions of either capillary equilibrium or negligible
capillary pressure, the saturation distribution can be found
analytically as demonstrated in Sections V and VI. It should

be noted that capillary equilibrium in this context is a state of a very small volume of a reservoir. It is not the same as vertical equilibrium as defined by Coats (4).

For all steady-state systems the following relationships must be satisfied:

$$\frac{dS_w}{dt} = \frac{dS_{nw}}{dt} = 0 \tag{3}$$

where subscript nw denotes the non-wetting phase. Consequently

$$\nabla \cdot q_w = - \frac{dS_w}{dt} = 0$$

and

$$\nabla \cdot q_{nw} = 0 \tag{4}$$

where q is the flow of a phase. Neglecting gravity these flows are given at position $\underline{r}$ by

$$q_w = - (K(\underline{r})k_{rw}(\underline{r},S)\nabla P_w(\underline{r})) / \mu_w \tag{5}$$

and

$$q_{nw} = - (K(\underline{r})k_{rnw}(\underline{r},S)\nabla P_{nw}(\underline{r})) / \mu_{nw} \tag{6}$$

where

K	is the permeability
k_r	is the relative permeability of a phase
P	is the pressure of a phase
μ	is the viscosity of a phase

S is the scaled wetting phase saturation given by

$$S = \frac{S_w - S_c}{S_{max} - S_c}$$

where S_c is the critical saturation at which the wetting phase first becomes mobile and S_{max} is the maximum wetting phase saturation which can be attained before the non-wetting phase becomes immobile.

V. CAPILLARY EQUILIBRIUM APPROXIMATION

Two conditions must be satisfied for the capillary equilibrium approximation to be valid within the small volume of interest. The first is that the viscous pressure gradients are small compared with capillary pressure gradients. The second condition is that capillary equilibrium must be achieved within a short time. For small-scale heterogeneity, capillary equilibrium should be achieved in less than 1 day, which is a short time compared with reservoir timescales.

If the limiting case of no flow is considered, then both P_{nw} and P_w must be constant if the permeability is finite. Therefore the volume is in capillary equilibrium with a constant capillary pressure difference, P_c, between the two phases. Therefore

$$P_c(\underline{r},S) = P_d \qquad (7)$$

where P_d is a constant pressure difference. If the function $P_c(\underline{r},S)$ is known, S can be found. Corey and Rathjens (5) used this relationship to calculate the relative permeability in a simple layered system obtaining similar qualitative results to those recently obtained by Kortekaas (2).

If there is flow, $P_c(\underline{r},S)$ will no longer be constant and the volume will not be in perfect capillary equilibrium. The wetting phase pressure will therefore be given by

$$P_w(\underline{r},S) = P_{nw}(\underline{r},S) - P_d - \Delta p_c(\underline{r},S) \qquad (8)$$

where $\Delta p_c(\underline{r},S)$ is a small perturbation in the capillary pressure difference between the two phases. This would then give a perturbation to the saturation given by

$$\Delta S(\underline{r}) = \Delta p_c(\underline{r},S) \ / \ \frac{dP_c(\underline{r},S)}{dS} \qquad (9)$$

If the average pressure gradient across the volume is ∇P_{av} then, as will be demonstrated numerically later, the root mean square (r.m.s.) value of $\Delta p_c(\underline{r},S)$ is typically about $L_H \nabla P_{av}$ where L_H is the correlation length of the heterogeneities. For a typical reservoir of permeability 100 mD with a Buckley-Leverett shock front advancing at 1 ft/day, ∇P_{av} would be about 0.2 psi/ft. If the typical length of a heterogeneity was about 3 inches, then the r.m.s. value of $\Delta p_c(\underline{r},S)$ would be about 0.05 psi. For a typical capillary pressure curve dP_c/dS might be about 5 psi, which would give a r.m.s. variation of the saturations from their equilibrium values of about 1%. The assumption that the system is in capillary equilibrium is therefore satisfactory for small scale heterogeneities in reservoirs with pressure gradients of around 0.2 psi/ft.

VI. NEGLIGIBLE CAPILLARY PRESSURE

It is useful also to consider the converse case when the capillary pressure is negligible and the pressure gradients in the reservoir are large. For the limit of zero capillary pressure

$$P_w(\underline{r}) = P_{nw}(\underline{r}) \qquad (10)$$

Using equations 4, 5 and 6 gives

$$\nabla \cdot (K(\underline{r})k_{rw}(\underline{r},S)\nabla P_w(\underline{r})) = 0 \qquad (11)$$

and

$$\nabla \cdot (K(\underline{r})k_{rnw}(\underline{r},S)\nabla P_{nw}(\underline{r})) = 0 \qquad (12)$$

For an arbitrary permeability distribution and non-zero $\nabla P_w(\underline{r})$ the above equations can both be satisfied if

$$\frac{k_{rw}(\underline{r},S)}{k_{rnw}(\underline{r},S)} = C \qquad (13)$$

where C is a constant. If the relative permeability is only a function of scaled saturation S, then S and the relative permeabilities must be constant. As shown in the next section this requires that the average relative permeability is equal to the rock relative permeability.

VII. CALCULATION OF AVERAGE RELATIVE PERMEABILITY

The average relative permeability $\bar{k}_{r\alpha j}$ of phase α, for a region of a reservoir in direction j is defined by

$$\bar{k}_{r\alpha j} = \frac{\bar{K}_{\alpha j}}{\bar{K}_j} \qquad (14)$$

where $\bar{K}_{\alpha j}$ is the average permeability for phase α and $\bar{K}_j$ is the average total permeability. This region can be divided into N small blocks. For each of these small blocks both the permeability K_j and relative permeability $k_{r\alpha j}$ can be assumed to be homogeneous and isotropic. The average permeability of the region in a direction j, $\bar{K}_j$, is a function, f_{Kj}, of the permeabilities of the N blocks; thus,

$$\bar{k}_{r\alpha j} = \frac{f_{Kj}(K_1 k_{r\alpha 1}, K_2 k_{r\alpha 2}, \cdots K_N k_{r\alpha N})}{f_{Kj}(K_1, K_2, \cdots K_N)} \qquad (15)$$

In general f_{Kj} is a complicated function which needs to be evaluated numerically. However if the heterogeneities are randomly oriented and are not too long and thin, f_{Kj} is approximately the geometric mean of the individual permeabilities (6). Also, when the heterogeneities are much longer in one direction than another and are all oriented in the same direction, the permeability to flow parallel to the longer dimension of the heterogeneities tends to the arithmetic mean of the individual permeabilities.

If $\bar{K}_j$ is calculated using equation 28 then f_{Kj} can be assumed to be a first order homogneous function, so

$$f_{Kj}(AK_1, AK_2, \cdots AK_N) = Af_{Kj}(K_1, K_2, \cdots K_N) \qquad (16)$$

where A is a constant. If the capillary pressure is negligible, equation 13 shows that the relative permeability is constant throughout the system; consequently,

$$\bar{k}_{r\alpha j} = \frac{f_{Kj}(K_1 k_{r\alpha}, K_2 k_{r\alpha}, \cdots K_N k_{r\alpha})}{f_{Kj}(K_1, K_2, \cdots K_N)} = k_{r\alpha} \qquad (17)$$

This result shows that, in the absence of gravity, a heterogeneous permeability distribution does not affect the average relative permeability if the system is otherwise homogeneous. This simple result agrees with many numerical studies that have shown that small-scale heterogeneities have little effect on recovery from a reservoir (1).

VIII. CORRELATION OF PERMEABILITY AND CAPILLARY PRESSURE

The calculation of average relative permeability is not trivial if the rock relative permeability varies with position. This will happen if the scaled saturation S

depends upon the rock properties. Calculations of average
relative permeability in the remainder of this paper therefore assume that the capillary pressure for a given saturation S is correlated with the permeability of the rock.
This provides a correlation between relative permeability and
permeability that can significantly alter the average relative
permeabilities.

In general the correlation between capillary pressure and
permeability could be very complex, involving cross-correla-
tions with porosity, relative permeability and rock type.
In this paper it will be assumed that all the properties other
than the magnitudes of the capillary pressure and the
permeability are homogeneous. It is however trivial to
generalise to variable porosity and end points by using a
weighted average to calculate the average saturation. For a
system containing N blocks the average wetting phase
saturation

$$\bar{S}_W = \frac{\sum\limits_{i=1,N} \phi_i \, (S_{ci} + (S_{max_i} - S_{c_i})S_i)}{\sum\limits_{i=1,N} \phi_i} \tag{18}$$

where ϕ_i is the porosity of the block. Changes in the
dependence of relative permeability and capillary pressure on
scaled saturation S would produce more complex effects.

The following analytic expressions from Corey (7) have
been used for the rock relative permeability

$$K_{rw} = S^\epsilon \tag{19}$$

and $K_{rnw} = (1 - S)^2(1 - S^\gamma) \tag{20}$

where $\epsilon = \dfrac{2 + 3\lambda}{\lambda}$

and $\gamma = (2 + \lambda)/\lambda$

The constant λ, which is related to the pore size distribution
is a fitting parameter, which in this paper is normally set to,
2. These rock relative permeability functions are illustrated
graphically for $\lambda = 2$ in a number of figures, for example
Figure 6. A simple capillary pressure function

$$P_c(S) = \frac{P_1}{\sqrt{S}} (1 - S)^{1/n} \tag{21}$$

from Corey (7) is used, where P_1 is a constant the value of

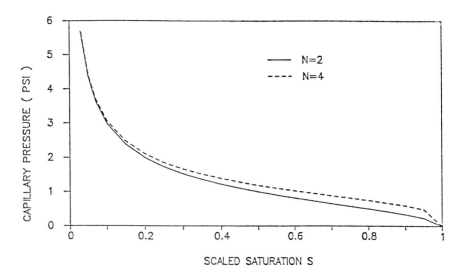

FIGURE 3.　CAPILLARY PRESSURE FUNCTIONS.

which is dependent on the permeability. In this paper the
constant n has values of either 2 or 4. The (1 - S) term is
used to ensure that the capillary pressure is zero at
residual oil saturation. Capillary pressure curves for n = 2
and 4 with P_1 equal to 1 psi are shown in Figure 3.

　　The correlation between permeability and capillary
pressure is modelled using the Leverett J function given by

$$J(S) = \frac{P_c(S)}{\sigma} (K/\phi)^{1/2} \tag{22}$$

where $J(S)$ is a function of scaled saturation only and σ is
the interfacial tension. As the porosity typically varies by
a factor of 2 whilst the permeability varies by several
orders of magnitude, it is possible to make the assumption
that the porosity is constant throughout a small volume of a
reservoir. Therefore

$$P_c(S) = \frac{P_2}{\sqrt{KS}} (1 - S)^{1/n} \tag{23}$$

where P_2 is a constant. For capillary equilibrium P_c is
constant throughout the volume so for n = 2, S as a function
of FK is

$$S(FK) = \frac{1}{1 + FK} \qquad (24)$$

and for n=4

$$S(FK) = \frac{-1 + (1 + 4(FK)^2)^{1/2}}{2(FK)^2} \qquad (25)$$

where $F = (\frac{P_c}{P_2})^2$

and the average scaled saturation

$$\bar{S} = \int S(FK)P(K)dK \qquad (26)$$

where $P(K)$ is the probability distribution function for the permeability. Thus for a given average scaled saturation it is possible to calculate F by iteration. Knowing F, the saturation distribution and hence the relative permeability distribution can be found throughout the volume, using equation (25).

IX. NUMERICAL METHOD

    In order to calculate the average relative permeability using the capillary equilibrium approximation in a reservoir with correlation between permeability and capillary pressure, the small-scale heterogeneous permeability distribution is required. As no suitable measured data were available, distributions have been generated from given probability distribution functions with specified correlation structure using a method of Farmer (8). Three-dimensional arrays of 5,000 to 200,000 blocks have been used in this paper with a relatively short correlation length of 2 blocks. A typical cross-section through such an array is shown in Figure 4. In order to model media with heterogeneities with different correlation lengths in different directions, the dimensions of the blocks are varied.
    Once the permeabilities have been allocated to the blocks in the model the saturation distribution can be found for a prescribed value of average saturation using equation 26 and then either equation 24 or 25. These scaled saturations are then used in equations 19 and 20 to calculate the relative permeabilities and the phase permeabilities of each block.

PERMEABILITY

mD

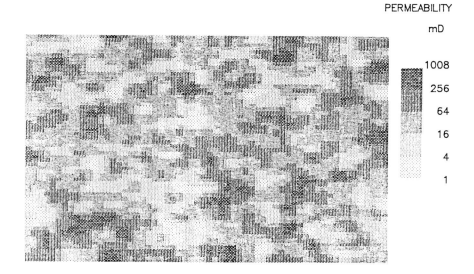

1008

256

64

16

4

1

FIGURE 4. PERMEABILITY DISTRIBUTION.

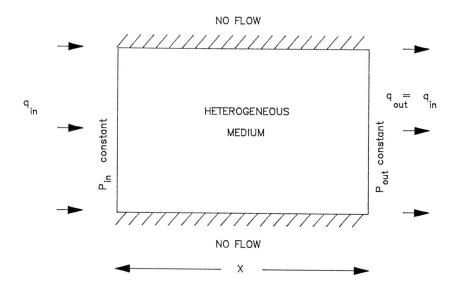

FIGURE 5. BOUNDARY CONDITIONS FOR STEADY-STATE CONDITIONS.

Calculation of the average relative permeabilities requires the solution of the equation

$$\nabla K_\alpha(\underline{r}) \nabla P_\alpha(\underline{r}) = 0 \qquad (27)$$

for each phase α and the total permeability. Equation 27 is solved in three dimensions using a standard 7-point finite difference scheme with the boundary conditions shown in Figure 5. Numerical experiment showed that errors due to the finite difference formulation could be approximately halved by refining each permeability block by two in the direction of the flow. There was however found to be little benefit from refinement perpendicular to the flow.

Having solved equation 27 the average permeability is given by

$$\bar{K}_\alpha = \frac{q_\alpha \mu_\alpha X}{A(P_{in} - P_{out})} \qquad (28)$$

where q_α is the flow through the system of length X and cross-sectional area A. If the permeability calculations are repeated for different values of average saturation over the range 0 to 1, a complete relative permeability curve can be generated.

The main computational expense is therefore the assembly and inversion of a single-phase matrix for each point on the relative permeability curves. This should be compared with the two-phase matrix which needs to be assembled and inverted several times for each timestep when a typical fully implicit reservoir simulator is used for the dynamic approach.

X. ANALYTIC CALCULATION OF RELATIVE PERMEABILITY

In order to check the numerical techniques it is useful to have an analytic test solution available. If the correlation length is much longer in one direction than the others, the system starts to behave as if it were layered. The permeability to flow parallel to the longer direction therefore tends to the arithmetic mean of the permeability distribution. For such a system analytic expressions for the relative permeabilities of the phases can be obtained if the phase permeability distributions can be integrated. This is possible if the Corey relative permeabilities given in equations 19 and 20 are used with $\lambda = 2$, n is set to 2 in equation 21 and a uniform permeability distribution is used such that

$$P(K) = \frac{1}{K_{max} - K_{min}} \qquad K_{min} \leqslant K \leqslant K_{max}$$

and $\quad P(K) = 0 \qquad\qquad\qquad K < K_{min}$ or $K > K_{max}$ (29)

Therefore the average permeability

$$\bar{K} = \frac{1}{2}(K_{max} + K_{min}) \tag{30}$$

Integrating equation 26 gives

$$\bar{S} = \frac{\ln(\dfrac{1 + FK_{max}}{1 + FK_{min}})}{F(K_{max} - K_{min})} \tag{31}$$

The average wetting phase relative permeability using equations 19 and 24 is given by

$$\bar{k}_{rw} = \frac{1}{\bar{K}} \int_{K_{min}}^{K_{max}} K(\frac{1}{1 + FK})^4 \, P(K) \, dK \tag{32}$$

and using equation 20 the average non-wetting phase relative permeability is

$$\bar{k}_{rnw} = \frac{1}{\bar{K}} \int_{K_{min}}^{K_{max}} K(1 - \frac{1}{1 + FK})^2 (1 - (\frac{1}{1 + FK})^2) P(K) \, dK \tag{33}$$

Both these integrals can be evaluated analytically. For the wetting phase for example

$$\bar{k}_{rw} = (\frac{-3FK - 1}{6\bar{K}(K_{max} - K_{min})F^2(1 + FK)^3}) \bigg|_{K_{min}}^{K_{max}} \tag{34}$$

The resulting relative permeabilities are shown graphically in Figure 6 for K_{max} = 100 mD and K_{min} = 10 mD. The rock curves in this and subsequent figures are plots of equations 19 and 20.

    Numerical calculations of relative permeability are shown for comparison with the analytic curves in Figure 6 and can be seen to be in good agreement. This demonstrates that the numerical calculations using the random heterogeneous distribution give results which tend to the correct limit for

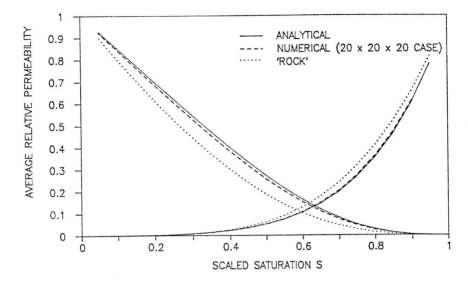

FIGURE 6. COMPARISON OF RELATIVE PERMEABILITIES CALCULATED NUMERICALLY
WITH THOSE OBTAINED ANALYTICALLY.

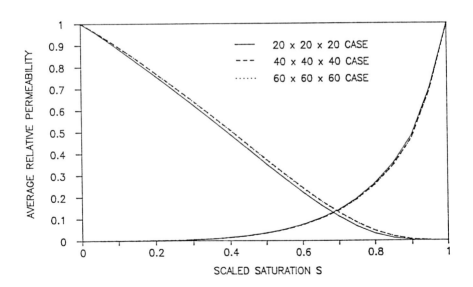

FIGURE 7. COMPARISON OF RELATIVE PERMEABILITIES CALCULATED USING DIFFERENT
SIZE PERMEABILITY ARRAYS.

a long correlation length. For the numerical calculation the
permeabilities were generated on a 60 x 60 x 60 grid. Each
grid block was assumed to be 20 units long by 1 unit high and
1 unit wide. In order to speed up the computation a 20 x 20
x 20 sample of the full array was used for the permeability
calculation. The permeability calculation used a 40 x 20 x
20 finite difference grid in order to minimise numerical
errors. Tests showed that even with the grid refinement
these errors dominated the differences seen between the
numerical and analytic calculations.

XI. EFFECT OF SYSTEM SIZE

Ideally a very large 3-dimensional system would be used
in order to minimise stochastic and edge effects on the
calculated relative permeabilities. However increasing the
system size makes the calculations much more expensive, as
the computer time required is at least proportional to the
number of grid blocks used.

To test the effect of system size on the results, a set
of relative permeabilities were generated from both 20 x 20 x
20 and 40 x 40 x 40 subsets of a 60 x 60 x 60 permeability
array. For these calculations the rock relative permeabilities and capillary pressure functions were defined by $\lambda = 2$
and $n = 4$. The grid blocks were given dimensions of DX = 4,
DY = 4, DZ = 1. A log-normal probability distribution
function was used for the permeability with a geometric mean
permeability of 31.6 mD with 10 and 100 mD being one standard
deviation of the log-normal distribution from the mean.

A comparison of the relative permeabilities calculated
using the different arrays is shown in Figure 7. This
illustrates that the 20 x 20 x 20 grid is giving good
agreement with the curves generated using the finer arrays.
This 20 x 20 x 20 case is used as a reference for the rest of
the paper and, unless stated otherwise, all the parameters
used for the calculations in the rest of this paper are
identical with those used for this 20 x 20 x 20 case.

XII. RATE DEPENDENCY AND THE EFFECT OF VISCOUS PRESSURE

The assumption of capillary pressure equilibrium within a
small heterogeneous volume, described in Section V, required
that the difference between the phase pressures was
approximately constant throughout the system. However, if

there are high flow rates and large viscous pressure drops across the system, this condition will not be satisfied and, as shown in equation 9, the saturations will not be equal to their capillary equilibrium values.

In Figure 8 the r.m.s. values of $\Delta p_c(\underline{r},S)$, calculated using equation 8, divided by $L_H \nabla P_{av}$ are plotted as a function of saturation for the 40 x 40 x 40 array case described in the previous section. Other cases were found to give roughly the same function. It can be seen that $\Delta p_c(\underline{r},S)$ has a minimum value approximately equal to $L_H \nabla P_{av}$. The function however does tend to increase if either of the phases has a low saturation. This is because the variance of the permeability of a phase increases dramatically at low saturations. The higher variance causes more distortion of the pressure distribution of that phase and therefore a higher r.m.s. value of $\Delta p_c(\underline{r},S)$.

A typical spatial variation of $\Delta p_c(\underline{r},S)$ is shown in Figure 9 for an X-Z cross-section using the 60 x 60 x 60 permeability distribution. It can be seen that $\Delta p_c(\underline{r},S)$ shows large scale structure. This structure will cause large regions of higher or lower saturation that will have a relatively small effect on the average relative permeabilities. Short-scale variations in $\Delta p_c(\underline{r},S)$ would be more significant, reducing the correlation between saturation and permeability, which is the main cause of the difference between the average and rock relative permeability curves.

If the heterogeneities are relatively large and the flow rates are high, the capillary pressure equilibrium assumption will break down. For very high flow rates the effects of capillary pressure can be ignored and the average relative permeabilities become equal to the rock relative permeabilities. This provides a mechanism for producing rate dependency.

Unfortunately, using the simple calculational techniques described in this paper it is not possible to determine the flow rates at which the relative permeabilities are most rate dependent. Some steady-state calculations at finite flow rates have been performed using the PORES reservoir simulator. These calculations are very expensive, even using relatively small 50 x 100 2-dimensional grids, and it was not possible to allow enough time for the system to fully reach steady-state conditions. The calculations did show that, for a constant average saturation, the fractional flow of the wetting phase increased with rate.

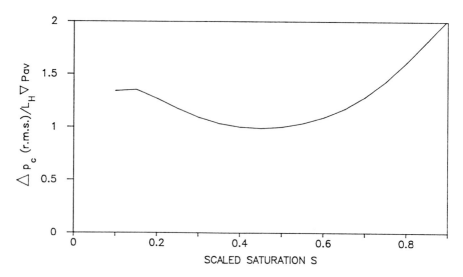

FIGURE 8. VARIATION OF THE r.m.s. VALUE OF $\triangle p_c$ AS A FUNCTION OF SCALED SATURATION S.

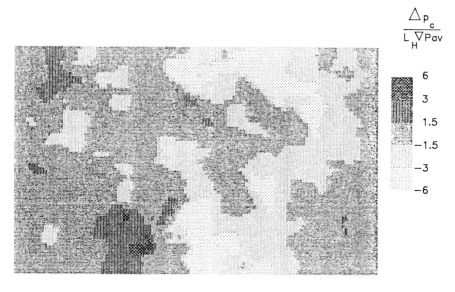

FIGURE 9. SPATIAL VARIATION OF $\triangle p_c / L_H \nabla Pav.$

XIII. NEED FOR 3-DIMENSIONAL CALCULATIONS

Most heterogeneity studies have considered 2-dimensional systems only. In order to compare the results from 2 and 3-dimensional systems, average relative permeabilities have been calculated using a 50 x 50 grid. Figure 10 shows that the 2-dimensional and 3-dimensional average relative permeabilities are very different. Comparison of the fractional flow curves for an end point mobility ratio of unity shows that at breakthrough the 2-dimensional system gives almost the same recovery as the rock curves whilst the 3-dimensional system gives about 10% better recovery. Consequently average relative permeabilities calculated using a 2-dimensional model may be seriously in error.

The main reason for the difference is the poorer connectivity in a 2-dimensional system compared with a 3-dimensional system. If for example the system was composed of a random distribution of two different types of rock, one of which was impermeable, a 2-dimensional system would require about 50% of the rock to be permeable for the system as a whole to be permeable. However for a 3-dimensional system, with flow in all 3 directions allowed, only 25% of the rock needs to be permeable, Kirkpatrick (9). This effect is most important for the non-wetting phase permeability at high wetting phase saturations when the permeability to the non-wetting phase of a large proportion of the system is very low.

XIV. DIRECTIONAL EFFECTS

In Figure 11 the average relative permeabilities calculated in both the X and Z directions using the default permeability block dimensions are compared. As can be seen, for flow in the Z-direction the average relative permeabilities are similar to the rock curves. The average relative permeabilities are therefore directional with the largest effects of heterogeneity seen for flow parallel to the direction of maximum correlation.

The non-wetting phase relative permeability is the most directional. This is because the variance of the non-wetting phase permeability tends to be significantly higher, causing a much larger difference between the arithmetic and geometric means.

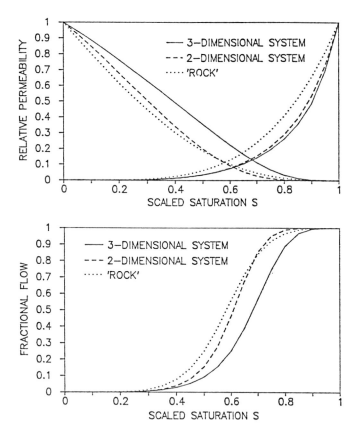

FIGURE 10. COMPARISON OF 2 AND 3—DIMENSIONAL SYSTEMS.

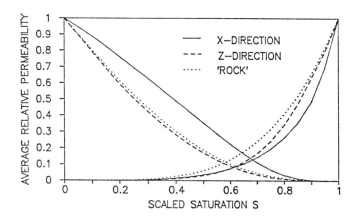

FIGURE 11. DIRECTIONAL DEPENDENCE OF AVERAGE RELATIVE PERMEABILITY.

The total average permeability is also directional. In the X-direction it is about 45 mD whilst for the Z-direction it is only 28.6 mD. Calculations with DX = 32, DY = 16, DZ = 1, gave a X-direction permeability of 52 mD and also showed even more favourable X-direction relative permeability curves than the default DX = 4, DY = 4, DZ = 1, system as shown in Figure 12.

XV. DIFFERENT ROCK CURVES

In Figure 13 the fractional flow curves calculated from the average relative permeability for both λ = 4 and λ = 2 are compared with the rock curves. Unit end point mobility ratio is assumed. This illustrates that the average curves show less difference than the rock curves.

Using the default value of λ = 2 the effect of changing n from 4 to 2 was investigated. This again made little difference, the effect being approximately half that seen changing λ from 2 to 4. Both these results show that the effects of the heterogeneous structure are more important than differences in the shapes of the rock curves.

XVI. EXPENSE OF CALCULATION

All the calculations were performed on a CRAY-2 computer. The standard 20 x 20 x 20 calculation using 19 saturation points required about 1 minute. Larger problems took significantly longer times, with the 60 x 60 x 60 calculation of only 5 points requiring about 15 minutes.

These times are significantly shorter than those which would be required by a reservoir simulator to model the dynamic flood of a heterogeneous medium. A typical reservoir simulator would require about 30 minutes to simulate a 50 x 50 system of permeability blocks using a 100 x 50 numerical grid.

XVII. SUMMARY OF POSSIBLE EFFECTS ON RESERVOIR PERFORMANCE

This paper has shown that small-scale heterogeneities can affect the average relative permeability if their permeability and capillary pressure functions are correlated. For an

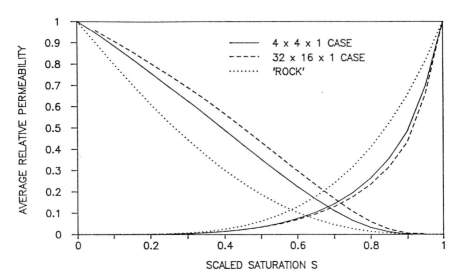

FIGURE 12. EFFECT ON AVERAGE RELATIVE PERMEABILITY OF INCREASING
HORIZONTAL CORRELATION LENGTH.

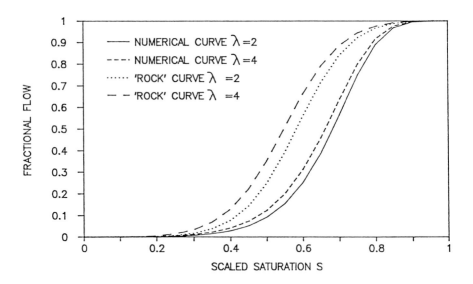

FIGURE 13. EFFECT ON FRACTIONAL FLOW OF CHANGING ROCK RELATIVE
PERMEABILITY.

oil reservoir which is water-wet, developed by water injection or with natural influx, the heterogeneities can significantly improve sweep. If the predominant flow direction is parallel to the long axis of heterogeneities, normally parallel to the bedding plane, the fractional flow curves show that scaled saturation at breakthrough can be increased for unit end point mobility ratio by about 10 saturation units. If the flow is perpendicular to this direction little or no benefit is seen.

If the reservoir is oil-wet the recovery will be reduced by the presence of heterogeneities. Welge analysis for the unit mobility ratio fractional flow curves shows a reduction in recovery at breakthrough from about 65% of the mobile oil down to 55%. Recovery could be increased if the flow is optimised to be perpendicular to the bedding planes or the production rate is increased.

XVIII. CONCLUSIONS

This paper has shown that the steady-state approximation is valid and useful for evaluating the effects of small-scale heterogeneity. Using this approximation it has been possible to show the following.

(1) Small-scale heterogeneities in the permeability alone do not affect the average relative permeabilities.

(2) If the capillary pressure function correlates with the permeability, the average permeability, and fractional flow curves are significantly altered.

(3) The average relative permeability can be directional and rate dependent.

(4) The calculation of average relative permeabilities using the steady-state approximation is very fast.

(5) 3-dimensional heterogeneity models should be used because 2-dimensional models can give very misleading results.

ACKNOWLEDGEMENTS

This work was funded by the UK Department of Energy. I should also like to thank my colleagues at Winfrith, especially Dr. C. L. Farmer, for their encouragement and help.

REFERENCES

1. Lasseter,T.J., Waggoner,J.R. and Lake,L.W. (1986). In "Reservoir Characterisation" (L.W.Lake, Ed.), p.545. Academic Press, New York.
2. Kortekaas,T.F.M. (1985). SPE J. 25,917.
3. Kyte,J.R. and Berry,D.W. (1975). SPE J. 15,269.
4. Coats, K H (1971), Soc. Pet. Eng. J., 63, 71.
5. Corey,A.T., Rathjens,C.H. (1956) Jour. Petr. Techn., Trans. AIME, Technical Note 393, December.
6. Warren,J.E., Price,H.S. (1961) SPE J. II, 153.
7. Corey,A.T. (1977), "Mechanics of Heterogeneous Fluids in Porous Media", p.95, Water Resources Publications, Fort Collins, Colorado.
8. Farmer,C.L. (1988). In "Mathematics in Oil Production" (S.Edwards and P.R.King, Ed.), p.244, Clarendon Press, Oxford.
9. Kirkpatrick, S. (1973), Reviews of Modern Physics, 45, 4.

CORE SAMPLE HETEROGENEITY FROM LABORATORY FLOW EXPERIMENTS

Ibrahim Bahralolom
John Heller

New Mexico Petroleum Recovery Research Center
New Mexico Institute of Mining and Technology
Socorro, New Mexico

I. INTRODUCTION

A major factor causing flow nonuniformity, and inefficiency of displacement in reservoirs, is the heterogeneity of rock properties at all scales. A problem that confronts engineers in the design of any enhanced recovery project is the evaluation of the effect of this variability of reservoir rock on the performance of the displacement process. Both hydrologists and petroleum researchers have directed considerable effort toward understanding the effect of nonuniformity on miscible displacements at field scale. Not as much work has been done on smaller-scale nonuniformities however, and their effect is not fully understood. In this study, our main concern has been to investigate the effect of the variation of rock properties on miscible displacement performance at the scale of laboratory experiments, where the interaction of nonuniform flow and dispersion is more likely to be important. Our objective is also to examine the geostatistical details of the small-scale variations themselves.

In the usual miscible displacement experiments, performed on reservoir core samples or packed columns, a design effort is made to minimize the effect of nonuniform flow. In those experiments, the flow from the entire outflow face of the core sample is collected and combined for analysis. The concentration curve obtained in this way is then matched with one or another of the existing theories that are based on a one-dimensional model of flow. The function of this mathematical match is to predict one or more parameters which presumably describe the effect of small-scale heterogeneity on mixing behavior. Whereas these fitting parameters are quite characteristic of the particular sample, the description may raise questions concerning both the importance of nonuniformity of the rock on mixing in the direction normal to average flow and of the scale of the sample used in the experiment.

Heller (1963), Dupuy et al.(1966), Withjack (1986), and perhaps other investigators have displayed visual observations of miscible displacements in

RESERVOIR CHARACTERIZATION II

relatively uniform media in which the effect of rock nonuniformity is nevertheless quite obvious. The results of these experiments clearly indicate the existence of channeling or other flow nonuniformity, even in Berea sandstone. It is clear that laboratory-scale variations of velocity and concentration result from nonuniformity of rock properties. The question to be considered here concerns the effect of rock heterogeneity on the details of miscible displacements in laboratory experiments performed at relatively low flow rates where dispersion may be important.

To measure and observe the nonuniformity of rock properties by the use of miscible displacements, two different experimental techniques are presented. Data from one of these methods, which affords useful information on the variations of porosity as well as of permeability, are examined in greater detail. The presentation and analysis of these data occupy the major part of our report.

In this paper, the nonuniformity of permeability and porosity, and the effect of this variability on displacement, are considered over a range of scales from only a few orders of magnitude larger than pore size to distances of ten centimeters or more. Furthermore, we investigate the spatial correlation of these properties in directions normal and parallel to the Darcy flow by utilizing geostatistical approaches. The statistical parameters used to characterize small-scale heterogeneity for the porous medium under consideration are discussed. In a concluding section, the influence of dispersion on these results is considered, and a calculation is made of the relative magnitudes of dispersive and convective fluxes in laboratory-scale experiments.

II. EXPERIMENTAL METHODS

The main goal of this study is to examine the influence of small-scale heterogeneity on displacement performance. Although the evaluation of laboratory-scale patterns of variability and its effect on displacement is by no means complete, preliminary results show that variations of velocity and concentration transverse to the average flow exert a relatively large effect on the outcome of miscible processes. Thus, the presence of permeability variations that are correlated in the direction of flow is particularly important.

In the following sections, brief summaries are given of two experimental techniques through which the impact of relatively small-scale heterogeneity can be observed and measured in miscible floods. In both, the mobility ratio was held at unity. The measurement of local concentration curves is a major feature of both of these experimental approaches, although it will be seen that the second method described below provides an additional advantage.

A. Output Face Concentration Technique

In experiments being performed at the New Mexico Petroleum Recovery Research Center, breakthrough curves are being observed at different points on the output face of core during miscible floods. The experimental technique by which this is done involves the use of a fluorescent dye tracer, with illumination by collimated light from a mercury vapor lamp, defined by a slit, and parallel to and grazing the output face of the sample. The illuminated plane of the fluid emerging from the output face of the rock sample is observed by a video camera and recorded. Periodically, image frames are digitized and stored by a computer. Figure 1 is a schematic of the apparatus used to measure flow heterogeneity. From the successive images, independent records can be constructed of the light brightness at different locations on the surface versus time.

Berea sandstone core is selected to validate the experimental technique. Figure 2 shows the breakthrough curves for different locations at the output face of a fired Berea core sample; the core is 12 cm long with almost a square cross section of 14.8 cm^2 (3.8 x 3.9 cm). The experiment is conducted at a relatively slow flow rate of 105 cc/hr(about 6 ft/day). It is possible to notice the existence of velocity variation along different paths in this relatively uniform porous medium. Even though the preliminary results indicate that small-scale heterogeneity can be revealed by using this technique, there are limitations as in other experimental techniques. The limitations of this technique are caused by adsorption of the fluorescent dye onto the rock, by transverse convection and diffusion of the dyed output fluid after emergence from the core, and by the complexities of obtaining and correcting the brightness/concentration calibration curve. In this study, the main focus will be the presentation and discussion of the results of observations by the second method in a similar miscible flood experiment.

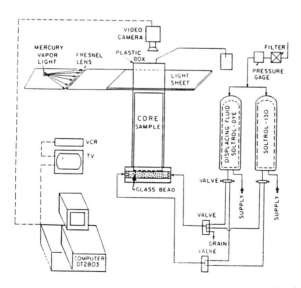

Figure 1. Apparatus for measurement of flow heterogeneity from output concentration.

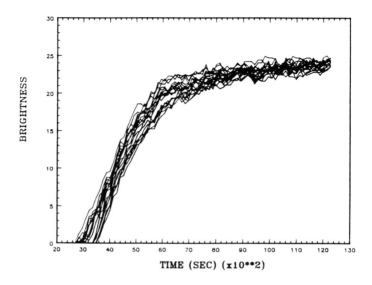

TIME (SEC) (x10**2)

Figure 2. Brightness vs. time, Berea sandstone: indicating dye concentration at 20 different locations on the output face of sample.

B. Flow Visualization Data Using CT-Scanner Technique

In addition to yielding the expected results provided by observation of the output concentration, the computed tomography (CT-Scanner) is able, by duplicate runs, to reveal the pore-to-bulk volume fraction (i.e., the porosity) averaged over small volumes, and many of the internal details of tracer material concentration during a miscible flood. A valuable set of scanned image files of an experiment on a sample of Antolini sandstone was graciously provided for additional research by Drs. Gary Pope and Bruce Rouse of the University of Texas at Austin. The data provided consist of two sets of scanned image files. One set of these images includes data from which the porosity values can be calculated at various points on different cross sections or " slices" of the core; the other set of images was taken consecutively of a slice near the output, while a miscible flood of a unit mobility ratio was being performed in the same core. Figure 3 shows a sketch of the core and the positions at which cross-sectional CT-Images were taken. Details of the experimental procedure in the use of the CT-Scanner and the calibration technique used to convert CT-number directly to density measurements at each picture element (pixel) are given by Pope et al. (1988). The porous medium used in these experiments is an outcrop sample of Antolini sandstone, 14.1 cm long, with a rectangular cross-sectional area of 27.44 cm^2 (5.6 x 4.9 cm). The porosity, measured by a material balance method, is 11.79% as reported by Pope et al. (1988). The core permeability with brine was reported to be 470 md.

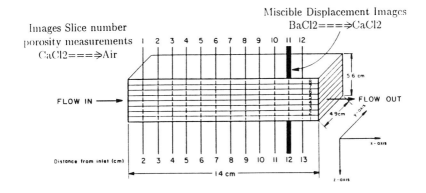

Figure 3. Sketch of Antolini sandstone core showing locations of image-scans for porosity and concentration during miscible flood.

The data file of each of the scanned images of the sandstone core sample was provided in a matrix of 256 x 256 pixels. The size of the image of the rock is 158 x 138 pixels, which contains the entire cross section of the core (5.6 x 4.9 cm). The individual pixel size is approximately 0.35 x 0.35 mm on the cross section, with a thickness in the longitudinal direction of 1 mm (slice thickness). Although the boundaries are not defined with perfect sharpness, the CT- Scanner thus samples a rectangular pixel-volume of about 0.1225 microliter of the sample.

As expected, there is natural fluctuation in the pixel-to-pixel density readings. Such fluctuation arises from two sources, both of them connected with the fact that the computed values of density-at-a-pixel result from the counting of random events. The first type of random events is the arrival of X-ray photons in the detectors used in the CT-Scanner, which leads to different measured values of density for repeat determinations at the same pixel. The second type leads to fluctuations that are characteristic of the individual pixels, and is associated with the randomness of the number of sand grains in each of the sampled pixel volumes. One would expect, here, that the variance (the mean square variation) of the measured density values would be roughly equal to the product $N\phi(1-\phi)$, where N is the number of sand grains in a pixel-volume and ϕ is the porosity. To reduce the effect of the natural fluctuation, the readings of the image were subjected to averaging over groups of 16 pixels arranged 4 x 4. In consequence, the individual 'group pixel' size is larger—approximately 1.44 x 1.44 mm, and the root mean square fluctuation is reduced by a factor of four. Figure 4 shows a schematic representation of individual and group pixel volume elements.

CT-SCAN IMAGE

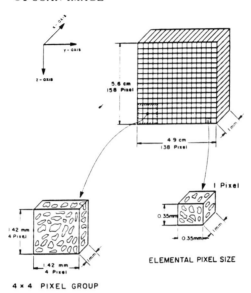

4 × 4 PIXEL GROUP

ELEMENTAL PIXEL SIZE

Figure 4.   Schematic representation of pixel and pixel-group volume elements of CT-scanner image.

III.   POROSITY DISTRIBUTION MEASUREMENTS

Two sets of scanned images were taken to provide quantitative measurements at different sections of the core. They were taken equally spaced along the core (10 mm apart) beginning 20 mm from the inlet as shown in Fig. 3. The first set consists of the images when the core sample was dry. The other set of images was taken with the core completely saturated with calcium chloride ($CaCl_2$) solution. The porosity was calculated by the formula

$$\phi_{i,j} = (A_{i,j} - B_{i,j}) / \rho_{CaCl2} , \tag{1}$$

where $\phi_{i,j}$ is the porosity at pixel (i,j), $A_{i,j}$ is the scanner-measured density of the wet sample (at 100% saturation with $CaCl_2$) at pixel (i,j), and $B_{i,j}$ is the scanner-measured density of the dry sample at the same place. The apparent density of the $CaCl_2$ solution is denoted by ρ_{CaCl2}.

Some kind of layering is clearly visible in this sample, and the cross-section CT-mages clearly show this as well. Alternate dark (low density) and light (high density) regions indicate lower and higher porosity layers. The local porosity distribution at location ("slice number") 11 is shown in Fig. 5. The term "local porosity" refers to the calculated porosity of "pixel-groups" from the CT-images, at the particular sample location. In this figure, the

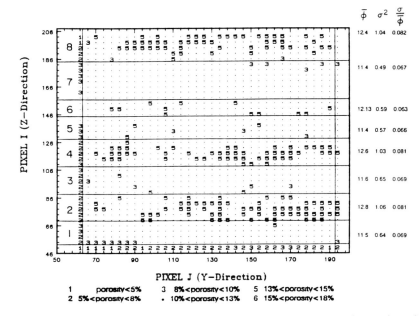

Figure 5. Local porosity values of Antolini sample cross section at location 11, 120mm from the inlet. (Obtained from CT-scanner)

coordinates represent the position of the pixel group in the z and y directions, and the porosity range at each pixel-group position is represented by a digit from 1 to 5. The places labelled with the number 5 have the highest porosity (about 13%). The most numerous pixel-groups, which would have been labelled with a 4, have been identified only with a dot in order to reduce clutter in the presentation. The effect of penetration of coating material around the core can be observed along the edges; clearly, it has reduced the porosity in these locations to the lowest values. In consequence, these boundary points have not been utilized in the calculations to follow.

Analysis of the porosity distribution across the core sample at location 11 confirms the initial visual classification, that the cross section can be divided into eight distinct strata or regions of alternating density. If the layer numbers are assigned as shown in the column at the left side of the figure, the even-numbered regions have the highest porosity. A statistical summary of the porosity measurements for each layer is shown at the right side of Fig. 5. In addition to the arithmetic mean and the variance, a coefficient of variation ($C\phi$) is calculated for each layer. At location 11, the coefficient of variation of porosity measurements is relatively constant for all layers, ranging from 0.063 to 0.082. That suggests two conclusions. The first is that the sample standard deviation increases with sample mean. Therefore the spread, or average deviation from the mean, of local porosity measurements in the more porous layers is relatively high compared to that of the low-porosity layers. The second observation is concerned with the porosity distribution within the layers themselves. For instance, layer 2 in the same figure shows that the spatial concentration of group-pixels labelled with number 5 increases gradually from the left to the right. This indicates the local mean porosity of layer 2 changes with position in the y direction.

The histogram of all local values of porosity on the location 11 cross section is shown in Fig. 6. Despite the obvious layering, which might have led one to expect a bimodal distribution, the curve closely matches a Gaussian distribution. One might speculate that the major depositional process that determined the porosity of a layer was the extent of sorting in the population of grains available for transport and deposition at the time that particular portion of the formation was laid down. In that case, it would be natural to expect that the overall porosity distribution should indeed be normal.

The porosity distributions at the other locations are similar to that at location 11, with only slight differences in the average and variance of the local porosity at each location. The only exception is the local porosity distribution at location 1, shown in Fig. 7, in which it is seen that the mean porosity over the entire cross section is significantly lower than the others. For this slice, it can also be noted that the layering in ϕ patterns is not so clear as in the other locations.

Departing, now, from the display of measurements at individual locations, Fig. 8 shows the average porosity distribution. The term "average porosity" used here refers to the arithmetic average of local porosity in the x direction, over eleven cross sections of the sample, but at the same pixel-group index (i,j) of each cross section. The averaging process reduces the porosity variance dramatically. As also shown in Fig. 8, the average porosities of the three most porous layers are almost the same.

A different view of the amount of detail which is "washed out" by the above averaging process is shown in Fig. 9. The eight curves in this figure represent the layer-averaged porosity values as a function of the x coordinate (the sample length) for each of the layers identified in Fig. 3.

IV. CALCULATIONS OF PERMEABILITY

In order to deduce permeabilities in the rock sample, we need to have information or data on the distributions of Darcy velocity, porosity, and output concentration gradient from the miscible flood conducted in the sample.

The CT-Scanner data allow us to compute the porosity over eleven evenly-spaced "slices" or cross sections, in extraordinary spatial detail, within the limitations of measurement error and of the statistical fluctuations of the values. Similarly, the concentration of $BaCl_2$ tracer can be obtained from CT-images taken over an entire cross section (near the output face) at different times during the $BaCl_2$-$CaCl_2$ miscible flood, within similar limitations. For both ϕ and C, the values calculated represent averages over small volumes of rock in the immediate neighborhood of the pixels measured.

On the other hand, values of permeability cannot be obtained so directly.

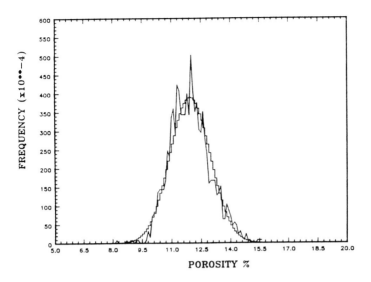

Figure 6.   Frequency distribution of local porosity values at location 11, and Gaussian distribution for comparison.

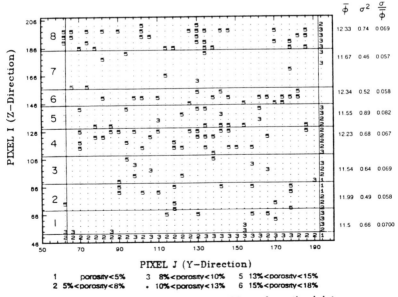

Figure 7.   Local porosity values at location 1, 20mm from the inlet.

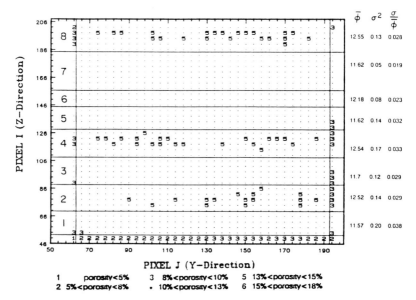

Figure 8. Lengthwise average porosity values (cross sections 1 through 11).

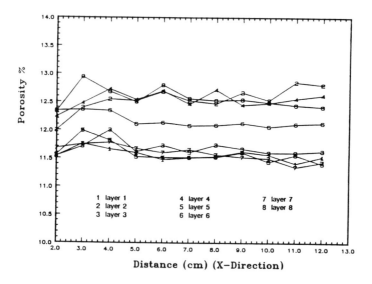

Figure 9. Variation along sample of average porosity in eight identified layers in Antolini core sample.

The data which are available, from the time-indexed concentrations of $BaCl_2$, give the distribution of travel times when the 50% concentration of $BaCl_2$ appeared at different locations over the output cross section. From these data, and from the porosities at corresponding locations in the different slices, values of the permeability can be computed as averages over the length of the flow path. But this computation cannot be performed without certain assumptions of regularity in the pattern of permeability variation throughout the sample.

In particular, we make the assumption that permeability variations are zero or negligible in the direction of average flow (that is, as selected by the original sampling cuts, and by the sealing of the "sides"). This is the x direction, as indicated in Fig. 3. With that assumption, and given the rectangular boundary conditions and an incompressible fluid, the solution of Darcy's equation is simply that the pressure will vary linearly with x, and that the Darcy velocity is in the x direction throughout the sample. Judging from the measured variation of porosity in the x direction, the assumption made about permeability is almost certainly in error. But because the measurements made do not provide enough information for a unique determination of permeability along the individual flow lines, we have chosen the simplest assumption among many possibilities.

Figure 8 shows that the major rock property variation in the University of Texas sample is the layering in the z direction. Although statistically significant variations of ϕ are perceptible in the y direction, their magnitude is considerably smaller than those in the z direction. Considering only the variation in the z-direction, the lengthwise average porosity of group pixel data is averaged out in the y-direction and presented in a form of histogram in Fig. 10a. There are 38 vertical bars in Fig. 10a, the heights of which are proportional to $\overline{\phi}(z)$, the average porosities of flat "tubes" or "stripes." These vertical bars are four pixels high and as wide in the y direction as the sample. The z axis is placed horizontally in this figure and in the two figures below it.

Directly under Fig. 10a, with the layer boundaries lined up, is Fig. 10b, a histogram displaying the travel times along each of the individual "stripes" or sub-layers. Each of the arrival times was estimated as the average of the times when the 50% concentration of $BaCl_2$ appeared at location 11 for the individual pixel-groups of a sublayer. The estimate was made by linear interpolation between time and the inverse error function of the argument (1-$2C_{(y,z,t)}$). The calculation of average displacement velocity was thus based on the assumption that all flow paths have equal lengths. The validity of this assumption is questionable except for "perfectly stratified" rocks, and should certainly be investigated further.

If these travel times are denoted by T(z), then one can compute the average permeabilities, K(z) (averaged not only over the width of the particular layer, but also over the length of the flow path, L, from the inlet face to location 11, at which the concentrations were observed). From Darcy's equation, the formula is:

$$\overline{k}(z) = (\mu L^2 \overline{\phi}(z))/(\Delta PT(z)). \tag{2}$$

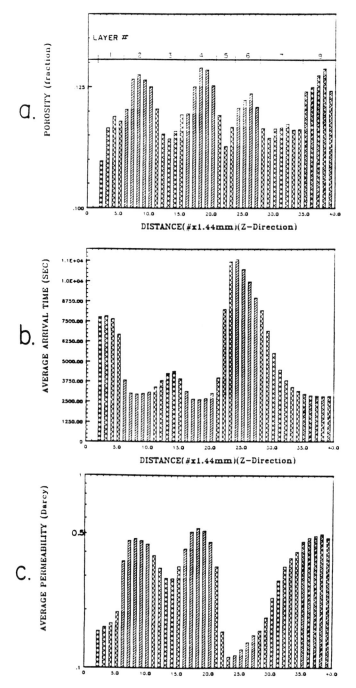

Figure 10.   Lengthwise averages of 4-pixel high sublayers: (a) porosity; (b) travel time; and (c) permeability.

These values are displayed in Fig. 10c, which is also lined up under the other two, so that the individual layers can be identified. It can be seen that, although there is some correlation of these calculated permeabilities with the porosities shown in Figs. 10a and 10c, the relationship is by no means exact. For instance, one can see that, whereas the porosity in layer 7 (toward the right side of these figures) is relatively uniform from top to bottom of the layer, the variation of the calculated permeability is almost linear between the values in layers 8 and 6. Whether this is real, or an artifact of the experiment, is discussed in a later section.

The permeabilities shown in Fig. 10c are averages over the entire width of the sample in the y direction, calculated with the use of porosities that are also averaged over the width. It is also possible to make the same calculations on a finer scale. This has been done for 4 by 4 pixel "streamtubes" running the length of the sample. The calculation has naturally resulted in a much larger number of permeability determinations, which can then be utilized for various statistical purposes. As an example, these individual permeability calculations were paired with the lengthwise average porosities over the same 4 x 4 streamtubes to plot the (k, ϕ) correlation scatter diagrams shown in Fig. 11. While there may be no good reason to expect that all of the porosity and permeability values should fall on a single correlation line, the great variation among the different layer correlations shown in Fig. 11 is noteworthy, and may suggest that significant differences existed in the depositional conditions for the layers. But before drawing further conclusions on this topic, it is appropriate to perform some geostatistical calculations.

V. SPATIAL VARIABILITY OF POROSITY AND PERMEABILITY IN ANTOLINI SANDSTONE

One of the purposes of this study is to investigate the spatial variability of rock properties on a small scale, and to decide whether these properties can be well represented at this scale by mathematically continuous functions. Properties such as permeability and porosity vary from place to place in real porous media. These variations have been observed on different scales from pore level to interwell distance. It has been noted that such variation of rock properties is not completely random, but possesses some kind of structure. The measured values of rock properties have some spatial continuity embedded within the randomness. However, it has been pointed out that the variation of permeability is large compared to the porosity variation, and in most petroleum literature the effect of porosity variation on fluid flow is considered to be small. In this study, the effect of small distance-scale variations of permeability and porosity on miscible displacement performance are considered. Geostatistical methods are used for the evaluation of the spatial correlation of rock properties.

One of the basic geostatistical tools used in this study to quantify the variability of rock properties is the semivariogram. The semivariogram is a graphical technique by which the correlation of the measured values at different distances can be displayed.

Ibrahim Bahralolom and John Heller

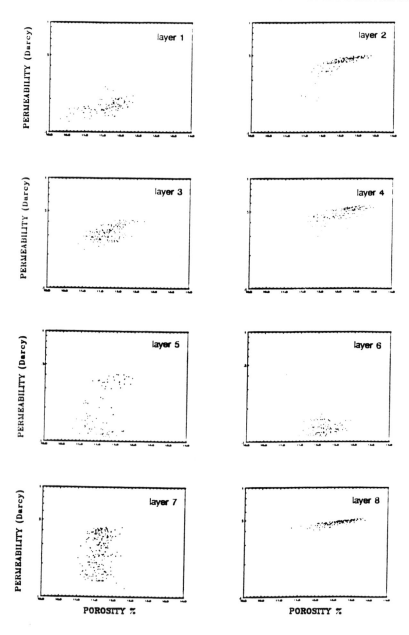

Figure 11. Permeability-porosity scatter plots by layers, Antolini sandstone sample.

Mathematically, this can be expressed as follows:

$$2\gamma(\underline{h}) = (1/N_{(h)}) \Sigma (X(i)-X(i+\underline{h}))^2, \tag{3}$$

where $N_{(h)}$ is the number of samples separated by the vector $\underline{h}$, $X(i)$ and $X(i+\underline{h})$ are the values of the samples, measured at positions a distance $\underline{h}$ apart, and the sum is over all such pairs. The lag distance, $|h|$, is the distance between two data points. Application of this technique provides information about the distance of the transition zone—the distance between the correlated and uncorrelated points. High correlation between the neighboring points is expected at small lags, while for large lags the sample values become independent of one another and the variability reaches a maximum equal to the sample variance. The distance at which the dependence between neighboring points vanishes is called the range of influence or the correlation length.

Two-dimensional semivariograms have been constructed from the porosity data at all locations to study the variability of porosity in both directions in the (z,y) plane perpendicular to flow. Comparison of those computed at different locations will give an idea about the variability in the flow (x) direction. Finally, examining the semivariogram of the average porosity may provide clues about the porosity variation within the system.

Figures 12 and 13 present the semivariograms versus lag distance for the local porosity at locations 11 and 1, respectively (Fig. 3). Due to the anisotropy of rock properties, the semivariogram is calculated for the two directions for each location. These directions are called the vertical direction (along the z axis, with the calculated points indicated by the digit 2) and the horizontal direction (in the y direction, indicated by the digit 1). In all semivariogram figures presented in this study, the abscissa represents the lag distance along either y or z axis of the core sample, and each pixel unit is equivalent to 0.35 mm. The ordinate unit of the semivariogram figures corresponds to the porosity being given in percent. In practice, the values of the semivariogram are not considered reliable (being based on insufficient samples) past about half of the total sampled extent. In this case, the limit is about 75 pixels (2.6 cm).

Two major features can be studied in these semivariograms--their behavior at the origin and at large values of lag spacing. The local porosity semivariogram at location 11 (Fig. 12) shows clear differences in the two directions, as would be expected from the known stratification. The vertical semivariogram of the local porosity rises to much higher γ values than that in the horizontal direction, reflecting greater variability due to the layering in the z-direction. In the vertical direction, the semivariogram values reach almost a maximum at about 0.65 cm and continue with some variation around this value. The magnitude rises and falls regularly, with a dip in the curve at 1.4 cm and another at 2.7 cm. As has been pointed out by geostatisticians (such as Journel and Huijbregts, 1978), this behavior is expected when there is a periodic trend or stratification present within the structure.

On the other hand, the semivariogram in the direction parallel to the strata showed different behavior. Here, the values of $\gamma(h)$ are not as high as

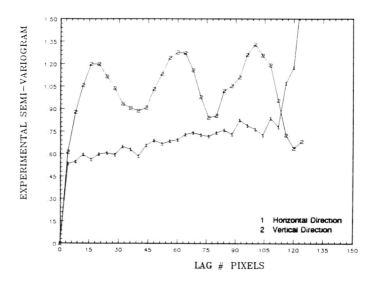

Figure 12.   Semivariogram of local porosity measurements at location 11.

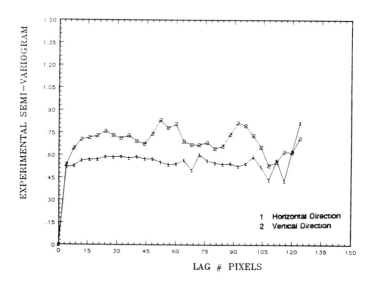

Figure 13.   Semivariogram of local porosity measurements at location 1.

those in the z-direction, and they continue to rise linearly for the largest reliable lag distances. This indicates a more regular variation in porosity than in the vertical direction. Comparison of the semivariogram behavior for the vertical and horizontal directions indicates that a high degree of anisotropy of the spatial porosity distribution exists at this location.

Figure 13 shows the semivariogram of local porosity at location 1, and shows less variability in the vertical direction, reflecting the lack of well-developed layers, as shown in Fig. 7. At location 1, the amplitudes of the vertical direction semivariogram peaks are smaller and closer to the computed values of the semivariogram for the horizontal direction. In the horizontal direction, the semivariogram values achieve almost a constant value at small lag distances. They remain at this constant level with little variation for the largest lag distances. Comparison of this variogram with the one at location 11 (shown in Fig. 12) indicates a considerable change in the geostatistical properties between these locations, particularly in the vertical direction.

The semivariogram of the lengthwise average porosity (that is, of the arithmetic averages of porosity at the same pixel-group, over all cross sections or locations) is presented in Fig. 14. The porosity shows similar behavior to those for the single-location variograms, particularly in the vertical direction. The lengthwise averaging process smoothed the porosity variations and reduced the sample variance. In the horizontal direction, the semivariogram-computed values do not reach the variance, but instead rise very slowly with increase of the lag distance. This behavior shows that the porosity values are much more variable in the vertical direction as a result of the layering.

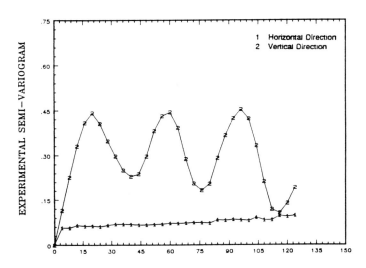

Figure 14.   Semivariogram of lengthwise average porosity of 11 cross sections (measurements at 11 different locations along flow direction).

The computed semivariogram values for the horizontal direction either achieve a constant value less than the variance of the local porosity measurements, as at sample location 1, or they continue to rise for large lag distances, as at location 11. Because of this, rough estimation of the correlation length is not easily obtained from these variograms. The autocorrelation function is, in this case, a more useful alternative to characterize spatial variability. This function is proportional to the complement of the semivariogram. The behavior of the autocorrelation functions for the local porosity values at different locations along the Antolini sandstone core sample has also been investigated, but it will not be presented here. The autocorrelation functions of the porosity measurements in the horizontal and vertical directions decrease rapidly in general. In the horizontal direction, the correlation scales for porosity at locations 2 and 3 are small compared to those estimated at locations 7 and 10. However, none of the auocorrelation functions, at any locations approach the zero autocorrelation line. Rather they tend to remain between 0.37 and 0.2. The autocorrelation functions of local porosity in the vertical direction at the same locations indicate that the stratification at locations 7 and 10 are more pronounced than at locations 2 and 3.

The variation of porosity in the direction of flow is also investigated in more detail. The autocorrelation function is calculated for two different strips of group-pixel porosity in the z-direction over 11 locations. These two strips are taken 3 cm apart in the y-direction. Figure 15 presents the autocorrelation function for each strip porosity over the entire length of the core sample. The auocorrelation functions indicate that there is spatial correlation in the porosity values along the flow direction, the correlation distance ranging between 1.7-3.5 cm. Also indicated is that in the x, as well as the y direction, the autocorrelation functions do not die; instead, they continue to remain between 0.375 and 0.3 for the larger lag spacing.

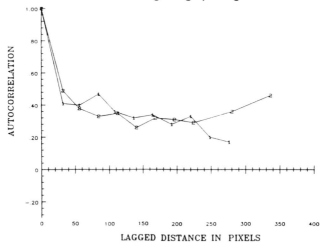

Figure 15.   Autocorrelation functions representative of the two strips of porosity in the x-direction.

In summary, the results of the local and average porosity distributions, the semivariograms, and the autocorrelation functions suggest that the spatial correlation of porosity in the horizontal direction, within the layer itself, is much greater than that in the vertical direction. Along the flow direction, the average correlation length is estimated to be about 2.5 cm.

VI. SPATIAL VARIABILITY OF PERMEABILITY

Because there are no spatially detailed permeability measurements available for the core sample, the permeability has been estimated as described earlier from the consecutive concentration curves measured at location 11. Using that procedure, the average permeability was estimated for each 4 x 4 pixel-group streamtube, and is shown in Fig. 16. In the figure, ten ranges of logarithm of permeability are denoted by the digits 1 through 10, where digit 10 is labelled with 0. Comparison of the lengthwise average porosities (Fig. 8) with the average permeability distribution shows that the high porosity layers generally have higher permeability as well. It is also clear that there is much more variation of permeability in the vertical direction than in the horizontal direction.

The spatial variability of the average permeability can be estimated from the logarithmic semivariogram shown in Fig. 17. Again, the directional dependency of the permeability variability is clear. The semivariogram in the vertical direction rises almost parabolically from the origin and, after remaining for several millimeters at a plateau, rises again and reaches a maximum at about 2.1 cm. It then drops off steadily to a minimum at a distance greater than 3 cm. The semivariogram of average permeability in the horizontal direction behaves differently, rising very slowly and regularly. It then rises at an increasing rate, up to the greatest lag distances available in this sample, suggesting that any limiting range of influence or correlation distance is beyond 3 cm. This is similar to the behavior of the semivariogram of the average porosity, and probably indicates a trend within the structure in the horizontal direction.

These results are similar to those of Dupuy et al., (1966). They measured the local permeabilities at the surface of a Fontainebleau sandstone core sample by using a micropermeameter device They characterized the spatial heterogeneity using the autocorrelation function of the micropermeability measurements. Their results showed that the autocorrelation function in the horizontal direction did not reach zero for relatively large lag distances. We speculate that this behavior might indicate that the parameter of interest does not follow the assumption of second-order stationarity, at least within this size of sample.

A separate question that is important to consider in this study concerns the relationship between average porosity measurements and the average estimated permeability--especially at the small scale available for study here. Although there can certainly be no general relationship, relatively simple linear correlations do exist between the average porosity and permeability of closely related rocks from the same formation. In addition, it has also been known

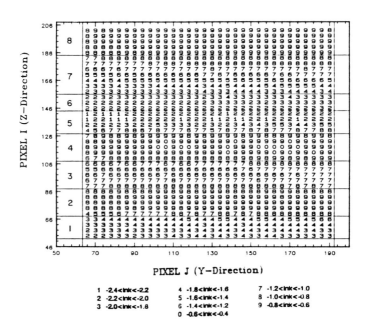

Figure 16.   Average permeability values in cross section.

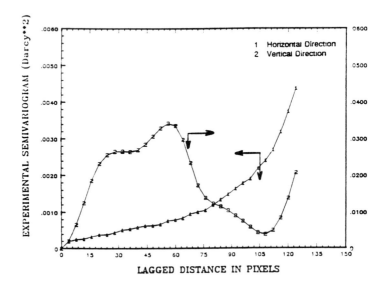

Figure 17.   Semivariogram of average permeability of Antolini sample.

(Krumbein and Monk, 1942) for unconsolidated sandstone or laboratory packs with narrow grain size distributions, that permeabilities are proportional to the square of the grain size, while porosity is little affected by it. But scatter diagrams of porosity and permeability data for large samples of consolidated porous media show a large degree of variability. Variation of the width of the grain size distribution, and of the amount of cement, are usually considered to be the main cause for this scattering. It has been suggested that the scatter may be due to the combination of many linear trends, each dependent on individual petrological variables (Wendt et al., 1985).

This wide variation is also evident by our analysis of the CT-Scanner data. Figure 11 shows the permeability-porosity correlations for each of the eight layers. The overall variation of permeability on these plots is not large, but it is clear that a considerable difference exists among the different layers. A simple linear relationship seems to hold for the high-permeability layers 4 and 8, and probably 2 as well. In these layers, we speculate that the grain sizes are comparable and the sand was well-sorted by the depositional process. However, a linear relationship can also be found in low-permeability layers such as layers 1 and 6. These layers may consist of finer grains with little variation in size. On the other hand, the scatter in layers 3, 5, and 7 shows a much greater variation in permeability, probably indicating considerable variation in grain size.

We would also speculate that the low porosity level of all of the layers indicates a large amount of cement or very fine material mixed into the primary grains. Petrographic analysis of these layers might be of great assistance in verifying or contradicting these speculations and in casting light on the geological processes during the deposition and subsequent diagenesis of such dune deposits as the Antolini sandstone.

VII. THE EFFECT OF DISPERSION

In the discussions above, we have tried to deduce information about variability of rock properties from the CT-Scanner data of the sample and from the concentration patterns and history that were observed near the outflow end of the core sample. The methods used consider a somewhat simplified Darcy flow (some assumptions about the regularity of the layering were necessary) and do not consider the effect of the rate at which the experiment was performed. It is clear that, simultaneously with Darcy flow, dispersion causes changes in the concentration gradients. The question can then be asked, might it therefore be necessary to make any corrections to our calculations of the permeability?

When the first one-dimensional miscible displacement experiments were performed in uniformly packed columns, it was observed that, over a wide range of flow rates, the measured transition zone length was independent of the rate. It was soon realized (see Blackwell, 1962; Perkins and Johnston, 1962; Heller, 1963 and 1972) that this occurred because the largest term in the longitudinal dispersion coefficient was proportional to the flow velocity and did not explicitly contain the molecular diffusivity. It was the nearly complete diffusive mixing within the pores that enabled miscible displacement to proceed with a relatively low dispersion coefficient, determined by the microscopic variability of the rock. Molecular diffusion averages out the extreme changes in concentration which would otherwise be caused by the sharp variation of velocity within the pores. In relatively uniform cores or packed columns, the remaining variability of the concentration pattern was due to the random variation of average velocities between adjacent flowpaths.

In more natural porous media, there is also larger-scale variability of rock properties, and consequently in the velocities during flow. During miscible displacement, the moving isoconcentration surfaces that describe the flood's progress become corrugated on many larger scales than merely that from pore-to-pore. In the larger distance scales, and within reasonable time scales, the concentration patterns that evolve during miscible displacement through large-scale rock heterogeneities will once again be independent of flow rate. The reason in this case is that dispersion cannot in the time available make a change that is appreciable over the larger distances. The time for approach to equilibrium, in any process described by a diffusion equation, is proportional to the square of the distance scale.

In miscible floods through nonuniform media at laboratory scales, the simplification of rate independence will not hold, and dispersion might produce significant effects. There are two sources of change in the concentration pattern during miscible displacement. The first is that due to the flow velocity (which, though it may vary from place to place, will be constant in time in a flood with unit viscosity ratio), and the second is that due to dispersion. There are several choices of a method to determine whether dispersion is significant in the data reported here, or in any other experiment. One would be to consider the general equation for the change in the concentration field, the so-called convection-dispersion equation. This equation gives the rate of change of concentration as a sum of two terms that can be compared in magnitude. Written in vector form, with U representing the Darcy velocity, C the concentration, and $\mathfrak{D}$ the dispersion tensor, the equation is

$$\partial C/\partial t = -(U/\phi) \cdot \nabla C + \nabla \cdot (\mathfrak{D} \cdot \nabla C). \tag{4}$$

Comparing the first (the convective) term to the second might require more data than are available. Instead, one could go back one step in the derivation, so to speak, and compare the convective and dispersive flux vectors directly. The convective flux is just the displacement velocity, U/ϕ, whereas the dispersive flux is $-(\mathfrak{D} \cdot \nabla C)$. This plan is followed in the next paragraphs.

The major way that dispersion can change the concentration pattern in a miscible flood in a nonuniform medium is by smoothing the irregularities in the isoconcentration surfaces. This is accomplished, in general, through dispersion transverse to the flow, in the manner discussed quantitatively by G.I.Taylor (1953) for capillary tubes and by Hirasaki and Lake (1981) in stratified porous rock. Thus, in this experiment, the important dispersion will be transverse to flow at the borders of the layers and in the z direction, and

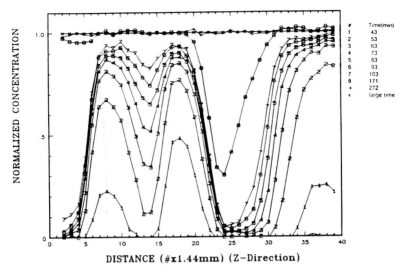

DISTANCE (#x1.44mm) (Z-Direction)

Figure 18.   Z-direction of $BaCl_2$ concentrationprofiles at location 11, at successive times.

we need only compare the displacement velocity with $D_T(\partial C/\partial z)$. Successive measurements of the concentration profile in the z direction are presented in Fig. 18, from which estimates of the concentration gradients at location 11 can be obtained.

The displacement velocity was approximately 5.1×10^{-3}cm/sec. From Fig. 18, the maximum gradient appears to be 2.2 concentration units per centimeter, between layers 1 and 2, at the scan made at 73 minutes. The molecular diffusion coefficient of $CaCl_2$ (which is probably somewhat greater than D_T at this low flow rate) is about 1.2×10^{-5}cm^2/sec, so the dispersive flux is no more than 2.6×10^{-5}cm/sec, and much less than the convective flux. It thus appears that the dispersive change in the concentration pattern in this experiment is negligible.

The above result also indicates that the calculations that were made of permeability probably do not need any corrections because of the omission of the dispersion term.   At worst, the permeabilities computed at the boundaries between layers might be in error.   In the neighborhood of the boundary, transverse dispersion would decrease sharpness of the apparent permeability contrast.

This decrease can be calculated as follows. If the boundary between the different permeability regions were "perfectly sharp" (i.e., if any graded transition region between them were thinner than the distances that can be resolved), and if permeabilities of the two layers differed by more than ten percent or so, then at the output face, the graded zone of concentration between the regions could be computed from the well-known formula for the distance between the 90% and 10% concentration levels on a diffuse front,

$$\Delta x_{90,10} = 3.62\mathrm{sqrt}(D_T T) = 3.62\mathrm{sqrt}(\phi D_T L/U).\qquad (5)$$

Squaring and solving the above equation for D_T, and using again the values above for ϕ, L, and U along with 0.42 cm for $\Delta x_{90,10}$ (read off Fig. 18, again between layers 1 and 2), we obtain the very reasonable value for D_T of $5 \times 10^{-6} cm^2/sec$. The result reinforces the conclusion that, in this case, any dispersive corrections for the permeability variation would be negligible.

VIII. CONCLUSIONS

This paper reports the analysis of CT-Scanner data from a laboratory-scale miscible displacement experiment performed at the University of Texas at Austin. From this analysis, we have been able to observe and calculate the variabilities of porosity and permeability in a visibly stratified sample of Antolini sandstone. The measurements at the level of pixels (0.035 cm square by 1mm thick) show considerable variation, but are at least partly significant, indicating real statistical variations of the porosity from one pixel to the next.

At only slightly larger scale, the variations of porosity do not appear to be random, but rather indicative of layering caused by the depositional conditions and subsequent cementation. These porosity variations are in the range from 11 to 13%.

Permeability could not be measured directly, of course. Instead, data which chronicled the concentration changes associated with a miscible flood were analyzed. The permeabilities were calculated from miscible front travel times, coupled with the measured values of porosity at eleven locations between the inflow face and the "slice" at which the concentrations were measured. These permeability values varied from 100 to 600 millidarcies. Generally, the higher porosity layers showed greater permeability, although the correlations, shown on the scatter diagrams, differ noticeably for the different layers.

Finally, an approximate calculation indicates that the effect of transverse dispersion in this experiment was not severe, and that permeability calculations by the method indicated do not stand in need of appreciable correction for this reason.

ACKNOWLEDGEMENTS

Great appreciation is expressed to Drs. Gary Pope and Bruce Rouse for their interest and generosity in supplying us with copies of the digitized data from an experiment performed with the CT-Scanner at the University of Texas at Austin. The authors also wish to acknowledge the support, on the larger project of which this work is a part, of the Department of Energy, the New Mexico Research and Development Institute, and the consortium of oil companies including Abu Dhabi National Reservoir Research Foundation, Amoco, Arco, BP America, Marathon Oil, OXY USA, and Texaco. Finally, we wish to acknowledge the assistance of K. Allbritton, Karen Bohlender, Toni Johnson, and Kevin Clower in preparing and editing the manuscript. In addition, appreciation is expressed to Dr. Graham Fogg for a critical review and many valuable suggestions.

REFERENCES

Blackwell, R.J. (1962), "Laboratory Studies of Microscopic Dispersion Phenomena," Trans., AIME 225, pp.1-8.

Dupuy, M., Morineau, Y., and Simandoux, P. (1966), "About The Importance of Small Scale Heterogeneity on Fluid Flow in Porous Media," paper SPE 1840 presented at 1966 SPE Annual Meeting, Dallas, Feb.8.

Heller, J.P. (1963), "The Interpretation of Model Experiments for the Displacement of Fluids Through Porous Media," AIChE J., 9, pp.452-459.

Heller, J.P. (1972), "Observations of Mixing and Diffusion in Porous Media," Proc., Second Symposium, "Fundamentals of Transport Phenomena in Porous Media," IAHR-1SSS, Ontario, pp.1-26.

Hirasaki, G.J. and Lake, L.W. (1981), "Taylor Dispersion in Stratified Porous Media", Soc. Pet. Eng. J., 21, pp.459-468.

Journel, A.G. and Huijbregts, C. (1978), Mining Geostatistics, Academic Press, NY, p.248.

Krumbein, W.C. and Monk, G.D. (1942), "Permeability as a Function of the Size Parameters of Unconsolidated Sand," AIMME, Petroleum Technology, Technical Publication 1492, pp.1-11.

Perkins, T.K. and Johnston, O.C. (1963), "A Review of Diffusion and Dispersion in Porous Media," Soc. Pet. Eng. J. (March) 3, pp.70-84.

Pope, G.A., Lake, L.W., and Sepehrnoori, K. (1988), "Modelling and Scale-up of Chemical Flooding," Annual report to U.S. Department of Energy, Report No. DOE/BC/10846-10 (November).

Taylor, G.I. (1953), "Dispersion of Soluble Matter in Solvent Flowing Slowly through a Tube," Proc. Roy. Soc.(A) 219, pp.186-203.

Wendt, W.A., Sakurai, S., and Nelson, P.H.(1986)," Permeability Prediction From Well Logs Using Multiple Regression," Reservoir Characterization, Academic Press, NY, pp.181-221.

Withjack, E.M. (1987), "Computed Tomography for Rock-Property Determination and Fluid-Flow Visualization," paper SPE 16951 presented at the 1987 SPE Annual Meeting, Dallas, Sept. 27-30.

QUANTIFYING SATURATION DISTRIBUTION AND CAPILLARY PRESSURES USING CENTRIFUGE AND COMPUTER TOMOGRAPHY

Helga Baardsen
Victor Nilsen
Jørgen Leknes
Arvid Hove

Production Laboratories, Statoil
Stavanger, Norway

1. ABSTRACT

X-ray Computer Tomography (CT) is used to investigate the centrifuge technique for capillary pressure measurements in core samples. The validity of the Hassler & Brunner boundary condition is studied, redistribution of water saturation is visualized and quantified from CT- images, and capillary pressures are estimated.

Samples of reservoir sandstone, Berea sandstone and aluminum oxide are saturated with brine and drained with gas or a light refined oil. After measurements of produced water at each angular velocity, the samples are scanned in the CT. The water saturation is quantified, and the change in the saturation profile after centrifuging is followed as a function of time. CT-images show that the water saturation changes towards a uniform distribution. However, a totally uniform distribution is not established in any of these samples within 12 hours. This change is faster for increasing capillary pressure.

A wax technique is used to study the Hassler & Brunner boundary condition. Having obtained equilibrium for a gas/wax drainage at an elevated temperature, the wax distribution is frozen during centrifuging. The CT-images show that the boundary condition is not valid for all capillary pressures.

A technique to estimate capillary pressures from CT-visualized saturation profiles is developed.

RESERVOIR CHARACTERIZATION II

2. INTRODUCTION

The centrifuge technique is a rapid method used to measure capillary pressure curves (1,2), relative permeability (3-5), wettability indices (6,7) and for the establishment of saturations for other analyses such as water flooding and electrical measurements (8). During centrifuging there exist a capillary pressure gradient and also a saturation profile within the sample. In a conventional centrifuge experiment the mean wetting phase saturation in the sample, $<S_w>$, and the rotation velocity, ω, are measured.

In the calculation of capillary pressure (2) and the simulation of relative permeability curves (4,5), it is assumed that the outer core end is 100% saturated with wetting phase during centrifuging, implying that the capillary pressure at this boundary is zero. This Hassler & Brunner boundary condition is studied in the literature (9 -12).

The application of an epoxy casting technique to study the fluid distribution during centrifuge capillary pressure measurements is described earlier (9). Two different coloured epoxy were used. Thin section studies of the samples after centrifuging indicated that the 100% saturation at the outlet end is not valid beyond a limited rotation velocity. Hence, the traditional Hassler & Brunner boundary condition is not valid for all capillary pressure levels.

A theoretical study has shown that the validity of assuming that a 100% water saturation at the outer core end exists is dependent on the magnitude of the displacement pressure for the particular rock sample (10). One of the conclusions was that "the failure of the Hassler assumption occurs only in very unusual reservoir situations."

When a centrifuge is used to establish water saturations for electrical measurements or before a waterflood, it is assumed by many laboratories that, after a certain time, a uniform saturation distribution is again established. For resistivity measurements, it was concluded from a theoretical study that accurate determination of saturation exponent requires uniformly saturated core, and that procedures for monitoring the saturation profiles are needed and must be established (13).

The purpose of our study is, by use of X-ray Computer Tomography, to check the validity of the Hassler & Brunner boundary condition and to visualize the wetting phase saturation during and after centrifuging. By applying an image processing system the CT-visualized saturations are quantified and capillary pressures estimated.

3. EXPERIMENTAL

Three samples of a North Sea reservoir sandstone, four Berea sandstone samples and four aluminum oxide samples were used in the experimental programme. They were cleaned in a Soxhlet apparatus (methanol, toluene), oven dried at 60 $°C$ and 40% relative air humidity and saturated with a NaI-solution or wax. Primary drainage was performed with a light refined oil or gas using a Beckman ultra centrifuge (pir 16.5 rotor). No modification of the coreholders was applied. The samples were wrapped in teflon tape to ensure longitudinal flow.

The Amott wettability test (6) was performed on selected samples after the oil/water drainage. The "water-wetting" indices (0.9 -1.0) indicate strongly "water-wet" systems. Petrophysical data and fluid properties are listed in Tables I and II.

Table I. Core sample properties. Ka and Kl are the permeabilities at $1/P_m$ equal to 1 and 0 respectively.

Sample	Mater.	Length (cm)	Diam. (cm)	Poros. (%)	Gr.dens. (g/cc)	Ka (mD)	Kl (mD)
A1	al.ox.	4.80	3.67	37.9	2.98	3890	3763
A2	al.ox.	4.75	3.82	43.0	3.16	53	41
A3	al.ox.	4.94	3.63	38.1	2.98	3721	3496
A4	al.ox.	5.02	3.75	43.3	3.16	55	42
B1	Berea	4.92	3.79	20.2	2.66	393	346
B2	Berea	4.87	3.79	19.9	2.66	356	320
B3	Berea	4.87	3.74	22.5	2.65	593	552
B4	Berea	4.89	3.74	22.4	2.65	751	691
R1	res.s.	4.90	3.78	17.9	2.66	78	69
R2	res.s.	4.89	3.78	19.4	2.66	127	115
R3	res.s.	4.92	3.79	17.4	2.66	114	105

Table II. Fluid properties

Fluid	Density (g/cc)	Viscosity (cP)
Water (NaI 100g/kg), 20 $°C$	1.0785	1.01
Light refined oil, 20 $°C$	0.743	1.31
Wax (*Therell type* 5006)[a], 70 $°C$	0.773	4.10

[a] Melting point for wax (droplets of 1,3 diiodpropane added) is about 50 $°C$.

A SIEMENS Somatom DRH X-ray CT-scanner was used, and the CT-technique is described elsewhere (14-16). A ContextVision

GOP-300 image processor has been applied to further filter, enhance and do calculations on the images.

Experimental programme for cases 1 (gas/water) and 2 (oil/water) :

1. CT- scan of dry, 100% water- and 100% oil-saturated samples
2. Drainage in centrifuge at ω_1 for about 12 hours
3. CT- scans of the samples (removed from the centrifuge) to visualize the redistribution of water saturation
4. Repeat steps 2 and 3 for higher ω

Experimental programme for case 3 (gas/wax):

1. CT- scan of dry and 100% wax-saturated samples
2. Centrifuge at elevated temperature and at 20 °C to freeze the equilibrium saturation distribution in the sample
3. CT- scan
4. Repeat steps 2 and 3 for higher ω

4. RESULTS AND DISCUSSION

4.1 Water Saturation Estimated from CT-Images

The CT-images show the distribution of water or wax saturation in an 8 mm slice of the sample, which corresponds to about 1/4 of the total bulk volume. The slice is illustrated in Fig. 1. The principle of determining saturations from CT measurements is described elsewhere (15, 16). The mean water or wax saturation estimated from the CT-images, $S_w(CT)$, and from material balance, $<S_w>$, are plotted in Figs. 2 and 3. Experimental conditions and results are listed in Tables III and IV which follow after Fig. 3.

In Fig. 2 the mean wetting phase saturation values are plotted for all the samples listed in Tables III and IV. The saturation values, based on the CT-images, agree with the values from material balance within 0.03 saturation units. For some cases the disagreement is greater. This can be attributed to sample heterogeneity (the slice is not representative), and uncertainties in the material balance as well as in the CT-quantification. Shrinkage of wax during the solidification is also a source of error. The correlation coefficient for these data is estimated to 0.974.

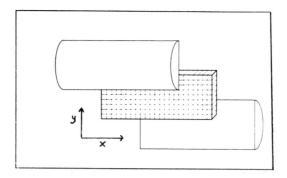

Fig. 1. The CT-slice in the longitudinal direction.

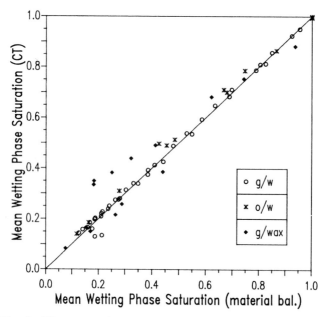

Fig. 2. Mean wetting phase saturation from material balance and estimated from CT-slices for different fluid systems in samples of sandstone and aluminum oxide.

To compare the CT-estimation to the material balance in further detail, data for three *homogeneous* Berea sandstone samples have been extracted and plotted in Fig. 3. In this case the correlation coefficient is increased to 0.999.

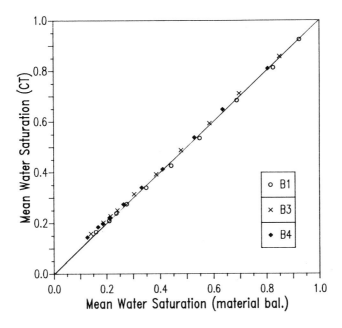

Fig. 3. Mean wetting phase saturation from material balance and estimated from CT-slices for gas/water system in Berea samples.

Table III. Experimental conditions and results for mobile fluids. The $P_c(r_i)$ is the capillary pressure at inner centrifuge radius. A hyphen (-) means that $S_w(CT)$ could not be estimated.

Sample	Fluids	RPM	$P_c(r_i)$ (bar)	$< S_w >$	$S_w(CT)$
A1	gas/water	1000	0.40	0.283	-
		5000	9.91	0.097	-
	oil/water	500	0.03	1.000	1.000
		750	0.07	0.667	0.710
		900	0.10	0.454	0.490
		9000	9.99	0.118	0.140
A2	gas/water	5000	9.84	0.307	-
A4	gas/water	1200	0.59	0.954	0.950
		5000	10.2	0.280	0.280
		7200	21.2	0.176	0.160
	oil/water	1000	0.13	1.000	-
		1800	0.41	0.896	-
		5000	3.18	0.384	-
		10000	12.7	0.248	-

(continued)

Table III. (Continued)

Sample	Fluids	RPM	$P_c(r_i)$ (bar)	$<S_w>$	$S_w(CT)$
B1	gas/water	500	0.10	0.924	0.923
		600	0.15	0.825	0.814
		700	0.20	0.688	0.683
		850	0.29	0.548	0.535
		1000	0.40	0.441	0.426
		1200	0.58	0.387	-
		1250	0.63	0.348	0.339
		1500	0.90	0.293	-
		1600	1.03	0.275	0.275
		1930	1.50	0.236	0.239
		2250	2.04	0.209	0.210
		2500	2.51	0.193	-
		2750	3.04	0.181	-
		3000	3.62	0.170	-
		3250	4.25	0.159	0.165
	oil/water	500	0.03	1.000	0.995
		750	0.07	0.866	0.865
		900	0.10	0.679	0.700
		9000	10.2	0.163	0.185
	(New exp.)	1800	0.40	0.425	0.497
B2	gas/water	600	0.14	0.789	0.787
		1200	0.58	0.384	0.375
		5000	10.0	0.212	0.135
		7200	20.7	0.186	0.130
B3	gas/water	500	0.10	0.849	0.857
		600	0.14	0.697	0.711
		700	0.20	0.585	0.594
		850	0.29	0.478	0.488
		1000	0.40	0.385	0.393
		1250	0.63	0.303	0.315
		1500	0.90	0.257	-
		1600	1.02	0.240	0.251
		1930	1.49	0.214	0.228
		2250	2.02	0.187	0.203
		2500	2.50	0.173	-
		2750	3.02	0.160	-
		3000	3.60	0.149	-
		3250	4.22	0.140	0.158
B4	gas/water	500	0.10	0.805	0.810
		600	0.14	0.634	0.648
		700	0.20	0.528	0.538
		850	0.29	0.409	0.413
		1000	0.40	0.331	0.340
		1250	0.63	0.263	0.274
		1500	0.90	0.225	-
		1600	1.03	0.210	0.222
		1930	1.49	0.185	0.198
		2250	2.03	0.167	0.185

(continued)

Table III. (Continued)

Sample	Fluids	RPM	$P_c(r_i)$ (bar)	$<S_w>$	$S_w(CT)$
B4	gas/water	2500	2.51	0.154	-
		2750	3.03	0.143	-
		3000	3.61	0.134	-
		3250	4.23	0.126	0.144
R1	gas/water	900	0.33	0.767	-
		5000	10.0	0.370	-
	oil/water	1000	0.13	0.853	-
		1800	0.41	0.544	-
		5000	3.13	0.316	-
		10000	12.5	0.291	-
R2	gas/water	900	0.33	0.741	-
		5000	10.0	0.266	-
	oil/water	1000	0.13	0.747	0.785
		1800	0.41	0.484	0.513
		5000	3.12	0.278	0.310
	(collapsed)	10000	12.5	0.278	0.310
R3	oil/water	900	0.10	0.910	-
		5000	3.14	0.410	-

Table IV. Experimental conditions and results for gas/wax drainage

Sample	RPM	$P_c(r_i)$ (bar)	$<S_w>$	$S_w(CT)$
A3	600	0.10	0.288	0.258
	1100	0.35	0.169	0.150
	5900	10.1	0.075	0.084
A4	700	0.14	0.937	0.882
	1200	0.42	0.743	0.753
	3200	2.99	0.322	0.439
	5900	10.2	0.182	0.350
B2	700	0.14	0.440	0.386
	1200	0.41	0.264	0.216
	3200	2.94	0.153	0.166
	5900	9.98	0.102	-
R1	700	0.14	0.621	0.682
	1200	0.42	0.412	0.491
	3200	2.95	0.250	0.383
	5900	10.0	0.181	0.335

4.2 The Hassler & Brunner Boundary Condition

The boundary condition at the outer core end is studied in the literature, and recently an equation to calculate the critical capillary pressure is suggested (10). This is the maximum capillary pressure at the inlet core end before the boundary condition at the outlet fails. It is a function of fluid combinations, permeability, porosity and pore structure among other parameters.

For Berea sandstone (air/oil), it was concluded that "for reservoir water saturation prediction the failure of the Hassler-Brunner boundary condition will be encountered only in very unusual reservoir situations. When using the centrifuge technique to provide pore size distribution data, the boundary condition limitation may arise more frequently" (10).

However, CT-images for the present samples indicate a breakthrough of nonwetting phase at outer core end at relatively low capillary pressures.

For the gas/water drainage of sample B1 the estimated and the observed critical capillary pressure can be approximated as suggested by Melrose (10). This approximation is however very uncertain. For the Berea sample B1 and the reservoir sample R1 with oil/water system, this model gives up to 10 times higher critical capillary pressure than observed from the CT-images.

The capillary pressure at a distance r from the centrifuge axis is given by

$$P_c(r) = \alpha \, \Delta\rho \, \omega^2 \, (r_e^2 - r^2) \tag{1}$$

where

$P_c(r)$ = capillary pressure at a distance r in the sample, (bar)
α = conversion factor (5.0 • 10^{-7})
$\Delta\rho$ = density difference, (g/cc)
ω = angular rotation velocity (2π RPM/60), (1/sec)
r_e = outer centrifuge radius, (here: 9.38 cm)

The critical capillary pressures listed in Table V are *rough* estimates. These values are based on observing whether or not a saturation gradient exists at the outer core end. Even with a small gradient, we believe that there has been a thin film at outlet during centrifuging, but a redistribution has already started within the 2-3 minutes before the first CT-scan. A breakthrough will probably occur if the angular rotation velocity is increased only a few RPM.

At the highest ω used, a nearly uniform distribution of wetting fluid phase is observed for most of the samples. Hence, a breakthrough of nonwetting fluid phase has occurred.

Table V. Estimated critical capillary pressures at inner centrifuge radius from CT-images. For some cases the critical pressure can not be more exactly determined due to few measurements.

Sample	Fluids	P_c(bar)
A1	gas/water	10-12
	oil/water	< < 10
A2	gas/water	> 10
A3	gas/wax	< < 10
A4	gas/water	20-22
	oil/water	> 12
	gas/wax	> 10
B1	gas/water	5-6
	oil/w..ter	8-10
B2	gas/water	< 10
	gas/wax	< 3
B3	gas/water	5-6
B4	gas/water	5-6
R1	gas/water	< < 10
	oil/water	3-4
	gas/wax	3-4
R2	gas/water	< < 10
	oil/water	3-4
R3	oil/water	< 3

The indicated critical capillary pressures are approximately the same for gas/water and oil/water systems within each sample, except for sample A1. For this sample in the oil/water drainage, the critical capillary pressure is uncertain. The value is based on the gas/wax study of a parallel sample, as illustrated in Fig. 4. This figure shows that, increasing the velocity from 1100 to 5900 RPM, a breakthrough has occurred. For the aluminum oxide samples, the critical capillary pressure seems to be higher for the lower permeable samples.

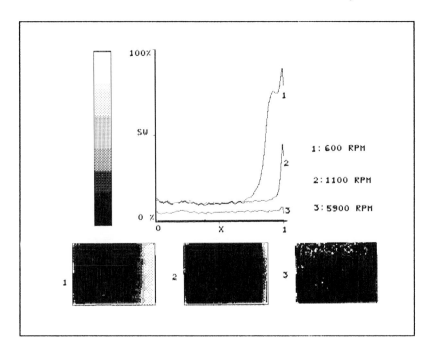

Fig. 4. The distribution of wax saturation in sample A3 in the study of gas/wax drainage. The corresponding CT-images are shown beneath the profiles.

The saturation profiles in Figs. 4 to 7 are plotted as a function of position along the sample, where the x-axis represents the positions from inner to outer centrifuge radius (left to right), see Fig. 1. The saturation values are averaged across the images, and the ordinate-axis represents the saturation values from 0 to 100%.

The CT-wax technique is nondestructive. Since the same sample can be used for all the measurements, this method is preferred to the combination of epoxy and thin section analysis described in the literature (9). Our results for Berea sandstones agree with the data from the "above mentioned" epoxy-study, but the critical pressures found from both studies are lower than theoretical predictions (10).

4.3 Redistribution of Water Saturation After Centrifuging

The images show that, at high mean water saturations, the redistribution of water is a slow process. The change in saturation distribution as a function of time for one measurement is visualized in Fig. 5 for an oil/water system. A similar change in saturation gradient is observed for gas/water system. A uniform distribution is not achieved for any of the samples within the observation time if the Hassler & Brunner boundary condition is still valid. To accelerate this process, the plug was mounted invertedly and centrifuged for a short period. The gradient was then reduced.

The loss of 100% water saturation at outlet, as for sample R2 at 1800 RPM in Fig.6, can also be caused by some redistribution during deceleration. During this period no wetting phase can imbibe into the sample from the outlet, and the 100% saturation can not be kept (especially in the case of a thin film at the outlet).

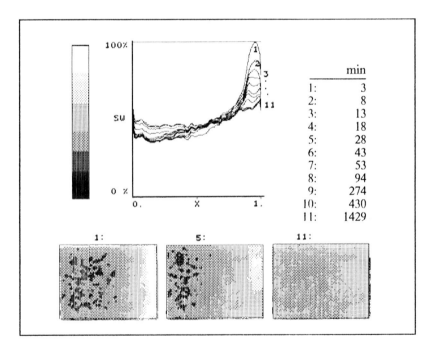

Fig. 5. The distribution of water saturation in sample B1 in the study of oil/water drainage at 1800 RPM. Three corresponding CT-images are shown. The time refers to minutes after the centrifuge was stopped.

By increasing the capillary pressure, the mean water saturation decreases and the redistribution process is faster. This is illustrated in Figs. 6 and 7.

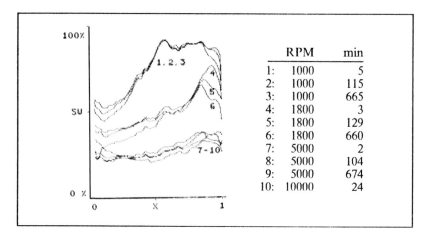

Fig. 6. The distribution of water saturation in sample R2 in the study of oil/water drainage.

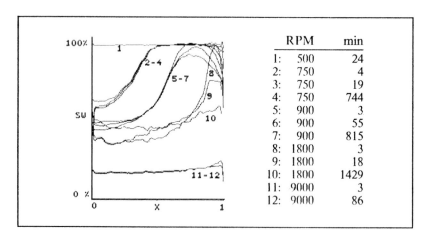

Fig. 7. The distribution of water saturation in sample B1 in the study of oil/water drainage.

Further analysis of the redistribution process is done. From the first saturation profile the maximum water saturation is located, see Fig. 5. Fixing the corresponding x-value, the water saturation is read as a function of time and plotted on a semilog scale in Fig. 8. Two trends are observed that can be explained by a change in the driving force, capillary pressure. At the outer core end the capillary pressure increases and elsewhere in the sample it decreases. The effect of viscous forces is also reduced as a function of time (viscous and capillary forces are probably more equal after 1 hour). A similar plot for aluminum oxide samples shows that the water saturation changes towards a uniform distribution linearly with the logarithm of time. The process also seems to be faster for high than low permeable aluminum oxide samples.

CT-images indicate that the same kind of redistribution occurs for gas/water and oil/water systems. The process seems to be faster for the gas/water than for the oil/water system in Berea samples, probably due to different capillary forces and viscosities.

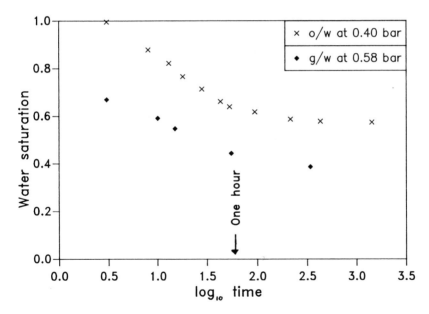

Fig. 8. The redistribution of water in sample B1. The saturation at a fixed position x in the sample is plotted as a function of time (time in minutes).

4.4 Capillary Pressures Estimated from CT-Images

Eq.1 is rewritten to include a distance l from outer core end to a point in the sample, and this gives

$$P_c(l) = \alpha \, \Delta\rho \, \omega^2 \, (2\,r_e - l)\,l \tag{2}$$

where

$l = r_e - r = r_e - (r_i + x)$
$l = 0$ at outer core end (r_e)
$l = L$ at inner core end (r_i)

The different variables are also defined in Fig. 9.

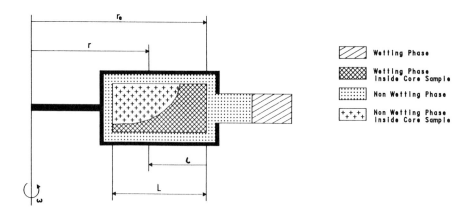

Fig. 9. A porous sample in a rotating system.

Reading pairs of (x, S_w) and using eq. 2, the capillary pressure can be obtained as a function of water saturation. The saturation profile, $S_w(x)$, is shown in Fig. 10 for the gas/water drainage of sample B1. Similar plots for samples B3 and B4 have been obtained. Fig. 11 illustrates the conversion to a capillary pressure curve for B1. The discrepancy between estimated curves and Hassler & Brunner curves for B3 and B4 are almost identical to this discrepancy for B1.

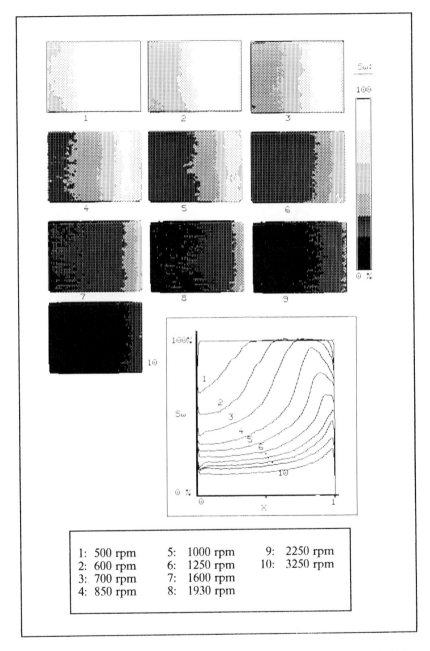

Fig. 10. The water saturation from CT-images for sample B1 plotted as a function of normalized core length; gas/water drainage. The CT-scans were obtained 1-2 minutes after stopping the centrifuge.

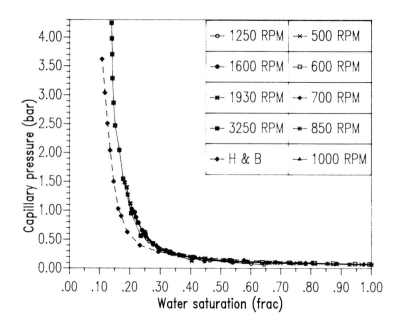

Fig. 11. Drainage capillary pressure curves for gas/water system
in sample B1. Data based on CT-images 1-2 minutes after stopping
the centrifuge are compared to a traditional Hassler & Brunner
solution (H & B).

The discrepancy between the capillary pressure curves calculated
from CT-images and from the traditional Hassler & Brunner method
is most pronounced for the transition zone. Uncertainties in the
H & B solution (about 2%) based on the material balance and in CT-
values (2 − 3%) are sources of error in saturation determination. The
possible redistribution of mobile fluids can also partially explain the
deviations.

A gas/wax drainage of sample B2 (Fig.12) indicates that the redistribution of fluids can be the explanation for the difference in the
transition zone. The capillary pressure curve for gas/wax system is
adjusted to the curve for gas/water system using measured interfacial
tensions (25.2 mN/m and 70.6 mN/m) and contact angles about zero
for both fluid systems.

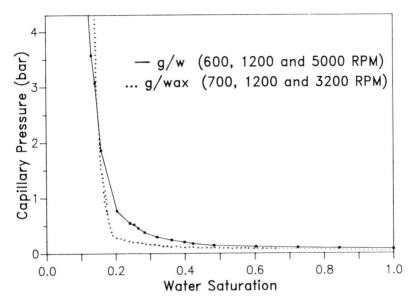

Fig. 12. Gas/wax and gas/water capillary pressure curves for sample B2 estimated from CT-images.

As illustrated, the capillary pressure curve can be estimated from CT-images. Great parts of the capillary pressure curve can be obtained from one centrifuge experiment (one ω) with a succeeding CT-visualization if a suitable choice of ω is made. The time before the CT-scan ought to be less than 1-2 minutes. To measure capillary pressures in a saturation interval corresponding to what is obtained in traditional flooding experiments, ω can be low. A capillary pressure level lower than 3 bar may be useful.

5. CONCLUDING REMARKS

In this study fluid distribution in "water-wet" core samples during and after centrifuging is visualized and quantified. Based on the results, we make the following remarks:

1. The Hassler & Brunner boundary condition at the outer core end seems to fail at relatively low capillary pressures.

The error in capillary pressure curves assuming that the boundary condition does not fail, ought to be estimated and equations modified if necessary.

A new outlet criterion may be used to control the boundary condition in simulations of relative permeabilities from centrifuge data.

2. In the gas/wax study, the CT-technique is an excellent method to visualize the saturation distribution in a core sample during centrifuging. The same sample can be used in all the measurements, which implies that the Hassler & Brunner boundary condition can be studied in detail.

3. The capillary pressure curve can be estimated from CT-images of water saturation distribution if CT-scans are made a short time after centrifuging.

4. For these samples the redistribution of water is a slow process at high mean saturations since the capillary pressure is low. The process is faster as the capillary pressure increases.

   Effects of a nonuniform saturation distribution in measurements of electrical resistivity, or as a base before a waterflood, ought to be investigated.

5. The redistribution process seems to be faster for the gas/water than for the oil/water system due to higher capillary forces and lower viscosity.

6. The water saturation gradient seems to exist for a longer period in the aluminum oxide samples than in the sandstones.

ACKNOWLEDGEMENTS

The authors acknowledge many helpful discussions with J. K. Ringen of Statoil. We would like to thank O. Vikane, M. Skarestad and L. M. Meling for comments to the manuscript, and Statoil for permission to publish this paper.

REFERENCES

1. Slobod, R.L., Chambers, A. and Prehn Jr., W.L. (1951). Trans., AIME, Vol. 192, pp. 127-134.
2. Hassler, G.L. and Brunner, E. (1945). Trans., AIME, Vol. 160, pp. 114-123.
3. Hagoort, J. (1978). SPE 7424.
4. O'Meara Jr., D.J. and Crump, J.G. (1985). SPE 14419.
5. Nordtvedt, J.E., Nilsen, V. and Kolltveit, K. (1988). Sci.Techn. report no.184, Univ. of Bergen.
6. Amott, E. (1959). Trans., AIME, Vol. 216, pp. 156-162.
7. Sharma, M.M. and Wunderlich, R.W. (1985). SPE 14302.
8. Szabo, M.T. (1974). SPE-journal, pp. 243-252.
9. Wunderlich, R.W. (1985). SPE 14422.
10. Melrose, J.C. (1988). The Log Analyst, pp. 40-47.
11. O'Meara Jr., D.J., Hirasaki, G.J. and Rohan, J.A. (1988). SPE 18296.
12. Firoozabadi, A., Soroosh, H. and Hasanpour, G. (1988). JPT, Vol. 40. no.7, pp. 913-919.
13. Lyle, W.D. and Mills, W.R. (1987). SPE 17143.
14. Hove, A., Ringen, J.K. and Read, P.A. (1987). SPERE, Vol. 2, no. 2, pp. 148-154.
15. Wellington, S.L. and Vinegar, H.J. (1987). JPT, Vol. 39. no.8, pp. 885-898.
16. Leknes, J., Hove, A. and Nilsen, V. (1988). Statoil report no. 88.49 (In Norwegian).

Session 2
Mesoscopic

The scale of bed boundaries, stratification types, or any other small-scale geological features, usually inferred from well logs

RESERVOIR CHARACTERIZATION AT THE MESOSCOPIC SCALE

Paul F. Worthington
BP Research, Sunbury-on-Thames TW16 7LN, England

ABSTRACT

The intermediate scales of measurement serve as a fulcrum for the extremes. Core-calibrated downhole measurements effectively fulfil this pivotal role in reservoir characterization. They provide a linkage, hitherto underutilized, whereby reservoir data from the micro- to the mega-scales can be merged into a unified, meaningful and internally self-consistent reservoir model in a way which overcomes problems due to anisotropy and heterogeneity.

Several key stages have been identified in the development of this model at the mesoscopic (bedding) scale. Firstly, the representativeness and compatibility of the measured parameters should be established, especially for core-log calibration. Secondly, reservoir zones must be identified within each of which there is no significant variation of the characteristic petrophysical algorithms and the characterizing parameters they contain: this procedure is termed *petrophysical characterization*. Thirdly, the downhole measurements must be integrated with data from pore studies, core analysis and geophysical surveys, through inter-scale reconciliation. Finally, the petrophysical interpretations should be related to and reconciled with geochemical, sedimentological, stratigraphic and structural information, through cross-discipline correlation.

Used in this way, downhole measurements have a substantial, quantitative and spatially continuous input to the static reservoir model upon which the dynamic reservoir model is founded. Their further input to the dynamic model is limited by the absence of an established log-derivable permeability indicator. This means that at the bedding scale, the direct data input to the dynamic model has to be either discontinuous or semi-quantitative. This shortcoming constitutes an important target area for improvement.

Today, reservoir evaluation is driving towards an improved understanding of basic reservoir science. This strategy seeks a better appreciation of the ordering effects of nature and their relationship to scale. With the data acquisition and processing technology currently at our disposal, there is every prospect that the objective will be achieved.

RESERVOIR CHARACTERIZATION II

1. INTRODUCTION

The brief of this paper is to focus on reservoir characterization at the mesoscopic (bedding) scale, broadly defined here as ranging from a few centimetres to a few tens of metres. As such, the subject matter is concentrated around core-calibrated downhole measurements, and is essentially concerned with vertical scaling, lateral (inter-well) scaling having been specified to be a complementary theme. Although mesoscale studies are generally perceived to constitute a mature scientific area, relative to the evaluation of larger heterogeneities at the inter-well and reservoir scales, the high reliance on empiricism has actually impeded scientific advancement. Further advances in understanding are needed if mesoscale data are to be used most effectively in controlling reservoir characterization at larger scales.

Reservoir characterization as a whole is progressively gaining a sounder scientific pedigree under the influence of three coexisting forces, all of which have been fuelled by substantial increases in computing power. The first of these is directed at making geology in general, and geological description in particular, much more quantitative and thereby much less subjective. The second represents the drive towards three-dimensional, high-resolution seismic data with a potential for lithology and porosity mapping at depth on a gross scale. The third reflects the increasing application of core-calibrated wireline logging data in geochemistry, sedimentology, stratigraphy and structural geology. These three forces are converging through a contemporary reservoir-description scenario that is very different from the classical approach of the past. Within this new scenario, reservoir boundaries are delineated by continuous geophysical measurement rather than inferred through interpolation between wells. Well evaluation of reservoir properties, taken to be representative of a drainage area, gives way to field-wide evaluation based on definable reservoir zones of petrophysical consistency. These zones are compatible with seismic information and with sedimentological and geochemical characteristics of facies and reservoir quality. Thus, the contemporary reservoir-description scenario encompasses a reservoir model which is internally self-consistent in that geological, geophysical and petrophysical data are reconciled.

The integration of these data into a physically equivalent, unified reservoir model is the process of reservoir characterization. A primary purpose is to take account of reservoir heterogeneity through synthesis of reservoir description as a basis for reservoir simulation. Thus, reservoir characterization provides a link between the static and the dynamic components of the reservoir model. A prerequisite for an effective link is a proper understanding of the controls on fluid flow in the porous media which comprise the reservoir unit, an objective which is not achievable without meaningful reservoir description. Thus, reservoir characterization cannot be considered in isolation from the input reservoir description. Reservoir description and characterization, taken together, comprise much of the advancing science of reservoir evaluation (Figure 1).

Not all aspects of this scenario can be enacted at the present time. Further advances in technology and scientific understanding are needed before a fully integrated reservoir model can be developed as a matter of routine. The process of

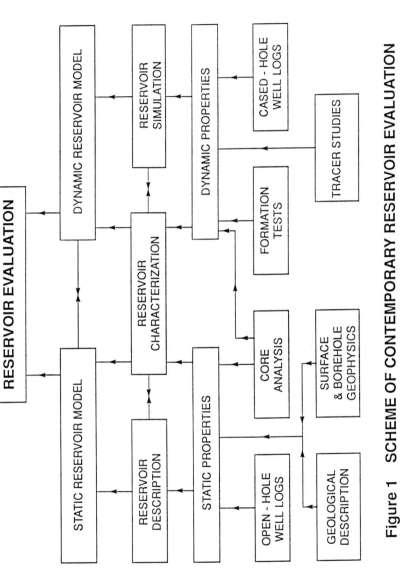

Figure 1   SCHEME OF CONTEMPORARY RESERVOIR EVALUATION

integration itself, with the associated problems of scale reconciliation, is by no means clear-cut. Issues such as these are strongly influenced by technical advances at the mesoscopic (bedding) scale, which is intermediate relative to the micro- and core scales, on the one hand, and the geophysical scales, on the other. An appraisal of the current status of reservoir characterization at the mesoscopic scale, and of those developments that are needed before self-consistent integrated reservoir models can be established routinely, forms the substance of this paper.

This appraisal is based upon a threefold approach. Firstly, the parameters needed for effective reservoir description are detailed. Secondly, the petrophysical algorithms that interrelate these parameters are discussed. Thirdly, the grossing up of these parameters to larger scales is considered against the robustness of the algorithms themselves. Throughout, the underlying contention is that there is little prospect of a fully integrated, internally consistent reservoir model without properly calibrated downhole measurements. These are fundamental to the reservoir characterization process. There is no other route to continuous characterization at the bedding scale, unless there is full core recovery throughout the reservoir, a situation which is rarely, if ever, realized.

2.   CENTRAL ROLE OF DOWNHOLE MEASUREMENTS

Downhole measurements have a pivotal role in reservoir characterization in three important respects: scale of measurement, evaluation strategy and reservoir technology.

The scales of the various types of investigation in exploration and production activity have been compiled in Figure 2. Downhole measurements, encompassing wireline well logs and formation tests, measurement-while-drilling and the newly emerging probe tools, are distinguished from interwell measurements and vertical seismic profiles for present purposes. In particular, the collective resolutions of most wireline well logs are at a scale (0.01 - 2 m) which encompasses the geometric mean of other scales represented in contemporary reservoir evaluation. This medial scale lies within the mesoscopic or bedding scale and it overlaps the core scale (Figure 2), an important observation since downhole measurements should be calibrated through core analysis if they are to be used quantitatively with confidence.

The reservoir-evaluation strategy of contemporary engineering geoscience is shown in Figure 3. Downhole measurements at the mesoscopic scale, with appropriate core calibration, have a direct input into most other areas of exploration and production activity. Figure 3 also conforms to the progressive calibration of data at larger scales by (higher-resolution) measurements at smaller scales. This is the key to cross-scale data reconciliation.

The technology of downhole measurements encompasses several branches of physics (Table 1), unquestionably the fundamental science of reservoir evaluation. The range of physical measurements that can be achieved downhole is more extensive than those of surface or subsurface geophysics and approximates that at the core scale. There is some correspondence between these four measurement domains. It is through this correspondence that downhole

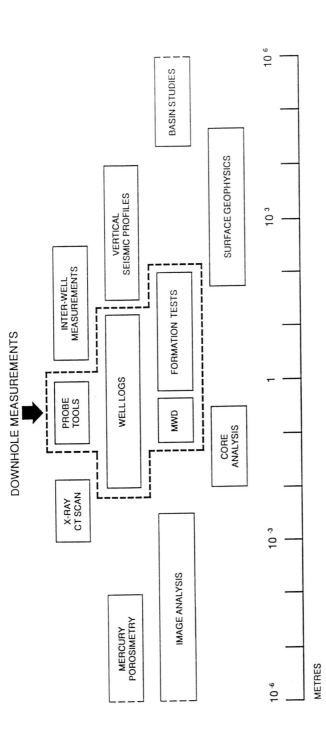

DOWNHOLE MEASUREMENTS

MERCURY POROSIMETRY

IMAGE ANALYSIS

X-RAY CT SCAN

PROBE TOOLS

INTER-WELL MEASUREMENTS

WELL LOGS

VERTICAL SEISMIC PROFILES

BASIN STUDIES

MWD

FORMATION TESTS

CORE ANALYSIS

SURFACE GEOPHYSICS

METRES

10^{-6} 10^{-3} 1 10^{3} 10^{6}

Figure 2   SCALES OF MEASUREMENT EXPRESSED IN TERMS OF RESOLUTION, EMPHASIZING DOWNHOLE MEASUREMENTS AS DEFINED IN THIS PAPER

(MWD - measurement-while-drilling)

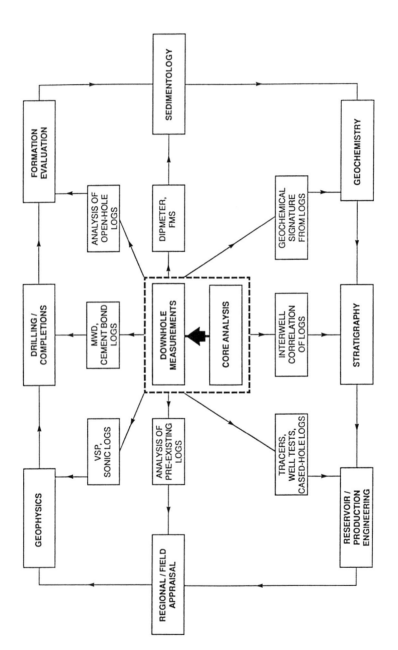

Figure 3   SCHEME OF ENGINEERING GEOSCIENCE IN RESERVOIR EVALUATION, EMPHASIZING THE ROLE OF DOWNHOLE MEASUREMENTS

VSP - Vertical seismic profiles;   MWD - measurement while drilling;   FMS - formation microscanner

TABLE 1

PHYSICAL MEASUREMENTS IN RESERVOIR EVALUATION

BRANCH OF PHYSICS	MEASUREMENT	CORE ANALYSIS	DOWNHOLE MEASUREMENTS	SUBSURFACE GEOPHYSICS	SURFACE GEOPHYSICS
Acoustics	Acoustic Velocity	•	•	•	•
	Ultrasonic Imaging		•		
Electricity	Dielectric	•	•		
	Resistivity	•	•	•	•
	Electrochemical Potential	•	•		
	Electrical Imaging	•	•		
Magnetism	Remanent Magnetism	•			
	Field Strength				•
	Magnetic Resonance	•	•		
Nuclear/Atomic	Natural Gamma Radiation	•	•		
	Neutron Moderation		•		
	Gamma Ray Attenuation	•	•		
	Induced Gamma Radiation	•	•		
	Neutron Activation	•	•		
Mechanics & General Properties	Fluid Pressures	•	•	•	
	Fluid Sampling	•	•	•	
	Gravity/Density	•	•		•
Thermodynamics	Temperature		•		
Optics	Imaging	•			

measurements are able to provide the linkage between high-resolution measurements at the micro- and core scales and the broader measurements at the geophysical scales.

The key to successful data integration is the effective reconciliation of those physical data that are relevant to reservoir evaluation. This reconciliation has two components, accommodating the change in each physical quantity with the scale of measurement and recognizing the degree of association of different physical properties, even from within traditionally separate disciplines. The former component depends strongly on the nature of the measured parameters, the latter on the relationships between them.

3. PARAMETRIC DESCRIPTION

Effective reservoir characterization at the bedding scale requires the meaningful integration of core and log data. This exercise has several important prerequisites. In the following, and subsequent, discussions the components of a reservoir unit will be described, in increasing order of vertical scale, as: laminae (millimetres), strata (several centimetres), beds (centimetres up to a few tens of metres), intervals (tens of metres up to several hundred metres and beyond). A reservoir zone is the largest region within which there is consistency of those physical parameters needed to effect an evaluation of downhole (and surface geophysical) measurements in terms of reservoir properties.

Depth Matching

Well log data are identified by wireline depths, core data by drillers' depths. The two are usually different. They must be integrated. The process of depth matching is more complicated where there is incomplete core recovery.

Separate logging runs are reconciled through the incorporation of a natural gamma log within all tool combinations. The merged wireline depths can be related to core depths through comparison with laboratory natural gamma data measured on whole core and/or with geological descriptions of shale boundaries in core. Where the total natural gamma count is not sufficiently diagnostic, recourse might be made to the measured spectral components associated with the uranium, potassium and thorium series. Natural gamma spectral data are recorded routinely downhole and in the laboratory.

Natural gamma data may be featureless over clean (i.e. clay-mineral-free) reservoir intervals and where the predominant clay mineral is not significantly radioactive, e.g. kaolinite. In these cases, precise core-log depth matching within the reservoir interval(s) will have to rely on the relationship between core description and some other petrophysical parameter, e.g. sonic and density. Since no other parameters are monitored routinely at the whole core stage, the depth-matching process might be tuned beneficially using laboratory data measured on core plugs.

Representative Sampling

There are two considerations. The first requires that the data to be integrated are actually representative of the reservoir rock at the scale in question. This means that logging tools must have sufficiently sharp vertical resolutions to record meaningful parametric values in target beds; data which contain a substantial representation from adjacent beds are not likely to be useful. Log-derived parameters should also be corrected for perturbations caused by the borehole environment. Thus, log data should faithfully represent the true parametric values within a target bed over a distance which corresponds to tool vertical resolution, in most cases around 0.6 m. On the other hand, core data should, where possible, be measured on samples which are known to be homogeneous, e.g. through core description supported by X-ray CT scanning, and which are sufficiently large to provide a meaningful average of the pore structure. Whole core is more likely to fail on the former count, core plugs on the latter. Preferred plug sizes are 1.5 inches diameter by 3 inches length (3.8 x 7.6 cm). Cores do, of course, afford the opportunity to observe heterogeneities that cannot be resolved by logs. These heterogeneities can take the form of sequences of laminae of contrasting character. Petrophysical tests on core plugs orientated parallel to the laminations represent average physical responses of these small-scale elements. The statistical significance of these averages will depend upon the thickness(es) of the laminae relative to plug diameter.

The second sampling consideration concerns the representativeness of the sample populations themselves. For a given scale of measurement, this can only be assessed by reference to the immediately larger scale. Thus, core plugs should be representative of the target beds within the reservoir interval for which core-log data integration is to be effected. The logs should, in turn, be representative of a reservoir block on the inter-well scale; satisfying this requirement depends partly on the criteria for siting the well. We would expect to see correlation between logs run in different wells, provided that the wells are sufficiently closely spaced, the logs are of properties relevant to interwell correlation, and there are no effects due to faulting, etc. If no correlation is evident, the wells might be spaced too far apart to be effective control points for reservoir characterization in that particular geological situation, without recourse to high-resolution 3D seismic surveys. If closely spaced wells show little correlation between logging suites, reservoir characterization at the bedding scale has no obvious relationship to the inter-well scale. In cases such as these, where each well samples a seemingly random distribution of reservoir rock, the question of representativeness of the logging data cannot be answered without establishing an association with larger-scale data. Fortunately, the ordering effects of gravity and sedimentary deposition often allow petrophysical correlation between closely spaced wells.

Equivalent Rock Fabrics

There is little point in integrating data that purport to describe the same porous medium at a given depth location but do, in fact, describe lithologies that are very different. This situation can arise if the core samples have been handled in such a way that their rock fabrics cease to represent the *in-situ* reservoir rock which is measured by the logging tools. Inappropriate core recovery, storage, preservation, cleaning and drying practices can all result in fabric alteration. Material that is designated for special core analysis must be quickly and properly preserved, non-destructively cleaned, and should never be fully dried until all fabric-sensitive measurements, such as permeability and resistivity, have been completed. There are special recovery problems in the case of unconsolidated reservoirs; these are the subject of ongoing research (Worthington *et al.*, 1987).

It is equally important that the well logs providing parametric input to the reservoir model do not sample predominantly those zones which have been damaged or altered by the drilling process. Logging tools that sample only the near-well-bore region are most at risk. Thus, for example, microresistivity devices do not have a direct application to reservoir characterization in terms of the parameters they measure. However, they have an indirect application through the correction for invasion effects on deeper sensing resistivity logs. Shallow sensing devices do have an important application in bed resolution.

Comparable Measurements

There are three aspects relevant to data integration: measuring the same parameter in the laboratory as is measured downhole, enforcing the same measurement conditions, and relating the measurements to a common parametric system. Examples of the first case are resistivity measurements in the laboratory, which should be made at frequencies similar to those used downhole with laterologs. If the measurement frequencies are different, it should be established that there is no frequency-dependence of resistivity. If there should be such a frequency-dependence, the measurements will be incompatible and core-log integration of resistivity data will be less meaningful than it might otherwise have been.

Enforcing the same measurement conditions for laboratory and log data requires core measurements at simulated reservoir conditions. Hitherto, the industry has frequently used core data measured at ambient conditions to calibrate log data measured *in situ*. This practice, which is sometimes necessary for financial reasons or because of technical shortcomings, is scientifically lacking, unless it can be shown that the core parameter(s) in question are unchanged when reservoir conditions are imposed. For many years the industry has taken the laboratory simulation of reservoir conditions to mean the application of overburden pressures. The question of temperature has usually been ignored. The reason is that core measurements at truly simulated reservoir conditions of pressure and temperature are prohibitively expensive and technically very difficult. Nevertheless, it is an avenue the industry may have to take.

A common parametric system requires that the same parameters describe the same physical characteristic. For example, porosities can be presented in the total or the effective porosity systems, as required for the evaluation of shaly (i.e. clay-mineral-bearing) sands by the cation-exchange-capacity and the shale-volume-fraction methods, respectively. The total porosity system relates to the entire interconnected pore space, the effective porosity system to that portion of the interconnected pore space which does not comprise chemically bound water. Since each system produces different numerical values for a shaly sand, there would be little point in integrating, for example, core porosities measured by helium expansion and conventionally reported in the total porosity system with porosities from neutron-density log combinations reported in the other. The same system should be used throughout a reservoir database.

Discussion

If the above prerequisites are fully satisfied, an effective integration of core and log data is more likely, because we will be relating compatible parameters. The integration is two-fold. In addition to providing calibration of log measurements, core data also allow the definition of interpretative algorithms used in log analysis. These algorithms constitute the foundation of reservoir characterization at the bedding scale.

4.   CHARACTERISTIC ALGORITHMS

All algorithms used to determine reservoir parameters from measured physical properties should either be applicable everywhere or be characteristic of the reservoir or a given reservoir zone. This means that algorithms of the latter category must be verified for the reservoir (zone) before application. These are the characteristic algorithms. There are two types of characteristic algorithm, those with a standard algebraic form for which certain coefficients and exponents are reservoir-specific (Type 1) and those which vary in algebraic form from reservoir to reservoir (Type 2). Algorithms of Type 1 can be conceptual, theoretical or empirical. Algorithms of Type 2 are mostly empirical. The algorithms are usually established and verified using core data as a basis for the subsequent application to well log data. The comments of the previous section relating to compatibility of measurements, parametric systems, etc., are therefore especially pertinent.

Algorithms in common use in reservoir evaluation are shown in Figure 4, which indicates their type and how they fit into the overall evaluation strategy. The establishment of reservoir-descriptive algorithms introduces a new type of parameter, the derived or secondary parameter. These parameters are established through the correlation of measured (primary) parameters. Both primary and secondary parameters can characterize a reservoir unit or a reservoir zone, on the one hand, or be spatially variable, on the other. In particular, secondary parameters are usually intended to be characterizing parameters at the mesoscopic scale, but they sometimes have to be admitted as spatially variable in the presence of variable (complex) lithologies or changing electrolyte characteristics. Certain parameters (e.g. ρ_{ma}, ρ_{sh}) can be either primary or secondary, according to how they were determined. Primary and secondary characterizing parameters that have an application at the bedding scale are identified in Table 2. Comments on specific algorithms now follow.

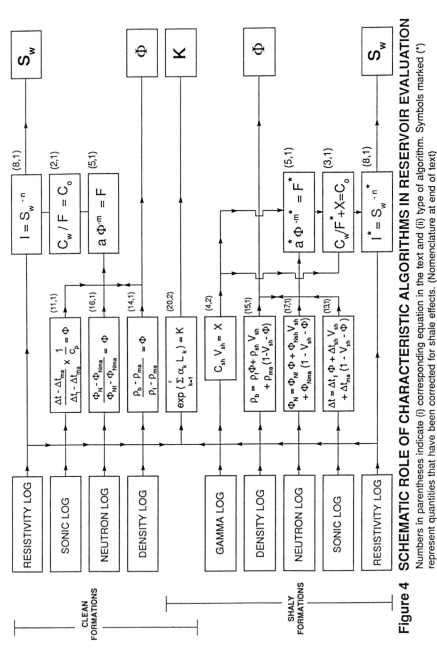

Figure 4 SCHEMATIC ROLE OF CHARACTERISTIC ALGORITHMS IN RESERVOIR EVALUATION

Numbers in parentheses indicate (i) corresponding equation in the text and (ii) type of algorithm. Symbols marked (*) represent quantities that have been corrected for shale effects. (Nomenclature at end of text)

134

TABLE 2

CHARACTERIZING PARAMETERS *

PHYSICAL NATURE OF MEASUREMENT	PRIMARY PARAMETERS	SECONDARY PARAMETERS	COMMENTS
Acoustic	Δt_f Δt_{ma} Δt_{sh}	c_p	provided c_p is not a function of depth
Electrical	C_w C_{sh}	a m n	provided C_w is constant within zone
Nuclear/Atomic	ρ_f ρ_{ma} ρ_{sh} ϕ_{Nf} ϕ_{Nma} ϕ_{Nsh} GR_{sand} GR_{shale}		corrected for mud radioactivity
Various		α_k	coefficients in permeability regression equation(s)
Various		c d e f	regression terms in Q_v vs ϕ relationships

* Nomenclature at end of text

Note: Certain primary parameters can be classified as secondary if derived from crossplots.

135

Definition of Formation Resistivity Factor

In terms of conductivity, formation resistivity factor, or formation factor, F is defined for clean sands as:

$$F = \frac{C_w}{C_o}$$

(1)

or

$$C_o = \frac{C_w}{F}$$

(2)

where C_o is the conductivity of a rock that is fully saturated with an aqueous (NaCl) electrolyte of conductivity C_w. F is a derived parameter that is spatially variable: it can be determined from single measurements of C_o and C_w.

For shaly sands equation (2) becomes

$$C_o = \frac{C_w}{F} + X$$

(3)

where X is the excess conductivity associated with the presence of dispersed clay minerals, herein loosely described as shaliness. F and X cannot be determined from single measurements: a crossplot of C_o vs C_w is required for several different saturating electrolyte salinities. X is a derived parameter that is spatially variable.

Several interpretations of X have been made in the literature in an attempt to facilitate its downhole measurement. For example, Simandoux (1963) adopted the specification:

$$X = V_{sh}\, C_{sh}$$

(4)

where V_{sh} is the shale volume fraction and C_{sh} is the shale conductivity. V_{sh} is a spatially variable parameter that can be determined from log shale-indicators downhole (Poupon & Gaymard, 1970). The V_{sh} approach to shaly-sand evaluation is used by an estimated 85 per cent of oil companies. V_{sh} is used to report shale content without discriminating between dispersed, laminated and structural shales, in cases where the shale elements cannot be separately resolved by the relevant logging tool. C_{sh} is often taken to be constant within a reservoir (zone). It can be determined using electric log response(s) opposite shale beds that are sufficiently thick to be fully resolved by the logging tool. In so doing, it is presumed that clay minerals dispersed within reservoir sands have a physical response which is identical to that of the intraformational shales, an assumption which is seen as tenuous. C_{sh} characterizes a reservoir (zone) and is not otherwise admitted to be spatially variable (Table 2). Therefore, equation (4) is a characteristic algorithm of Type 2 whereas equation (3) is not. This statement is not appropriate if a constant value of C_w is specified throughout a reservoir (zone), instead of C_w being determined on a depth-by-depth basis; in the former case equations (2) and (3) become characteristic algorithms, too.

First Archie Equation

The empirical relationship between F and fractional porosity ϕ, noted by Archie (1942), takes the form

$$F = \frac{a}{\phi^m} \qquad (5)$$

where the parameters a and m are usually derived from a (well-defined) bilogarithmic correlation of the spatially variable parameters F and ϕ for several samples. The exponent m, originally termed the *cementation exponent* and subsequently seen as a grain shape exponent, is now regarded as an indicator of pore shape. The coefficient a is regarded by many as having to conform to unity, in order to satisfy the boundary condition that F = 1 when ϕ = 1. The opposing view is that this boundary condition at 100% porosity is not relevant to an empirical relationship established over the range of porosities in a reservoir (zone). When established by correlation, the parameters a and m characterize a reservoir (zone) and are usually not allowed to be spatially variable (Table 2).

The assumption of constancy of a and m may not be justified; in such cases a "variable-m" approach is sometimes pursued by assuming that a = 1 and calculating m from the *definition*

$$m = - \frac{\log F}{\log \phi} \qquad (6)$$

where F and ϕ are obtained from single measurements and the derived parameter m is now spatially variable. This approach is not usual. Equations (5) and (6) can only be applied in shaly sands if F has been corrected for shale effects or if the latter can be ignored. Equation (5) is a characteristic algorithm of Type 1 : Equation 6 is not.

Definition of Resistivity Index

The resistivity index I is defined for clean sands (Archie, 1942) as:

$$I = \frac{C_o}{C_t} \qquad (7)$$

where C_o is the conductivity of a fully electrolyte-saturated reservoir rock and C_t is the rock conductivity when partially saturated with the same electrolyte. The term I represents a derived parameter which is spatially variable : it can be determined from single measurements of C_o and C_t.

The quantity I is affected by shaliness, and complex correction procedures have been developed. These use the same physical interpretations of X as in the fully saturated case. Thus, the correction procedures use characteristic algorithms of Type 1.

Second Archie Equation

The empirical relationship between I and fractional water saturation, S_w, proposed by Archie (1942), takes the form

$$I = S_w^{-n} \tag{8}$$

where the exponent n is usually derived from a bilogarithmic correlation of the spatially variable parameters I and S_w. The sum of S_w and the fractional hydrocarbon saturation, S_h, is unity. The quantity n is termed the *saturation exponent*. A value of n can be established for several different desaturation levels of a single sample in which case the derived value relates only to that sample. Where a composite plot of data from several samples shows a consistent data trend, the value of n, established by correlation, characterizes the reservoir zone and is not allowed to be spatially variable (Table 2). Equation (8) then becomes a characteristic algorithm of Type 1. Where such a composite plot reveals a data cloud, a variable-n approach may be necessary either through single-sample correlations or through the *definition*

$$n = - \frac{\log I}{\log S_w} \tag{9}$$

where I and S_w are obtained from single-desaturation measurements and n is spatially variable. Equations (8) and (9) can only be applied in shaly sands with appropriate corrections for shale effects. Equation (9) is not a characteristic algorithm.

Transit Time - Porosity Relationship

The time-average equation for clean sands (Wyllie *et al.*, 1956, 1958), relates reciprocal acoustic velocity or sonic transit time in the bulk rock, Δt, to the porosity:

$$\Delta t = \Delta t_f \, \phi + \Delta t_{ma} \, (1 - \phi) \tag{10}$$

where Δt_f and Δt_{ma} are the transit times in the pore fluid and rock matrix, respectively. (The word *matrix* is used in the petroleum engineering sense to mean those solid constituents that do not comprise clay minerals.) Equation (10) is based on the concept of acoustic mixing and requires perfect acoustic coupling between grains, i.e. the formation must be compacted or, perhaps, silica cemented. If it is not, a lack-of-compaction correction is applied to the computed porosity as indicated in the following rearrangement of equation (10):

$$\phi = \frac{\Delta t - \Delta t_{ma}}{\Delta t_f - \Delta t_{ma}} \times \frac{1}{c_p} \tag{11}$$

where c_p is a dimensionless correction factor for lack of compaction. Equation (10) is a characteristic algorithm of Type 2 for which Δt_f and Δt_{ma} have to be specified for a reservoir (zone) and are not allowed to be spatially variable (Table 2). The quantity c_p can be evaluated by comparing porosities calculated from equation (10) with those measured on core plugs or derived from other well logs. The parameter c_p can either characterize a reservoir (zone) and show no intrazone variations (Table 2) or show significant changes with depth within a reservoir (zone). In both cases, equation (11) is a characteristic algorithm of Type 1.

Other equations have been proposed for the relationship of acoustic velocity or transit time to porosity at the mesoscopic scale. Some of these contain depth as an input parameter (e.g. Faust, 1951), in which cases the equations cannot be regarded as characteristic algorithms. Of those alternative equations that might be regarded as characteristic algorithms, the equation of Raymer *et al.* (1980) is particularly interesting. This equation, written

$$\frac{1}{\Delta t} = \frac{(1 - \phi)^2}{\Delta t_{ma}} + \frac{\phi}{\Delta t_f} \quad , \tag{12}$$

was claimed to describe the transit time - porosity relationship for sedimentary rocks without the need for a compaction correction. Yet, Hartley (1981) provided examples for which equation (12) did not fit the data distributions. At present, the time-average equation remains the primary transit time - porosity relationship and equation (12) is not sufficiently used to be regarded as a characteristic algorithm of Type 2.

In the race towards porosity prediction away from the well site, through the integration of petrophysics and geophysics, equations such as (10) - (12) are being supplanted by more complex expressions which involve elastic moduli (e.g. Gassmann, 1951). This process will not be complete until there is a better understanding of those additional characterizing parameters that are needed to apply meaningfully the more complicated equations.

For shaly formations the time-average equation can be written:

$$\Delta t = \Delta t_f \; \phi + \Delta t_{ma} \; (1 - \phi - V_{sh}) + \Delta t_{sh} \; V_{sh} \qquad (13)$$

where Δt_{sh} is the interval transit time in shale. The parameter Δt_{sh} is characteristic of a reservoir (zone) and is not allowed to be spatially variable (Table 2). It can be determined from the response of a sonic log opposite intraformational shale beds of sufficient thickness to be fully resolved, subject to the earlier assumption of equivalent physical responses of the shale beds and those shales contained within the reservoir sands. Equation (13) is a characteristic algorithm of Type 1.

Density - Porosity Relationship

This conceptual relationship, which is analogous to Equation (10), can be written for clean sands as:

$$\rho_b = \rho_f \; \phi + \rho_{ma} \; (1 - \phi) \qquad (14)$$

where ρ_b is the bulk density of the rock, and ρ_f, ρ_{ma} are the densities of the pore fluid and rock matrix, respectively. For shaly sands

$$\rho_b = \rho_f \; \phi + \rho_{ma} \; (1 - \phi - V_{sh}) + \rho_{sh} \; V_{sh} \qquad (15)$$

where ρ_{sh} is the shale density, which can be determined from the response of a density log opposite thick intraformational shales, subject to the earlier assumption of physical equivalence. The parameters ρ_f, ρ_{ma} and ρ_{sh} are characteristic of a reservoir (zone) and are not allowed to be spatially variable (Table 2). A possible exception would be if ρ_f was allowed to vary with temperature.

Neutron Porosity - Porosity Relationship

The response of the neutron porosity log is calibrated empirically in limestone porosity units because the primary calibration standard comprises test pits in limestone formations. Neutron log response ϕ_N (in limestone porosity units) is related to porosity through the conceptual expression

$$\phi_N = \phi \; \phi_{Nf} + (1 - \phi) \phi_{Nma} \qquad (16)$$

where ϕ_{Nf}, ϕ_{Nma} are neutron log responses (in limestone porosity units) for the fluid and matrix, respectively. ϕ_{Nf} is unity for water and moderately heavy oils. ϕ_N is equal to ϕ only for pure limestone (for which $\phi_{Nma} = 0$ in limestone porosity units). The corresponding equation for shaly formations is:

$$\phi_N = \phi \ \phi_{Nf} + (1 - \phi - V_{sh}) \phi_{Nma} + V_{sh} \ \phi_{Nsh} \qquad (17)$$

where the ϕ_{Nsh} is the neutron log response (in limestone units) for shale. ϕ_{Nsh} can be determined from the response of the neutron porosity log in thick intraformational shales, with the usual assumption of equivalent physical responses of the shale beds and those shales contained within reservoir sands. The parameters ϕ_{Nf}, ϕ_{Nma} and ϕ_{Nsh} are characteristic of a reservoir (zone) and are not allowed to be spatially variable (Table 2). Equations (16) and (17) are characteristic algorithms of Type 1. True porosity ϕ can be calculated from these equations by inserting appropriate values of the characterizing parameters.

Composite Relationships

Equations (1) - (17) can be combined in various ways for log analysis purposes. Two examples for clean sands follow. The combination of Equations (1), (5), (8) and (10), for application of the Hingle crossplot method, furnishes

$$\Delta t = \Delta t_{ma} + (\Delta t_f - \Delta t_{ma}) \left[\frac{a \ C_t}{C_w \ S_w^n} \right]^{1/m} \qquad (18)$$

The combination of equations (10) and (14), for the generation of sonic - density crossplots, gives

$$\rho_b = \rho_{ma} \frac{(\rho_f - \rho_{ma}) \ \Delta t_{ma}}{(\Delta t_f - \Delta t_{ma})} + \frac{(\rho_f - \rho_{ma}) \ \Delta t}{(\Delta t_f - \Delta t_{ma})} \qquad (19)$$

Equations (18) and (19) are characteristic algorithms as are all conventional combinations of equations (1) - (17). It is worth noting that although C_w can be either a characterizing or a spatially variable parameter, Δt_f, ρ_f and ϕ_{Nf} are usually regarded as characterizing parameters because the effects of moderate changes in fluid composition are less pronounced.

Permeability Relationships

There is no established procedure for the evaluation of permeability from wireline logs (Worthington, 1988). Most investigators have relied upon an empirical correlation of core permeability with a variety of log responses such as GR, ϕ_N, ρ_b, Δt, and self-potential E_{sp} (e.g. Wendt et $al.$, 1986). One example, due to Allen (1979) and based on single-well considerations, takes the form

$$\log_e K = \alpha_1 + \alpha_2 \, \phi + \alpha_3 \, \phi_N + \alpha_4 \, GR \qquad (20)$$

where GR is the response of the natural gamma log.

In equations such as (20) the coefficients α_k, k=1,...,4, might characterize the reservoir (zone) over which the correlation was established if the equation could be extended to multi-well applications. In such a case equation (20) would be a characteristic algorithm of Type 2. However, such a multi-response application is highly dependent on a knowledge of lithofacies, in terms of the lateral uniformity of mineralogy, fabric, pore structure, etc. Since permeability is variably sensitive to all these characteristics, a zone indicated by one approach might not be the same as that defined by another.

Shale Indicator Relationships

The natural gamma log is often used as a shale indicator. This procedure assumes a linear increase in V_{sh} with gamma response from a clean sand reference level, GR_{sand}, up to a shale reference level, GR_{sh}. The conceptual equation is

$$V_{sh} = \frac{GR - GR_{sand}}{GR_{sh} - GR_{sand}} \qquad (21)$$

The parameters GR_{sand} and GR_{sh} are characteristic of a reservoir (zone) provided that they can be measured without degradation due to radioactive mud systems. Under these conditions equation (21) is a characteristic algorithm of Type 1. Difficulties with this approach are the lack of integrity of the end points as truly clay-mineral-free, at one extreme of tool response, and 100 per cent clay mineral composition, at the other, variations in clay mineralogy, and the complicating natural-radioactivity characteristics of kaolinite, micas, feldspars, heavy minerals, and organics.

The downhole prediction of another shale parameter, cation exchange capacity per unit pore volume Q_v, has come to rely on a somewhat unsatisfactory relationship with porosity. This relationship has taken several forms. Two empirical versions are

$$Q_v = c\phi^d \qquad (22)$$

and

$$Q_v = \frac{e}{\phi} + f \qquad (23)$$

where the parameters c, d, e, f, determined by regression, are characteristic of a reservoir (zone). Equations (22) and (23) are characteristic algorithms of Type 2.

Reservations about equations (22) and (23) are that they assume all changes in porosity to be caused by changes in clay mineral content and that the data distributions are highly scattered.

Discussion

Evaluation of the parameters in Table 2 leads to the establishment of the characteristic algorithms described above. There are other algorithms which have a role in environmental corrections but these are not discussed here. The process of establishing the characteristic algorithms is termed *petrophysical characterization*; it is not complete until sufficient algorithms have been characterized to allow a complete level-by-level well log analysis in terms of the reservoir parameters that are determinable. These reservoir parameters are ϕ, S_h, and K. Reservoir evaluation is much less meaningful without proper petrophysical characterization. Extrapolation of relationships from other reservoirs is no substitute.

The classification of algorithms into Types 1 and 2 is time-variant. As our knowledge advances, a Type 2 relationship might emerge as dominant and justify reclassification as Type 1. On the other hand, a Type 1 relationship might cease to predominate as alternative forms emerge and might therefore be reclassified Type 2. The classification is merely a guide to the status of contemporary thinking.

It is clear that characteristic algorithms contain at least two types of parameters, measured parameters and characterizing parameters. The former are input to algorithms characterized by the latter. The aim is usually to calculate a third type of parameter, the reservoir parameter. An exception is Equation (19), developed for crossplot generation.

Reservoir characterization requires that parametric distributions be grossed up to describe larger blocks of reservoir in forms that are physically equivalent to the basic data. From a mesoscopic standpoint the grossing up procedure is especially important from core to log scales and from conventional logs to long-spacing logs and VSPs. It should be considered in the light of the characteristic algorithms themselves.

5.   INTEGRATION OF SCALES

Scaling methodology in petrophysics is a function of the nature of the reservoir and the orientation of the borehole axis relative to bedding planes or to other specified directions of spatial correlation. In this paper, it will be assumed that boreholes are vertical and bedding is horizontal. In terms of the reservoir itself, important considerations are the degrees of anisotropy and heterogeneity at the relevant scales. It is not sufficient to describe a reservoir as anisotropic or heterogeneous without reference to the scales of measurement.

Anisotropy

The relationship between anisotropy and scale is illustrated by Figure 5, which shows a stratified sandstone. The (smaller scale) strata form part of larger beds that appear petrophysically uniform at the logging scale where the strata cannot be resolved. The beds form part of a reservoir zone that appears physically consistent at the geophysical scale where the beds cannot be resolved. The strata might, for all practical purposes, appear isotropic within themselves at the core scale and yet create through their relative differences significant anisotropy at the bedding scale. In reality, reservoir rocks of the type illustrated are likely to be anisotropic at all three scales. It is frequently assumed that reservoir rocks are isotropic in the horizontal (bedding) plane but anisotropic in the vertical plane, at all scales of measurement. Media which satisfy these conditions are known as *transversely isotropic*. Experience has shown that this assumption can be fairly well approximated at the core scale from a petrophysical standpoint (Sawyer *et al.*, 1971), although departures must be expected.

Considerations of anisotropy are important because petrophysical measurements vary in their directional dependence. For example, sonic logs measure a vertical transit time, resistivity logs provide a horizontal electrical resistivity (conductivity), and nuclear logs respond to various parameters, such as electron density and hydrogen concentration, which show no directional dependence. At the core scale, petrophysical measurements are usually made on horizontal plugs (parallel to the bedding). With this orientation, acoustic and contact resistivity (conductivity) measurements will be horizontally directed, with measured porosities and water saturations showing no directional dependence. Therefore, at both core and log scales the characterizing parameters within, for example, equation (5) must be directionally dependent for they (a and m) are established through a correlation of a quantity that is directionally dependent (F) with one that is not (ϕ).

Heterogeneity

Reservoir characterization increases in complexity as the reservoir becomes increasingly heterogeneous. At one extreme, if a reservoir unit is homogeneous, a supposition that is at variance with most studies of rock systems, reservoir characterization is trivial because data measured at the core scale in one locality are representative of the entire field. At the other extreme, if a reservoir unit is randomly heterogeneous with no spatial correlation, a supposition that defies natural ordering effects, it cannot be characterized in terms of definable zones. Further, the scaling procedures will be different from those appropriate to a layered distribution of reservoir properties, requiring a statistically significant data sample that is sufficiently large to represent the entire reservoir unit. Mixed conditions can occur in which reservoir zones might be randomly heterogeneous but, after being characterized individually, they comprise a layered distribution of physically equivalent pseudo-homogeneous zones that can be effectively integrated into a reservoir unit.

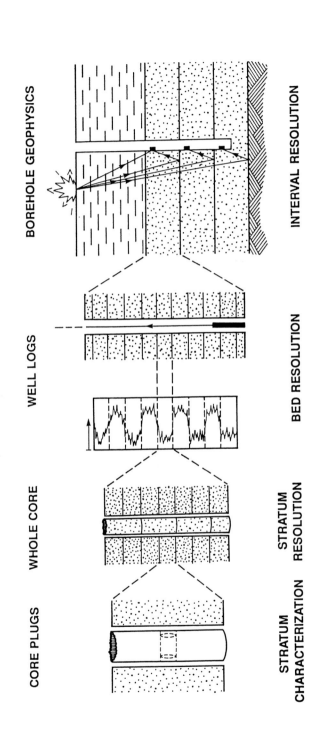

CORE PLUGS   WHOLE CORE   WELL LOGS   BOREHOLE GEOPHYSICS

STRATUM
CHARACTERIZATION

STRATUM
RESOLUTION

BED RESOLUTION

INTERVAL RESOLUTION

Figure 5 INTERRELATIONSHIP OF DIFFERENT SCALES OF MEASUREMENT THROUGH
PROGRESSIVELY CHANGING RESOLUTION

Strata can be isotropic or anisotropic.
Beds that contain strata are anisotropic. Intervals that contain beds are anisotropic.

In general, reservoirs tend to be heterogeneous within a larger ordered framework in a way which is a function of scale (Figure 6). It is essential that the scale of ordering be identified. For example, at the pore- and microscales both the packing and the sorting of rock grains can be markedly variable. But if the variations are random within a relatively large volume of interest, these heterogeneous elements can be sufficiently small for that volume to be considered homogeneous at the larger scale, e.g. the lamina, the stratum or the bedding scale. Again, if the randomly distributed contrasting elements are at the mesoscale, well-log data do not lend themselves to inter-well correlation for reservoir zonation. In such a case, ordering and apparent homogeneity will have to be sought at a still larger scale, e.g. the interval scale, and reservoir zonation might be based solely on borehole or surface geophysics. Reservoirs can therefore appear to be both homogeneous and heterogeneous; the assumption of either one of these conditions depends on whether heterogeneities are averaged or separately resolved at the scale of measurement being considered.

Although heterogeneity precludes the absolute cross-scale integration of laboratory, borehole and surface-geophysical data, the reservoir system is amenable to systematic data reconciliation if the ordering is known. This involves using the high-resolution data at the core scale to refine those at the log scale, and then progressively working up to the geophysical and reservoir scales. The first stage is core-to-log scaling.

Core-to-Log Scaling - Parameters

The primary function of core-to-log scaling is to calibrate log response. The calibration is frequently undertaken by relating a point core measurement to log response at the same (depth merged) level. An example is the calibration of neutron log response (in limestone porosity units) through core-plug porosities for a sandstone reservoir rock. This practice is scientifically unacceptable for it does not reconcile scales. A more correct procedure would be to average core porosities, measured at sufficiently close sampling points, over vertical distances that correspond to log resolutions. The sampling interval would depend on the scale of the vertical variations within the bed, for this would determine the number of samples needed for the population to be statistically representative. The approach would require pre-selection of key intervals over which more detailed plug sampling could be made. Key intervals would have to show effectively constant log responses. For the case of the neutron tool, which has a vertical resolution of about 0.6 m (2 ft), key intervals would need to be at least this thick, to be fully cored, and to have plug porosities measured every 15 cm (6 inches) to allow a five-point arithmetric mean to be determined for correlation with the log. A more detailed level of plug sampling might be needed where there is a high degree of vertical variability. A possible refinement might be to weight the mean in accordance with the shape of the tool response function, being careful to take account of any "black-box" smoothing undertaken by the logging contractor.

A similar procedure can be followed for core-to-log scaling of conductivity. However, since conductivity is directionally dependent, the method of averaging plug values must take account of anisotropy (Figure 7). Conductivities C_i measured

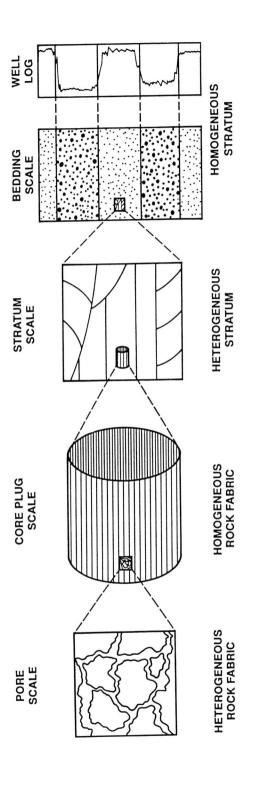

Figure 6 SCHEMATIC ILLUSTRATION OF HETEROGENEITY AS A FUNCTION OF SCALE

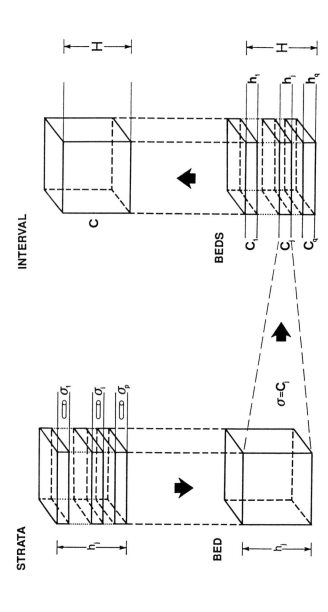

Figure 7   SCALE INTEGRATION OF CONDUCTIVITY

Integration of transversely-isotropic conductivities σ_i, $i=1, 2, \ldots, p$, into an electrically-equivalent longitudinal conductivity σ which is the longitudinal conductivity C_j of the j th bed in the interval of overall longitudinal conductivity C.

The symbol σ is used in this figure only.

by contact methods on regularly-spaced horizontal plugs should be averaged using the parallel conductor equation which can be written in terms of conductivity as follows:

$$C = \frac{1}{p} \sum_{i=1}^{p} C_i \tag{24}$$

where p is the number of samples. Note that equation (24) assumes a regular sampling interval for core plugs. In using this equation for core-to-log scaling, each plug is presumed to be representative of a stratum of thickness equal to the sampling interval and with a lateral extent that is at least as great as the diameter of investigation of resistivity logging tools. Some of these conceptual strata will be petrophysically similar. Values of C are determined over depth ranges that correspond to tool vertical resolutions for comparison with the deep laterolog or induction tool responses. These log responses are also in the form of horizontal resistivities (conductivities) but the vertical resolutions are very different, about 0.6 m for the laterolog, 1.5 m for the induction log in conductive beds, and larger still for the induction log in resistive beds. Therefore the depth-averaging of core plug data should relate to different vertical distances for correlation with these two logs. This calibration procedure will only be feasible for key intervals in the water zone for which the reservoir fluid properties can be properly simulated in the laboratory. A similar equation can be adopted for the formation resistivity factor since C_o is directionally dependent but C_w is not, i.e.

$$F = p \left[\sum_{i=1}^{p} \frac{1}{F_i} \right]^{-1} \tag{25}$$

where F_i is the formation factor of the i-th stratum.

It is sometimes useful to scale up water saturations measured directly on oil-base-mud cores for comparison with S_w values determined from resistivity logs. For p plugs spaced equally over a vertical distance that corresponds to resistivity-log resolution, the scaling equation is

$$S_w = \frac{\sum_{i=1}^{p} \phi_i \, S_{wi}}{\sum_{i=1}^{p} \phi_i} \tag{26}$$

where ϕ_i, S_{wi} are the porosity and water saturation, respectively, of the i-th plug.

Core-to-log scaling of interval transit time can be less meaningful if the plug data relate to horizontal measurements, as is conventional. In a transversely isotropic bed, where the individual strata are also transversely isotropic, horizontal core compressional velocities (or transit times) can be significantly greater (less) than their vertical equivalents as a consequence of pore aspect and grain shape. If we define δ as the ratio of the vertical to the horizontal transit times in the strata, so that

$$\delta = \frac{\Delta t_{vi}}{\Delta t_{hi}} , \qquad (27)$$

the mean vertical interval transit time Δt_v within the bed can be calculated from the expression

$$\Delta t_v = \frac{\delta}{p} \sum_{i=1}^{p} \Delta t_{hi} \qquad (28)$$

provided that δ is known. Equation (28) can also be used within key intervals to calculate a mean core-measured vertical transit time over a distance that corresponds to the vertical resolution of a sonic tool, i.e. about 0.6m (2 ft). This calibration procedure will not be valid where the interstitial fluids used in the laboratory have different acoustic properties from those present *in situ*.

Density can be scaled from the core scale to the log (bedding) scale in a straightforward manner using the equation

$$\rho_b = \frac{1}{p} \sum_{i=1}^{p} \rho_{bi} \qquad (29)$$

Note that the density log measurement is collimated and that there may not always be repeatability between runs in heterogeneous media.

The scaling of permeability is analogous to the scaling of electrical conductivity, the reciprocal of resistivity. For horizontal beds comprised of horizontal strata, the scaling equation for permeabilities measured on horizontal plugs is

$$K = \frac{1}{p} \sum_{i=1}^{p} K_i \qquad (30)$$

where K is the horizontal permeability of a bed calculated from measurements on p equi-spaced constituent core plugs, each of which is presumed to be representative of a stratum of thickness equal to the plug spacing and with a lateral extent that is at least as great as the diameter of investigation of formation testers.

Core-to-Log Scaling - Algorithms

The second function of core-to-log scaling is to apply to log responses characteristic algorithms that are defined by characterizing parameters established through laboratory measurement. This raises two important questions. (i) Are the methods of parametric scaling described in the previous subsection compatible with the characteristic algorithms? (ii) Are the characterizing parameters established at one scale also valid at another? The questions are interrelated.

In order to address these questions we consider some of the characteristic algorithms of Figure 4. Taking equation (5), with a = 1 for simplicity, and rewriting for individual core-plug values, i, we have

$$F_i = \frac{1}{\phi_i^{\ m}} \tag{31}$$

The exponent m has been established for a core sample population and it is presumed to be constant. Substitutions in equation (25) for F, using equation (5) with a = 1, and for F_i, using equation (31), give the expression

$$\phi = \left(\frac{1}{P} \sum_{i=1}^{P} \phi_i^{\ m} \right)^{1/m} \tag{32}$$

where ϕ is the mean porosity of the core plug population. But equation (32) is generally not the same as the expression for the arithmetic mean, viz.

$$\phi = \frac{1}{P} \sum_{i=1}^{P} \phi_i \tag{33}$$

which was earlier identified as the conventional approach. Equation (33) actually furnishes the correct porosity for a volume of rock provided that p samples are sufficient to represent the variability. This is because porosity is an intrinsic property at both stratum and bedding scales. It does not depend on spatial distribution and can therefore be averaged arithmetically. For the Archie-proposed case of m = 2, equation (32) provides a root-mean-square porosity. Equations (32) and (33) become the same when m = 1, the case of parallel capillaries. The only ways in which equations (32) and (33) might generally yield the same values of ϕ would be if the exponents m are different in the source equations (5) and (31), i.e. if the value of m at the tool-response (bedding) scale is fortuitously different from that, or those, at the core-plug (stratum) scale. Such a scale-dependence would preclude the establishment of log-applicable characteristic algorithms, e.g. equation (5), using core data.

The paradox can be resolved by recalling that equations (5) and (31) are empirical expressions which are not valid outside the range of values of porosity

and formation factor for which they were established. In fact, equation (5), written
with a = 1, is an approximation to a more fundamental scientific equation

$$F = \frac{T}{\phi}$$

(34)

where T is the electrical tortuosity. Writing equation (34) in terms of core-plug
parameters, we have

$$F_i = \frac{T_i}{\phi_i}$$

(35)

Substitution for F and F_i in equation (25) yields

$$\frac{\phi}{T} = \frac{1}{P} \sum_{i=1}^{P} \frac{\phi_i}{T_i}$$

(36)

If the core plugs all show the same tortuosity, and this is representative of the bed
as a whole, equation (36) provides an arithmetic mean porosity and the paradox is
resolved.

The first Archie equation serves as an approximation to equation (34). The
incompatibility of the conventional scaling procedures and the first Archie equation
is a consequence of the shortcomings of this approximation. A value of m
established through regression of core data should be compared with that derived
from a crossplot of resistivity (conductivity) and porosity log data in the water zone.
While there is no theoretical reason for these values to be the same, because of the
empirical nature of the equations, a broad agreement is observed in practice.

A similarly directed discussion can be developed for all the algorithms in
Figure 4, but there are distinctive differences. For example, in the time-average
equation (10) the characterizing parameters are not directionally dependent, unlike
the quantities a and m in the Archie equation, and they are not a function of scale.
For the core (stratum) scale

$$\delta \quad \Delta t_{hi} = \Delta t_{ma} + \left(\Delta t_f - \Delta t_{ma} \right) \phi_i$$

(37)

by analogy with equation (10). Summing for p equi-spaced samples, we have

$$\frac{\delta}{p} \sum_{i=1}^{P} \Delta t_{hi} = \Delta t_{ma} + \frac{1}{p} \left(\Delta t_f - \Delta t_{ma} \right) \sum_{i=1}^{P} \phi_i$$

(38)

and equation (28) can be incorporated so that

$$\Delta t_v = \Delta t_{ma} + \frac{1}{p} \left(\Delta t_f - \Delta t_{ma} \right) \sum_{i=1}^{P} \phi_i \qquad (39)$$

Direct application of equation (10) to the log-response (bedding) scale furnishes

$$\Delta t_v = \Delta t_{ma} + \left(\Delta t_f - \Delta t_{ma} \right) \phi \qquad (40)$$

Comparison of equations (39) and (40) shows that through this simplistic approach, the conceptual time-average equation is compatible with the core-to-log scaling technique and the characterizing parameters are common to both scales. Similar conclusions can be drawn for the density - porosity (equation 14) and the neutron porosity - porosity (equation 15) relationships. In those cases, there is no directional dependence of the petrophysical parameters.

Scaling from Logs to Larger Vertical Intervals

The equations of the previous two subsections can applied to the calibration by logs of larger scale measurements such as ultra-long-spacing electrical logs (ULSEL), vertical seismic profiles (VSP), and perhaps borehole gravimetry. The characteristic algorithms can be applied with appropriate modifications to these larger-scale data provided that our prerequisites are satisfied and that the scaling interval is chosen to match the vertical resolution of the larger scale measurement. The algorithms can also be used to determine the effective properties of a given lithological zone.

For example, equation (24) would be interpreted in terms of log responses rather than core plugs. The logs indicate q beds of different thicknesses h_j within an interval of thickness H, so the grossing-up procedure to determine the effective horizontal conductivity of a reservoir interval must allow for this variation. Equation (24) would then be written

$$C = \frac{1}{H} \sum_{j=1}^{q} h_j C_j \qquad (41)$$

and equation (25) becomes

$$F = H \left(\sum_{j=1}^{q} \frac{h_j}{F_j} \right)^{-1} \qquad (42)$$

where C_j, F_j relate to beds rather than constituent strata. The quantities C and F can be compared with long-spacing electrical log measurements which actually record horizontal resistivities (conductivities) following the so-called *paradox of anisotropy* (Keller & Frischknecht, 1966). In so doing, it is presumed that the beds extend horizontally throughout the zone of investigation of the long-spacing log. When F is related to a bulk porosity through the first Archie equation, there is again an incompatibility between the conventional depth-averaging procedures and those implied by the characteristic algorithms.

The appropriate expression for the average transit time for a reservoir interval of thickness H comprising q log-resolved beds of thickness h_j is given by

$$\Delta t_v = \frac{1}{H} \sum_{j=1}^{q} h_j \ \Delta t_{vj} \qquad (43)$$

where Δt_{vj} is the transit time for the j-th bed and Δt_v now relates to the entire interval. Equation (43) does not include a term analogous to δ for we assume that all measurements are vertical. The quantity Δt_v can be correlated with vertical VSP data. The transit time Δt_{vj} is related to the porosity of the bed through an equation of the form of (10). The quantity Δt_v is related to the bulk porosity ϕ of an interval through a characteristic algorithm that is unaffected by the parametric scaling.

Scaling of the density log for comparison with borehole gravity data will be impeded in heterogeneous reservoirs. This is partly because of the collimated measurement with pad-type density tools. Another factor, which also affects other tool comparisons, is the potentially greater depth of investigation of the borehole gravimeter relative to that of the density tool. In fact, the spacing between borehole gravity stations can be tuned so that a target volume of interest can be studied more effectively. Density log data should be integrated over intervals that correspond to these spacings if the logs are to be used for control.

Permeability data are often integrated in a way that omits the bedding scale, i.e. core permeabilities are averaged for direct comparison with production test data. Because core recovery is usually incomplete, creating a data deficit that is exacerbated by the scale disparity, core permeability data in a single well are sometimes interpreted as random samples of a property that shows a logarithmic normal distribution and can therefore be represented by a geometric mean. Thus, the degree of data deficiency is allowed to govern data integration procedures. In fact, the geometric mean can provide a highly pessimistic measure of permeability relative to that obtained from a production test. This disparity has been noted even where core plugs have been sampled and measured at regular intervals and where these data are representative of a horizontal stratification. In such cases equation (30) should be used to generate a mean horizontal permeability. In other cases the disparity between permeabilities from core analyses and well tests can be accentuated by the inclusion in the core sample data of low permeabilities that are incorrectly assumed to persist laterally through the larger volume of reservoir rock sampled by the well test. Here, a useful rule-of-thumb might be to average geometrically the core-derived permeabilities associated with each bed type and then to average arithmetically the geometric means.

The debate on averaging procedures might be clarified if characteristic algorithms, which do not contain logarithmic functions, were used to infer a vertical distribution of horizontal permeabilities from the log data. This approach would provide a more complete insight into the spatial distribution of permeability, especially in the case of extensive horizontal layering. These log-derived permeabilities can be averaged through an equation of the form

$$K = \frac{1}{H} \sum_{j=1}^{q} h_j K_j \qquad (44)$$

where K_j is the inferred permeability of the j-th log-resolved bed, of large lateral-correlation length, within the test interval. Although the characteristic algorithms for permeability represent no more than an empirical approximation, the averaging process can render the resulting mean permeability from equation (44) more useful than other approaches that do not draw upon log data, especially if wireline logs indicate an interwell correlation of those beds within the test interval. That having been said, there is no single approved method for averaging or scaling permeabilities. The optimum approach changes with the nature of their spatial continuity. It is a lack of knowledge of this aspect that constitutes the major impediment to effective reservoir characterization. In this respect, continuous well logs used in conjunction with inter-well measurements have a vital role to play.

6.   THE FUTURE

The foreseeable future will continue to be dominated by the oil-price regime and its inherent uncertainty, coupled with a development scenario for smaller fields. Tight financial constraints will inevitably impact on management decision-making at the exploration, appraisal and development stages. In particular, the nature of the fiscal regime means that higher risks will be associated with decisions of commerciality and with the design of production programmes. Reservoir characterization has a key role to play in assessing and reducing those risks.

High Resolution Seismics

The principal driving force will be high-resolution seismic prospecting. Advances in 3D seismic reflection technology and processing developments such as amplitude-versus-offset analysis bring nearer the ultimate goal of broad surface mapping of reservoir porosity, lithology and fluid type (e.g. Gelfand & Larner, 1984; King et al., 1988). This goal will not be attained without progressive calibration of the (mega-scale) seismic data by (macro-scale) interwell measurements which, in turn, are controlled by core-calibrated downhole measurements at the bedding scale. Acoustic measurements are fundamental to data integration for they can be made routinely at all scales from the laboratory to surface geophysics. A major area of future activity will be directed at an improved understanding of the interrelationship of elastic (acoustic) data measured at different scales.

Within the framework of the seismically defined reservoir model, petrophysical zoning must be effected so that it is compatible with definitive reflecting horizons. The zones must be regions of petrophysical consistency, i.e. the characterizing parameters should be effectively constant within them. A seismic reflector indicates a change in acoustic properties but not necessarily in the characterizing parameters. For this reason, some reflectors will be intrazonal whereas others will define major changes in lithology and thence in characterizing parameters. It is important to be able to distinguish between definitive and intrazonal reflectors. A second area of projected future activity will therefore be concerned with the relationship of acoustic properties to other physical parameters, both primary and secondary, that can be measured or determined in the laboratory and downhole.

Geological Information from Wireline Logs

The physical model of a reservoir unit must be compatible with geological data, otherwise it is not meaningful. The bridgeheads between physical measurements and geological interpretation are indicated in Figure 8. Two of these merit special mention. The as yet unestablished geochemical logging tool uses nuclear measurements to infer a geochemical signature in terms of elemental abundances within a reservoir rock : the elemental concentrations are then used to predict mineralogy (Hertzog *et al.*, 1987). This approach relies strongly on the integrity of element-to-mineral transforms. Much more work needs to be done to establish whether these transforms are universally applicable or reservoir-specific, or have little meaning because of the non-uniform chemical composition of naturally occurring minerals. The other topical area is high-resolution electrical imaging of the borehole wall to investigate bedding, structure, and thence depositional facies. The formation microscanner provides conductivity images with a vertical resolution of the order of a centimetre. This has important implications for the characterization of thinly-bedded reservoirs (McGann *et al.*, 1988). Different methods of data processing emphasize different structural details, e.g. interformational vs intraformational. Logging for sedimentological purposes will advance strongly as the microscanner concept displaces the more conventional dipmeter approach. We can therefore expect to see a great deal of research interest in the acquisition, processing, resolution and utilization of downhole electrical images.

Downhole Permeability Estimation

Refined continuous estimates of permeability at the bedding scale, for input to dynamic models, will be sought through some of the sophisticated logging tools that are either just becoming established in terms of oil-industry applications or are still under development. For example, the analysis of Stoneley waves from sonic waveform logs has already provided some promising indications of permeability prediction (Cheruvier & Winkler, 1987). The development of proton magnetic resonance tools that are not excessively affected by the wellbore environment (Jackson, 1984) might lead to more meaningful estimates of pore (surface) characteristics for better permeability prediction *in situ*. Yet again, a renewal of

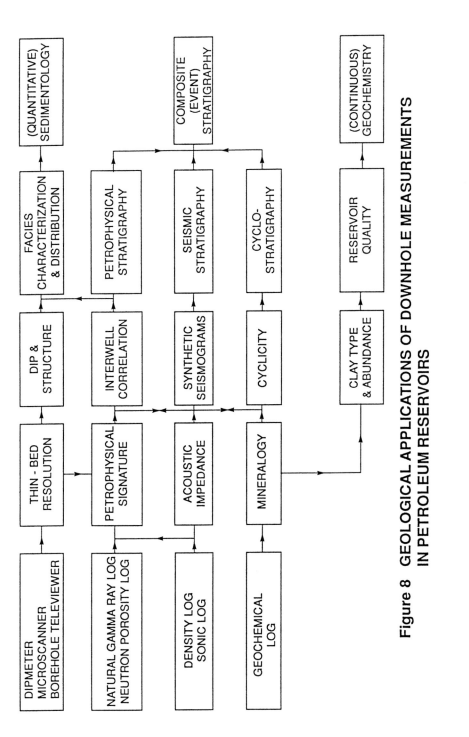

**Figure 8   GEOLOGICAL APPLICATIONS OF DOWNHOLE MEASUREMENTS
IN PETROLEUM RESERVOIRS**

interest in induced polarization logging (Vinegar *et al.*, 1985) might eventually lead to commercial downhole measurements of excess conductivity; these could be used conjunctively with porosity log data in pursuance of the permeability-prediction strategy favoured by the author's hierarchical study of algorithm performance (Worthington, 1976). Success in one of these areas would remove a major data deficiency, that of no continuous, quantitative permeability data at the bedding scale for input to dynamic models.

Discussion

There are many areas of projected future activity that have not been mentioned, e.g. fracture characterization, but all are governed by two basic issues. The first is formation anisotropy, an appreciation of which is essential for realistic 3D models. An important aspect is the influence of anisotropy upon log measurements in dipping beds and/or deviated wells. The second issue concerns uncertainty about the scaling procedures themselves, with particular regard to the heterogeneity of the system. Because measurements at different scales never sample the same volume and are rarely unambiguous, we cannot expect perfect data reconciliation. Yet, we do need to know the extent to which a perceived mismatch can be attributed to physical phenomena through uncertainties in the spatial distribution of the properties at the relevant scales, on the one hand, or to deficiences in the scaling algorithms, on the other. Successful reservoir characterization requires that both these issues be resolved.

7. CONCLUSIONS

Reservoir characterization, as defined here, is fundamental to contemporary reservoir evaluation. It is an integrating process whereby static and dynamic properties of reservoir rocks are molded into an integrated reservoir model with static and dynamic components. The static model draws upon characterizing information at the bedding scale, benefiting from calibration at the core scale. If the static model is inadequate, the dynamic model cannot be properly representative of the reservoir. Because of the fundamental role of the static model, and its closer ties to the bedding scale through continuous downhole measurements, this paper has focussed most strongly on the contribution of core-calibrated log data to the generation of the static reservoir model. The dynamic model is based upon comparatively little information at the bedding scale, i.e. logging input to Sw - height curves, repeat formation tests, and production logging to monitor reservoir performance; this shortfall has been identified as a weakness.

The key to successful reservoir characterization, in the presence of formation anisotropy and heterogeneity, is the effective reconciliation of precise and representative data measured at different scales. This approach allows high-resolution data at smaller scales to be used for the calibration of related data at larger scales. The attainment of this objective requires solutions to problems associated with the different physical natures of measurements in routine use, from the microscale through to reservoir scales. In the aftermath of strong advances in data acquisition technology during the 'Seventies and early 'Eighties, the emphasis is now on an improved understanding of these data and how they can contribute to more meaningful reservoir evaluation. Reservoir characterization is at the forefront of this "return to basics".

Effective data reconciliation also requires the removal of traditional inter-disciplinary barriers. It should no longer be appropriate to refer to a person as, for example, a sedimentologist or a petrophysicist. The required role is that of an *engineering geoscientist* with specialization in, for example, sedimentology or petrophysics. In other words, the need for lateral thinking and cross-discipline fertilization is acknowledged. Geoscientists concerned with downhole measurements at the mesoscopic (bedding) scale have a special responsibility because of the high degree of interaction of their subject with other traditional disciplines. In particular, wireline logs constitute vital control data for reservoir characterization at larger scales; their use encourages reservoir models that are factual as opposed to those that have been allowed to cross the thin dividing line into the realm of fiction.

Reservoir characterization at the bedding scale is essentially concerned with core-calibrated well logs. A major objective is the identification of reservoir zones within each of which those petrophysical algorithms used in log analysis do not change. In structurally complex reservoirs, the attainment of this objective requires an especially close association of the petrophysical data with the interpretations of high-resolution seismic surveys. The identification of the characteristic algorithms, achieved by evaluating the characterizing parameters that they contain, has been termed *petrophysical characterization*. The systematic procedures described here allow petrophysical characterization to be effected in the presence of both anisotropy and heterogeneity and in a way that permits the end-product to be related to data at different scales and from other disciplines. The entire process is dependent upon recognizing the ordering effects of nature.

The German philosopher, Immanuel Kant (1724-1804) stated:

"Everything in nature acts in conformity with law".

Belief in this statement might well be the ultimate driving force of reservoir characterization.

8.   ACKNOWLEDGEMENTS

The author is grateful to BP Research for sanctioning the publication of this work. Special thanks are due to those friends and colleagues who have provided helpful comments during the preparation of the manuscript: they are Steve Begg, Jon Bellamy, Bob Ehrlich, Kes Heffer, Sue Raikes and Dick Woodhouse.

9. NOMENCLATURE

C_i conductivity of i-th plug or stratum [S m^{-1}]

C_j conductivity of j-th bed [S m^{-1}]

C_o conductivity of fully water-saturated reservoir rock [S m^{-1}]

C_{sh} conductivity of intraformational shale [S m^{-1}]

C_t conductivity of partially water-saturated reservoir rock [S m^{-1}]

C_w conductivity of saturating aqueous electrolyte [S m^{-1}]

E_{sp} electrochemical self-potential [mV]

F (mean) formation resistivity factor of strata or beds
               (= R_o/R_w = C_w/C_o) [dimensionless]

F_i formation resistivity factor of i-th plug or stratum [dimensionless]

F_j formation resistivity factor of j-th bed [dimensionless]

GR   natural gamma log response [API units]

GR_{sand} natural gamma log response to clean sand [API units]

GR_{sh} natural gamma log response to intraformational shale [API units]

H   thickness of reservoir interval [m]

I   resistivity index (= R_t/R_o = C_o/C_t) [dimensionless]

K   (mean) intergranular permeability of strata or beds [mD]

K_i intergranular permeability of i-th plug or stratum [mD]

K_j intergranular permeability of j-th bed [mD]

L_k k-th log response in permeability regression equations [variable units]

Q_v cation exchange capacity per unit pore volume [eq. litre^{-1}]

R   average resistivity of strata or beds [Ω m]

R_i resistivity of i-th plug or stratum [Ω m]

R_j resistivity of j-th bed [Ω m]

R_o resistivity of fully water-saturated reservoir rock [Ω m]

R_{sh} resistivity of intraformational shale [Ω m]

R_t resistivity of partially water-saturated reservoir rock [Ω m]

R_w resistivity of saturating aqueous electrolyte [Ω m]

S_h fractional hydrocarbon saturation (=$1-S_w$) [dimensionless]

S_w fractional water saturation (=$1-S_h$) [dimensionless]

S_{wi} fractional water saturation of i-th plug or stratum [dimensionless]

T (mean) electrical tortuosity of strata or beds [dimensionless]

T_i electrical tortuosity of i-th plug or stratum [dimensionless]

V_{sh} fractional shale volume [dimensionless]

X excess conductivity associated with shale effects [S m^{-1}]

a coefficient in the Archie formation factor - porosity equation [dimensionless]

c_p sonic correction factor for lack of compaction [dimensionless]

h_j thickness of j-th bed within an interval [m]

m exponent in the Archie formation factor - porosity equation [dimensionless]

n exponent in the Archie resistivity index - water saturation equation [dimensionless]

p number of strata within a bed

q number of beds within a reservoir interval

Δt sonic transit time (reciprocal acoustic velocity) [μs m^{-1}]

Δt_f sonic transit time for pore fluid [μs m^{-1}]

Δt_{hi} horizontal sonic transit time for i-th plug or stratum [μs m^{-1}]

Δt_{ma} sonic transit time for rock matrix [μs m^{-1}]

Δt_{sh} sonic transit time for intraformational shale [μs m^{-1}]

Δt_v mean vertical sonic transit time for strata or beds [μs m^{-1}]

Δt_{vi} vertical sonic transit time for i-th plug or stratum [μs m^{-1}]

Δt_{vj} vertical sonic transit time for j-th bed [μs m^{-1}]

α_k k-th characterizing coefficient in permeability regression equation [variable units]

δ ratio of vertical to horizontal sonic transit times for plugs or strata [dimensionless]

ϕ (mean) fractional porosity of strata or beds [dimensionless]

ϕ_i fractional porosity of i-th plug or stratum [dimensionless]

ϕ_N neutron log response (limestone porosity units)

ϕ_{Nf} neutron log response to pore fluid (limestone porosity units) [dimensionless]

ϕ_{Nma} neutron log response to rock matrix (limestone porosity units) [dimensionless]

ϕ_{Nsh} neutron log response to intraformational shale (limestone porosity units) [dimensionless]

ρ_b (mean) bulk density of strata or beds [g cm^{-3}]

ρ_{bi} bulk density of i-th plug or stratum [g cm^{-3}]

ρ_f density of pore fluid [g cm^{-3}]

ρ_{ma} density of rock matrix [g cm^{-3}]

ρ_{sh} density of intraformational shale [g cm^{-3}]

10. REFERENCES

ALLEN, J.R., 1979.
Prediction of permeability from logs by multiple regression.
Trans. SPWLA 6th Eur. Formation Evaluation Symp., pp M1-17.

ARCHIE, G.E., 1942.
The electrical resistivity log as an aid in determining some reservoir characteristics.
Trans. Am. Inst. Min. Metall. Eng. **146**, 54-62.

CHERUVIER, E. & WINKLER, K.W., 1987.
Field example of *in situ* permeability indication from full acoustic wavetrains.
Trans. SPWLA 28th Ann. Logging Symp., pp NN1-15.

FAUST, L.Y., 1951.
Seismic velocities as a function of depth and geologic time.
Geophysics **16**, 192-206.

GASSMANN, F., 1951.
Elastic waves through a packing of spheres.
Geophysics **16**, 673-685.

GELFAND, V.A. & LARNER, K.L., 1984.
Seismic lithologic modeling.
The Leading Edge **3**, 30-35.

HARTLEY, K.B., 1981.
Factors affecting sandstone acoustic compressional wave velocities and an examination of empirical correlations between velocities and porosities.
Trans. SPWLA 22nd Ann. Logging Symp., pp PP1-21.

HERTZOG, R., COLSON, L., SEEMAN, B., O'BRIEN, M., SCOTT, H., McKEON, D., WRAIGHT, P., GRAU, J., ELLIS, D., SCHWEITZER, J. & HERRON, M., 1987.
Geochemical logging with spectrometry tools.
SPE Paper 16792, Society of Petroleum Engineers, Dallas, Texas.

JACKSON, J.A., 1984.
Nuclear magnetic resonance well logging.
The Log Analyst **25** (5), 16-30.

KELLER, G.V. & FRISCHKNECHT, F.C., 1966.
Electrical methods in geophysical prospecting.
Pergamon Press, Oxford, 517 pp.

KING, G.A., DUNLOP, K.N.B. & GRAEBNER, R.J., 1988.
Surface seismic monitoring of an active water flood.
SEG Preprint, Society of Exploration Geophysicists, Tulsa, Oklahoma.

McGANN, G.J., RICHES, H.A. & RENOULT, D.C., 1988
Formation evaluation in a thinly bedded reservoir - a case history:
Scapa Field, North Sea.
Trans. SPWLA 29th Ann. Logging Symp., pp. V1-20.

POUPON, A. & GAYMARD, R., 1970.
The evaluation of clay content from logs.
Trans. SPWLA 11th Ann. Logging Symp., pp. G1-21.

RAYMER, L.L., HUNT, E.R. & GARDNER, J.S., 1980.
An improved sonic transit time-to-porosity transform.
Trans. SPWLA 21st Ann. Logging Symp., pp. P1-13.

SAWYER, W.K., PIERCE, C.I. & LOWE, R.B., 1971.
Electrical and hydraulic flow properties of Appalachian petroleum
reservoir rocks. U.S. Bur. Mines Rept. Inv. 7519, 22 pp.

SIMANDOUX, P., 1963.
Dielectric measurements on porous media: application to the
measurement of water saturations: study of the behaviour of
argillaceous formations.
Revue de l'Institut Francais du Petrole 18, supplementary issue,
193-215. (Translated text in Shaly Sand Reprint Volume, SPWLA,
Houston, pp IV 97-124).

VINEGAR, H.J., WAXMAN, M.H., BEST, M.H. & REDDY, I.K., 1985.
Induced polarization logging : borehole modeling, tool design and field
tests.
Trans. SPWLA 26th Ann. Logging Symp., pp AAA1-62.

WENDT, W.A., SAKURAI, S. & NELSON, P.H., 1986.
Permeability prediction from well logs using multiple regression.
In: Reservoir characterization (Lake, L.W. & Carroll, H.B., Eds.),
Academic Press, pp 181-221.

WORTHINGTON, A.E., GIDMAN, J. & NEWMAN, G.H., 1987.
Reservoir petrophysics of poorly consolidated rocks: I. Well-site
procedures and laboratory methods.
Trans. SPWLA 28th Ann. Logging Symp., pp BB1-17.

WORTHINGTON, P.F., 1976.
Hydrogeophysical properties of parts of the British Trias.
Geophys. Prospect. **24**, 672-695.

WORTHINGTON, P.F., 1988.
Permeability evaluation.
The Technical Review **36** (1), 1.

WYLLIE, M.R.J., GREGORY, A.R. & GARDNER, L.W., 1956.
Elastic wave velocities in heterogeneous and porous media.
Geophysics **21**, 41-70.

WYLLIE, M.R.J., GREGORY, A.R. & GARDNER, L.W., 1958.
An experimental investigation of factors affecting elastic wave velocities
in porous media.
Geophysics **23**, 459-493.

PERMEABILITY VARIATIONS IN SANDSTONES AND THEIR RELATIONSHIP TO SEDIMENTARY STRUCTURES

Andrew Hurst

Statoil, Forus, N-4001 Stavanger, Norway*

Kjell Johan Rosvoll**

Geology Department, University of Bergen, N-5000 Bergen, Norway

ABSTRACT

A high precision laboratory minipermeameter is used to make more than 16000 measurements on cores of lightly consolidated shallow marine sandstones. Measurements are collected on 2mm or 5mm orthogonal grids which provide a sufficiently dense data set to identify the permeability variations present in sedimentary structures. From ANOVA testing it is inferred that greater permeability variation is present within sedimentary structures than between them. Permeability anisotropy is identified in whole core samples and quarter-cuts and its relation to sedimentary structures illustrated.

Optimal sample density is evaluated for each lithofacies by N_0 testing. Even the most "homogeneous" large-scale cross stratified and massive lithofacies examined require permeability measurements to be made at, on average, a 5mm spacing if permeability variation is to be known to within a $\pm$ 5% level. However, optimal sampling densities vary considerably, from approximately 2mm to 10mm, within the homogeneous lithofacies. More heterogeneous lithofacies need denser measurements to optimalise permeability sampling, heterolithic facies often requiring sampling at sub-mm spacing.

Permeability contrasts which correspond to mm-scale sedimentary layering are resolved with the minipermeameter. Variation of minipermeameter probe geometry changes the resolution of measurement, smaller probe radii giving better resolution of fine permeability layering. Comparison of measurements made with different radii probes on the same sample area shows that similar mean permeabilities are derived for each measurement series but, that the smallest radius probe resolves a higher permeability heterogeneity.

Average permeabilities derived from minipermeameter measurements on a preserved core are similar to the whole-core permeability. Average permeability derived from core plug measurements taken subsequently are considerably higher. The core plugs thus give an over optimistic impression of the permeability.

Present addresses, * UNOCAL UK, 32 Cadbury Rd., SUNBURY-ON-THAMES, TW16 7LU, UK and ** Imperial College, Dept. of Mineral Resources Engineering, Prince Consort Road, London SW7 2BP, U.K.

I. INTRODUCTION

It is acknowledged that sedimentary heterogeneities affect the recovery efficiency of reservoirs over a wide range of scales of investigation (Haldorsen 1986; Weber 1986). Several recent studies address the problem of metre-scale or greater permeability distribution by examining outcrops (Jones et al. 1984; Stalkup and Ebanks 1986; Stalkup 1986; Goggin et al. 1986; Lewis et al. 1990). Others (Weber et al. 1972; Pryor 1973) have attempted to examine the sub-metre scales of permeability heterogeneity caused by sedimentary structures, which are usually identified during geological core description.

Pryor's (1973) comprehensive work uses core plugs to characterise permeability variations and concludes that a greater permeability variability exists within bedding and lamination packets than between them. The size of the core plugs (2.5cm diameter) limited the resolution at which Pryor could investigate permeability variation within sedimentary structures, which often have dimensions less than 2.5 cm; the data have no geological validity for examining the permeability variations within individual sedimentary structures.

Distinction of sediment bodies provides a basis for prediction of the well-to-well continuity of reservoir units. However, such well-to-well correlations use a coarse scale of facies definition, the "megascopic" (Haldorsen 1986) or "medium scale" (Lasseter et al. 1986) of reservoir heterogeneity, and many scales of permeability heterogeneity may be present at successively smaller scales of examination (Lewis 1988), all of which may have some significance for recovery efficiency (Weber 1986). If both the geometry and permeability distribution of different sandbodies within a reservoir are known, it is possible to use sedimentological data predictively in reservoir modelling (Pryor and Fulton 1978).

As contrasts in depositional energy are preserved as variations in grain size and sorting, and are visible as sedimentary structures, it is reasonable to expect that permeability variations are associated with primary depositional features (Mast and Potter 1963; Weber 1982). However, can the permeability contrast be confirmed statistically and can the effects of sedimentary structures on the orientation of the principle axes of permeability, or more generally the presence of permeability anisotropy, be identified?

Sedimentary lithofacies define intervals with similar sedimentary structures, grain size and sorting, and they may be useful for defining reservoir zones and petrophysical characterisation of cores. As different lithofacies have contrasting degrees of internal heterogeneity, it is reasonable to assume that permeability characterisation of each lithofacies will require a different density of permeability measurements. This assumption is tested and predictions are made of optimal sampling densities for different lithofacies. In addition, the effects of varying the volume of investigation of permeability measurements are quantified both with respect to estimation of the net permeability of samples and with respect to detection of heterogeneities. Here, a laboratory minipermeameter is used to investigate the distribution of permeability within sedimentary structures in a cored section.

II. EXPERIMENTAL METHODS AND MATERIALS

A. Minipermeameter Measurement

All measurements are made using a laboratory minipermeameter similar in basic design to that described by Eijpe and Weber (1971). Nitrogen is used as the injection gas at a controlled injection pressure of 0.025 bar. Minipermeameter probes have rubber tips to ensure sealing against sample surfaces. Several probes were tested with a variety of internal and external diameters, different rubber seals and different contact pressures. Three probes were chosen for routine measurements in this study (Table 1). Where appropriate, experimental conditions, in particular probe diameter and gas pressure, are adjusted to give optimal measuring ranges for specific samples. For all series of measurements we choose to operate with constant gas pressure for each probe type, and where possible, the same measuring probe for any one series of measurements (Halvorsen and Hurst 1990). P_a , the force of application, is the downwards force exerted by a pneumatic cylinder which presses the minpermeameter probe against the rock surface. P_c is the contact pressure between the rubber tip and surface of the sample, which of course decreases for a given P_a with increased area of contact.

Testing of the equipment has proved that permeability can be measured with a satisfactory degree of accuracy and a high degree of precision. One aspect of the instrumentation, which is of particular importance in this study, is the capability of positioning the measuring probe on the surface of samples to within 0.1mm accuracy. It is thus possible to construct subcentimetre grids of data points where the measuring probe can be placed, and re-placed, within thin sedimentary laminae. Subsequently, the equipment has been up-graded to fully automated status where probe positioning is possible to within 0.001mm accuracy (Halvorsen and Hurst 1990).

Calibration of minipermeameter flow rates (cm^3 min^{-1}) with permeability (mD) is made by following the procedure of Cadman (1984). Standard "homogeneous" core plugs (3.8cm diameter) are selected which cover a wide range of permeabilities, in this case from < 1mD to > 2000mD. Here, sample homogeneity is evaluated by visual inspection and CAT-scanning. Approximately 50% of the plugs are, after CAT-scanning, rejected because of the heterogeneities detected. Permeability of selected homogeneous samples is then measured using Hassler-sleeve apparatus. Three minipermeameter measurements are made on each end of the core plugs. The net flow rate of each plug is calculated by taking the geometric mean of the measurements from the plug ends and then, the harmonic mean of those values.

Permeability (Hassler-sleeve) is plotted against flow rate for each plug, the relationship between the two values defined and the correlation coefficients calculated (Fig. 1). For the probes used in this study (Table 1) the following relationships between permeability (K) and flow rate (S) are defined:

$$\text{for Probe 1D2,} \quad K = 13.2 . S - 20.1$$

$$\text{for Probe 2,} \quad K = 15.1 . S + 13.6$$

$$\text{for Probe 4,} \quad K = 26.0 . S - 35.1$$

Table 1. Minipermeameter probes. d_i = internal diameter (mm), d_x = external diameter (mm), P_a = application pressure (bar), P_c = contact pressure at sample surface (bar).

Probe	d_i	d_x	P_a	P_c	Application
1D2	6	24	2.5	6.8	poorly consolidated samples
2	4.5	11	3.0	43.5	consolidated medium-coarse sands
4	3	6	1.4	69.2	consolidated silty-fine sands

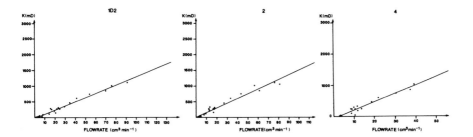

Fig. 1. Calibration curves for minipermeameter (flowrates) and Hassler-sleeve (mD) measurements on standard homogeneous core plugs.

B. Core Material

Samples are taken from shallow marine sandstones of the Fangst and Båt Groups, in the Haltenbanken area of the Norwegian continental shelf (Dalland et al. 1988). The sandstones are subarkoses, with occasional quartz arenites (> 95% quartz) and varied but low (< 2% to 10%) mica contents (Bjørlykke et al. 1986). Mica and clay contents tend to be greatest in the finer grained sandstones. All samples are lightly consolidated and it is assumed that any variations in permeability reflect primary depositional characteristics rather than the effects of cementation.

Unless stated otherwise, quarter-cut cores are used. Thus, measurements are made on two flat, approximately orthogonal slabbed faces. Each face is approximately 7.5cm wide and in long section is broken only by the fragmentation caused by coring and the holes left by routine sampling of core plugs. Care was taken to avoid possible errors caused by measuring too near sample edges. Experiments proved that it was advisable to measure no nearer a sample edge than 15mm.

No core cleaning is made prior to measurement apart from a gentle brushing of the surface to remove dust. The residual oil saturation of the cores is known to be very low from independent analyses.

Samples are coded using a lithofacies designation (Fig. 2) and two numbers, the first denotes a core section and the second denotes a sedimentologically defined part of the cored section.

C. Sedimentology

Before permeability distribution is evaluated, a sedimentological classification of lithofacies is made. Lithofacies units, which are directly related to specific depositional processes, are the building blocks used by geologists to build 3D sedimentary reservoir models. Lithofacies are similar sedimentary rocks, in so far as within a particular geological succession, a given lithofacies will always look approximately the same. Similarity of appearance is a direct consequence of similar grain size and sorting, which are produced by specific physical regimes during deposition (Allen 1984). Thus, we expect lithofacies should provide a basis for describing and comparing permeability distributions in sedimentary rocks.

1. Sedimentological Classification

Five sandstone lithofacies are identified: large-scale cross stratified, small-scale cross stratified, horizontally stratified, massive and heterolithic (Fig. 2). The sequence is dominated by the large-scale cross stratified lithofacies that are known to have good reservoir characteristics and high (> 2D) permeabilities. In the cross stratified units, laminae and sets are identified and used to sub-divide the samples. In the massive units, where structures are scarce or poorly defined, and in heterolithic units, which are often thin-stratified and contain irregular structures, sample populations are divided into "parts" of similar size which have no specific geological association but may reflect a source of permeability variance.

III. RESULTS

A. Minipermeameter Data

More than 16,000 minipermeameter measurements were made on approximately 15 m of core. Examples of each lithofacies together with permeability trends and distributions are given in Figs. 3, 6 and 8. In general, the data give better approximations to normal distributions when plotted on log rather than a linear scale. Good log normal fits are not always obtained, particularly in the case of heterolithic sandstones. Permeability data from core plugs usually have an approximately log normal distribution (Havlena 1966), so it is significant that the minipermeameter data, which have a much smaller volume of investigation (Daltaban et al. 1989) and a higher measuring density (2mm and 5mm spacing) than core plugs, also approximate log normal distributions.

Vertical permeability profiles show clearly the range of variation present and the small scale at which the variations occur. Grain size variation is minimal in most of the samples and permeability varies independent of grain size. Even in the apparently "homogeneous" large-scale cross stratified units and the "structureless" massive units permeability heterogeneity is obvious.

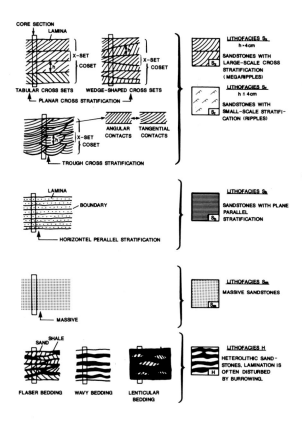

Fig. 2. Sandstone lithofacies classification.

B. Source of Permeability Contrast

Permeability variations in cores are readily identified by either minipermeameter or core plug measurements. To attain a full understanding of the dynamic properties of a reservoir, in this case with respect to permeability distribution in cores, it is necessary to discover where the contrasts in permeability originate. Geological description of cores identifies lithofacies, which then form a basis for defining flow units for modelling recovery efficiency in reservoir simulation models. The sedimentary structures within each lithofacies provide a logical focus for examining permeability contrast within units and here, we assess whether permeability contrast is attributable to variations within or between sedimentary structures. This is examined using analysis of variance (ANOVA, Davis 1986). A summary of methods and all results are found in Rosvoll (1989).

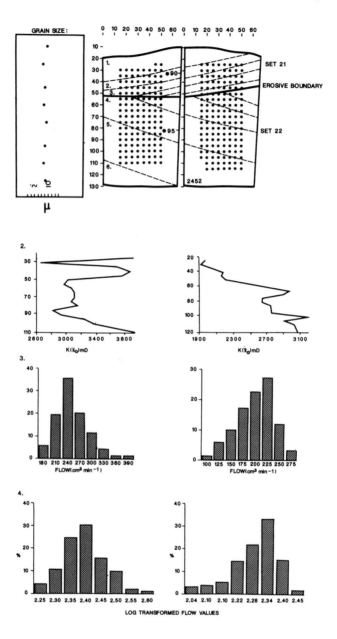

Fig. 3. Sample S_X - 1.18, a large-scale cross stratified sandstone. Two scales of sedimentary bedding are present, laminae (1 - 5 cm thick) and coarser sets which comprise several laminae. 1 - core sample with measuring grid, dimensions in mm; 2 - vertical profiles of permeability (K), $\bar{x}_G$ = geometric mean; 3 - flow rates for each side of sample; 4 - log transformed versions of 3.

1. Large-Scale Cross Stratification

A typical sample containing large-scale cross bedding, lithofacies S_X , is shown in Fig. 3. When examined collectively, the samples have uniformly high permeabilities (Fig. 4). In each data set (laminae from the same lithofacies) permeabilities tend to group together, and only in one data set, lithofacies S_X-1.21, is a clear divergence observed from the general trend. There, four laminae have permeability variations between 2000 and 2600mD, whereas the other two laminae have values between 5750and 6000mD, a 1:3 contrast of permeability between laminae within the same lithofacies. The maximum permeability variation between successive laminae varies from approximately 250mD to around 2000mD.

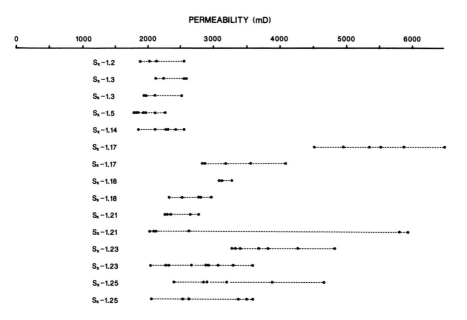

Fig. 4. Mean permeabilities $(\overline{k}_A)$ of laminae within each sample of lithofacies S_X. Where two samples have the same sample code, data from both sides of a sample are presented.

ANOVA tests are made on fifteen cross stratified units. In fourteen of the fifteen examples, 70 to 95% of the total variation for each unit can be attributed to variations within laminae (Fig. 5). F testing is used to test whether there is any statistical basis for assuming that all laminae have similar permeability. In thirteen of the examples the 5% confidence level is not satisfied so there is little reason to assume that the permeabilities of the laminae are similar.

Results of the ANOVA analysis indicate that, in general, the greatest permeability contrast is present within the individual laminae of cross stratified sets. This confirms the impression given by a more cursory examination of the data (Fig. 4) where average permeabilities for laminae within specific units are seen to cluster.

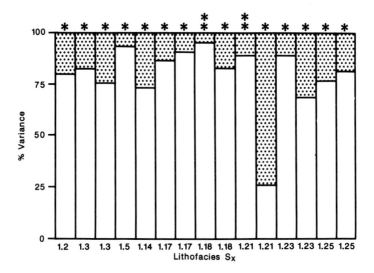

Fig. 5. Results of ANOVA testing on lithofacies S_X. The stippled area denotes the percentage of the permeability variance not attributable to variations present within individual cross sets. Single * denotes that F test failed to satisfy both 1% and 5% errors. Two *'s denotes that the 1% condition is accepted.

2. Small-Scale Cross Stratification

An example of small-scale cross stratified sandstone is shown in Fig. 6. ANOVA is made on data from from one sample, S_R 1.8, which contains both laminae and sets. Additional data from S_R - 3.7a are used to compare permeability variance between approximately equal parts of the sample, as no clear sedimentological divisions could be made.

Results of the ANOVA analysis show that the source of the permeability contrast is attributable to variance within the sedimentary laminae (Fig. 7a). F testing indicates that a high degree of homogeneity may be present within some laminae, the 1% confidence interval being accepted. In the same sample, comparison of the permeability variance is greater within sets than between. Random division of data (parts) from another sample (S_R 3.7a) shows also that the source of the permeability variance lies mainly within each part of the unit.

3. Horizontally Stratified Sandstones

The characteristics of the horizontally stratified sandstones are shown in Fig. 6b. Only one example of horizontally stratified sandstone has a sufficiently high data population to warrant statistical analysis. The sample is divided into two approximately equal parts, for which the permeability variance belongs almost exclusively (99.9% of the total variation) within each part of the unit (Fig. 7b). It is implied that the horizontal laminae contain a distinct permeability variation and that each of the laminae in this sample are very similar. This is confirmed by the F test.

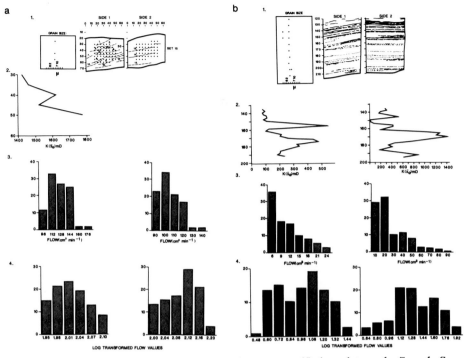

Fig. 6. a- Sample S_R - 1.12, a typical small-scale cross stratified sandstone. b- Sample S_H - 3.7b, a horizontally stratified sandstone. Annotation as in Fig. 3.

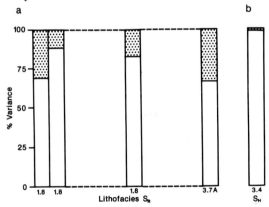

Fig. 7. Results of ANOVA for, a- lithofacies S_R, small-scale cross stratified, and b-lithofacies S_H, horizontally laminated sandstones.

4. Massive Sandstones

The absence of clearly defined sedimentary structures in massive sandstones (Fig. 8a) makes analysis of the source of variance dependent on dividing each unit into "parts" that have no specific geological affinity but have approximately similar areas and populations. The source of permeability variance is mainly attributable to variations within each part of the massive units (Fig. 9a). Results of F testing show that statistically three of the parts can be considered to have uniform permeability distributions.

5. Heterolithic Sandstones

Heterolithic units are interstratified sandstones and shales, which in detail may have very varied bedding characterisitics (Figs. 2 and 8b). Their permeability distribution is controlled largely by the arrangement and dimensions of shale interbeds, which also restrict the vertical movement of fluids. Sampling of heterolithic units using conventional core plugs causes problems when the the diameter of the plugs approaches, or is less than, the thickness of the laminae under investigation. Minipermeameter measurement is far less limited by this constraint and may be used to investigate the permeability of very small scale features (Weber 1986, p.515).

Because of the complexity of the stratification in the heterolithic samples, the data are divided into approximately equal parts (as for the massive and horizontally stratified sandstones). In general, a high proportion of the variance present originates within each part, $> 96\%$ (Fig. 9b), i.e. the permeability distributions of the parts are very similar. In the two cases where a lower proportion of the variation belongs within the parts, there are markedly higher sand contents in one part, which causes a bimodal distribution of the highest permeabilities between parts.

C. Permeability Anisotropy

Permeability anisotropy is known to be a major factor in determining the recovery efficiency of reservoirs (Weber 1986). However, conventional core analysis and special core analysis programs are rarely designed to make a quantitative evaluation of permeability anisotropy. In this study, permeability anisotropy is evaluated by two different methods: 1- by measurement along vertical sections of a whole core sample; 2- by analysis of the source of variance between the two sides of core quarter-cuts.

1. Whole Core

a. Sample Description. A core of fine-grained sandstone, approximately 20cm long and 12cm in diameter, typical of the preserved cores used for special core analysis, is examined. Four flat surfaces, approximately 4cm wide, were cut orthogonally on the perimeter of the sample (Fig. 10). Minipermeameter measurements were made in vertical columns at 5mm intervals on two of the surfaces, and at 15mm intervals on the other surfaces. Three distinct sedimentological units are identified on the basis of their sedimentary structures: an upper cross stratified unit, a unit containing stylolites and a lower, massive unit (Fig. 11).

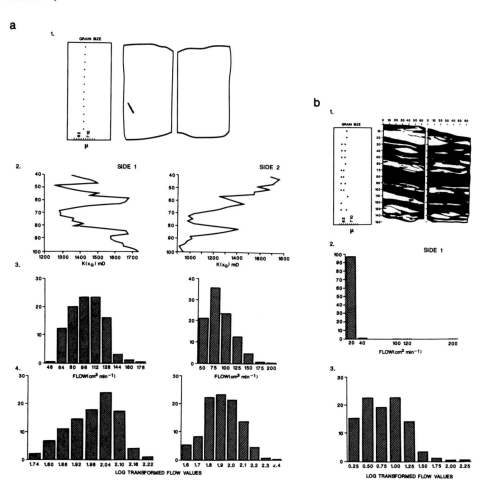

Fig. 8. Minipermeameter data from, a- sample S_M - 9.9, a typical massive sandstone, and b- sample H - 9.9, a typical heterolithic sandstone. Annotation as Fig. 3.

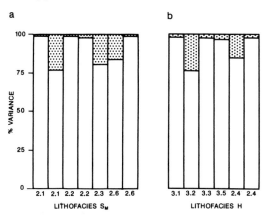

Fig. 9. ANOVA results from parts of, a- massive sandstones, and b- heterolithic sandstones. Data from orthogonal faces of the same sample are prese nted.

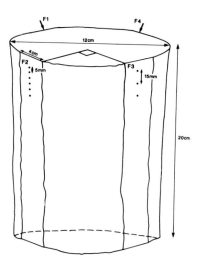

Fig. 10. Whole core sample.

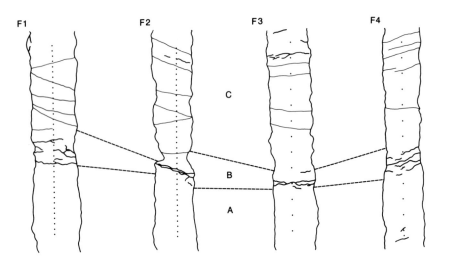

Fig. 11. Sedimentological division of whole core sample based on description of each vertical section.

b. Results. Both the permeability trends along the vertical profiles (Fig. 12) and the average values for each zone (Table 2) reveal the the presence of considerable permeability variation. From Table 2 it is also clear that columns F1 and F3, and columns F2 and F4, have similar average permeabilities, F1 and F3 being significantly higher.

Table 2. Mean permeabilities (mD) for geological zones A, B and C (Fig. 11) for each vertical section. $\bar{x}$ = geometric mean, σ = standard deviation, n = number of measurements.

		F1	F2	F3	F4
	$\bar{x}$	809	674	797	646
C	σ	177	202	287	191
	n	18	22	9	7
	$\bar{x}$	192	277	7	101
B	σ	65	91	-	-
	n	5	4	1	2
	$\bar{x}$	687	425	622	443
A	σ	132	123	207	216
	n	14	11	3	4

It is unlikely that the vertical sections by chance have coincided with the maximum and minimum directions of permeability in zones A and C, just as it is unlikely that the axes of permeability anisotropy are coincident in the two different zones. The data are however, indicative of the existence of directional variations in permeability.

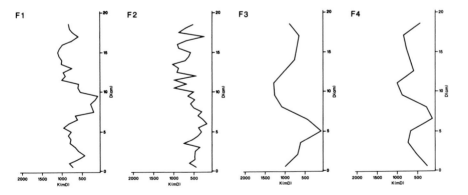

Fig. 12. Vertical profiles of permeability from the orthogonal sections shown in Figs. 10 and 11.

As minipermeameter measurements have a very small volume of investigation and lack directional constraints (relative to core plug measurements), the data should be considered only as proof of the presence of permeability anisotropy. The quantitative significance of the anisotropy is evaluated by taking 1 in. diameter core plugs from each face. The core plugs give directional permeabilities, perpendicular to the cut surfaces, and should confirm to what extent the minipermeameter data can be used to define anisotropy. In zone C the plug data confirm the minipermeameter data but in zone A the confirmation is less clear. As is shown in Table 3, very few plug data are obtained, thus, their statistical validity is less than that of the minipermeameter data, particularly in view of the variations of permeability known to exist in vertical section (Fig. 12).

Table 3. Comparison of mean permeabilities from minipermeameter data (M) and core plugs (P) for zones A and C for each face. Terminology is otherwise as in Table 2.

		F1		F2		F3		F4	
		M	P	M	P	M	P	M	P
C	$\bar{x}$	809	1018	674	755	797	978	646	662
	n	18	2	22	2	9	2	7	2
A	$\bar{x}$	687	526	425	588	622	-	443	709
	n	14	1	11	2	3	0	4	2

Zone B, which contains stylolites, is more cemented than the other zones and has lower permeability (Table 2). Minipermeameter measurement of permeability in zone B gives evidence of very low permeabilities, several measurements of < 20 mD. Because of the thin, irregular nature of zone B, no assessment of permeability anisotropy is made.

c. Flow Characteristics. If this sample was used for special core analysis, several problems can be expected in the light of the minipermeameter data. First, considerable fine-scale variation in the magnitude of permeability is recognised along vertical sections. In particular, zone B has several areas of low permeability which would be expected to split the sample into two separate horizontal flow units (approximately zones A and C) and to severely limit vertical flow through the sample. Second, the sample shou ld not be treated as a homogeneous isotropic material as distinct permeability anisotropy is identified in two parts of the core. The anisotropy is confirmed by core plug measurements.

2. Orthogonal Faces

a. Sample Description. Nine of the samples are considered to have sufficient data points on both faces of the quarter-cut core to make statistical testing of their similarity possible. As in the preceeding section where the source of permeability contrast was examined, the ANOVA method is used to evaluate the source of variance. Although a quarter-cut core does not make possible identification of the directions of permeability anisotropy, analysis of the data will at least determine whether the two cuts have any significant permeability contrast, from which it may be inferred that anisotropy exists.

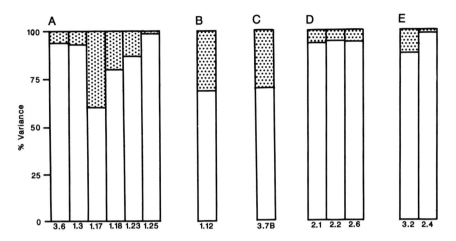

Fig. 13. Results of ANOVA for comparison of two sides of the same sample. Lithofacies as follows: A- S_X, B- S_R, C- S_H, D- S_M, E- H.

Table 4. Comparison of mean (geometric) permeabilities (mD) of orthogonal faces, $\bar{x}_{S1}$ and $\bar{x}_{S2}$, n = number of measurements on each face.

Facies	$\bar{x}_{S1}$	$\bar{x}_{S2}$	n
S_X-3.6	1044	1332	900
S_X-1.3	2389	2139	167
S_X-1.17	5347	3115	395
S_X-1.18	3193	2769	206
S_X-1.23	3687	2811	408
S_X-1.25	3329	2912	300
S_X-1.12	1614	1354	99
S_H-3.7b	204	513	660
S_M-2.1	1477	1253	900
S_M-2.2	1845	1672	870
S_M-2.6	427	298	1725
H -3.2	60	24	1080
H -2.4	33	27	900

b.   Results.   In all cases ANOVA proves that greater variation exists within the permeabilities on any one side of a sample than between two sides of the same sample (Fig. 13). F testing however, shows that the permeability distributions on each side of any one sample are significantly different, even within a 5% tolerance level. The presence of permeability anisotropy in all samples is thus inferred. Differences in permeabilities are summarised in Table 4. Although it is not surprising that the samples are shown to exhibit anisotropy, it is significant that despite what amounts to random orientation of each pair of faces when cutting the quarter cores, detection of anisotropy is nevertheless possible even in the most homogeneous of lithofacies. These data give no information regarding the orientation of the principle axes of permeability anisotropy. It would be expected that orientations of the principle axes of perm eability coincided with the dip of the sedimentary structures.

D. Sampling Density

There is little doubt that the routine sampling intervals used for core plugs are of little relevance to the actual scale of permeability variation in most formations (Allen et al. 1988). Similarly, though the physical importance of the presence of small-scale features such as shale laminae and other sedimentary structures on recovery is apparently well-known (e.g. Richardson et al. 1987), no permeability data are published which can be used to make quantitative evaluations of these effects. As shown in previous sections minipermeameter measurements can be made on a very fine-scale grid such that it is possible to make statistically significant comparisons between the permeability distributions of individual laminae in cross stratified units.

For most practical purposes it is probably unreasonable to embark on routine minipermeameter studies of cores using the sampling densities of 2mm and 5mm square grids employed in this study. It is however, appropriate to test which sampling density is suitable for a particular facies so that one can at least be aware of the statistical validity of the data and hopefully, define statistically homogeneous rock volumes which are suitable for inclusion in quantitative models (J.J.M.Lewis pers.comm. 1988). Intuitively, one would expect well-sorted "homogeneous" sandstones, such as the cross stratified lithofacies (Fig. 2, S_X), to require fewer measurements to give a statistically acceptable data base than poorly sorted "heterogeneous" sandstones, such as our heterogeneous lithofacies (Fig. 2, H). An evaluation of the optimal sampling density is made using a N_0 test (Appendix). Calculations are made both to test the validity of the grid size chosen for the experiments and to predict optimal grid sizes (measuring densities) for each sample.

It should be noted that the N_0 test assumes that the data are statistically independent. This assumption is almost certainly untrue for most lithologies, as may be inferred from the ANOVA testing, where individual sedimentary structures are shown to affect permeability distribution. A more rigorous statistical testing of appropriate sample density should take account of any natural ordering present, for example by using variogram input (Barnes 1988).

1. Analysis of Experimental Data

The N_0 testing is made by varying the range of acceptability about the average permeabilities (k) of each sample. By increasing the range one expects fewer total observations to be necessary to satisfy the confidence limits defined for the test. For example, in facies S_X, large-scale cross stratification, a 5mm sampling density is used. Both visually and from prior knowledge of the reservoir characteristics, S_X is assumed to be a very homogeneous facies. However, over half the samples require a higher density of sampling if the permeability distribution is to be constrained to within $\bar{k} \pm 5\%$. If constrained to within $\bar{k}$ $\pm 25\%$. all samples have acceptable sampling densities.

In terms of defining permeabilities for use in reservoir simulation models, $\pm 5\%$ about $\bar{k}$ is probably of little consequence when describing bulk flow properties, certainly in high-permeability
facies such as S_X. More alarming from a geological standpoint however, is that the apparently homogeneous sandstones which belong to one lithofacies give so varied results. By comparing N_0 with N_x it is apparent that ranges of acceptable sampling density vary from 0.2 to 5 times the experimental sampling densities. Even when the variation about $\bar{k}$ is increased to $\pm 10\%$, 15% of the samples still require increased sampling to attain statistical validity, although the increase of sampling to satisfy the pre-defined confidence interval is now always less than 1.5%.

No relationship is found between the average permeability of samples and N_0 (Fig. 14). However, the lithofacies which intuitively are expected to require the highest measuring densities, such as the heterogeneous facies, do so.

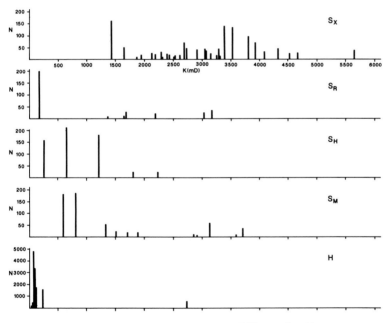

Fig. 14.   Plot of mean permeability against N_0.

2. Prediction of Optimal Sampling Density

Values of N_0 may be used to derive the dimensions of sampling grids which give optimal sampling densities for specific samples. Optimal sampling densities (d_I) for each sample are calculated for a rectangular grid from N_0 as follows:

$$d_I = \sqrt{\frac{N_x}{N_0}} \cdot d_x$$

where d_x is the size of grid used in the actual data gathering (2mm or 5mm). If the variation about $\bar{k}$ is increased, a proportional increase in the optimal dimensions of the grid size is obtained, i.e. a $\pm 20\%$ of $\bar{k}$ has a grid twice the size of $\pm 10\%$ of $\bar{k}$. The greater the size of d_I, the greater the homogeneity of a sample.

Optimal sampling densities (d_I) for the various lithofacies are summarised in Table 5. Mean values for d_I are given for each lithofacies, however, only for lithofacies S_X are there sufficient data to make any generalisations about d_I. In Table 5 two values are given for $\bar{d}_I$, A and B, where $\bar{k}$ falls within ranges of acceptability of $\pm 25\%$ and $\pm 5\%$ respectively. As to be expected for a range of acceptability of $\pm 25\%$, grids with larger dimensions provide a more satisfactory sampling of the permeability variation than for $\pm 5\%$. Similarly, the most geologically homogeneous lithofacies, S_X and S_M, require the least dense grids to give an optimal sampling.

Table 5. Mean optimal sampling densities ($\bar{d}_I$) in mm for each lithofacies, where the range of acceptability about $\bar{k}$ is, A- $\pm 25\%$,and B- $\pm 5\%$. n = number of samples.

Lithofacies	Case A			Case B			n
	min	$\bar{d}_I$	max	min	$\bar{d}_I$	max	
S_H	6.4	6.5	6.6	2.6	2.6	2.6	2
S_R	5.6	7.6	11.1	2.3	5.0	8.2	6
S_X	5.2	9.7	17.2	1.7	4.6	10.1	32
S_M	3.7	7.3	11.3	2.2	4.6	9.4	11
H				0.3	0.8	1.5	7

Of particular interest from a geologic standpoint is that within a specific lithofacies there is a large variation of d_I (Fig. 15). For the large-scale cross stratified lithofacies (S_X), which appears homogeneous, there is apparently more variation of d_I than there is difference between S_X and the other lithofacies (Table 5). This comparison cannot be given any statistical significance as too few data are available from the other lithofacies. In none of the samples can the permeability variation be characterised (with a minipermeameter) with a sampling interval of 25cm which is typical of the sampling density used for core plugs.

Another feature of variations in d_I is that orthogonal surfaces of the same sample may have quite different values (Fig. 16), e.g. samples S_X 1.02 and 1.21. The variation in d_I from side to side reflects the heterogeneities present within sedimentary structures, such as imbrication and sorting of grains, which have orientation-dependent effects on permeability. The significance of orientational facets of sedimentary structures is particularly important when evaluating the results of any directional measurements, for example interpretation of core floods or relating downhole measurements of electrical conductivity to permeability anisotropy.

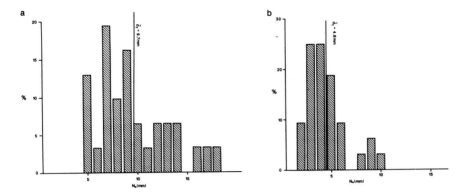

Fig. 15. Variation of optimal sampling density (d_I) in the large-scale crossstratified lithofacies (S_X). a- $\bar{k} \pm 25\%$, b- $\bar{k} \pm 5\%$.

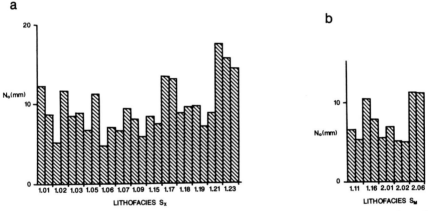

Fig. 16. Comparison of optimal sampling densities (d_I) from orthogonal sections of lithofacies (a) S_X and (b) S_M . d_I, are for a $\pm 25\%$ range of acceptab ility about $\bar{k}$.

E. Quantitative Effects of Measurement Resolution

Increasing the volume of investigation will, in heterogeneous media, always tend to smooth out any permeability variation detected. With respect to minipermeameter probes, increasing the contact area gives an increased volume of investigation. The actual relationship between contact area and volume of investigation is probably complex and strongly dependent on the permeability anisotropy of samples. Here, an experiment is conducted to investigate how use of different diameter probes affects the resolution of the minipermeameter and to quantify those effects with respect to prediction of the average permeability.

A sample of parallel laminated sandstone is examined which contains 4-12 mm thick laminae of fine-grained sand interstratified with 0.3-2 mm thick micaeous, silty laminae

Andrew Hurst and Kjell Johan Rosvoll

(Fig. 17). A 3 x 3 cm area was selected for analysis, and measurements were made on a 2mm grid using the three probes described in section II, A.

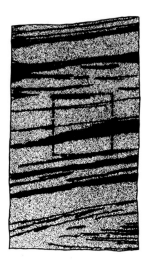

Fig. 17. Plane laminated, fine grained sandstone with 3 x 3 cm area selected for minipermeameter analysis.

All three probes were able to detect a similar permeability distribution and the location of the silty laminae correlates with zones of low permeability (Fig. 18). As expected the probe with the largest contact area, probe 1D2, gives the smoothest plot of permeability variation. However, the ranges of permeabilities measured with each probe varies considerably,

Probe 1D2   264 - 1769 mD

Probe 2   112 - 2052 mD

Probe 4   43 - 2864 mD

the summary statistics for which are given in Table 6.

Although the average permeabilities and the spatial permeability distribution are similar for each probe, the capability of the probes to detect the range of permeability present varies (Fig. 19).

The area studied is similar in size to the end of a standard 3.8cm core plug. Assuming that the sample is homogeneous, the $\bar{x}_A$'s for the area of examination are believed to approximate to a core plug permeability for the sample. Clearly, by just examining the bulk data, i.e. mean values, all impression of the the sample heterogeneity is lost. The significance of retaining the information about sample heterogeneity probably depends on how the permeability data are to be applied. Certainly, the variations among $\bar{x}_A$ for the different probes are unlikely to be significant in evaluation of the flow properties of a reservoir zone. However, the detection with probe 4 of areas of permeability almost twice as high and seven times as low as with probe 1D2 are of significance when evaluating the permeability of core samples, both with respect to interpretation of core flood experiments and with respect to the averaging of core data when defining flow units.

Table 6. Summary statistical data for the data presented in Fig. 18. $\bar{x}_A$ = arithmetic mean permeability, σ = standard deviation.

Probe	$\bar{x}_A$	σ	median
1D2	837	408	785
2	835	545	655
4	901	703	641

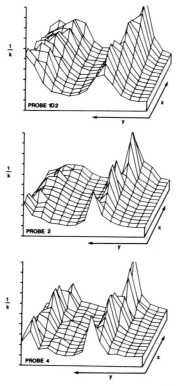

Fig. 18. 3D presentation of permeability (K mD) variations detected within the 3 x 3 cm area of the plane laminated sandstone shown in Fig. 17.

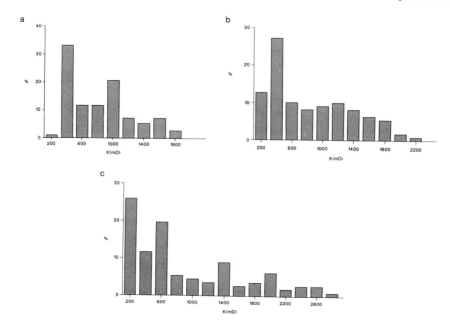

Fig. 19. Permeability distribution of the data shown in Fig. 18. a - probe 1D2, b - probe 2, c - probe 4.

F. Estimation of Mean Permeability

Estimation of mean permeability measured on core samples is inevitably done as part of the process of calibration of core, wireline log and test data when defining permeability values for grid blocks in reservoir simulation models (Dubrule and Haldorsen 1986). Application of a minipermeameter to core analysis instead of, or in addition to, core plugs will in itself give no better data for this estimation. If however, an enlightened program of minipermeameter sampling is made using the geological description of the core material as a basis for the sampling program, rather than using a routine (e.g. 25cm) vertical sampling interval, it should be possible to both test and quantify the significance of geological heterogeneities on recovery factors by evaluating different averaging techniques for each lithofacies.

A comparison of the estimation of mean permeabilities, both K_H and K_V, is made using the data from a whole core. It should be noted that, in this example, the minipermeameter data have a much lower sampling density than elsewhere in this study. Furthermore, core plugs have been taken at a high density, without any consideration of the location of possible geological features. As such, the density of core plug data is much greater with respect to the density of minipermeameter data than is normal.

Prior to minipermeameter measurement, axial (K_V) and radial (K_H) permeabilities were measured on the whole core. Following minipermeameter measurement, the core was cut into two approximately equal parts, so forming four serial sections on each of which were made 20 minipermeameter measurements. Two core plugs were taken normal to each face. Axial permeabilities are calculated from the plug and minipermeameter data by calculating the geometric mean for each face and the harmonic mean of those values. The permeabilities obtained are as follows,

$$\text{from whole core,} K_V = 100 \text{ mD,}$$

$$\text{from minipermeameter,} K_V = 155 \text{ mD,}$$

$$\text{from core plugs,} K_V = 377 \text{ mD,}$$

which in this case shows that a considerable inconsistency is present between the core plug data and the other data.

Here, the same methods for averaging the permeability data were used for both core plugs and minipermeameter data, the difference in the results being solely attributable to the actual measurements made. The minipermeameter was able to pick up more information about the presence of low permeability zones within the sample. Had the core plugs been selected with a view to sampling the full range of geological heterogeneities present, rather than maximising the total number of plugs, averaging of the plug data may have produced a result more similar to the minipermeameter and whole core data.

IV. DISCUSSION

A. Analytical Considerations

Although minipermeameters are easy and cheap to use, the actual measurement they make is difficult to interpret, either in terms of volume of investigation or in terms of the relationship between flowrate and gas pressure. No results are known from experiments designed to examine flow paths. Results from numerical simulations are published (Goggin et al. 1988), including a simulator which allows input of rock heterogeneities on a mm-scale (Daltaban et al. 1989). However, the simulations cannot include the effects of tortuosity or grain surface heterogeneities, so their value for estimating volume of investigation etc. should be treated circumspectly until experimental data are available. Input of heterogeneities into mm-sized grid cells (Daltaban et al. 1989) may improve our understanding of flow patterns in heterog eneous media, but assignment of realistic transmissibilites in x, y and z directions fo r such small grid cells presents severe problems, i.e. one must know more about the rock on a mm-scale than is tenable experimentaly.

On a more encouraging note, the excellent correlations obtained between minipermeameter measurements and homogeneous core plugs (Halvorsen and Hurst, 1990) at least allow application of minipermeameters in reservoir characterisation to continue in the knowledge that the individual data are at least as good as plug data, and more representative. Certainly, the advantages of being able to obtain statistically meaningful data sets with a minipermeameter allow quantitative evaluation of, for example, the significance of geological characteristics and optimal sampling densities for different formations, advantages which should not be ignored in reservoir studies.

B. Quantitative Aspects of Sedimentary Facies Analysis

Ultimately, the value of sedimentary facies analysis of cores is judged by the extent to which it makes a direct contribution to the definition of both in-place and recoverable reserves. On a well-to-well scale, correlation and mapping of sedimentological trends has become an accepted and valuable method applied in reservoir zonation. On a well-scale sedimentological data are probably less well tried. Sedimentological data may be complementary in several areas including special core studies, geological applications of wireline logs (electrofacies), definition of "h", and thus interpretation of permeability, in the interpretation of well-test data, etc. (Hurst 1987). Major weaknesses with all sedimentological data are the difficulty of assigning values to the range of observations made and the lack of continuous measurements. A lack of continuous quantitative measurements, such as grainsize or sorting, stems from the manual methods employed by geologists for core description.

If the quantitative significance of sedimentary features on flow characteristics can be proven by, for example, minipermeameter measurement, then application of sedimentological data to the evaluation of reservoir flow characteristics becomes a reality. Similar to Pryor (1973), but on on finer scale, we have shown that the origin of permeability contrast can be identified in different sedimentary structures; almost always more contrast is present within individual structures, laminae, sets or even randomly assigned parts, than between the individual structures. Similarly, the presence of permeability anisotropy is readily identified and its relationship to sedimentary features illustrated.

As the physical conditions for the formation of sedimentary structures are well-known (Allen 1984) the geological interpretation of sand deposition can be given significance in terms of permeability distribution. Cross stratification is interpreted to have formed by migration of large-scale bedforms, such as megaripples, and has a degree of cyclicity which tends to form approximately parallel laminae. Each lamina represents a similar sedimentary event that records a change in the hydraulic conditions during deposition, assuming that the supply and characterisitics of the detritus are approximately constant. In response to changing hydraulic conditions the sorting and grain size within a lamina vary in vertical section; each lamina may approximate to a small fining-upwards sequence. Implicit in the sedimentological reasoning, and confirmed by the ANOVA analysis, is that the petrophysical characteristics of each lamina are expected to vary in vertical section, whereas some degree of similarity is to be expected between adjacent laminae.

Cross stratification is here divided in two different lithofacies which are size -dependent, S_X and S_R, large- and small-scale cross stratification respectively. As cross bedding is not unique to one depositional environment, and size of the structures is not environmentally significant, one cannot expect all cross beds to have similar permeability characteristics. Additional information about the environment of deposition would be helpful when interpreting the permeability distributions. In tidal channel facies we have observed that cross stratified intervals have more uniform permeability distributions than in fluvial facies. This is consistent with observations made on modern sand bodies using core plugs (Pryor 1973), and reflects a difference of sorting in the two environments.

None of the sandstones examined in this study have undergone more than very local cementation, for example the presence of isolated carbonate nodules. Thus, the permeability variations are interpreted to reflect primary depositional features. In sandstones where the effects of diagenesis have reduced porosity and changed the saturation characteristics by growth and dissolution of minerals, there need not be such clear-cut relationships between sedimentary structures and permeability variations. Additional data are needed to test to what extent sedimentary structures determine permeability distribution independent of diagenetic effects.

Minipermeameters or Core Plugs

Minipermeameter measurements should not be seen as a replacement for core plugs. Indeed, the two forms of permeability measurement are complementary; calibration of minipermeameters is made using core plugs! Clearly, any choice between the two methods should depend on the eventual application of the data. Core plugs, because of their size, never reveal a comparable amount of information about permeability variation as a minipermeameter (Halvorsen and Hurst, 1990). The results of the N_0 testing of optimal sampling densities for different sedimentary facies show that, even in geologically homogeneous lithofacies, high-density sampling with a minipermeameter is required to give a statistically valid sampling of the permeability variation. Core plugs of course have a much greater volume of investigation than even the largest diameter probe (1D2) used in this study. Therefore, they will tend to smooth out the presence of many permeability heterogeneities detected by the minipermeameter and consequently, fewer core plugs will be needed to give a statistically valid sampling of the observed permeability variation.

On the scale of laboratory measurements on cores, the minipermeameter has many advantages over core plugs. As minipermeameter measurement is basically non-destructive, the limited volumes of core material available will escape the destruction caused by plugging. Current routines for core plugging can be readily modified so that any 1 in. diameter plugs can be taken after minipermeameter measurement is completed. More relevant in terms of comparison with special core analysis is that minipermeameters can gather data on a scale similar to the scal e of investigation in core experiments (Giordano, et al. 1985). Thus, permeability ma ps be constructed from samples examined by CT-scanning or other tomographic methods (Halvorsen and Hurst, 1990). If a goal of the collection of permeability data is to tie together wireline log data, sedimentological data and sub-surface pressure data, the density of data available from minipermeameters is very attractive, if only in terms of providing a statistically meaningful data set for calculation of mean permeabilities.

Although the example of identification of permeability anisotropy is hardly exhaustive, it does show the ease with which experiments can be made on whole, or part, cores to examine orientational heterogeneity. Plug measurements are of course ultimately required to confirm directional trends revealed by the minipermeameter, but plugs alone will never produce a sufficiently dense data set to make the initial identification of anisotropy at this scale of investigation.

Perhaps most significant with respect to optimisation of core measurement programs is the possibility of defining optimal sampling densities for different intervals of core, based on a traditional sedimentological description and minipermeameter measurements. The confidence intervals required in practice may be considerably wider than 95% with a range of acceptability for $\bar{k}$ of $\pm$ 25% as used in this study and will depend on how the data are to be applied. An approach for description of permeability variation in cores is to select several "type" lithofacies defined from the sedimentary description of the core and, with minipermeameter measurements, to evaluate the permeability variance present in each lithofacies. Thus, permeability variation is quantified and statistically valid uncertainties may be assigned when defining flow units.

Definition of flow units requires quantification of both the continuity of each unit and their internal organisation. With respect to the latter, the permeability distribution is of critical importance when evaluating the transmissibilities used in simulation grid cells. To what extent flow units can be considered to be statistically homogeneous rock volumes is largely a question of at which scale the data are gathered. "Homogeneous" will nevertheless be expressed as mean values with associated variances. From the results of the N_0 testing it seems unlikely that present-day routines gather sufficient data to allow statistically valid evaluations of permeability variation and distribution. Increasing the density of core plugs may improve this situation, but this would probably cause a more-or-less total destruction of the core, which is unlikely to be condoned as good practise in core analysis. It seems appropriate to conduct N_0 tests on presently available plug data from different lithofacies if only to quantify the statistical validity of the data.

V. CONCLUSIONS

Laboratory minipermeametry can provide non-destructive permeability measurements with comparable precision to core plugs, at a sampling density which allows statistically valid characterisation of the permeability distribution of sedimentary structures. The main source of the permeability variation in sandstones is confirmed to lie within sedimentary structures.

Permeability anisotropy is readily identified on both slabbed and unslabbed cores, and its directional significance is quantifiable with core plug measurements.

Determination of optimal sampling densities for resolving the permeability variation of various lithofacies shows that even the most geologically homogeneous rocks need sampling densities of approximately 10cm. One can probably never make a sufficiently dense sampling of the most heterogeneous lithofacies with a minipermeameter. Increasing volume of investigation, either by increasing contact area or by using core plugs, smooths out the presence of any permeability heterogeneities present. If the volume of investigation is increased lower sampling densities are required to characterise permeability variation, but the capability for identifying both high and low permeable laminae diminishes.

Varying the resolution of minipermeameter measurements has little effect on resultant mean permeabilities. However, the measured heterogeneity of samples decreases with decreased resolution thereby losing information about the heterogeneity.

VI. ACKNOWLEDGMENTS

Wojtek Nemec is thanked for his advice, criticism and enthusiasm. Erling Siring gave valuable support and wrote the programs for the statistical analysis. Adolfo Henriquez, Jerry Jensen and John Lewis are thanked for reading and criticising a preliminary version of the manuscript. Den norske stats oljeselskap a.s. (Statoil) is acknowledged for supporting and encouraging the publication of the paper.

VII. REFERENCES

Allen,D., Coates,G., Ayoub,J. et al. (1988). Probing for permeability: an introduction to measurements. The Technical Review 36, 6-20.

Allen,J.R.L. (1984) Sedimentary structures - their character and physical basis. Developments in Sedimentology 30, Elsevier, 663p.

Barnes,R.J. (1988) Bounding the required sample size for geologic site characterization. Mathematical Geology 20, 477-490.

Bjørlykke,K., Aagaard,P., Dypvik,H., Hastings,D.S., and Harper,A.S. (1986). Diagenesis and reservoir properties of Jurassic sandstones from the Haltenbanken area, offshore mid Norway. In "Habitat of Hydrocarbons on the Norwegian Continental Shelf" (A.M.Spencer et al., eds.), p.275-286. Graham & Trottman, London.

Cadman,M. (1984). Non-destructive permeability measurement. M.Eng. thesis Heriot-Watt University (unpublished).

Dalland,A. Worsley,D. and Ofstad,K. (1988). A lithostratigraphic scheme for the Mesozoic and Cenozoic succession offshore mid- and northern Norway. Norwegian Petroleum Direktorat Bulletin No.4.

Daltaban,T.S., Lewis,J.J.M. and Archer,J.S. (1989). Field minipermeameter measurements - their collection and interpretation. 2nd. International Conference on EOR, Budapest.

Davis,J.C. (1986). Statistics and data analysis in geology (2nd. edition). John Wiley & Sons Inc., 646 p.

Dubrule,O., and Haldorsen,H.H. (1986). Geostatistics for permeability estimation. In "Reservoir Characterisation" (L.W.Lake and H.B.Carroll,Jr. , eds.), p.223-247. Academic Press, New York.

Eijpe,R., and Weber,K.J. (1971). Mini-permeameters for consolidated and unconsolidated sands. Amer.Assoc.Petrol.Geol.Bull. 55, 307-309.

Giordano,R.M., Salter,S.J. and Mohanty,K.K. (1985) The effects of permeability variations on flow in porous media. SPE 14365.

Goggin,D.J., Chandler,M.A., Kocurek,G.A. and Lake,L.W. (1986) Patterns of permeability in eolian deposits. SPE/DOE 14893.

Goggin,D.J., Thrasher,R.L. and Lake,L.W. (1988). A theoretical and experimental analysis of minipermeameter response including gas-slippage and high-velocity flow effects. In Situ 12, 79-116.

Haldorsen,H.H. (1986). Reservoir simulator parameter assignment and the problem of scale in reservoir engineering. *In* "Reservoir Characterisation" (L.W.Lake and H.B.Carroll Jr., eds.), p.293-340, Academic Press, New York.

Halvorsen,C. and Hurst,A. (1990). Principles, practice and applications of laboratory minipermeametry. *In* "Advances in core evaluati on: accuracy and precision in reserves estimation," (EUROCAS I) (P.F.Worthington, ed.). Gordon and Breach, London (in press).

Havlena,D. (1966). Interpretation, averaging and use of the basic geological - engineering data. J.Canadian Pet.Tech. 5, 153-164.

Hurst,A. (1987). Problems of reservoir characterisation in some North Sea sandstone reservoirs solved by the application of microscale geological data. *In* "North Sea Oil and Gas Reservoirs" (J.Kleppe, E.W.Berg, A.T.Buller, O.Hjelmeland and O.Torsæther, eds.), p.153.167, Graham & Trottman.

Jones,J.R.Jr., Scott,A.J., and Lake,L.W. (1984). Reservoir characterisation for numerical simulation of Mesaverde meanderbelt sandstone, northwestern Colorado. SPE 13052

Lasseter,T.J., Waggoner,J.R. and Lake,L.W. (1986). Reservoir heterogeneities and their influence on ultimate recovery. *In* "Reservoir Characterisation" (L.W.Lake and H.B.Carroll,Jr., eds.), p.545-559. Academic Press, New York.

Lewis,J.J.M. (1988). Outcrop-derived quantitative models of permeability heterogeneity for genetically different sandbodies. SPE 18153, p.449-463.

Lewis,J.J.M, Hurst,A. and Lowden,B. (1990). Permeability distribution and measurement of reservoir-scale sedimentary heterogeneities in the Lochaline Sandstone (Cretaceous). Field Guide, 13th International Sedimentological Congress, Nottingham, U.K., 26-31 August, 1990, (in preparation).

Mast,R.F. and Potter,P.E. (1963). Sedimentary structures, sand-shape fabrics, and permeability, pt.2. Jour.Geology 71, 548-565.

Pryor,W.A. (1973). Permeability-porosity patterns and variations in some Holocene sand bodies. AAPG Bull. 57, 162-189.

Pryor,W.A. and Fulton,K. (1978). Geometry of reservoir-type sandbodies in the Holocene Rio Grande delta and comparison with ancient reservoir analogs. SPE 7045.

Richardson,J.G., Sangree,J.B., and Sneider,R.M. (1987). Permeability distributions in reservoirs. J.Pet.Tech. 39, 1197-1199.

Rosvoll,K.J. (1989). Small-scale permeability variation in reservoir sandstones: laboratory analysis and quantitative evaluation. Can.Scient. thesis, Univ.Bergen (in Norwegian).

Stalkup,F.I. (1986). Permeability variations observed at the faces of crossstratified sandstone outcrops. *In* "Reservoir Characterisation" (L.W.Lake and H.B.Carroll,Jr., eds.), p.141-179. Academic Press, New York.

Stalkup,F.I. and Ebanks,W.J.Jr. (1986). Permeability variation in a sandstone barrier island - tidal channel - tidal delta complex, Ferron Sandstone (Lower Cretaceous), Central Utah. SPE 15532

Weber,K.J. (1982). Influence of common sedimentary structures on fluid flow in reservoir models. J.Pet.Tech. 44, 665-672.

Weber,K.J. (1986). How heterogeneity affects oil recovery. In "Reservoir Characterisation" (L.W.Lake and H.B.Carroll,Jr., eds.), p.487-544. Academic Press, New York.

Weber,K.J., Eijpe,R., Leynse,D. and Moens,C. (1972). Permeability distribution in a Holocene distributary channel-fill near Leerdam (the Netherlands) - permeability measurements and in situ fluid-flow experiments. Geologie en Mijnbouw 51, 53-62.

VI. APPENDIX

N_0 Testing

If one assumes a standard error in the calculated average (tolerance level) and a chosen significance level, an N_0 test is designed to calculate the optimal number of data points required to satisfy these predetermined levels of tolerance and significance.

On each sandstone sample a number (not predetermined) of minipermeameter measurements (N_x) are made on a measuring grid with either a 2mm or 5mm density. Estimation of how many measurements (N_0) are required to have a mean value (as calculated from N_x), with chosen tolerance and significance levels is done as follows: let $\bar{k}$ be the calculated mean permeability (e.g. geometric mean) of a sample, σ is the standard deviation of the "infinite" population (of permeability measurements with "true" mean value, μ), N_x is the number of measurements made, and S is the calculated standard deviation of the N_x permeabilities. $\bar{k}$ is normalised by dividing by $\dfrac{S}{\sqrt{N_x}}$. It is assumed that,

$\dfrac{\bar{k}\sqrt{N_x}}{S}$ has a t-distribution where $N_{0-1,0.975}$ is the number of degrees of freedom (Wonnacott and Wonnacott 1977).

The aim is to find how many observations are required such that $\bar{k} \pm p$ percent can fulfill the predetermined 95% confidence interval. It is assumed that the t-statistics are based on N_0 obsevations rather than N_x, such that the following relationship is satisfied, $\dfrac{\sqrt{N_x}\,kp}{S100}$ be equal to half of the length of the confidence interval such that,

$$\frac{\sqrt{N_0}\,kp}{S100} \simeq t_{N0-1,0.975} \tag{1}$$

When N_0 is large, $t_{N0-1,0.975} \simeq \Phi_{0.975}$, where $\Phi = 0.975$ quantile in the normal distribution $(0,1)$. $\Phi_{0.975} \simeq 2$ therefore $t_{0-1,0.975}$ can be assumed to equal 2. Thus,

$$\frac{\sqrt{N_0}\,\overline{k}p}{S100} = 2 \tag{2}$$

which implies that,

$$N_0 = \left[\, \frac{2V \text{ sub } k100}{p} \,\right]^2 \tag{3}$$

where $V_k = S/\overline{k}$. V_k is the coefficient of variation and is a measure of the spread present in the distribution.

The resultant value N_0 gives a direct indication of whether too many or too few measurements were made for any one sample and, provides a measure of the heterogeneity of samples for a specific grid size.

Reference
Wonnacott,T.H. and Wonnacott,R.J. (1977). Introductory Statistics (3rd edition). John Wiley & Sons, New York, 650p.

GENERATION OF EFFECTIVE RELATIVE PERMEABILITIES FROM DETAILED SIMULATION OF FLOW IN HETEROGENEOUS POROUS MEDIA

A.H.Muggeridge

BP Research, Sunbury Research Centre,
Sunbury-on-Thames, Middx TW16 7LN
ENGLAND

ABSTRACT

The amount of oil recovered from waterflooding an oil reservoir can be significantly altered by variations in the reservoir rock permeability. However the typical grid block dimensions used in conventional reservoir simulation cannot generally resolve the details of the permeability variations and hence cannot accurately model the displacement. Instead the average properties of flow are represented by replacing the rock relative permeabilities with pseudo functions.

Very often a successive scaling up procedure has to be applied in order to represent the many scales of permeability variation present in a complex heterogeneous system. In this work, the detailed simulation methods described by Christie (1989) are used to study how well pseudo functions actually represent the average properties of fluid flow through heterogeneous porous media. This is achieved by examining the properties and variability of pseudo functions generated to represent flow through three different types of permeability distribution, namely uncorrelated random, highly correlated and a braided stream environment typical of a channel sand.

This paper demonstrates the importance of defining appropriate reservoir volume elements and of relating this to the correlation length of the permeability distribution, so that the pseudo functions accurately represent the detailed fluid flow through the reservoir. It is necessary to select the dimensions of the coarse grid so that each block contains a representative element of the permeability distribution.

1. INTRODUCTION

The recovery of oil can be significantly altered by the presence of geological heterogeneities in the reservoir rock. In waterflooding, spatial variations in absolute and relative permeabilities may mean the water finds preferred flow paths, resulting in an early breakthrough and a reduction in the amount of oil contacted.

The typical grid block dimensions used in conventional reservoir simulation cannot generally resolve the details of the permeability variations and hence cannot accurately model the displacement. One way of addressing this problem is to alter the input data used by the reservoir model so that simulations reproduce the average fluid flow observed in the heterogeneous reservoir.

The most widely used method of representing the effects of heterogeneities is to replace the measured rock relative permeabilities in a reservoir model with pseudo relative permeabilities (Haldorsen (1986), Lasseter (1986), Kossack et al. (1989), Smith (1989)). These pseudo functions are chosen so that they reproduce the altered fluid flow in a coarse-grid model. An alternative approach, proposed by White (1987), uses a tensor formulation for coarse-grid transmissibilities to model the altered fluid flow. This method appears to be successful in modelling the average fluid flow, but does not attempt to compensate for the increased levels of numerical diffusion present in the coarse-grid model.

Pseudo functions can be derived analytically for flow through layered reservoirs (Pande et al. (1987), Reznik et al. (1984), Tompang and Kelkar (1988)) but for more complex permeability distributions they have to be generated numerically by matching coarse-grid and fine-grid simulations (Kyte and Berry (1975)). The problem with generating pseudos by this method is that most simulation codes cannot use the very fine grids required to represent all scales of heterogeneity.

A successive scaling up procedure has to be applied in order to represent all scales of permeability variation in a complex heterogeneous system. A fine-grid simulation is used to generate pseudos to represent flow through a 'typical' section of a permeability distribution with variations on the scale of inches. This small-scale model is then replaced by a single grid block in a model representing permeability variations on the medium scale. The average properties of flow on the small scale are represented by pseudo functions. Again simulation is used to generate pseudos to represent the flow on the medium scale. These pseudos can then be used to represent the average flow in a single grid block in a reservoir model representing the largest scale of permeability variation. This approach has been used by Lasseter *et al.* (1986) and Kossack *et al.* (1989) to enable them to investigate how different types of permeability distribution altered the fluid flow.

The problem with this approach is that, in the absence of suitable fine-grid simulation techniques, it is not possible to check how accurately the pseudos actually reproduce the average properties of the real displacement. The aim of this paper is to use the detailed simulation methods described by Christie (1989) to study how well effective relative permeabilities actually represent the average properties of fluid flow through heterogeneous porous media. This is achieved by examining the properties and variability of effective curves generated to represent flow through three different types of permeability distribution.

2. METHOD

The problem with most numerical methods for generating pseudo functions is that they seek to represent a number of different effects which makes it difficult to determine how the functions depend upon any one variable, such as the permeability distribution. Clearly the number of variables represented can be reduced by eliminating physical effects, such as gravity and capillary pressure.

The pseudoization process also compensates for the increased levels of numerical diffusion present in the coarse-grid model. The pseudo functions will depend upon the relative sizes of the fine-grid model used in their generation and the final coarse-grid model on which they will be applied. However the effect of permeability variations on the pseudo functions may also change with the relative sizes of the coarse- and fine-grid models. For example, suppose we want to represent flow through a highly correlated permeability distribution using pseudo functions. Changing the averaging volume from one in which the dimensions are of the order of one correlation length to one in which the dimensions are perhaps two or three correlation lengths will alter the mean fluid distribution and hence the pseudos.

In this paper we have used a multi-stage approach for generating pseudo relative permeabilities which enables us to separate the effects of permeability variations from those of numerical diffusion in the coarse-grid model. We use *effective* relative permeabilities to represent the effects of heterogeneities and then generate pseudo functions which compensate for numerical diffusion as well. The effective relative permeabilities are derived from simulations in the same way that rock curves are derived from displacement experiments in the laboratory. We can then use these effective functions to represent a displacement through a two-dimensional, heterogeneous system on a one-dimensional, homogeneous, fine-grid model. This is equivalent to generating pseudo functions for a one grid-block model without having to compensate for numerical diffusion. The method has the advantage that the effective relative permeabilities can then be compared directly with the original rock curves.

2.1 High-Resolution Simulation

High-resolution reservoir simulation was used to understand the effects of the permeability variations on the detailed flow through the different permeability distributions and to generate effective relative permeabilities which represented the average properties of the flow. The numerical methods used in the simulation program have been fully described by Christie (1989) and will not be repeated here. The program was originally developed to model viscous fingering in miscible displacements, but the combined requirements of both speed and accuracy necessary to model this type of flow apply equally well to the simulation of immiscible flow in heterogeneous media. Conventional black oil simulators would not have been sufficiently accurate to model the flow through permeability distributions containing small-scale variations and extreme permeability contrasts.

We have verified that the code is accurate when modelling immiscible displacements by comparing its results with analytical solutions (for simplified cases) and with the predictions of a commercially available black oil simulator.

2.2 A Method of Generating Pseudo Relative Permeabilities

We used a three-stage approach to generate pseudo relative permeabilities from simulations of a line drive through different permeability distributions. This enabled us to separate the effects of the permeability distribution and the averaging volume on the pseudo functions from those of numerical diffusion. We applied this method to each distribution that we wished to represent by a single homogenous grid block. We illustrate the technique in figure 1, where we show how to reduce a two-dimensional fine-grid model to either a single grid block or a row of four grid blocks.

The first stage was the computation of the effective absolute permeability for single-phase flow using the pressure-solver method described by Begg, Carter and Dranfield (1987). This technique uses a finite difference solution of the pressure equation to calculate the effective permeability. We used this method to ensure that the two-phase displacement would satisfy the correct limits for single-phase flow. The average absolute permeabilities calculated by the Kyte and Berry pseudoization method do not necessarily have this property as they are obtained from simple harmonic weighting.

The second stage used the method of Jones and Roszelle (1978) to compute effective relative permeabilities. This method is generally used to derive rock relative permeability curves from displacement experiments through cores. It is used for dynamic rather than steady-state displacements and is based on the analytical method described by Johnson, Bossler and Naumann (1959). A fractional flow of one at the inlet is assumed. The method uses the pressure drop along the model, the flow rate through the model and the volume of oil recov-

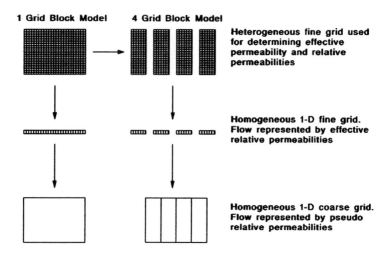

Figure 1:   Procedure for reducing two-dimensional fine-grid to one-dimensional coarse grid.

ered as a function of the volume of water injected to derive relative permeability curves which describe a continuous saturation distribution along the model.

The effective relative permeabilities and effective absolute permeabilities generated during these two steps were then used to represent the flow through the heterogeneous, two-dimensional, permeability distribution on a one-dimensional, homogeneous, fine-grid model.

The third stage was to apply the pseudoization technique described by Kyte and Berry to this one-dimensional, fine-grid model to give pseudo functions which could then be used in a coarse-grid simulation. These incorporate the effects of the permeability variations and compensate for numerical diffusion. They describe the change in mean saturation in each coarse grid block rather than the saturation distribution between the inlet and outlet faces of the grid block. This third stage is not use in this work.

2.3 Effective Relative Permeabilities

We generated the effective relative permeabilities from a detailed simulation of a constant-rate line drive through the chosen permeability distribution, maintaining a constant pressure along the outlet face. The volume of water injected, the volume of oil produced, the injection rate and the mean pressure drop across the model were recorded as a function of time. We then combined these data using the method of Jones and Roszelle (1978) to generate the *effective* relative permeabilities for the displacement. These curves are not the same as the *pseudo* functions used by Lasseter *et al.* and Kossack *et al.* They only represent the effects of the heterogeneities on the average fluid flow. They cannot be used directly in a coarse-grid model as they do not compensate for the increased levels of numerical diffusion present in such a model.

The shape of the effective curves was determined exactly for all water saturations up to about 0.6. This saturation corresponded to having injected 2 pore volumes of water into the model. The shape of the curves between this saturation and the endpoint was obtained by linear interpolation. This did not significantly affect the results as the amount of oil displaced towards the end of a waterflood is small.

The results of the detailed simulations gave us an understanding of how the fluid flow was altered by the presence of heterogeneities. We compared the results of simulations through different permeability distributions to discover the way that the properties of the permeability distribution (such as correlation length) altered the fluid flow.

We used the effective relative permeabilities that we generated from the detailed simulations to model the average flow through the heterogeneous systems. These effective curves allowed us to represent the flow through the two-dimensional, heterogeneous permeability distribution on a one-dimensional, homogeneous, fine-grid model.

We compared the results of these simulations with the original simulations in which we modelled the detailed fluid flow. This enabled us to determine how accurately the effective curves reproduced the average fluid flow.

In a successive scaling-up procedure, fine-grid simulation is used to generate pseudos to represent flow through a 'typical' section of a permeability distribution. These pseudos are then used in a coarse-grid simulation in which the flow through each grid block represents the average flow through that section of the permeability distribution.

It is important to determine whether the accuracy of these pseudo functions is affected by the properties of the section of the permeability distribution from which they were generated. It is also important to determine whether the pseudos can be used to model flow in displacements different from the one in which they generated.

We investigated how accurately a scaling-up procedure might reproduce the average flow through a heterogeneous reservoir by generating effective relative permeabilities for the whole of a heterogenous, permeability distribution and also for sections of that distribution. We chose to divide the distributions longitudinally into four sections, giving an inlet and an outlet section and two central sections (see figure 1). We then simulated the flow through the permeability distribution using:

a)   a homogeneous, one-dimensional, fine-grid simulation using the effective permeabilty and effective relative permeability curves for the whole distribution.

b)   a one-dimensional model in which the flow through each section of the permeability distribution was represented by the effective absolute and relative permeabilities corresponding to each section.

We compared these results with those obtained from the original fine-grid simulations in order to determine how accurately the effective properties represented the mean flow through the permeability distribution.

3. PERMEABILITY DISTRIBUTIONS

We modelled the flow through three different permeability distributions, each with a different correlation length. Each distribution was modelled on a 100×100 grid with an overall system length $1\frac{1}{2}$ times the width.

The first model reservoir consisted of an uncorrelated, log-normal permeability distribution (figure 2). A random number generator was used to give the permeability in each grid block. The distribution was isotropic with a geometric mean permeability of 100mD and a variance in $\log_e K$ of 20%. This corresponds to a standard deviation of about 50mD.

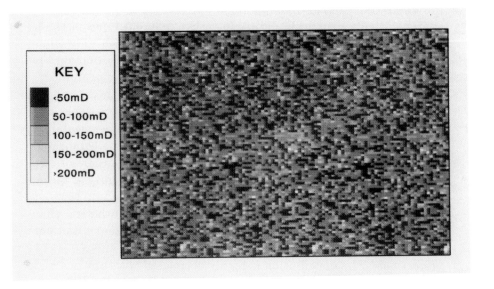

Figure 2: Log-normal, uncorrelated permeability distribution

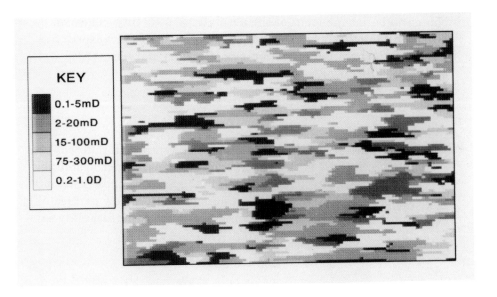

Figure 3: Sandbody permeability distribution

The second model reservoir (figure 3) was generated to represent a sandbody distribution containing five different types of sand, each with a different permeability and correlation length. The highest permeability sand has a correlation length in the x-direction that approaches the length of the model, while the lower permeability sands have correlation lengths of the order of one-eighth the reservoir length. The y-direction correlation lengths of all the rock types are of the order of one-tenth the width of the model. The permeability distribution within each sand-type was isotropic and the total range of permeabilities in the whole model covers some four orders of magnitude. This compares with a standard deviation of 52mD in the first model.

The third distribution (figure 4) represents a hypothetical channel sand made up of two rock types, each with a different mean permeability and ratio of permeabilities to flow in the x- and y-directions. The distribution is highly correlated with obvious high permeability channels through the reservoir, although the difference in permeabilities (500mD in the high permeability sand and 50mD in the low permeability sand) between the two sands is much less than that in the sandbody distribution.

In the first two models the rock relative permeability curves were chosen to be independent of the absolute permeability. They were based on those observed in high permeability sands in a North Sea reservoir. Using a power law parameterization and normalized water saturations they can be written as,

$$k_{rw} = 0.45 S_w^2 \tag{1}$$
$$k_{ro} = \left(1 - S_w\right)^5 \tag{2}$$

In the third model there were two rock types with different permeabilities and these were assigned different relative permeabilities. The high permeability rock curves were the same as those used in the first two permeability distributions(eqs. (1,2)). The low permeability rock curves were based on those observed in a low permeability sand from the same North Sea reservoir. They are given by,

$$k_{rw} = 0.3 S_w^3 \tag{3}$$
$$k_{ro} = \left(1 - S_w\right)^3 \tag{4}$$

We assumed that the critical water saturations and residual oil saturations were constant throughout each reservoir in order to simplify the problem. In practice different rock types would also have

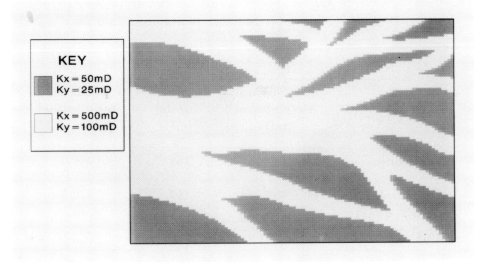

Figure 4: Channel sand permeability distribution

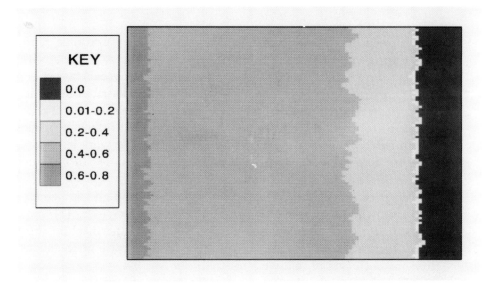

Figure 5:   Saturation distribution in log-normal permeability distribution at 0.4 (PVI) pore volumes injected.

Table I: Effective permeabilities for different models.

	Log-normal distribution (mD)	Sandbody model (mD)	Channel sand model (mD)
Whole	99.4	134.	249.5
1st quarter	96.9	152.	310.5
2nd quarter	96.7	141.	161.
3rd quarter	96.7	161.	149.
4th quarter	97.3	149.	197.6
Harmonic mean	96.9	150.	260.1

different critical water saturations and residual oil saturations. Similarly there should be different capillary pressure curves in rocks with different permeabilities, but again as we wished to concentrate on how the pseudo curves changed with permeability distribution we ignored this variation and assumed that capillary pressure was negligible. However capillary effects may have a significant influence on the flow in heterogeneous systems (Smith 1989). Further work is required to assess their impact. A constant porosity was assumed throughout each permeability distribution.

We used a constant oil/water viscosity ratio of 6 giving a shock front saturation for the high permeability rock curves (eqs. (1) and (2)) of 0.38. The shock front saturation for the low permeability rock curves (eqs. (3) and (4)) was 0.6 and the mobility ratio at the shock front for both sets of relative permeabilities was 0.4.

The x-direction effective permeabilities for each of the subsections obtained when each model is divided into four longitudinally, as shown in figure 1, are given in Table I. The effective permeabilities of each of the four regions are virtually identical in the random distribution and are very similar in the sandbody distribution, but there is quite a large difference between the regions taken from the channel sand distribution. An examination of the models (see figures 2-4) shows that for the random and sandbody distributions the structure of the distribution in each section is very similar, but in the channel sand the amount of high and low permeability rock changes noticeably between sections.

Table I also shows that the harmonic mean of the effective permeabilities calculated for each section is very close to the effective value for the whole model in the random distribution; however for the sandbody and channel sand distributions there is a significant difference between the two values. The harmonic mean permeability will only be the same as the effective permeability when the flow is one-dimensional through a series of different permeabilities (cf. an electrical current through a series of resistors).

The difference between the harmonic mean and the effective permeability for the whole sandbody and channel sand models means that the detailed flow is only approximately one-dimensional. The effective absolute permeabilities were derived from the mean flow rate through the permeability distribution and the pressure drop across it. This forces the flow at the outlet to be perpendicular to the outlet face. However when the effective permeability is derived for the whole distribution, the flow within that distribution may not be perpendicular to the boundaries between the sub-sections that make up that distribution. White (1987) has attempted to address this problem by using tensor effective permeabilities.

4. RESULTS
4.1 Random Permeability Distribution

Figure 5 shows the water saturation distribution predicted by high-resolution simulation after 0.4 pore volumes of water have been injected into the random permeability distribution. Some small-scale fingering of the water into the oil is evident.

Comparing the oil recovery with that predicted for the same displacement through a homogeneous reservoir (figure 6), we can see that the overall flow pattern is scarcely altered. There is a slightly earlier water breakthrough, but otherwise the permeability variation has very little effect on the recovery process.

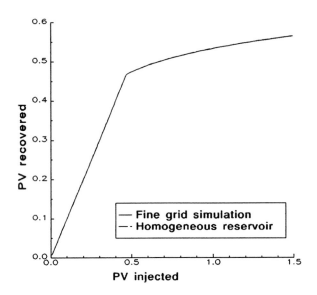

Figure 6:   Oil recovery curves for log-normal permeability distribution.

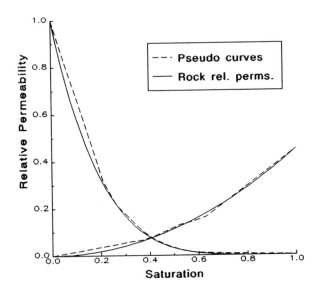

Figure 7:   Effective relative permeability curves for log-normal permeability distribution

The effective relative permeability curves that we generated for this displacement are compared with the rock curves in figure 7. They are virtually identical. Differences arise for saturations lower than the shock front saturation, as there is insufficient information in a single displacement to determine the correct shape for the curves below this point. There is also no information concerning the shape of the curves for saturations greater than 0.5, as the displacement was only simulated until 2 pore volumes had been injected.

The uncorrelated, log-normal, permeability distribution does not significantly alter the flow compared with that observed in a homogeneous reservoir with the same effective permeability. This is still true even when the variance in $\log_E K$ is increased to 200%. The oil-water interface is perturbed by the permeability variations, but these perturbations do not grow because the correlation length of the distribution is small and the shock front mobility ratio is less than 1.

4.2 Sandbody Distribution

The fingering of water through the higher permeability flow channels in the sandbody distribution is evident in figure 8, which shows the saturation distribution obtained at 0.4 pore volumes injected. These preferred flow channels are a result of the larger correlation length of this distribution and the more extreme permeability contrasts.

Comparison of the oil recoveries obtained from the heterogeneous model with those obtained in a homogeneous model (figure 9) show the effect of these flow channels. There is an earlier water breakthrough and a lower overall oil recovery when waterflooding the sandbody distribution.

The average saturation profiles observed in the sandbody model at different times are compared in figure 10. The distance moved by each saturation has been normalized so that the area under each average saturation profile is the same. This enables us to examine how the shape of the profile changes with time. The profile is very spread out when compared with that predicted for a homogeneous reservoir, as the water is tending to finger through the high permeability paths in the rock. The profile shape is essentially constant with time – there are some small variations but each saturation moves with approximately constant speed. This suggests that the average motion

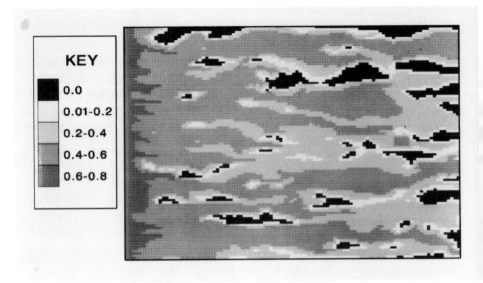

Figure 8:   Saturation distribution in sandbody model at 0.4 PVI.

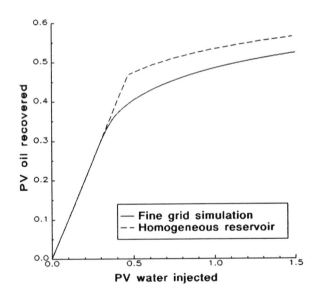

Figure 9: Oil recovery curves for sandbody model

can be satisfactorily matched with a fractional flow model, *i. e.* a hyperbolic first-order equation.

Figure 11 shows the effective water and oil relative permeability curves calculated for the whole sandbody distribution. The oil curve is remarkably similar to the rock curves, considering the large difference in the mean saturation profiles for homogeneous and heterogeneous media seen in figure 10. The largest difference occurs between the rock and effective curves for water. The effective water relative permeability is greater at all saturations for which the curves were calculated.

We compare the effective relative permeability curves calculated for each of the four subsections of the reservoir in figure 12. They are very similar. This reflects the small differences in effective absolute permeability that were observed in Table I.

The oil recovery curves obtained from one-dimensional, fine-grid simulations using *a)* a single homogeneous region and *b)* four regions representing the four subsections of the reservoir are compared with the results of the two-dimensional, heterogeneous simulation in figure 13. The results agree very well although there is a small underprediction of oil recovery when the reservoir is divided into four regions.

4.3 Channel Sand Distribution

The saturation distribution predicted by fine-grid simulation of a waterflood through the channel sand is shown in figure 14. The effect of the high permeability channels on the displacement is even more obvious than in the simulation through the sandbody distribution. The patches of higher water saturation in the picture show the buildup of water in the lower permeability sands. This buildup occurs because of the different rock relative permeability curves that we used in the high and low permeability sands. There is a higher shock front saturation moving with a lower velocity in the lower permeability rock.

The oil recovery obtained for this displacement (figure 15) is intermediate between that which would be obtained for a homogeneous reservoir using the relative permeabilities assigned to the higher permeability sands (lowest recovery curve) and that obtained using the curves for the lower permeability rock (highest recovery). There is a much earlier water breakthrough in the heterogeneous model corre-

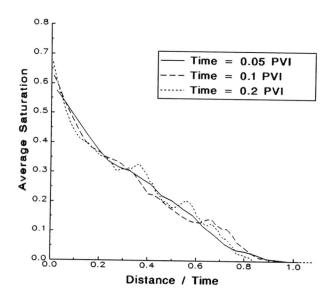

Figure 10: Saturation profiles along the sandbody model.

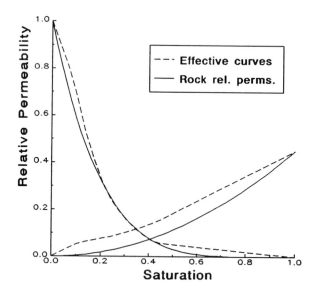

Figure 11:   Effective relative permeability curves for whole sand-
                body model.

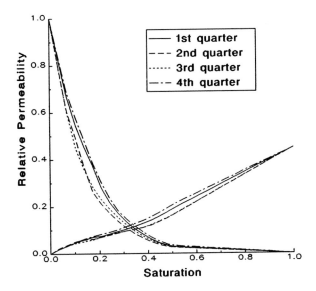

Figure 12:   Effective relative permeability curves for quarters of sandbody model.

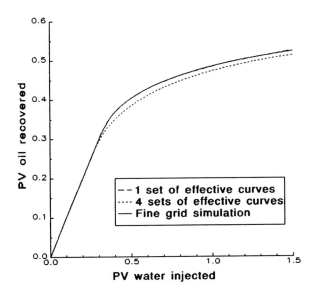

Figure 13:   Oil recovery curves using effective relative permeabilities for sandbody model

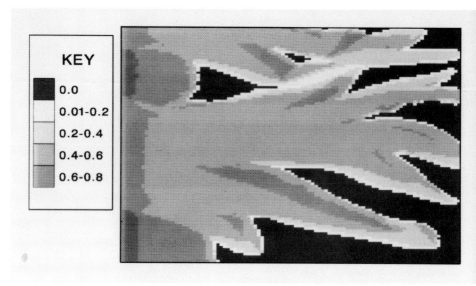

Figure 14: Saturation distribution in channel sand at 0.4PVI.

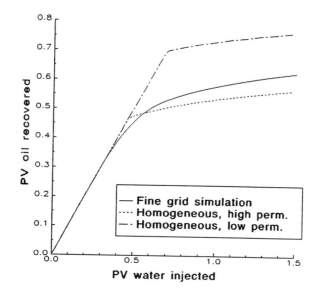

Figure 15: Oil recovery curves for channel sand model

sponding to the water finding high permeability paths through the reservoir.

Figure 16 compares the shapes of the average saturation profiles along the model at different times. The distance scale has been normalized in the same way as in figure 10. Although the shapes of the curves are broadly similar at different times, the different mean saturations do not move with a constant speed. Saturations between 0.3 and 0.6 slow down as the front progresses along the model, while lower saturations (below 0.3) seem to speed up.

The effective relative permeability curves generated for the whole channel sand distribution are compared with the rock curves used for the two different rock types in figure 17. The effective curves are very similar to those used in the higher permeability sand and bear little resemblance to those for the lower permeability sand. The same higher water relative permeability at lower saturations is observed as was noted in the sandbody effective relative permeabilities. The endpoint of the effective water relative permeability curve was calculated so that the effective permeability of the system to water at residual oil saturation was correct.

The effective relative permeability curves obtained for the four subsections of this distribution are shown in figure 18. There is a greater variation in shape between the four regions than was observed in the sandbody distribution, reflecting the variability between the effective permeabilities of the regions that is seen in Table I.

Figure 19 compares the recoveries obtained from one-dimensional simulations using a single region and the effective relative permeabilities shown in figure 18 with those obtained when the model is divided into four regions, each with different effective permeabilties and relative permeabilities. If the single-region model is used then we can reproduce the oil recoveries obtained with the fine-grid model very well, but when we subdivide the reservoir into four regions the one-dimensional model predicts a very much worse oil recovery. Thus it is important to generate effective relative permeabilities at the scale length for which we want to represent the average flow accurately.

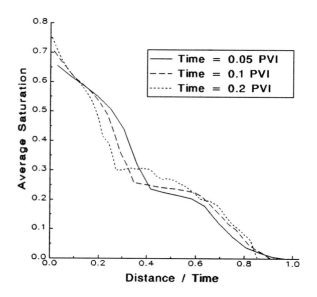

Figure 16: Saturation profiles along channel sand model.

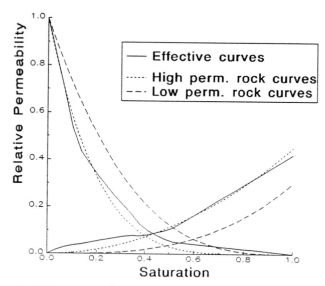

Figure 17: Effective relative permeabilities for whole channel sand

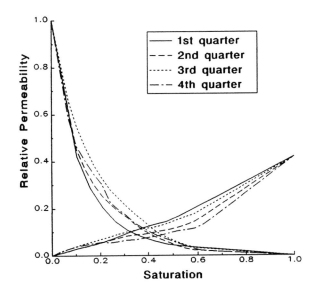

Figure 18:   Effective relative permeabilities for quarters of channel
              sand model.

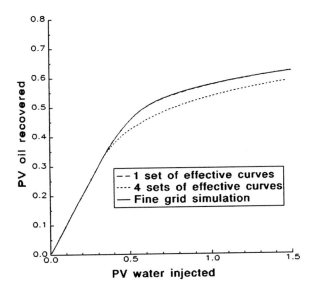

Figure 19:   Oil recovery curves using effective relative permeabili-
              ties for channel sand model

5. DISCUSSION

The fine-grid simulations showed that the different permeability distributions modified the flow in a line drive in different ways and this was reflected in the different effective relative permeabilities generated.

The uncorrelated random permeability distribution did not really alter the flow at all despite the relatively high variance because the correlation length of the distribution was negligible. As a result the effective relative permeability curves for the displacement were virtually indistinguishable from the rock curves.

The waterflood through the sandbody model was significantly less efficient because of the greater correlation length and the more extreme range of permeabilities. The high permeability channels through the distribution meant that the mean saturation profile along the model was more diffuse. The effective relative permeability curves for the distribution are consistent with this – the oil curve is very similar to the rock curve, while the effective water relative permeability is increased with respect to the rock curve, particularly at low saturations. This means that the water is more mobile, especially at low saturations, and hence moves more quickly through the system. The total mobility of the two fluids is also greater at low saturations in order to model the flow through the high permeability channels.

The channel sand also altered the displacement efficiency, again because of the greater correlation length of the distribution and the large difference in permeabilities between the two rock types. The effective relative permeability curves are very similar to the rock curves for the higher permeability sand because the flow occurs principally through this rock type. The higher effective relative permeability for water at low water saturations is again needed to represent the more diffuse mean water saturation profile.

A displacement is linearly scalable if the saturation at any point along the mean flood front is a function only of the number of pore volumes injected with respect to that point (Rapoport (1955)). This means that the shape of the mean saturation profile in any given displacement does not change with time. The effective relative permeabilities that we generate from such a displacement should be accurate when applied to problems other than the particular displacement from which they were generated. In particular they will not depend upon the inlet and outlet boundary conditions used in the displacement.

From an examination of the mean saturation profiles shown in figures 10 and 16 we can infer that the displacement through the sandbody distribution is approximately linearly scalable whilst that through the channel sand is not. The shape of the mean saturation profile in the channel sand changes as the displacement progresses whilst the shape of the profile in the sandbody distribution does not. This corresponds to the way the channel sand permeability distribution changes along the model while the sandbody permeability distribution does not.

Therefore we should be able to generate effective relative permeabilities from displacements either through the whole of the sandbody distribution or through separate sections of it and still reproduce the results of the fine-grid simulation. This is because the flow through large subsections of the model is very similar to the flow through the whole reservoir. In other words large sections of the sandbody model are representative of the complete permeability distribution. This is not the case for the channel sand model.

Indeed there is reasonable agreement when we compare the recovery curve obtained when we represented the distribution by four homogeneous subsections with that obtained by fine grid simulation (figure 13). The agreement is not exact because the effective relative permeabilities generated for each subsection are slightly dependent on the inlet and outlet boundary conditions (cf the difference between the harmonic mean of the effective permeabilities of the four sections and the total effective permeability for the whole model noted in section 3).

In contrast the effective relative permeabilities generated for subsections of the channel sand will not reproduce the flow observed in the fine grid model as the flow changes along the model. The mean saturation at any given position is not solely dependent upon the number of pore volumes injected with respect to that point, but also on the position of that point with respect to the inlet. Hence the one-dimensional, fine-grid model using the effective properties for four different regions of the channel sand model underpredicts the amount of oil recovered from the displacement (figure 19).

In summary the averaging volume used when generating the pseudo functions must be chosen so that it contains a representative section of the permeability distribution through which flow is to be modelled. If this is not the case then the pseudo relative permeabilities will be dependent upon the inlet and outlet boundary conditions in the displacement from which they were derived. Kossack *et al.* attempted to overcome this dependency by incorporating suitable permeability distributions at the inlet and outlet faces of the distribution for which they were attempting to derive pseudo functions; however they did not show whether this was successful. It would probably give a good approximation as long as the combination of inlet, centre and outlet sections contained a representative section of the whole distribution.

6. SUMMARY AND CONCLUSIONS

We have used high-resolution simulation to simulate a waterflood through three permeability distributions with different correlation lengths, thus giving us an understanding of how heterogeneities altered the detailed flow. The data from these simulations were used to generate effective relative permeability curves for the whole of each model, reducing each heterogeneous two-dimensional reservoir model to a homogeneous one-dimensional problem.

Each permeability distribution was also divided longitudinally into four smaller distributions. High-resolution simulation was then used to generate effective relative permeability curves for each region. We have used both the single sets of effective curves and the sets of four pairs of effcetive curves to simulate waterfloods through each of the models, using a homogenous one-dimensional grid and compared these results with those of the original fine-grid simulations.

From this work we can draw the following conclusions:-

1) A fine, uncorrelated log-normal permeability distribution with a log variance of 20% does not substantially alter the fluid flow. The effective relative permeability curves generated for such a displacement are very similar to the rock curves, so the displacement can be adequately modelled in one dimension using the rock curves and a single effective permeability.

2) The shape of effective relative permeabilities differs significantly from the rock curves for more correlated permeability distributions. In particular the water relative permeability is increased at low satu-

rations because of the channelling of the displacing fluid through the high permeability paths.

3) Linear scaling can be used to determine whether the effective (and pseudo) relative permeability curves derived from a particular displacement can be applied to other problems without a significant loss of accuracy.

4) The reservoir volume used to generate effective (and pseudo) relative permeability curves must contain a representative section of the permeability distribution. This can be checked by comparing the values of effective permeability obtained for different averaging volumes. If the distribution changes considerably between coarse-grid blocks, then the effective relative permeabilities generated from displacements through separate grid blocks will not accurately represent the flow through a sequence of such grid blocks.

In summary, different permeability distributions alter the flow in a line drive in different ways and this is reflected in the properties of the effective and pseudo relative permeabilities. In particular the averaging volume used to generate the effective properties must be chosen so that it contains a representative section of the permeability distribution through which flow is to be modelled. If this is not the case then the effective relative permeabilities generated will not necessarily reproduce the flow observed in a similar waterflood through a series of such averaging volumes, perhaps representing the whole reservoir. This work did not investigate how different realizations of the models affected the pseudo functions. Further work is also needed to investigate the implications for two-dimensional flow and the successive rescaling of permeability distributions.

ACKNOWLEDGEMENTS: Permission to publish this work has been given by the British Petroleum Company plc. I would like to thank Dr. Steve Begg for providing the sandbody permeability distribution and the channel sand model used in this paper. I would also like to thank Dr. Mike Christie and Dr. John Fayers for helpful discussions during this work.

REFERENCES

Begg S.H., Carter R.R and Dranfield P., 'Assigning Effective Values to Simulator Grid Block Parameters in Heterogeneous Reservoirs', SPE 16754, presented at the 62nd Annual Technical Conference and Exhibition of the Society of Petroleum Engineers, September $27\text{-}30^{th}$ 1987.

Christie M.A., 'High-Resolution Simulation of Unstable Flows in Porous Media', SPE *Reservoir Engineering*, August 1989.

Johnson E.F., Bossler D.P. and Naumann V.O., 'Calculation of Relative Permeability from Displacement Experiments', *Trans* AIME, *216*, pp370-372, 1959.

Jones S.C. and Roszelle W.O., 'Graphical Techniques for Determining Relative Permeability from Displacement Experiments', *Journal of Petroleum Technology*, pp 807-817, May 1978.

Kossack C.A., Aasen J.O. and Opdal S.T., 'Scaling up Laboratory Relative Permeabilities and Rock Heterogeneities with Pseudo Functions for Field Simulations', SPE 18436, presented at the SPE Reservoir Simulation Symposium, Houston, Texas, $6\text{-}8^{th}$ February 1989.

Kyte J.R. and Berry D.W., 'New Pseudo Functions to Control Numerical Dispersion', *Journal of the Society of Petroleum Engineers*, pp 269-276, August 1975.

Lasseter T.J., Waggoner J.R. and Lake L.W., 'Reservoir Heterogeneities and their Influence on Ultimate Recovery', NIPER Reservoir Characterization Conference 1986.

Pande K.K, Ramey Jr., H.J., Brigham W.E. and Orr Jr., F.M., 'Frontal Advance Theory for Flow in Heterogeneous Porous Media', SPE 16344, 1987.

Rapoport L.A., 'Scaling Laws for Use in Design and Operation of Water-Oil Flow Models', *Trans.* AIME, *204*, pp 143-150, 1955.

Reznik A.A., Enick R.M. and Panvelkar S.B., 'An Analytical Extension of the Dykstra-Parsons Vertical Stratification Discrete Solution to a Continuous Real-Time Basis', *Society of Petroleum Engineers Journal*, pp 643-655, December 1984.

Smith E.H., 'The Influence of Correlation between Capillary Pressure and Permeability on the Average Relative Permeability of Reservoirs containing Small Scale Heterogeneity', 2nd NIPER Reservoir Characterization Conference 1989.

Tompang R. and Kelkar B.G., 'Prediction of Waterflood Performance in Stratified Reservoirs', SPE 17289 presented at the SPE Permian Basin Oil and Gas Recovery Conference, March 10-11th 1988.

White C.D., 'Representation of Heterogeneity for Numerical Reservoir Simulation', Ph.D. Thesis, Dept. of Petroleum Engineering, School of Earth Sciences, Stanford University, Stanford, CA, June 1987.

CHARACTERIZING SHALE CLAST HETEROGENEITIES
AND THEIR EFFECT ON FLUID FLOW

David L. Cuthiell, Stefan Bachu,
John W. Kramers, Li-Ping Yuan

Alberta Research Council
Edmonton, Alberta, Canada

I. ABSTRACT

A study has been performed to characterize shale clast zones and to predict the effects of this type of heterogeneity on permeability at the simulator grid block scale. Shale clast zones are a common feature of channel deposit reservoirs and represent partial but significant barriers to fluid flow. Unlike the larger shales investigated previously, individual shale clasts can be directly observed in core. Typically, they exhibit a much larger range of sizes and shapes than do larger shales. The present study is based on a data set of nearly 7000 digitized clast outlines taken from core in an Upper Mannville heavy oil pool in east central Alberta, Canada.

Shale clast zones are characterized by several scales of heterogeneity. At the largest scale the zones are composed of layers of clasts interbedded with layers of sand. Each clast layer is in turn composed of one or more distinct beds and each bed contains a large number of clasts. These heterogeneity scales provide a framework for characterizing both the geometry and spatial correlations of the clasts and the resulting flow properties. Finite element modeling has been used to simulate single-phase fluid flow through representative core scale regions and to obtain effective vertical and horizontal permeabilities for such regions. Based on observed and inferred properties of beds and layers, approximate methods are proposed for scaling these effective flow properties up to the scale of the entire shale clast zone.

RESERVOIR CHARACTERIZATION II
226

II. INTRODUCTION

The presence of discontinuous shales has been recognized as a significant factor controlling reservoir performance, and considerable research has been directed toward quantifying the effects of shales on fluid flow (Haldorsen and Lake, 1984; Haldorsen and Chang, 1985; Begg and King, 1985; Begg et al., 1985; Desbarats, 1987; Deutsch, 1989). The goal of such work is usually to predict the effective properties of a region of the reservoir, such as a simulation grid block, containing a large number of the shales. An important conclusion has been that the effects of very high contrast heterogeneities are poorly predicted by theories designed to deal with more continuously varying properties (Desbarats, 1987).

Shale studies reported to date have dealt (either explicitly or implicitly) with relatively large shale lenses with dimensions of at least meters. Typically, these lenses were modeled as rectangular in cross section and as having a relatively small ratio of thickness to width.

The present study concerns shale clasts, which are much smaller (dimensions mm to cm) than the previously studied shale lenses and have a more complex range of shapes. Shale clasts are expected to significantly reduce the effective permeability of the zones in which they occur and therefore have a potential impact on any recovery project undertaken. For example, Richardson et al. (1987) have recently advocated investigation of the effects of such core scale heterogeneities on gravity drainage processes.

The effect of shale clasts at the scale of a reservoir simulation grid block is more difficult to predict than that of larger, randomly distributed shales since the clasts are usually not distributed through the block in a statistically homogeneous manner. Instead, they are concentrated in distinct layers and the layers are in turn distributed in some fashion within the grid block. The prediction of effective permeabilities for a simulation grid block thus requires a scaling up process whereby effective permeabilities are calculated for individual clast layers and then for some configuration of layers within the block.

This paper is based on a reservoir case study, which supplied the clast shapes and distributions for our flow simulations. The paper first briefly describes the geology of the reservoir, then the characterization of both the geometry and flow properties of the clast zones at different heterogeneity scales.

III. CASE STUDY

A. Geology of the Reservoir

Shale clast zones were investigated as part of a larger
project analyzing the Provost Upper Mannville B Pool, a heavy
oil reservoir in east central Alberta. The geology of the
reservoir has been presented in greater detail elsewhere
(Kramers et al., 1989); here we summarize only the major
features.

The reservoir is contained in McLaren Formation sands of
Lower Cretaceous (Late Albian) age and occurs in valley fill
sequence deposits, which can be subdivided into a number of
lithologically distinct facies representing different
depositional environments. Figure 1 is a well log with
lithology, showing four of the five facies. The reservoir is
contained entirely within three of these lithofacies: the
blocky channel, shale clast and channel margin facies.

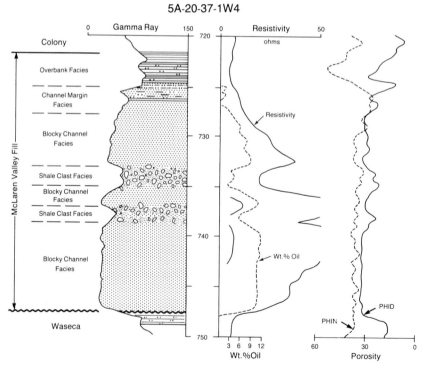

Figure 1.   Well log from the Provost Upper Mannville B Pool
            (well 5A-20-37-1W4) showing four of the five
            lithofacies of the McLaren Valley fill sequence.

The blocky channel lithofacies forms the main part of the
reservoir. It consists of trough and planar cross-bedded,
well-sorted, quartz-rich sands. The channel margin
lithofacies is found overlying and transitional with the
blocky channel facies and forms the top of the reservoir.

The shale clast facies occurs within the blocky channel
facies and could be considered a subfacies. It consists of
shale, silty shale or carbonaceous siltstone clasts in a
matrix of fine- to medium-grained sands (Figure 2). Some of
the clasts are structureless, whereas others show laminations
or soft sediment deformation structures. Individual clasts
exhibit a wide range of shapes and sizes. The largest clasts
can usually be distinguished from shale beds by their
contacts or orientation of contained sedimentary structures.
Gross thickness of the shale clast facies is up to 14 m, with
the thickest continuous layer being 3 m. Lateral continuity
of intervals with shale clast zones is expected to be on the
order of tens to hundreds of meters. In an EOR pilot in the
reservoir, the shale clast zone intervals can be correlated
across the pilot in an east-to-west direction a distance of
approximately 400 meters. In the north-to-south direction,
the same interval can only be correlated between two wells
25 m apart.

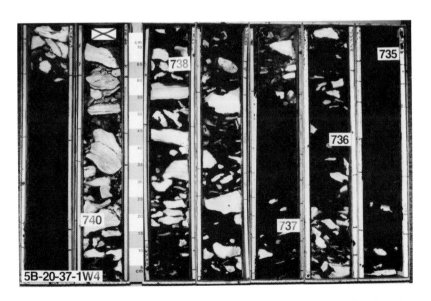

Figure 2.   Cored section through the shale clast facies in
             well 5B-20-37-1W4. The top of the core is in the
             upper right and each core tube is approximately
             75 cm long.

The shale clast facies is interpreted to have been deposited in a channel complex with the clasts coming predominantly from erosion of the channel cutbank. Because of the nature of deposition, the shale clast zones are expected to have their greatest dimension parallel to the paleocurrent direction. Shale clast zones are not unique to the Provost Upper Mannville B Pool. They have been observed in other reservoirs and can occur in any reservoir of channel origin.

B. Data for the Shale Clast Study

Data for the case study were obtained from three wells in the reservoir. Two of the wells (5A-20-37-1W4 and 5B-20-37-1W4) were located in the same experimental pilot pattern and had a separation of about 200 m. These two wells will be referred to as wells A and B respectively. The third well (14-29-37-1W4), located several km from wells A and B, will be referred to as well C.

Clast data were captured by manually tracing from core the outlines of all observed clasts in the clast zone for each of the three wells. The clast outlines were subsequently digitized, thus being represented as polygons with as many as 100 sides. The final data set comprised some 7000 clast outlines, representing approximately 20 m of core.

In principle, similar data could be captured using computerized image analysis techniques. This would have been difficult with the core available to us because of the very poor visual contrast between some of the clasts and the sand, resulting from absorption of oil by the clasts. However, in any large scale efforts to capture heterogeneity data of this type, automated data capture techniques would have to be developed.

The data used for the study are two-dimensional and therefore only partially representative of the real three-dimensional clasts. The implications this has for our results will be discussed later. In principle, three-dimensional clast data could be obtained from core by computed tomography.

IV. SCALES OF HETEROGENEITY

A clast zone exhibits heterogeneity at several distinct scales. The fundamental heterogeneity is created by the contrast in properties between individual shale clasts and

the surrounding sand matrix. However, the clasts occur in discrete layers within the zone and the spatial distribution of layers represents another scale of heterogeneity.
Figure 3 shows the interpreted outline of a shale clast zone in the Provost Upper Mannville B Pool case study. The zone is correlated between four closely spaced wells, and layers of clasts are shown as they were observed in core. Unlike the zone, individual layers of clasts cannot be correlated between wells and therefore represent a "stochastic" rather than a "deterministic" heterogeneity, in the terminology of Haldorsen and Lake (1984). Based upon core observation, the layers also display systematic internal variation. A typical example of a layer taken from our digitized data set is shown in Figure 4. It is apparent that there are distinct sub-layers or "beds" within the layer. The transition between one bed and its neighbor is marked by a sudden change in the size distribution and/or shapes of the clasts, while within the bed these properties change gradually if at all. These features are expected geologically since the different beds are created by individual events, which are not likely to result in similar clast characteristics.

The different heterogeneity scales must be taken into account when estimating effective flow properties of the shale clast zone. This is done by a scaling up procedure that calculates effective flow properties at successively larger scales, beginning with a scale smaller than bed size and ending with the entire zone. At each scale, the geometry

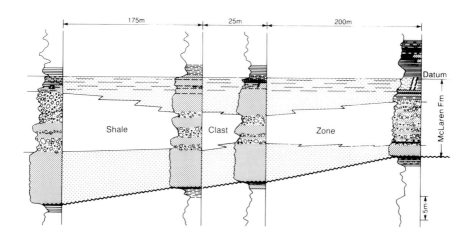

Figure 3. Stratigraphic cross section through four closely spaced wells with core control showing geometry of a shale clast zone (not to scale).

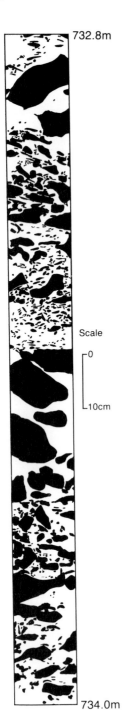

732.8m

Scale

0

10cm

734.0m

Figure 4. Digitized version of a
typical shale clast
layer.

and other characteristics of the heterogeneity must be known or estimated in order to predict the effective flow properties.

V. SHALE CLAST GEOMETRY

The "clasts" in our data set are in fact two-dimensional cross sections of clasts and are defined mathematically as polygons. The properties of the polygons (e.g. area) are readily calculated by standard numerical techniques. To define an orientation for a clast, it is useful to compute a "moment of inertia" tensor (Tough and Miles, 1984), where this terminology comes from mechanics. The tensor has associated with it a principal axis which may be taken to define the direction of orientation of the clast. This objective mathematical definition agrees well with visual intuition in real cases. The "length" and "width" of the clast are then defined as the dimensions of the rectangle oriented in the principal direction which just encloses the clast (Figure 5). An "aspect ratio" is defined as (width/length), and a "shape factor" as the area of the clast divided by (length x width). The latter parameter is a

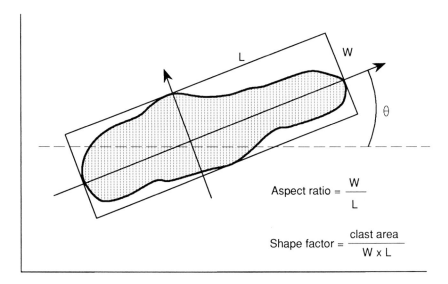

Figure 5.   Geometric parameters characterizing a shale clast. Arrows illustrate the principal axes of the moment of inertia tensor.

measure of how nearly the clast fills the enclosing
rectangle; it has the value 1 for a rectangular clast and $\pi/4$
for an elliptical clast.

For the purposes of analysis, we treat the clasts from
the three different wells separately. Figure 6 shows
distributions of clast length, orientation, aspect ratio and
shape factor for each of the three wells. Statistical
properties of the distributions are given in Table I. The
length distributions are approximately lognormal with a peak
around 5 mm. There are significant differences between
wells; in particular, well C, which is more distant from the
other two wells than the latter are from each other, contains
significantly smaller clasts. However, the widths of the
three distributions are very similar. Since only complete
clasts were analyzed to produce the distributions, there is a
bias: large clasts are more likely than small clasts to be
cut off by the edges of the core and thus excluded from
analysis.

The orientation distributions (Figure 6b) show that the
clasts are predominantly sub-horizontally aligned, although a
significant number are as much as 30° and a few up to 90°
from horizontal. The angular distribution is considerably
broader for well C. Aspect ratios (Figure 6c) are broadly
distributed with a mode somewhat less than 0.5. Again, the
clasts from well C have a different distribution of aspect
ratios than those from the other two wells. The shape

TABLE I. Means and standard deviations (in parentheses) of
the clast characteristic distributions shown in
Figure 6.

Clast Parameters	Well A	Well B	Well C
Length (log scale)	-0.13 (0.33)	-0.21 (0.32)	-0.32 (0.31)
Orientation (degrees)	0.2 (36.2)	1.9 (35.3)	3.5 (46.4)
Aspect Ratio	0.52 (0.18)	0.52 (0.18)	0.58 (0.18)
Shape Factor	0.68 (0.09)	0.73 (0.06)	0.72 (0.07)

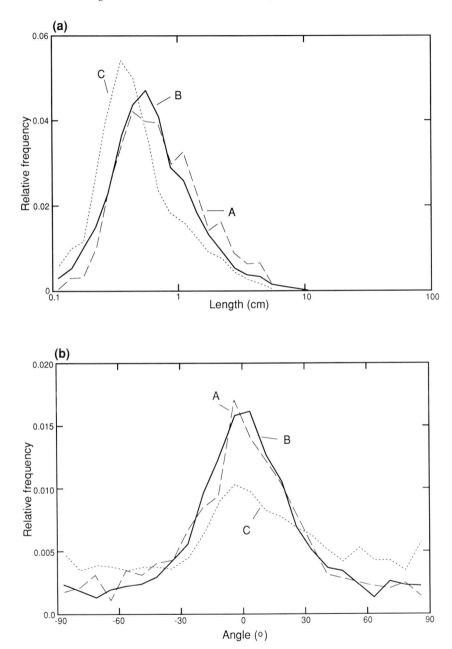

Figure 6.   Frequency distributions for clast parameters in
             wells A, B, and C: (a) length; (b) orientation;
             (c) aspect ratio; and (d) shape factor.

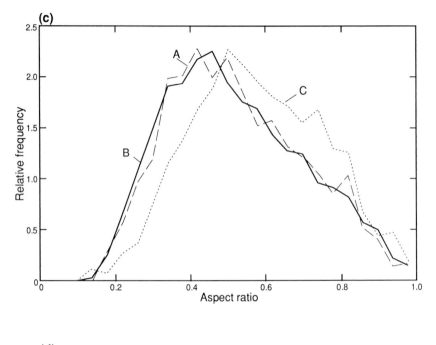

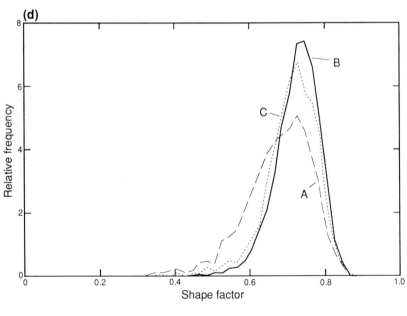

distributions (Figure 6d) have modes of about 0.75 and
indicate very few clasts with shape factor near 1. This
shows that the clasts are closer to being ellipses (shape
factor 0.785) than rectangles (shape factor 1). In the case
of shape factor, there is more difference between the nearby
wells (A and B) than between B and the more distant well C.

VI. CHARACTERIZING SHALE CLAST LAYERS

What affects flow within a clast layer is not only the
distribution of clast shapes and sizes but also the spatial
relationships among clasts. In principle, the effective flow
properties could be inferred by performing numerical flow
simulations for the entire layer. However, there are
typically about 1000 clasts of a wide variety of shapes and
sizes visible within one layer in core and this represents
only a tiny fraction of the entire layer.
It is not feasible to numerically simulate flow in a
region that is so large relative to the smallest
heterogeneities. For practical reasons, one initially
characterizes regions that are small enough to permit flow
simulations yet that contain a large enough sample of clasts
to be representative. Such regions must lie entirely within
a bed, since abrupt changes in clast distributions take place
at the boundaries of beds (Figure 4). This limits their size
to less than about 20 - 30 cm in the vertical direction and
in practical cases they are chosen even smaller, perhaps
10 cm long. We refer to such regions as "core scale"
regions. Once the flow properties are characterized for core
scale regions, the results may be scaled up to determine
properties for an entire layer.

A. Core Scale Flow Simulations

The ranges of sizes and shapes of clasts mean that flow
simulations for even small core scale regions require
substantial computer resources. Therefore, by practical
necessity, we chose to perform core scale flow simulations
for only five test regions selected from our data set. The
regions, ranging in shale fraction from 0.15 to 0.49, are
shown in Figure 7. Larger shale fractions were very
difficult to find in the data set except in areas of core
dominated by a single large clast, which were in many cases
completely blocked to flow.

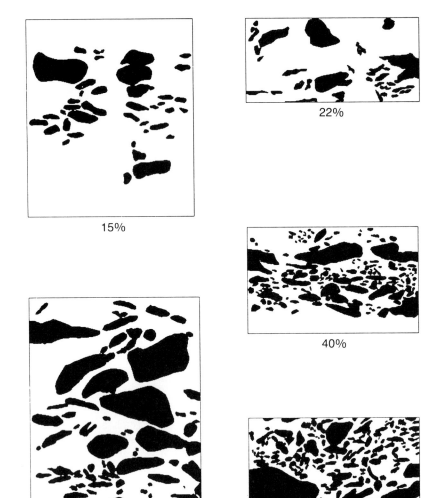

15%

22%

40%

35%   49%

Figure 7. Digitized test regions used in core scale flow
          simulations. Labels indicate shale fractions.

The test regions were chosen to contain reasonably uniform distributions of clasts. This requirement was imposed subjectively, based upon visual examination of the clast configurations. Ideally, one would like the test regions to satisfy objective criteria for statistical homogeneity. However, because of the large size of some of the clasts relative to the dimensions of the regions, this goal was unachievable in practice. Greater statistical homogeneity could be obtained by constructing synthetic clast regions based upon properties of the real clasts. Such a process, however, has its own shortcomings. For the simulations reported here, we have adopted the philosophy that the benefits resulting from the use of real data outweigh the statistical compromises.

Considering the range of clast sizes, accurate simulation of flow at core scale using a finite difference model with a uniform grid would have required of the order of 10^5 grid nodes. It was therefore decided to perform the simulations using a finite element model, which could more efficiently capture the complex geometry of the clasts. The model FE3DGW (Gupta et al., 1984a, 1984b) was used for this purpose.

The predominant orientation of the clasts is horizontal (Figure 6b) and therefore the principal directions of the effective permeability tensor are expected to be vertical and horizontal. With this assumption, vertical and horizontal effective permeabilities were computed for each of the test regions. A further assumption was made that shale and sand were separately homogeneous and isotropic with respect to flow. It has been shown for other sand/shale heterogeneities (Desbarats, 1987) that small spatial variations in the properties of the sand matrix and shale separately have relatively little impact on the final effective permeability, compared with the impact of a large permeability contrast between the sand and the shale.

To calculate effective vertical permeability, a constant pressure difference was imposed between the top and bottom of the region and the sides were assumed impermeable to flow. The numerical model was used to predict the resulting steady-state fluid flow. Effective permeability was then defined as the permeability of a homogeneous region of the same size having the same flow rate for the same pressure difference. Effective horizontal permeability was computed in a similar manner.

Large clasts on the boundaries of the flow regions might be expected to cause errors of unknown magnitude in the simulation results. Ideally, such boundary effects could be reduced by performing simulations for larger regions, but the physical limits of the core preclude doing so directly. A similar objective could be achieved by periodically

replicating a core region to create a larger region. Such a
technique presents computational challenges because of the
large grid which it requires, and also produces some
anomalies at the boundaries of the repeated regions. For the
present, we have assumed that boundary effects can either
decrease or increase the computed permeabilities and have not
attempted to mitigate these effects.

The calculated permeabilities for the five test regions
are shown in Figure 8. The results are presented in a
dimensionless form referred to as "permeability reduction",
defined as effective permeability divided by the permeability
of sand. Effective permeabilities decrease with increasing
shale fraction, with vertical permeability reduction reaching
0.163 at shale fraction 0.49. For any shale fraction,
effective vertical permeability is smaller than horizontal
despite the assumption of homogeneity and isotropy of the
sand and shale components. This core scale anisotropy is a
natural consequence of the predominantly flat shape and
sub-horizontal orientation of the clasts.

A power average function has been proposed (Journel et
al., 1986) as a fit to effective permeability for flow
through rectangular shale lenses. The power average fit has

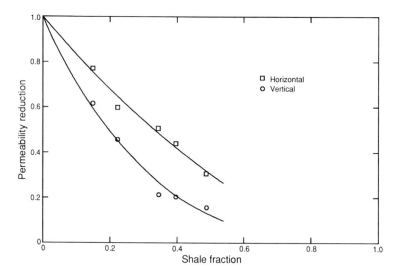

Figure 8. Vertical and horizontal effective permeabilities
          (relative to sand=1) for five test regions. Curves
          are power average fits (Equation (1))

the form

$$K_{eff} = \left[fK_{sh}^{\omega} + (1 - f)K_{sa}^{\omega} \right]^{\frac{1}{\omega}} \tag{1}$$

where f is shale fraction, K_{sa} and K_{sh} are sand and shale permeabilities, respectively, and ω is an empirical parameter. This function provides excellent fits ($R^2 > 0.99$) to the clast results, using values $\omega_v = 0.260$ for vertical permeability and $\omega_h = 0.574$ for horizontal permeability (Figure 8). The power average expression has the property that it interpolates between the conventional arithmetic, geometric and harmonic means, which correspond to $\omega = +1$, 0 and -1 respectively. Our results thus lie between the geometric and arithmetic means for both vertical and horizontal permeability and are poorly predicted by either of these conventional means.

The results shown in Figure 8 assumed a ratio of sand-to-shale absolute permeability of 1000:1, which is believed to be realistic for the shale clast system. To test sensitivity to this ratio, similar calculations were made for ratios ranging from 2:1 to 100,000:1. It was found that effective permeabilities changed very little for ratios greater than about 100:1, a result in agreement with similar calculations made for shale lens systems (Desbarats, 1987; Begg et al., 1985). The results for small sand:shale permeability ratios have intrinsic interest since they are relevant to the mathematically identical problem of thermal conduction.

It was not practical to significantly refine the finite element grid in order to test the sensitivity of our results to the gridding. Therefore, validation of the numerical results was obtained by means of electrical analog experiments. These were similar to experiments which have been reported for systems of shale lenses (Dupuy and Lefebvre du Prey, 1968). They exploit the mathematical analogy between fluid flow and the flow of electrical current. A core scale clast region is represented by a sheet of electrically conducting paper, with holes cut in it to represent clasts and equipotentials created at each end using highly conducting paint. The effective conductivity is obtained in the analog experiment by simply comparing the resistance of the paper with "clasts" (i.e. holes) to that of the original sheet with no "clasts". Since the holes in the paper have zero electrical conductivity, the analog results correspond to an infinite ratio of sand-to-shale permeability. However, as noted earlier, effective permeability depends very little on this permeability ratio when it is greater than about 100:1.

For the core scale region having shale fraction 0.15, the electrical analog experiments gave values of 0.771 and 0.613 for effective horizontal and vertical permeabilities; these were in excellent agreement with the corresponding simulation values of 0.772 and 0.620. For the region with shale fraction 0.35, the analog effective vertical permeability was measured to be 0.181, in comparison with the simulation value 0.213. Effective horizontal permeability was 0.484 by the analog method compared with 0.509 by the simulation method. These differences can be ascribed to small inaccuracies in either or both of the numerical and analog methods and are not large enough to significantly affect our conclusions.

The results of the core scale simulations demonstrate that shale fraction has a dominant impact on effective permeabilities. The difference between horizontal and vertical permeabilities shows that orientation and aspect ratio have important secondary effects. Effective permeabilities are also expected to depend on the dimensionality of the flow region, with three-dimensional permeabilities larger than two-dimensional for the same shale fraction (Matheron, 1967; Gutjahr et al., 1978; Desbarats, 1987). The study reported here has considered only two-dimensional flow. Three-dimensional flow simulations have been reported for simple rectangular shale shapes but are probably not feasible for clasts. Physical flow simulations through artificial (but realistic) clast systems are another option for obtaining three-dimensional results, and we are presently pursuing this alternative.

Our limited set of numerical simulations was not sufficient to fully explore the dependence of effective permeabilities on different properties of the clast distributions. Our present (labor-intensive) method of creating finite element grids does not permit a great many simulations to be performed. We are currently exploring automatic finite element generation or finite difference simulations using greater computer resources as means of obtaining more numerical results. In order to carry out a range of sensitivity tests it will be necessary to be able to create artificial regions with realistic configurations of clasts. Although the geostatistical method of indicator simulation has been successfully used to simulate distributions of rectangular shales (Desbarats, 1987), this method will probably not create geometries as complex as that of the clasts. Another potential method for creating artificial clast systems is the "Boolean sets" method (Matheron, 1975), which creates entire clasts and distributes them randomly in the matrix. This method has the advantage

of automatically incorporating the known geometrical
properties of the clasts, but cannot readily account for the
appropriate spatial relationships among them.

B. Flow Properties of Clast Layers

In order to quantify the flow properties of a shale clast
layer, it is necessary to scale up the core scale results
described in the previous section. Part of the purpose of
scaling up is to reduce the amount of detail at each stage of
calculation. Accordingly, at the layer scale, the detailed
clast distributions are replaced by continuous variables such
as shale fraction. At any point within the layer, such
variables have values representative of a core scale region
surrounding the point. Their relationship with the clasts is
comparable with the relationship between macroscopic porosity
and individual sand grains.

The small width of the core provides no sampling of the
lateral variation of mesoscale parameters like shale
fraction. Therefore, the data from one well provide only a
one-dimensional vertical sampling of variations in shale
fraction. Using the core data set, the vertical distribution
of shale fraction can be determined by sampling equal-sized
intervals of core through the clast zone. Such
distributions, using a 20 cm sampling interval, are shown for
wells A and B in Figure 9. Individual layers may be
recognized as contiguous groups of 20 cm intervals with
nonzero shale fraction. The results for well B (which has
the most data) appear to suggest that average shale fraction
is different in different layers. However, this appearance
is illusory. An F-test shows no significant difference in
shale fraction between layers. This means that the values of
shale fraction within any particular layer may be regarded as
a random sample from a common population. In the absence of
data describing horizontal variations in shale fraction
within a layer, a reasonable assumption is that they are
sampled from the same distribution.

Thus, the pooled set of shale fractions for all 20 cm
intervals within layers provides an experimental shale
fraction distribution which may be used to create
hypothetical shale clast layers. A hypothetical layer is
constructed by dividing it into a large number of core size
blocks and assigning a shale fraction to each using the
experimental distribution. Each of the core size blocks can
then be assigned effective horizontal and vertical
permeabilities based on the relationships shown in Figure 8.

Effective permeabilities for the entire layer will
clearly depend on the manner in which shale fractions are

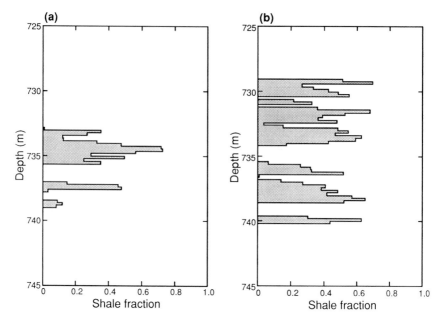

Figure 9.   Shale fraction sampled for 20 cm intervals
            through the shale clast zone in well A ((a)) and
            well B ((b)).

assigned to the blocks within the layer. The geostatistical
method of conditional simulation (Journel and Huijbregts,
1978) is well suited to the task of creating realistic
spatial distributions of shale fraction conditioned to well
data, but requires a model for the spatial correlations
within the layer. Since only one-dimensional data are
available, such a model cannot be constructed. In the
absence of a geostatistical model, several extreme
possibilities may be considered. In the case that shale
fraction is perfectly horizontally stratified, effective
vertical permeability for the layer will be given simply by
the harmonic mean of the effective vertical permeabilities of
the blocks constituting the layer. Effective horizontal
permeability in this case is given by the arithmetic mean of
the block horizontal permeabilities. In the opposite extreme
of complete vertical stratification, effective vertical
permeability for the layer will be the arithmetic mean of the
block vertical permeabilities, and effective horizontal
permeability for the layer (perpendicular to the
stratification) will be the harmonic mean of the block
horizontal permeabilities.

The true effective permeabilities must lie between these extremes. It is important to note that, for the distributions of shale fraction in the case study, the range of possibilities represented by the extreme cases is not too great. For example, based on the shale fractions in well B, effective vertical permeability of layers must lie between 0.12 and 0.18, and effective horizontal permeability of layers must lie between 0.31 and 0.36. A practical approximation is to estimate both vertical and horizontal effective permeabilities for layers by the geometric means of the corresponding block values. These predictions lie roughly in the middle of the range of possibilities and have been shown to be suitable for random, anisotropic permeability variations by others (Begg et al., 1987). In the case of well B, the geometric mean predictions are 0.15 and 0.34 for layer effective vertical and horizontal permeabilities respectively.

VII. CHARACTERIZING SHALE CLAST ZONES

It is likely that a typical reservoir simulation grid block located in a shale clast zone will contain a number of shale clast layers interspersed with layers of sand. Therefore the geometry and spatial distribution of shale clast layers within the grid block is important to the block effective permeability. This constitutes another scale of heterogeneity, which we will refer to as "zone scale", and requires another stage of scaling up.
    Only the intersections of clast layers with cored wells were available in the case study. Individual layers are generally not correlatable between wells. In this sense, the layers are analogous to the stochastic shales discussed by Haldorsen and Lake (1984). Accordingly, realistic distributions of layers within a shale clast zone can be created by the same method used by Haldorsen and Lake. This method requires an empirical distribution for lateral layer dimensions that in practice will have to be obtained from outcrops representing a similar depositional environment.
    Although the layer geometry and distribution can, in principle, be synthesized by existing methods, the flow properties cannot be computed as easily as for a distribution of large shales. Unlike large shales, the shale clast layers are relatively permeable and, furthermore, their internal effective permeability is anisotropic. In practice, the effective permeabilities of different shale clast layers are likely to be different, but the approximation described in

the last section gives all layers intersecting a particular
well the same effective permeabilities. This is similar to
the assumption at core scale that both clasts and sand matrix
have uniform and isotropic permeability. The major effect on
permeability at the zone scale is caused by the contrast
between sand layers and shale clast layers, rather than
internal variations in either of these subsystems. The
detailed geometry of the layers (like clasts at core scale)
is also important to the zone scale flow, but this geometry
is not known in the case study. Consequently, estimation of
zone scale effective permeabilities can only be approximate.

The thickness of clast layers observed in core is
typically one to two meters. For geological reasons, their
horizontal dimensions are expected to be much larger than
this. Thus, a zone is likely to have a highly stratified
structure over moderate horizontal distances. In the extreme
of complete stratification, zone effective permeabilities are
simple to calculate. Effective vertical permeability is
given by the harmonic mean of the effective vertical
permeabilities of the shale clast and sand layers
constituting the zone (the mean being weighted by the
thickness of the layers). Effective horizontal permeability
for the zone is given by the arithmetic mean of the
horizontal clast and sand layer permeabilities.

Another extreme for shale clast layer geometry is one in
which the lateral dimensions of layers are not much greater
than their thicknesses. We are not aware of any studies
dealing with a randomly distributed, binary heterogeneity of
this type for which one of the components (here the shale
clast layers) has anisotropic permeability. However, in the
similar anisotropic case with continuously variable
permeability the effective vertical and horizontal
permeabilities of the region are close to the corresponding
geometric means of the spatial distributions (Begg et al.,
1987). We assume that this result would also hold in the
case of very discontinuous shale clast layers.

The actual (unknown) layer geometry must lie between the
two extremes discussed above. Accordingly, we predict that
effective vertical permeability for the zone should lie
between the geometric and harmonic means of the layer values,
and that effective horizontal permeability should lie between
the arithmetic and geometric means of the layer values.
Applying these predictions to the data for well B discussed
in the last section and using the layer properties estimated
there, we predict vertical permeability reduction for this
zone between 0.23 and 0.31 and horizontal permeability
reduction between 0.52 and 0.59.

VIII. SUMMARY AND CONCLUSIONS

Shale clast zones are commonly found in channel deposit reservoirs and represent a hierarchical type of heterogeneity, since zones consist of a number of layers, each of which in turn contains one or more beds that finally contain large numbers of shale clasts. Because of the very small size of the clasts, repeated scaling up is required to obtain the effective flow properties of a single reservoir grid block in a clast zone. We have presented a case study showing that a reasonable strategy for scaling up follows the sequence: clast to core scale, core to layer scale, and finally layer to zone scale.

Based upon real clast distributions from core, geometric properties of the individual clasts have been characterized statistically. We have argued that a reasonably sized sample of clasts that is representative, yet small enough to be numerically tractable, is a core-sized region of about 10 cm on a side which lies entirely within one bed. A finite element model was used to simulate steady-state fluid flow for five such regions having a range of shale fractions. Effective vertical permeabilities are always smaller than horizontal at any shale fraction as a result of the flat shape and horizontal orientation of the clasts.

The scaling up from core to layer scale ideally requires a knowledge of the three-dimensional variations of mesoscale variables like shale fraction within a layer. However, core data provide only a small, one-dimensional sample of these variations. To characterize shale clast layers based on core alone, one computes shale fractions at 20 cm intervals through the shale clast zone. The set of non-zero shale fractions is assumed to be representative of the three-dimensional variations within any particular layer. These shale fractions can be converted to effective vertical and horizontal core scale permeabilities using the results of the core scale simulations. Effective vertical permeability for any layer is then estimated as the geometric mean of the set of vertical core scale permeabilities. Effective horizontal permeability for a layer is estimated as the geometric mean of the set of horizontal core scale permeabilities. These approximate predictions assign the same permeabilities to each of the layers sampled in a well.

At the zone scale, shale clast layers are analogous to the stochastic shales considered by others and their distribution within the zone can be simulated using existing stochastic methods. The layer length statistics needed to employ these methods were not available in the case study and we therefore propose an approximate method to estimate zone

permeabilities. Effective vertical permeability for the zone
is predicted to lie between the space-averaged geometric and
harmonic means of the vertical permeabilities of clast and
sand layers making up the zone. Effective horizontal
permeability for the zone is predicted to lie between the
space-averaged geometric and arithmetic means of horizontal
permeabilities for the layers.

The three stages of scaling up, clast to core, core to
layer and layer to zone, are significantly different. The
clast heterogeneity is essentially binary and of very high
permeability contrast. As a result of this contrast,
permeabilities at core scale are highly dependent on clast
shapes and spatial distributions. The core to layer scaling
up is characterized by more gently varying heterogeneity and
the results are accordingly less sensitive to the details of
spatial variation within the layer. The final scaling up
from layer to zone scale is intermediate between the first
two stages in the sense that the heterogeneity consists of
moderate permeability contrast between sand and shale clast
layers. Consequently, results are moderately sensitive to
the geometry of the layers. Considering these sensitivities
and present uncertainties, an order of priority to improve
the predictions presented here might be (1) to obtain core
scale results in 3D and for a wider range of shale clast
distributions, (2) to better characterize the layer geometry
at zone scale, and (3) to better characterize spatial
variability within both shale clast and sand layers.

ACKNOWLEDGEMENTS

The authors wish to express their great appreciation to Joe
Olic, whose prodigious efforts in digitizing the shale clasts
and in other aspects of the project, have made this work
possible. The authors also wish to acknowledge the Alberta
Research Council, the Alberta Oil Sands Technology and
Research Authority and the Alberta Department of Energy for
funding the research reported in this paper.

REFERENCES

Begg, S.H., R.R. Carter and P. Dranfield (1987). "Assigning Effective Values to Simulator Grid-block Parameters in Heterogeneous Reservoirs". SPE paper 16754 presented at 62nd Annual Technical Conference, Dallas, Sept. 27-30.

Begg, S.H., D.M. Chang and H.H. Haldorsen (1985). "A Simple Statistical Method for Calculating the Effective Vertical Permeability of a Reservoir Containing Discontinuous Shales". SPE paper 14271 presented at the 60th Annual Technical Conference, Las Vegas, Sept. 22-25.

Begg, S.H. and P.R. King (1985). "Modelling the Effects of Shales on Reservoir Performance: Calculation of Effective Vertical Permeability". SPE paper 13529 presented at the Reservoir Simulation Symposium, Dallas, Feb. 10-13.

Desbarats, A.J. (1987). "Numerical Estimation of Effective Permeability in Sand-shale Formations". Water Resour. Res., 23, 273-286.

Deutsch, C. (1989). "Calculating Effective Absolute Permeability in Sandstone/Shale Sequences". SPE paper 17264 presented at the Formation Evaluation Symposium.

Dupuy, M., and E. Lefebvre du Prey (1968). "L'anisotropie d'Ecoulement en Milieu Poreux Presentant des Intercalations Horizontales Discontinues". Presented at 3rd Colloquium, Assoc. de Rech. sur les Tech. de Forage et de Prod., Pau, France, Sept. 23-26.

Gupta, S.K., C.R. Cole, F.W. Bond, and A.M. Monti (1984a). "Finite-element Three-Dimensional Groundwater (FE3DGW) Flow Model Formulation, Program Listings and Users' Manual". Pacific Northwest Laboratory, Richland, Washington.

Gupta, S.K., C.R. Cole, and G.F. Pinder (1984b). "A Finite-Element Three-Dimensional Groundwater (FE3DGW) Model for a Multiaquifer System". Water Resour. Res. 20, 553-563.

Gutjahr, A.L., L.W. Gelhar, A.A. Bakr and J.R. McMillan (1978). "Stochastic Analysis of Spatial Variability in Subsurface Flows, 2. Evaluation and Application". Water Resour. Res. 14, 953-959.

Haldorsen, H.H., and D.M. Chang (1986). "Notes on Stochastic Shales: From Outcrop to Simulation Model" in Proceedings of the Reservoir Engineering Technical Conference, April 29 - May 1. Dallas, TX, edited by L.W. Lake and H.B. Carroll, Academic Press, Orlando, Fla., 1986.

Haldorsen, H.H., and L.W. Lake (1984). "A New Approach to Shale Management in Field Scale Simulation Models". Soc. Pet. Eng. J., 24, 447-457.

Journel, A.G., C. Deutsch, and A.J. Desbarats (1986). "Power Averaging for Block Effective Permeability". SPE Paper 15128, presented at the 56th California Regional Meeting of the Society of Petroleum Engineers, Oakland, CA., April 2-4.

Journel, A.G. and C. Huijbregts (1978). "Mining Geostatistics". Academic Press, Orlando, Fla.

Kramers, J.W., S. Bachu, D.L. Cuthiell, M.E. Prentice and L.P. Yuan (1989). "A Multidisciplinary Approach to Reservoir Characterization: the Provost Upper Mannville B Pool". J. Can. Petr. Techn., 28, no. 8, 48-58.

Matheron, G. (1967). "Composition des Permeabilites en Milieu Poreux Heterogene: Methode de Schwydler et Regles de Ponderation". Rev. Inst. Fr. Pet., 22, 443-466.

Matheron, G. (1975). "Random Sets and Integral Geometry". John Wiley and Sons (Interscience), New York.

Richardson, J.G., J.B. Sangree, and R.M. Sneider (1987). "Permeability Distributions in Reservoirs". J. Pet. Tech., October, 1197-1199.

Tough, J.G. and R.G. Miles (1984). "A Method for Characterizing Polygons in Terms of the Principal Axes". Computers and Geosciences 10, 347-350.

ANALYSIS OF UPSCALING AND EFFECTIVE[1] PROPERTIES IN DISORDERED MEDIA

Yoram Rubin [2]
J. Jaime Gómez-Hernández [3]
André G. Journel

Department of Applied Earth Sciences
Stanford University
Stanford, California

Abstract

Two methods to determine the effective properties of finite-size domains are presented and discussed. The first method is built around a normal-score transform of the conductivity, while the other calls for a non-parametric approach that requires an indicator-transform of the data. Both methods are discussed, demonstrated and compared.

[1]Supported by the Stanford Center for Reservoir Forecasting
[2]Now at Department of Civil Engineering, University of California, Berkeley, California
[3]Now at Departmento de Ingeniería Hidráulica y Medio Ambiente, Universidad Politécnica de Valencia, 46071 Valencia, Spain

251

I. INTRODUCTION

Random media are generally characterized by many scales of heterogeneity. When applying quantitative tools to analyze physical phenomena in random media, one needs a common yardstick to relate and compare processes that occur at the different scales. This need is particularly evident when addressing the problem of reservoir characterization. Flow processes observed at a given scale of heterogeneity display different features from those observed at other scales (Lasseter, Waggoner and Lake, 1986, hereafter referenced as LWL). When facing the actual task of characterizing a reservoir, one must use all available measurements that stem from a wide range of scales: from the core scale of laboratory measurements, to the intermediate scale of well test data and on to the largest scale of production data and basin description. A methodology is required for transporting information from one scale to another. In this study, we will adopt a somewhat widely accepted hierarchy of scales, and we will show how effective conductivities can be assigned to blocks of any given size, making use of measurements taken at different scales.

II. ON THE DEFINITION OF SCALES

Two methods for scaling heterogeneity were discussed recently that should be viewed as complementary.

The first approach (Hewett, 1986) suggests modeling geological formations with the tools of self-similarity and fractal dimensionality. This approach assumes that there is no disparity between scales; rather a continuum of successive new scales of heterogeneity is found as the size of the observation domain increases, all such scales of variability being self-similar.

The second approach assumes some disparity between scales. Similar hierarchies of scales were suggested independently by four different studies, approximately at the same time. LWL define the small scale, the medium and the large scale. The medium scale is the scale of the depositional unit, which is the body formed as a result of a single major depositional event. The small scale is the subdepositional unit scale and the large scale is an ensemble of many depositional units.

Haldorsen (1986) and, Krause and Collins (1984) define four scales: the microscopic scale which is the scale of pores and sand grains; the macroscopic scale, corresponding to the scale of conventional core plugs; the megascopic scale which is the scale of grid blocks in simulation models, and the gigascopic scale, which is the formation or regional scale.

Dagan (1986) distinguishes between three scales: laboratory scale, local scale and regional scale, which correspond to Haldorsen's (1986) microscopic, megascopic and gigascopic scales and LWL's (1986) small, medium and large scales, respectively. Note that Haldorsen's additional scale, the macroscopic, does not have any counterpart on either Dagan's or LWL's hierarchy of scales.

From a mechanistic point of view, following LWL, the pore scale can be considered as corresponding to the viscous-capillary regime, where gravity forces are generally unimportant, the local scale corresponds to the viscous-capillary-gravity regime, while the regional scale is the viscous-gravity regime.

In our study, we adopt the approach with disparity of scales and will focus on the flow variables at the local (LWL's medium) and regional (LWL's large) scales and their averaging process. A good discussion and demonstration of the flow regimes at each scale is given by LWL and by Dagan (1986).

Starting with the local scale the following regionalized variables (ReV) with quasi "point" support are considered: the specific flux, $\mathbf{q}(\mathbf{x})$ (where boldface letters denote vectors); the conductivity $k(\mathbf{x})$, defined through Darcy's law; and the piezometric head $\phi(\mathbf{x})$, with $\mathbf{x} = (x_1, x_2, x_3)$. Combining Darcy's law and the continuity equation leads to the steady-state relation:

$$\nabla \cdot (k \nabla \phi) = 0 \tag{1}$$

where:

$$\mathbf{q} = -k \nabla \phi$$

The previous variables are defined in the three-dimensional space at the local scale of, say, core plug measurements. They vary erratically over the domain and over the scale of the simulation grid blocks. At the local scale, the flow is considered as 3-D.

Based on the previous definition of the variables at the local scale, we derive the variables at the next scale, the regional. For that, we will average the ReV's over the reservoir depth. This averaging could be done over any chosen area resulting in a new ReV characterized at a different scale; thus there could be an infinite number of scales, depending on the volume of averaging. This aspect was demonstrated in Rubin and Gómez- Hernández (1989).

The regional scale is particularly useful in any flow problem where the horizontal dimension is much larger than the vertical one. This is the case often encountered in aquifers and thin oil reservoirs, Rosenzweig *et al.* (1986). The vertical averaging is justified if vertical flow is deemed negligible. In such case the flow is assumed to be strictly horizontal and we can define the new point variables in 2-D: head (equivalent to pressure) $H(x_1, x_2) = (1/L) \int_0^L \phi(x_1, x_2, x_3) dx_3$; flux $\mathbf{Q}(x_1, x_2) = \int_0^L \mathbf{q}(x_1, x_2, x_3) dx_3$ and transmissivity, which is defined as $T(x_1, x_2) = \int_0^L k(x_1, x_2, x_3) dx_3$ where L is the aquifer depth. These three variables are related through a relation similar to (1):

$$\nabla \cdot (T \nabla H) = 0 \qquad (2)$$

where:

$$\mathbf{Q} = -T \nabla H$$

The new variables $\mathbf{Q}$, H and T will be considered point variables at the regional scale since their "support" is much smaller than the regional integral scale.

III. EFFECTIVE PROPERTIES

The problem of upscaling or determination of effective properties, is a research topic in many areas of physics, where one is interested in averaging processes allowing estimation of an equivalent parameter that produces an output variable that replicates some a priori defined mean behavior of the system. For instance, starting with "point" transmissivity data $T(x_1, x_2)$ as defined in (2) one may want to define the single-valued effective transmissivity that matches the "average" flow characteristics over a given horizontal domain S. It is beyond our scope to review the wide body of literature on this subject, and the reader is referred to Landauer (1977). In order to conform with accepted terminology, it is useful to mention that effective properties refer to domains infinitely larger than the scale of heterogeneity. When dealing with domains of the order of the heterogeneity scale, the term block effective properties is more appropriate. While effective properties are devoid of any notion of domain size, block properties are functions of the block size.

All variables in (2) are now ReV's in 2-D defined as spatial integrals of the corresponding 3-D ReV's over the depth L (not necessarily constant over the horizontal extent of the aquifer). The probabilistic approach consists in treating heterogeneous soil variables as if they were samples drawn from a population of correlated random variables. This approach allows for both spatial variability and prediction uncertainty to be analyzed with the tools of probability theory.

One then interprets the previous ReV's as realizations of random functions $\tilde{H}(x_1, x_2)$, $\tilde{T}(x_1, x_2)$ and $\tilde{Q}(x_1, x_2)$ where the tilde is used to distinguish random functions from their realizations. From relation (2) comes the classical defintion of effective transmissivity:
$T_{ef} = E\{\tilde{Q}(x_1, x_2)\}/E\{-\nabla \tilde{H}(x_1, x_2)\}$.

Instead of the expected values defined on the random function models, one could define spatial integrals of the ReV's over any finite horizontal domain S:

$$\frac{1}{S} \int_S \mathbf{Q}(x_1, x_2) dx_1 dx_2 \quad \text{and} \quad \frac{1}{S} \int_S \nabla H(x_1, x_2) dx_1 dx_2 \qquad (3)$$

The ReV's defined in (3) are related through relation (2), leading to the integral relation:

$$\frac{1}{S} \int_S \mathbf{Q}(x_1, x_2) dx_1 dx_2 =$$
$$= -\frac{1}{S} \int_S T(x_1, x_2) \nabla H(x_1, x_2) dx_1 dx_2$$
$$= -T_S \frac{1}{S} \int_S \nabla H(x_1, x_2) dx_1 dx_2 \qquad (4)$$

with the block (S) effective transmissivity defined as:

$$T_S = -\frac{\frac{1}{S} \int_S \mathbf{Q}(x_1, x_2) dx_1 dx_2}{\frac{1}{S} \int_S \nabla H(x_1, x_2) dx_1 dx_2} \neq \frac{1}{S} \int_S T(x_1, x_2) dx_1 dx_2 \qquad (5)$$

The previous integrals and T_S itself are ReV's dependent on the geometry, size and location of the domain S. However, if interpreted as realizations of **stationary and ergodic** stochastic integrals, their limits, as S tends towards infinity, are equal to the expected values of the corresponding stochastic integrals, i.e., $T_S \rightarrow E\{\tilde{T}_S\} = T_{ef}$ as $S \rightarrow \infty$. However for practical domain size $S < \infty, T_S \neq E\{\tilde{T}_S\} \neq T_{ef}$.

Stationarity is a modeling decision rather than a property of the actual data or an assumption; it cannot be proved or disproved from data. It is a model property, a decision allowing to pool data from a given domain into statistics deemed representative of the whole domain. Without some form of stationarity in space and/or time, there would be no pooling allowed, hence no possibility of statistical inference.

The ergodic limit T_{ef} is unique and would not provide any spatial resolution over the entire horizontal extent of the averaging area S. This value is of little use except for global "accountability type" studies, for which one is only interested in the mean flux going through the aquifer.

The marginal expected value $E\{\tilde{T}_S\}$ can be seen as the average of all possible realizations T_S, obtained by moving the area S, fixed in size and geometry, throughout the entire aquifer–of size much larger than S. Although $E\{\tilde{T}_S\}$ does depend on the size and geometry of S, it does not depend any more on the particular location of S within the aquifer, i.e., it does not allow anymore spatial resolution than T_{ef}.

How then can we get spatial resolution short of actually sampling T_S at each location (x_1, x_2) of the centroid of S? An answer lies in noting that what makes an area $S(x_1, x_2)$ different from an area $S(x_1', x_2')$ is the actual information available within and in the vicinity of each area. Thus, the solution is to condition the expected value of $\tilde{T}_S$, or better, its probability distribution, to that local information. Consider the conditional expectation:

$$E\{\tilde{T}_S|(n_S)\} \tag{6}$$

with (n_S) being the information available within S and its neighborhood. Some remarks:

1. Even if (n_S) is extended to include all information (N) available within the aquifer, the two following conditional expectations are different

$$E\{\tilde{T}_S(x_1', x_2')|(N)\} \neq E\{\tilde{T}_S(x_1'', x_2'')|(N)\} \tag{7}$$

since conditional distributions, and hence their conditional expectations and variances, are never stationary, even if the original random function $\tilde{T}_S$ is strictly so.

2. The conditioning by (n_S) of the expected value $E\{\tilde{T}_S\}$ amounts to averaging only those realizations $T_S(x_1'', x_2'')$ that have the same data environment. This restriction allows accounting for local information and allows spatial resolution according to (7). The same line of reasoning can be used with regard to higher moments of the probability distribution function of the RV. Thus, conditional distributions are far more informative than the unconditional ones.

Derivation of either the marginal or conditional distributions can be done by each of the following techniques:

1. Make the hypothesis that the multivariate pdf of $\tilde{T}_S$ (or a normal score transform of it) is multinormal, and then derive the marginal or conditional distributions desired.

2. Use a Monte Carlo-type algorithm to generate the conditional distribution of $\tilde{T}_S$ given a model of spatial auto-correlation for $\tilde{T}_S$ and conditioned to actual data values.

The two approaches will be explored in the next two sections.

IV. MULTINORMAL PARAMETRIC STOCHASTIC APPROACH TO UPSCALING

As mentioned above, one possible approach to upscaling is to model the block effective property as a random function and to assume that its multivariate pdf is multinormal. Then, derivation of the conditional pdf of T_S is relatively easy. This statement succintly describes the method developed by Rubin and Gómez-Hernández (1989). In this section we summarize the principles and major results of the method.

The following assumptions are at the foundation of the approach:

1. $\tilde{Y}$, the logtransmissivity, is a random function of multinormal pdf. Some field studies (Hoeksema and Kitanidis, 1985; Sudicky, 1986;

Stalkup, 1986) have found Y to display a univariate normal distribution. This does not imply a multinormal probability distribution function for Y, which is adopted here as a constitutive hypothesis.

2. $\tilde{Y}$ is adequately described by a first-order expansion:

$$\tilde{Y}(\mathbf{x}) = \langle \tilde{Y}(\mathbf{x}) \rangle + \tilde{Y}'(\mathbf{x}) \quad \text{with } \mathbf{x} = (x_1, x_2) \tag{8}$$

where $\tilde{Y}'$ is the local random fluctuation of $\tilde{Y}$ around its stationary expected value $\langle \tilde{Y} \rangle$. $\langle \tilde{Y} \rangle$ is assumed to be constant in space.

3. We further define the random function

$$\tilde{H}(\mathbf{x}) = \langle \tilde{H}(\mathbf{x}) \rangle + \tilde{H}'(\mathbf{x}) \tag{9}$$

i.e., the head at any location is composed of its expected value and a local random fluctuation. We assume a constant, unidirectional mean head gradient, whose direction, for ease of mathematical manipulation, coincides with that of the x_1 axis. This means that the head residual, rather than the head itself, is stationary.

The preceding assumptions are used in conjunction with (4) and (2) in Rubin and Gómez-Hernández (1989) to obtain closed-form analytical expressions for the first and second moments of $\tilde{T}_S$.
The expected value of $\tilde{T}_S$ is:

$$\langle \tilde{T}_S \rangle = T_G \left\{ 1 + \frac{\sigma_Y^2}{2} - \frac{1}{J_1} \left[C_{YdH}(0) - \frac{1}{S^2} \bar{C}_{YdH}(S, S) \right] \right\} \tag{10}$$

where T_G is the geometric mean of the transmissivity field; $\sigma_Y^2 = \langle \tilde{Y}'^2 \rangle$ is the variance of $\tilde{Y}$; $J_1 = -\partial \langle \tilde{H} \rangle / \partial x_1$, is the constant expected value of the head gradient; $\bar{C}_{YdH}(S, S)$ is the regularization of the logtransmissivity and head-derivative cross-covariance over the area S. In the derivation of (10), it is assumed that the medium is isotropic and that the flux and head gradient are colinear.

In order to evaluate (10), an explicit model is needed for the point-support logtransmissivity covariance. For application we adopt for $\tilde{Y}$ the isotropic exponential covariance

$$C_Y(\mathbf{x}; \mathbf{x}') = \sigma_Y^2 \exp(-r/I) \tag{11}$$

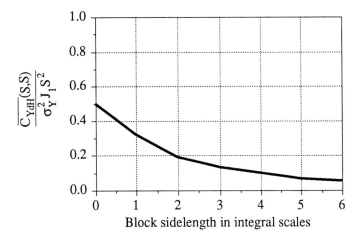

Figure 1: Block cross-covariance between Y and dH.

where I is the integral scale of the logtransmissivity and $r = |\mathbf{x} - \mathbf{x}'|$. Rubin and Gómez-Hernández (1989) obtained explicit expressions for $\bar{C}_{YdH}$ by using (11) and the linearized flow equation for an unbounded flow domain. This last assumption amounts to neglecting the influence of boundaries, and was shown by Rubin and Dagan (1988, 1989) to be acceptable at distances larger than $1I$ from the boundaries.

The covariance value $C_{YdH}(0)$ in 2-D was shown to be equal to $-\sigma_Y^2 J_1/2$, which is in conformity with the results of Gutjahr (1979). The results for $\bar{C}_{YdH}$ for equal size blocks as function of their non-dimensional side-length are depicted on Figure 1. In the linear development adopted here, $\bar{C}_{YdH}$ appears as a linear function of J_1, thus according to (10), $\langle \tilde{T}_S \rangle$ is not a function of the magnitude of the gradient. However, it is a function of the logtransmissivity variability and of the various assumptions concerning the flow configuration.

For $S = 0$, $\langle \tilde{T}_S \rangle = T_G(1 + \sigma_Y^2/2)$, which is the transmissivity expected value as it should be. On the other hand, for $S \to \infty$ $\langle \tilde{T}_S \rangle = T_{ef} = T_G$, which is a classical result (Dagan, 1979; Gutjahr et al., 1978).

Considering a first-order expansion of $\tilde{T}_S$ in $\tilde{Y}'$ and $\tilde{H}'$, the block transmissivity covariance is derived:

$$C_{T_S}(\mathbf{r}) = \text{Cov}\left\{\tilde{T}_S(\mathbf{x}); \tilde{T}_S(\mathbf{x} + \mathbf{r})\right\} = T_G^2 \bar{C}_Y(S, S_{+\mathbf{r}}) \tag{12}$$

$\bar{C}_Y(S, S_{+\mathbf{r}})$ is the regularized logtransmissivity covariance between two

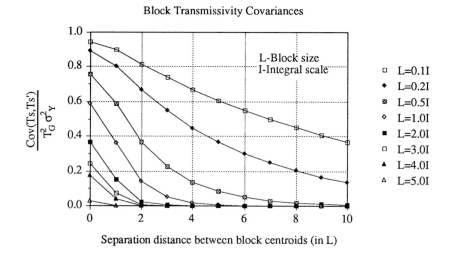

Figure 2: Block transmissivity covariances. Computed for two equal sized square blocks S and S' whose centroids are located along a line parallel to the block sides

blocks, with centroids distant of $\mathbf{r}$. The blocks S and $S_{+\mathbf{r}}$ need not be of the same size. Since $\tilde{Y}$ was assumed multinormal, and $\tilde{T}_S$ is found to be a linear function of $\tilde{Y}'$, it turns out that $\tilde{T}_S$ is also multinormal. Consequently, its pdf is defined entirely by (10) and (12). Figure 2 depicts C_{T_S} for equal size blocks as a function of their non-dimensional side length. As expected, volume averaging leads to a reduction of variance. However, even for relatively large side lengths, the block conductivity variance can not be neglected, and block conductivity can not be considered to be deterministic at practical block sizes. Different models for C_Y other than (11) would apparently show different rates of decrease after regularization. According to the regularization tables in Journel and Huijbregts (1978), the decrease would be much more dramatic for a covariance model with nugget effect, and less pronounced for a Gaussian covariance model.

An important outcome of the derivation of the multinormal pdf of the size-dependent block conductivity random function is the immediate extension to conditional probabilities. Using Bayes theorem, the conditional pdf of $\tilde{T}_S$ is obtained by Gaussian conditioning. The corresponding explicit relationships are given in Rubin and Gómez-

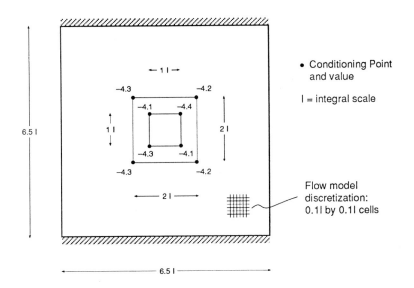

Figure 3: Plan view of the aquifer. The location and values of the conditioning points are also shown

Hernández (1989) and in Journel (1980). The conditional pdf ensures a conditioning of $\tilde{T}_S$ on transmissivity measurements taken over varying supports, as well as head (pressure) measurements, since all that is required to condition on such T measurements is the interscale covariance (12). As to the conditioning on head measurements one needs the logtransmissivity-head cross-covariance, given in Rubin and Gómez-Hernández (1989). In essence, the role of conditioning is to inform $\tilde{T}_S$ about the presence of various data in S or in the vicinity of S. In practice, conditioning capitalizes on the spatial structure to constrain the range of values that $\tilde{T}_S$ can adopt. The possibility to condition on both head and T measurements is important since head measurements are cheaper to obtain, and in general much more abundant than T measurements.

The practice of conditioning on data of different types, or data of different supports, is commonly referred to as cokriging (Journel and Huijbregts, 1978) in the geostatistical terminology. We shall now describe an application of the methodology.

Figure 3 depicts a 2-D domain. T_G has the value of $e^{-4.0}$. Also given are the locations and values of eight arbitrary logtransmissivity measurements considered as having quasi point support and located near the center of the domain. All distances are normalized by the integral scale. The exponential covariance model given by (11) is adopted with different values for σ_Y^2. Using Gaussian conditioning the conditional mean block conductivity and the conditional variance for any block size can be derived analytically. These results can also be obtained by Monte-Carlo (MC) simulation. The essence of the MC approach is that, by viewing the parameters of (2) as random functions, this equation turns out to be a stochastic partial differential equation, and its solution is also a random function characterized by a pdf. This pdf is inferred from the relative frequency distribution of the different solutions to (2) obtained numerically for different realizations of the random input transmissivity field. This procedure is described in greater detail in Rubin and Gómez-Hernández (1989) for unconditional simulations only. Its extension to conditional simulations is done following Journel and Huijbregts (1978). The major steps of the numerical conditional simulation are outlined here:

1. The MC simulations were carried out over a rectangular aquifer of size $6.5I$ that was divided into rectangular cells of size $0.1I$. The boundary conditions imposed are no flow along the boundaries parallel to the x_1 axis and constant head boundaries parallel to the x_2 axis. The constant head values were uniform along each side of the aquifer and introduced a hydraulic gradient that drives the flow across the aquifer.

2. The input random fields were generated by the turning band random generator TBM (Journel, 1974; Mantoglou and Wilson, 1981). The unconditional realizations were then made conditional to the data by kriging the differences between the measured values and the random values obtained at their locations and subtracting the kriged differences from the unconditional realizations (Journel and Huijbregts, 1978). Figure 4 depicts one of the unconditional realizations among the many that were generated for $\sigma_Y^2 = 1$ and $\langle \tilde{Y} \rangle = -4$, and Figure 5 shows another simulation that has been made conditional to the data configuration given in Figure 3. The conditional realization appears smoother than the unconditional one, since the conditioning measurements were little dispersed around the mean logtransmissivity.

3. The numerical solution for each realization under the specified

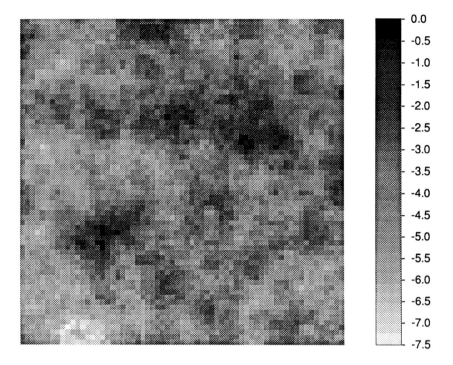

| 0.0 |
| -0.5 |
| -1.0 |
| -1.5 |
| -2.0 |
| -2.5 |
| -3.0 |
| -3.5 |
| -4.0 |
| -4.5 |
| -5.0 |
| -5.5 |
| -6.0 |
| -6.5 |
| -7.0 |
| -7.5 |

Figure 4: An unconditional simulation of logtransmissivity values

boundary conditions was obtained by the finite-difference code of Trescott, Pinder and Larson (1976).

4. Various blocks of different size S were outlined around the center of the domain. For each realization and after the flow simulation, the average value T_S was computed for each block using relation (4). This value is then considered as a realization of the random function $\tilde{T}_S$. The mean and variance of $\tilde{T}_S$ were then obtained by averaging over all such realizations.

Two hundred realizations were generated for each variance σ_Y^2 considered. It was checked that the magnitude of the mean head gradient did not influence the results. Also, it was checked that the random generator produced the desired input statistics, namely an exponential covariance model of given variance and integral scale and the correct mean of $\tilde{Y}$.

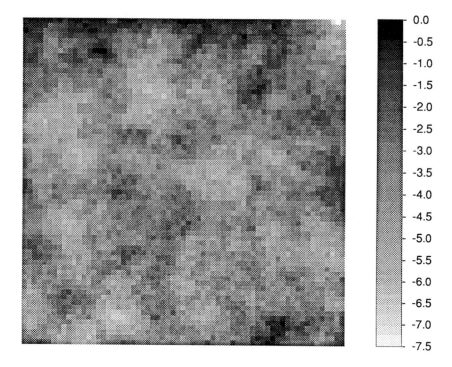

| 0.0 |
| -0.5 |
| -1.0 |
| -1.5 |
| -2.0 |
| -2.5 |
| -3.0 |
| -3.5 |
| -4.0 |
| -4.5 |
| -5.0 |
| -5.5 |
| -6.0 |
| -6.5 |
| -7.0 |
| -7.5 |

Figure 5: A conditional simulation of logtransmissivity values. The conditioning points appear in Figure 3

Figure 6 gives statistics corresponding to the simulations with $\sigma_Y^2 = 1$ and the theoretical mean block transmissivity for various square block sizes. The theoretical mean conforms quite well with numerical simulations. For the smallest block, the experimental distribution of $\tilde{T}_S$ appears skewed as indicated by the difference between the experimental values of the mean and median. This skewness reveals a departure from normality of the experimental results and, therefore, questions the validity of the theoretical results for such small blocks and for the adopted value for σ_Y^2. The skewness disappears, and the experimental distribution approaches a normal one, as the block gets larger. In standard modeling practice, the value chosen for the flow simulator blocks would be the geometric mean of the data, which, in this particular case, corresponds to $0.78T_G$, clearly underestimating the "true" value as obtained with the MC approach.

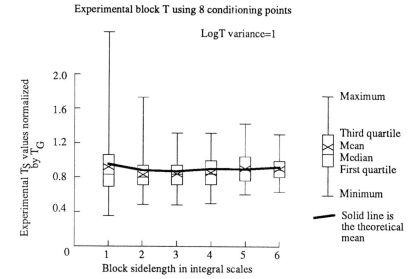

Figure 6: Summary statistics of the experimental block transmissivity values obtained using the conditioning points depicted in Figure 3 for $\sigma_Y^2 = 1$

Figure 7 gives the conditional block variances corresponding to these simulations, as well as their theoretical values, and the agreement is good. The shape of the conditional variance curve is a consequence of the particular location of the data relative to the different blocks. For example, for the smallest block of side-length $1I$, the four inner data are most consequential, 'screening' the other data. For blocks of side-lengths $2I$ and $3I$ the data configuration is more informative, there is no redundancy among the data since they are located evenly within the block. For larger blocks, the effects of screening and data redundancy are again observed. Upon comparing the conditional and unconditional variances for the largest block, it appears that conditioning becomes less important. This is intuitive, since for very large blocks, a data of quasi point support does not carry much information.

The exercise described above was repeated for $\sigma_Y^2 = 2$, with the results given on Figures 8 and 9. Although the variance value $\sigma_Y^2 = 2$ is beyond the limits of linear theory, the agreement between theory and experiments remains good. One explanation lies in the practice of

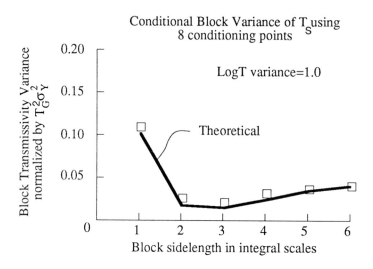

Figure 7: Comparison between the experimental block conditional variance and the theoretical value for the unconditional variance $\sigma_Y^2 = 1$

conditioning. The conditional variance is much smaller than the unconditional one, making the linear theory applicable over a wider range of
variability than assumed a priori. This robustness was demonstrated in
some previous studies, e.g., Rubin and Dagan (1988).

The assumption of multilognormality for T is essential if the first
two moments of $\tilde{T}_S$ are considered as sufficient to describe the entire
conditional pdf of $\tilde{T}_S$. However, the derivation of the stationary block
covariance (12) requires only the assumption of second-order stationarity
of Y. If the assumption of multinormality is withheld, that covariance
can be still used in a simple kriging formalism to obtain estimates of
the block transmissivities, but such estimates would not any more be
qualified for uncertainty; recall that the kriging variance is not a measure
of uncertainty but a mere ranking of data configuration (Journel, 1986).

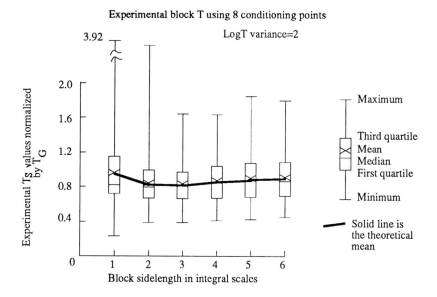

Figure 8: Summary statistics of the experimental block transmissivity values obtained using the conditioning points depicted in Figure 3 for $\sigma_Y^2 = 2$

V. A STOCHASTIC NON-PARAMETRIC APPROACH TO UPSCALING

The method described in the previous section corresponds to a parametric approach in the sense that a multi-Gaussian related model for the distribution of the point attributes is assumed a priori. The difficulties in accepting the basic assumption may be alleviated: (1) by requiring that the assumption is checked before being applied, using careful diagnostics of the data; (2) by noting that its parsimony is most attractive when data are scarce; (3) by noting that this assumption is less consequential when data are abundant and conditioned upon; and (4) by acknowledging the gain stemming from the possibility to condition on head measurements.

Intuitively, the multi-Gaussian assumption may be more relevant at the regional scale, since then the quasi smaller scale processes are already a volume average of the point process. In some formations, the

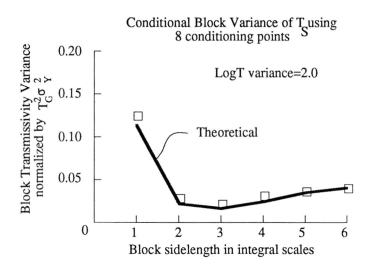

Figure 9: Comparison between the experimental block conditional variance and the theoretical value for the unconditional variance $\sigma_Y^2 = 2$

point process at the local scale may display a more pronounced spatial variability which can depart considerably from a unimodal distribution. Bimodal distributions such as sand-shale formations may sometimes be encountered, although the low mode (shales) or high mode (sand or fractures) may be difficult to sample. This situation calls for a somewhat different approach, where the above-mentioned assumptions are relaxed, allowing a wider range of applicability. One such approach, which is based on non-parametric statistics, will be described here only in principle.

In favor of a non-parametric approach, it may be noted that: (1) Gaussian-related models for stochastic simulation of random functions do not offer the flexibility to reproduce patterns, possibly critical, of spatial connectivity specific to extreme values (Journel and Alabert, 1988); (2) most often the attributes display a highly skewed distribution with large coefficient of variation and multi-modality not conducive to a Gaussian model.

In flow simulations the single most influential input is the permeability (or conductivity) field that conditions the flow paths. In a heterogeneous reservoir involving sequences (e.g. shales, sands and fractures) with permeability values differing by several orders of magnitude, flow is mostly conditioned by connected paths of either high or low permeability values (flow paths and barriers, respectively). Neither the histogram shape, whether lognormal or not, nor the proportions of extreme permeability values matter as much as the spatial connectivity of these extreme values. Randomly disconnected small fractures may not generate flow paths, whereas a minute volume proportion of connected high permeability values may wholly condition flow. In such situations, reservoir characterization and upscaling should strive to detect patterns of connected values, and account for their effect in a most realistic manner. In this section, an extension of the method to the upscaling problem will be described.

A. Characterization of the Random Field

Denote by $k(\mathbf{x}_l), l = 1, .., N$ the distribution over a given area of conductivity measurements. Define the indicator transform of any variable $k(\mathbf{x})$ by:

$$i(\mathbf{x}; k) = \begin{cases} 1 & \text{if} \quad k(x) \leq k \\ 0 & \text{otherwise} \end{cases} \tag{13}$$

The proportion of data valued less than or equal to a threshold k, is the arithmetic mean of the N indicator data:

$$\begin{aligned} F_A(k) &= \text{Proportion within } A \text{ of } k(\mathbf{x}_l) \leq k \\ &= \frac{1}{N} \sum_{l=1}^{N} i(\mathbf{x}_l, k) \end{aligned} \tag{14}$$

Similarly, consider the $N(\mathbf{r})$ pairs of data locations separated by the same vector $\mathbf{r}$; the $\mathbf{r}$-bivariate distribution is defined as the proportion of such pairs with simultaneously $k(\mathbf{x}_l)$ and $k(\mathbf{x}_l + \mathbf{r})$ valued less than or equal to the threshold k:

$$\begin{aligned} F_A(\mathbf{r}; k) &= \text{Proportion within } A \text{ of pairs} \\ &\quad k(\mathbf{x}_l) \leq k \text{ and } k(\mathbf{x}_l + \mathbf{h}) \leq k \\ &= \frac{1}{N(\mathbf{r})} \sum_{l \in \{N(\mathbf{r})\}} i(\mathbf{x}_l; k) i(\mathbf{x}_l + \mathbf{r}; k) \end{aligned} \tag{15}$$

with $N(\mathbf{r})$ being the number of pairs of locations such that $\mathbf{x}, \mathbf{x} + \mathbf{r}$ are in A. For a given separation vector $\mathbf{r}$ and a low threshold value k the bivariate cdf $F_A(\mathbf{r}; k)$ appears as a measure of connectivity between two low values. The greater $F_A(\mathbf{r}; k)$, the higher the probability to find a low value $k(\mathbf{x} + \mathbf{r})$ a vector $\mathbf{r}$ apart from an initial low value $k(\mathbf{x})$. The measure $F_A(\mathbf{r}; k)$ is taken in average over A and may change from one threshold value k to another. Similarly, a measure of high connectivity can be obtained.

Most traditional automatic interpolation algorithms do not account for any connectivity measure of type (15), and the information at the bivariate level, as expressed only by the covariance, does not distinguish between low, medium or high k-values. The Sequential Indicator Simulation (SIS) algorithm, described in Journel and Alabert (1988) and coded by Gómez-Hernández and Srivastava (1989), offers one solution since its essence is the ability to assign different connectivity patterns to different cut-offs.

B. Upscaling of Conductivities Using Indicator Coding

When facing a highly heterogeneous domain with the data available coming from different scales, a fine-scale reservoir model could be built and upscaled into the block effective properties; reservoir simulation based on such a grid would then provide a single answer. Unfortunately in most cases neither the information available nor the computer resources allow such detailed deterministic exercise.

Lack of information is usually confronted by building a stochastic model for the reservoir flow properties at the smallest scale of the data available. But then, one must deal with a distribution of reservoir models at that smallest scale rather than a single model. In order to avoid repetitive and computer-intensive upscaling of each alternative model, it is suggested to draw directly grid block properties from an upscaled stochastic model. The statistics (covariances) of that upscaled model are derived from a few synthetic block data generated by simulating flow on limited areas of the small-scale reservoir models. Both stochastic models at the smallest scale and at the final grid block scale are characterized by a series of indicator variograms reflecting spatial connectivity patterns of the flow properties.

Rather than a closed-form solution for upscaling, as described in the previous section, an experimental methodology is suggested. The approach would be to infer the parameters of the conditional distribution of the block values conditioned to the point values, and eventually to any information available. In this way, one could simply draw from this conditional distribution to obtain a likely representation of the block values, rather than simulate the reservoir at the smallest scale then upscale it. The methodology consists in the following steps:

1. First, the stochastic model of local-scale ReV is to be inferred. In the model described in the former section, this amounts to inferring the mean and covariance of the local scale permeability. In the indicator approach, this amounts to inferring the indicator means and spatial covariances at the various cut-offs, and possibly, the cross-covariances between different cut-offs.

2. Once the mean and covariance of the local-scale indicator transform are inferred, repeated realizations of the reservoir at the smallest scale can be generated. Rather than fill-in values for all local scale blocks, it is sufficient to fill in the values for a limited number of blocks or "patches" of given size at pre-designed locations scattered over the domain. This is particularly easy using SIS, which has the unique feature of partial generation along any pre-assigned path. The size of the blocks corresponds to the size of the grid blocks.

3. Rather than subjecting the whole domain to boundary conditions in a very demanding numerical flow simulation, the values of the block effective conductivities will be obtained by solving numerically the flow equation for each of the blocks separately, subject to appropriate boundary conditions, and using equation (4). The result will be a network of block conductivities.

4. The mean and covariances of the block conductivity indicator transforms will be inferred by using the network of block effective conductivities obtained in step 3. Also, one can infer at this stage the block-point covariance or the covariance of its indicator transforms.

5. The statistics inferred in step 4 constitute an upscaled stochastic model and allow to generate at random the grid block conductivity field, conditional to data at the grid block scale and the local scale. Repeated generations of that field and their corresponding numerical flow simulations are much less demanding numerically than a full-scale random generation and its numerical solution.

Step 3 is critical in this approach. When solving numerically the flow equation for each block separately, one can derive the spatial averages of flow and gradient and evaluate T_S using equation (4). However, there are two problems related to such evaluation: first, the value of T_S will depend on the particular boundary conditions applied to the block sides; second, the assumption of colinearity between flow, gradient and grid orientation will be violated for many of the blocks. To overcome the first problem we compute the value of T_S so that it minimizes the sum of squares of the differences between the mean flow obtained from the flow simulation for different boundary conditions and the flow obtained using the block effective value. The different boundary conditions are constant head boundaries defined by rotating a tilted planar head surface around the center of the block. The second problem amounts to replacing the single-valued T_S by an effective tensor. In this case, for each boundary condition applied to the block sides we will compute the mean flow in the x and y directions $(\bar{Q}_x, \bar{Q}_y)$ and the mean gradients in the x and y directions $(\overline{\nabla_x H}, \overline{\nabla_y H})$. The block effective transmissivity tensor:

$$T_S = \begin{bmatrix} T_{xx} & T_{xy} \\ T_{xy} & T_{yy} \end{bmatrix}$$

will have three unknowns that must satisfy the two equations:

$$\bar{Q}_x = T_{xx} \cdot \overline{\nabla_x H} + T_{xy} \cdot \overline{\nabla_y H}$$

$$\bar{Q}_x = T_{xy} \cdot \overline{\nabla_x H} + T_{yy} \cdot \overline{\nabla_y H}$$

When several boundary conditions are used the number of equations will be larger than the number of unknowns, thus defining an overdetermined system that can be solved using standard least-squares techniques.

VI. SUMMARY AND DISCUSSION

The discussion presented in this paper was aimed at exposing some concepts, some of them new, which are found useful when dealing with the problem of upscaling.

We adopt a stochastic approach to the description of the reservoir properties, and by doing so we also recognize that it is "...neither desirable nor practical to observe every detail of the hydraulic conductivity field. Rather, it is appropriate to characterize this highly complex spatial structure in terms of statistical quantities." (Bakr et al., 1978). This approach is widely accepted and applied in many branches of the earth sciences, as well as in physics.

An immediate question is how the random-like properties average in space. In order to provide a physically plausible solution, the flow equation must be accounted for in the averaging process. Past works along that avenue usually resulted in the derivation of effective properties that, by definition, pertain to very large domains and neccesitate the assumption of ergodicity; also the resulting effective properties evaluated are not qualified for uncertainty.

In the present work, we considered a definition of block effective properties which pertain to domains that are of the order of the heterogeneity scale. Instead of assigning single, deterministic (ergodic) values to the block effective properties, they are viewed as random functions dependent upon the stochastic properties of the point random process. This is a practical approach, since for most applications one needs to average flow properties over finite domains. This approach is richer in information content and constitutes one of the novel features of the present study. While the "classic" effective value allows only estimation of the mean flux, the block conductivity random function allows for a wider range of applications. For example, one can estimate which are the confidence intervals associated with the estimates of flux over a given domain, allowing for decision-making under uncertainty.

An advantage of the present approach is that it allows conditioning of the block effective properties on local data. Furthermore, data of different support and different quality can be used for that purpose.

An immediate gain from the upscaling procedure proposed is the reduction in computational burden. Rather than generating the random field and simulating it at the local scale, the field can be generated and simulated at the grid block scale, still being conditional to local data.

Conditioning is of importance for the following reasons. First, it leads to consistent models, since measurements are honored at their locations. Second, conditioning imparts robustness with regard to the assumption of stationarity. As already mentioned, stationarity is a modeling decision and can never be proved or disproved a priori. It is, however, a cornerstone in any attempt to derive effective properties, since stationarity and ergodicity lead to finite, well-defined limits for averages such as given in (4). When deriving block effective properties for finite domains, the assumption of stationarity is nothing but a prior decision, and conditioning serves to adjust that decision to local conditions.

Two models were presented in this study. The first is a parametric model, calling for multivariate normal hypothesis, and the second relies on non-parametric statistics. These models have different features, and each can be applied in different situations: (1) The parametric model is suitable for unimodal well-behaved distributions, while the non-parametric one is especially suited for multimodal or highly-skewed distributions; (2) The parametric model in the present study was developed under the assumption of limited conductivity variability, while the non-parametric model is not limited in that sense; (3) The inference of the parametric model is much easier than that of the non-parametric model, since only one covariance model needs to be inferred, while the non-parametric model requires inference of a much larger number of covariances, depending on the number of cut-offs and the decision to account or disregard the cross-covariances. The non-parametric approach requires a large amount of data either on the field under study or on some outcrops possibly available in petroleum applications; (5) the parametric model does not require the relatively demanding numerical computations associated with the non-parametric model.

ACKNOWLEDGMENTS

Financial support for this work was provided by the Stanford Center for Reservoir Forecasting (SCRF).

REFERENCES

Alabert, F.G., Stochastic imaging of spatial distributions using hard and soft information, M.Sc thesis, Stanford University, 1987.

Bakr, A. A., L. W. Gelhar, A. L. Gutjahr and J. R. McMillan, Stochastic analysis of spatial variability of subsurface flows, *Water Resour. Res. 14*, 263–270, 1978.

Dagan, G., Models of groundwater flow in statistically homogeneous porous formation, *Water Resour. Res., 15*(1), 47–63, 1979.

Dagan, G., Statistical theory of groundwater flow and transport: pore to laboratory, laboratory to formation and formation to regional scale, *Water Resour. Res., 22*(9), 120S-135S, 1986

Gómez-Hernández J.J., and R.M. Strivastava, ISIM3D: a three dimensional multiple indicator conditional simulation program, *Computers and Geosciences*, 1989 (submitted for publication).

Gutjahr, A. L., L. W. Gelhar, A. A. Bakr, and J. R. MacMillan, Stochastic analysis of spatial variability in subsurface flows, 2, Evaluation and application, *Water Resour. Res., 14*(5), 953–959, 1978.

Haldorsen, H., Simulator parameter assignment and the problem of scale in reservoir engineering, *in* Reservoir Characterization, Eds. L.W. Lake and H.B. Carroll, Academic Press, 1986.

Hoeksema, R.J., and P.K. Kitanidis, An application of the geostatistical approach to the inverse problem in two-dimensional groundwater modeling, *Water Resour. Res., 20*(7), 1003-1020, 1985.

Hewett, T., Fractal distribution of reservoir heterogeneity and their influence on field transport, SPE paper #15386, 1986.

Journel, A.G., Geostatistics for conditional simulation of orebodies, *Economic Geology, 69*(5). 673-687, 1974.

Journel, A.G., The lognormal approach to predicting local distributions of selective mining unit grades, *Math. Geolog., 12*, 285–303, 1980.

Journel, A. G., Geostatistics: models and tools for the earth sciences, *Math. Geol. 18*(1), 119–140, 1986.

Journel, A. G., and Ch. J. Huijbregts, *Mining Geostatistics*, Academic Press, 1978.

Journel, A. G., and F. Alabert, Focusing on spatial connectivity of extreme-valued attributes: Stochastic indicator models of reservoir hetergoneities, SPE paper # 18324, 1988.

Krause, F.F., and H.N. Collins, Pembina Cardium: recovery efficiency study, a geological and engineering synthesis, Volume I, II, Petroleum Recovery Institute, Calgary, Canada, 1984.

Landauer, R., Electrical conductivity in inhomogeneous media, *in* Electrical transport and optical properties of inhomogeneous media, Eds. J.C. Garland and D.B. Tanner, American Institute of Physics, 1978.

Lasseter, T.J., J.R. Waggoner and L.W. Lake, Reservoir heterogeneities and their influence on ultimate recovery, *in* Reservoir Characterization, Eds. L.W. Lake and H.B. Carroll, Academic Press, 1986.

Mantoglou, A. and J. L. Wilson, Simulation of Random Fields with the Turning Bands Method, Ralph M. Parsons Laboratory, Department of Civil Engineering, MIT, Report no. 264., 199 pp., 1981

Rosenzweig, J.J., N.A. Abdelmalek and J.R. Gochnour, The development of pseudo functions for three phase block oil simulators, *in* Reservoir Characterization, Eds. L.W. Lake and H.B. Carroll, Jr., Academic Press, London, 1986.

Rubin, Y. and G. Dagan, Stochastic analysis of boundary effects on head spatial variability in heterogeneous aquifers: 1. Constant head boundary, *Water Resour. Res.* 24(10), 1698-1697, 1988.

Rubin, Y. and G. Dagan, Stochastic analysis of boundary effects on head spatial variability in heterogeneous aquifers: 2. Impervious boundary, *Water Resour. Res.* 25(4), 707–712, 1989.

Rubin, Y. and J.J. Gómez-Hernández, A stochastic approach to the problem of upscaling of transmissivity in disordered media. 1. Theory and unconditional simulations, Submitted for publication to *Water Resour. Res.*, 1989.

Stalkup, F.I., Permeability variations observed at the faces of crossbedded sandstone outcrops, *in* Reservoir Characterization, Eds. L.W. Lake and H.B. Carroll, Jr., Academic, London, 1986.

Sudicky, E.A., A natural-gradient experiment on solute transport in a sand aquifer: spatial variability of hydraulic conductivity and its role in the dispersion process, *Water Resour. Res.,* 22(13), 2069-2082, 1986.

Trescott, P.C., G. F. Pinder and S.P. Larson, Finite-Difference Model for Aquifer Simulation in Two Dimensions with Results of Numerical Experiments, Chapter C1, Book 7, *Techniques of Water-Resources Investigations of the USGS*, 116 pp., 1976.

ROCK VOLUMES: CONSIDERATIONS FOR RELATING WELL LOG AND CORE DATA

Milton B. Enderlin
Diana K. T. Hansen
Brian R. Hoyt

Halliburton Logging Services
Fort Worth, Texas

I. INTRODUCTION

When cores, whether whole or sidewall, are taken from a well, the first task is to tie the core(s) to the openhole logs. When the observed core data correlate marginally with the log data, a great deal of interest is generated about the quality and validity of the log in question and/or the core analysis technique. The difficulty encountered in log-to-log and log-to-core integration can, in part, be attributed to an unclear sense of the volume of rock investigated by downhole logging tools and core analysis.

In the borehole environment, scale and perspective take on both two-dimensional and three-dimensional aspects: vertical, horizontal and volumetric. In 1987, Enderlin and Hansen discussed scale and perspective in the vertical sense as regards to sedimentary features recognizable on dipmeter logs. Aziz et al. (1987) have stressed the importance of scale in predicting three-dimensional reservoir behavior.

RESERVOIR CHARACTERIZATION II
Copyright © 1991 by Academic Press, Inc.
All rights of reproduction in any form reserved.

Jageler (1976) discussed volumetric considerations in log analysis, while Haldorsen and Lake (1984) discussed volumetric considerations in simulation models. It is important also to realize that sampling strategy is crucial to all forms of problem solving.

II. SCALE

Haldorsen (1983) and Haldorsen and Lake (1984) describe four levels of scale in geologic problems, specifically dealing with volumes: microscopic (grains and pores), macroscopic (measured rock properties), megascopic (reservoir simulators) and gigascopic (pattern of wells in a reservoir). Log and core analyses fall in the microscopic to the low end of the megascopic range.

III. VOLUMETRIC CONSIDERATIONS

Volumetric considerations when comparing logs or logs and cores are not a new concept and have been addressed by a number of authors (Jageler, 1976; Basan et al., 1988; Haldorsen and Lake, 1984). Figure 1 is an illustration of terms used in the discussion of rock volume. Figure 2 is a crossplot of radial depth of investigation versus vertical resolution. The farther a point or field lies from the origin the larger the rock volume investigated by that device. Figure 3 is a crossplot of low volume investigations. Table 1 provides a summary of vertical resolution, radial depth of investigation, angular dispersion and the resulting rock volume for a number of devices and/or sampling techniques. These values are not intended to represent absolute values, rather ballpark figures, since the rock volumes investigated are a function of where the observer chooses the 50%, 85% or 90% line, as well as a function of the porosity, grain density, mineralogy, sidebed effects, borehole fluids, invasion profile, etc.. Table 2 is a blank table which has been included here for input by anyone taking exception to the numbers we have used.

ILLUSTRATION OF TERMS

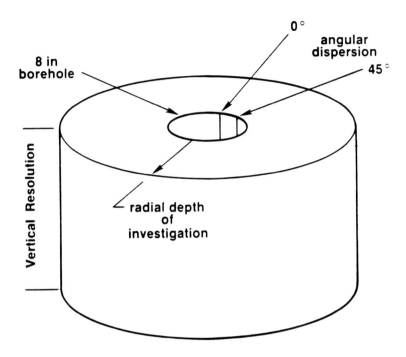

Figure 1. Graphic illustration of terms referred to in rock volume discussion.

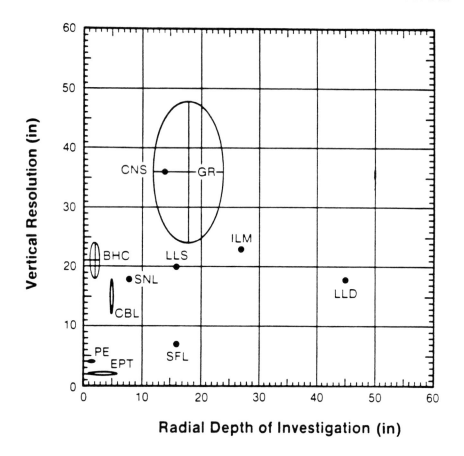

Radial Depth of Investigation (in)

Figure 2. Crossplot of radial depth of investigation versus vertical resolution.

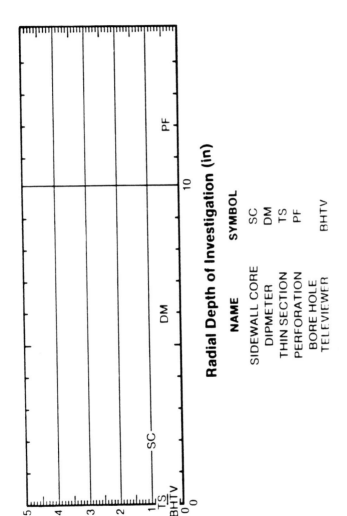

Figure 3. Crossplot of low volume investigation methods.

**Rock Volumes
(8 Inch Borehole)**

Observation	Vertical Resolution (IN)	Radial Depth of Investigation (IN)	Angular Dispersion (Degrees)	Rock Volume Investigated (IN³)
Density	15	5	30	245
Neutron (Mandrel)	24	16	360	28,952
Micro Electrical Log	2	1 5 (50%)	30	7
Gamma Ray	20	15	360	21,676
Unaided human eye focusing on a 1/3 slab of 4" whole core from a distance of 24"	2	N/A	N/A	N/A
Dipmeter (1 Button)	5	6 (90%)	14	5
Borehole Televiewer	3	1	360	7/Scan
Correlation core Gamma Ray of a 4" core using 3" Diameter detector	6	4	N/A	75
6FF40	96	50 (90%)	360	824,619
1' of 4" Core	12	N/A	360	150
1" by 2" Plug	1	N/A	360	1 6
MWD Resistivity	6"	32" (90%)	360	24,127
MWD Bit Resistivity (OBM)	1"	32" (90%)	360	4,021

Table 1. Summary of vertical resolution, radial depth of investigation, angular dispersion and the resulting rock volume for a number of devices and/or sampling techniques.

Rock Volumes
(8 Inch Borehole)

Observation	Vertical Resolution (IN)	Radial Depth of Investigation (IN)	Angular Dispersion (Degrees)	Rock Volume Investigated (IN³)
Density				
Neutron (Mandrel)				
Micro Electrical Log				
Gamma Ray				
Unaided human eye focusing on a 1/3 slab of 4" whole core from a distance of 24"				
Dipmeter (1 Button)				
Borehole Televiewer				
Correlation core Gamma Ray of a 4" core using 3" Diameter detector				
6FF40				
1' of 4" Core				
1" by 2" Plug				
MWD Resistivity				
MWD Bit				
Resistivity (OBM)				

Table 2. Blank table similar to Table 1 which can be filled in with information by the reader.

IV. HAZARDOUS GEOSTATISTICAL CONCEPTS

Logs and cores provide data useful in estimating reservoir parameters. These parameters have a spatial aspect which changes, although neighboring samples may not be independent. These parameters are described as regionalized variables and are actual functions taking on definite values at each point in space. Variation of such functions occurs within a geometric field of regionalization. The spatial aspects of sampling (shape, orientation and volume) interact with this geometric field of regionalization, resulting in a possible lack of continuity between samples of differing geometrical character. This type of variable is most correctly investigated using Markov chains, the variogram and kriging techniques (Matheron,1963; Dacey and Krumbein, 1970).

V. EXAMPLES

Examples of formation homogeneity and heterogeneity are included here to illustrate the method. In these examples RSC represents a small sidewall plug drilled from the borehole wall with a rotary diamond drill bit. WHOLE represents 1-inch plugs removed for analysis from a full diameter core taken with a core bit and barrel. Samples taken at the same depth will be seperated by at least several inches in a plane normal to the axis of the borehole.

A. Example 1

Example 1, an example of formation homogeneity, is from a sandstone reservoir in which whole core and rotary sidewall core samples were taken. Table 3 is a summary of the whole core and sidewall core derived parameters.

Example 1

HOMOGENEOUS

Perm. *SS* (md)		Poros. ϕHe(%)		Grain Density (g/cc)	
RSC▲	WHOLE✳	RSC▲	WHOLE✳	RSC▲	WHOLE✳
0.42	0.63	7.1	7.3	2.60	2.60
178	201	10.5	9.4	2.66	2.65
1050	852	23.5	24.6	2.64	2.65
XXX	XXX	25.4	27.6	2.65	2.68
0.41	0.1	9.1	7.6	2.66	2.65
1.61	4.19	10.0	13.1	2.64	2.64

SS   Steady State Method (Nitrogen Gas)

▲ $^{15}/_{16}$-inch Rotary Sidewall Core Plug

✳   1-inch Whole Core Plug

XXX   Unsuitable sample

Table 3. Example 1, an example of formation homogeneity, is from a sandstone reservoir in which whole core and rotary sidewall core samples were taken. This table summarizes the parameters which were derived from both types of core.

B. Example 2

Example 2, an example of formation heterogeneity, is from a carbonate reservoir in which logs, whole core and rotary sidewall cores were taken. Table 4 is a summary of core and log derived reservoir parameters. The log values listed in Table 4 were taken directly from field logs. The whole core was gamma logged in the laboratory and this log was correlated with the openhole gamma log. It was possible to depth correlate, with confidence, the rotary sidewall cores and the whole core for comparison. As can be seen in Table 4, the grain density varies greatly between core and log derived data. Porosity measurements can also vary greatly.

VI. CONCLUSIONS

- It is important to use common sense when comparing various kinds of data from the borehole.
- The user of log and core data needs to be aware of the rock volumes investigated by each method.
- A good correlation between diverse measurements with differing rock volumes provides insight into the degree of homogeneity of the larger volume. Conversely, poor correlation indicates heterogeneity in the larger volume.

Example 2

HETEROGENEOUS

Depth	Perm. SS* (md) RSC	Perm. SS* (md) WHOLE●	Poros. φHe(%) RSC▲	Poros. φHe(%) WHOLE●	Grain Density (g/cc) RSC▲	Grain Density (g/cc) WHOLE●	Log Data ρ$_b$ (g/cc)	φCDL (%)	φCNS (%)
4734	0.2167	0.7258	9.0	11.0	2.836	2.845	2.69	1.5	9.0
4736	0.7393	0.9213	14.1	13.4	2.845	2.845	2.60	6.0	11.5
4738	0.913	1.759	15.5	16.5	2.841	2.843	2.57	8.0	12.5
4740	0.3101	0.8836	13.0	16.3	2.838	2.834	2.57	8.0	13.5
4750	25.84	XXX	15.9	XXX	2.859	XXX	2.58	7.5	9.0
4804	608.0	0.2003	23.9	4.0	2.870	2.877	2.68	2.0	8.0
4806	99.28	348.4	18.8	20.2	2.887	2.875	2.52	11.5	18.5
4808	38.99	XXX	15.1	XXX	2.883	XXX	2.52	11.0	17.5
4810	10.54	85.09	13.1	17.7	2.874	2.886	2.60	6.5	14.0

SS   Steady State Method (Nitrogen Gas)

▲ $^{15}/_{16}$-inch Rotary Sidewall Core Plug

●   1-inch Whole Core Plug

XXX   Unsuitable sample

Table 4. Summary of core and log derived reservoir parameters. Example 2, an example of formation heterogeneity, is from a carbonate reservoir in which logs, whole core and rotary sidewall cores were taken.

REFERENCES

Aziz, K., A. Journel, and A. Nur, 1987, An integrated approach to forecasting the performance of petroleum reservoirs (Draft): Stanford Center for Reservoir Forecasting, Stanford University, 66 pages.

Basan, P., J.R. Hook, K. Hughes, J. Rathmell, and D.C. Thomas, 1988, Measuring porosity, saturation and permeability from cores: an appreciation of the difficulties: The Technical Review, v. 36, p. 22 - 36.

Dacey, M.F., and W.C. Krumbine, 1970 , Markovian models in stratographic analysis: Math. Geol. v. 2, p. 175-191.

Enderlin, M.B., and D.K.T. Hansen, 1987, Fundamental approach to dipmeter analysis: 11th Annual Canadian Well Logging Symposium, Calgary, Alberta.

Haldorsen, H.H., 1983, Reservoir characterization procedures for numerical simulation: Ph.D. dissertation, University of Texas.

Haldorsen, H.H. and L.W. Lake, 1984, A new approach to shale management in field-scale models: SPEJ, August, p. 447 - 452.

Jageler, A.H., 1976, Improved hydrocarbon reservoir evaluation through use of borehole-gravimeter data: Journal of Petroleum Technology, v. 28, p. 709-718.

Matheron, G., 1963, Principles of geostatistics: Econ. Geol., v. 58, p. 1246-1266.

THE DERIVATION OF PERMEABILITY-POROSITY TRANSFORMS
FOR THE H.O. MAHONEY LEASE, WASSON FIELD,
YOAKUM COUNTY, TEXAS

Douglas E. Craig

Production Geosciences Department
Mobil Exploration & Producing U.S. Inc
Midland, Texas

ABSTRACT

The H. O. Mahoney lease produces from two intervals within the Upper Permian San Andres Formation: the First Porosity and the underlying Main Pay. These intervals are predominately subtidal and intertidal in origin. Permeability-porosity transforms were derived by subdividing core analysis data from four wells by facies and type of analysis, whole core versus plug. The plug data were not used to derive the final transforms because the plugs were too small to adequately sample reservoir heterogeneities. Whole core data for each facies were analyzed by a pc-driven Mobil computer program that lists the relationship between permeability, porosity, percentage of cumulative flow capacity, percentage of cumulative pore volume, sweep efficiency at breakthrough, Dykstra-Parsons parameter, and Lorenz coefficient of heterogeneity. The permeability value that corresponded to 90% of the cumulative flow capacity was selected as the cutoff for each facies. Those values are: First Porosity intertidal 0.22 md, First Porosity subtidal 0.92 md, and Main Pay subtidal 6.2 md.
    Estimates show the First Porosity intertidal will contribute only 0.36% of the total cumulative pore volume; therefore, a transform was not computed for this facies. Furthermore, because both subtidal intervals are perforated in all injection wells and fluid injection profiles show fluid entering both facies, a single value of 1.0 md was selected as the cutoff for both subtidal facies. This cutoff describes the least permeable reservoir rock that can effectively be flooded under current operating conditions. The transforms for these

two facies were derived by entering all values above the
cutoff into a pc-driven Core Laboratories computer program
that crossplots permeability versus porosity on a semilog plot
and calculates the reduced major axis (RMA) line.

I. INTRODUCTION

The purpose of this report is to describe the procedure
used to derive the permeability-porosity transforms for the
productive interval of the San Andres Formation in the H. O.
Mahoney lease, Wasson(San Andres)Field, Yoakum County, Texas.
This was done as part of an on-going effort to update the
original reservoir description (Ballard, 1984). The derivation
of reliable permeability-porosity transforms was an important
part of the characterization process because the transforms
were used to calculate porosity-feet and permeability-feet,
two parameters used to quantify the amount of reservoir rock
present and the transmissibility of that rock. When the
updated reservoir description is complete, the data will be
used to evaluate the current injection pattern, evaluate the
need for infill wells, and computer-simulate the reservoir
performance.

II. GENERAL FIELD INFORMATION

Wasson Field is located on the Northwestern shelf,
adjacent to the Midland basin in Yoakum and Gaines Counties,
Texas (Figure 1). The field is a combination structural-
stratigraphic trap formed by the updip pinchout of porous
dolomite. The H. O. Mahoney lease lies on the NE flank of the
field approximately 400 feet below the crest (Figure 2).
Hydrocarbons are produced from two Permian zones in the
field: the San Andres Formation and Clearfork Wichita-Albany
Formations. This report will focus only on the Wasson (San
Andres) Field.
Wasson Field was discovered April 15, 1936 when the
Honolulu Oil Corp. and C. J. Davidson #1-678 L. P. Bennett was
drilled. Mobil drilled and completed its first well, the H.
O. Mahoney #1, on September 29, 1939. Most of the field was
developed on 40-acre spacing by the mid-1940's and continued

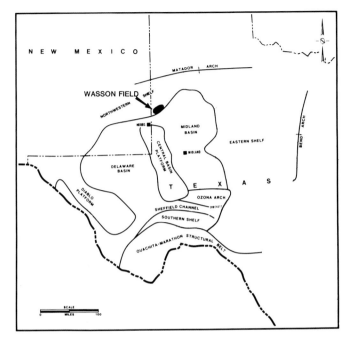

Figure 1. Location of Wasson Field.

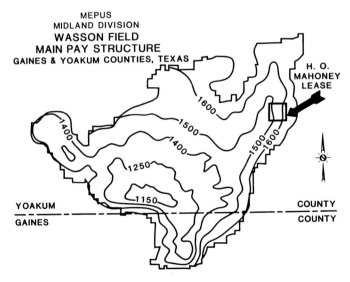

Figure 2. Subsea structure map of the top of the "Main Pay" interval of the San Andres Formation in the Wasson Field.

on primary production until the mid-1960's, when waterflood
operations began. Waterflood operations began when the field
was unitized into seven San Andres units (Figure 3): Shell
Denver, Amoco ODC, ARCO Williard, Cornell Cornell, Texas
Pacific (now Shell) Bennett Ranch, Texaco Roberts and Mobil H.
O. Mahoney lease. Mobil maintains 100% interest in the 640-
acre Mahoney lease and has a smaller interest in four of the
other six units.

The Shell Denver unit was the first to begin waterflooding
in 1964; Mobil began in November 1966. During the early 1980's,
several CO_2 feasibility studies, reservoir descriptions, and
pilot studies were conducted which showed CO_2 to be a viable
enhanced oil recovery method for the field. Shell began
injecting CO_2 into the Denver Unit in 1984. Mobil commenced
CO_2 injection on the Mahoney lease October 17, 1985 and is
currently in the fifth injection cycle. An increase of approximately 500 BOPD has been realized since CO_2 injection began
on the Mahoney lease. The Cortez pipeline supplies CO_2
to the Wasson field.

III. GEOLOGIC SETTING

The Midland Basin and associated Northwestern Shelf margin
were formed by the uplift of the Central Basin Platform at the
end of the Pennsylvanian Period (Figure 1). Regional Permian
sedimentation is characterized by the deposition of carbonates
on the shelf areas and clastics in the basins (Galley, 1955).
The Permian Period ended with the deposition of evaporites on
the Northwestern Shelf as shelf prograded and the sea regressed
toward the southeast.

The Permian stratigraphy at Wasson Field consists
predominantly of a thick section of dolomite and anhydrite up
through the San Andres Formation (Figure 4). Unconformably
overlying the San Andres Formation is a sequence of interbedded
dolomite and anhydrite that grades upward into evaporites and
clastics. The unconformity at the top of the San Andres
Formation is a regional feature characteristic of areas that
were depositionally high and underwent subaerial exposure and
erosion.

At the H. O. Mahoney lease, the San Andres Formation is
approximately 1400 feet thick and varies little in thickness.
Production is from a 300 to 400 foot interval that has been

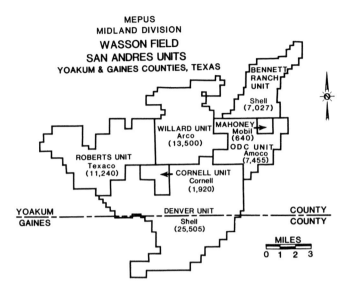

Figure 3. San Andres Units in the Wasson Field.

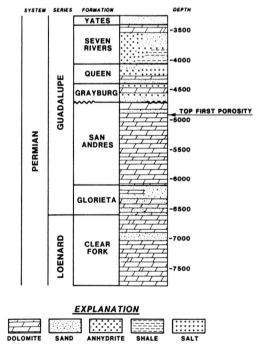

Figure 4. Generalized stratigraphic column for the Wasson Field (from Ballard, 1984).

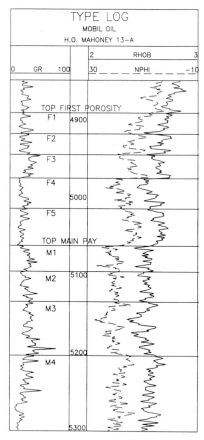

Figure 5. Type log of the San Andres Formation on the H.
O. Mahoney lease showing subdivided zones in the First
Porosity and Main Pay intervals. Wireline curves shown
are Gamma Ray (GR), Bulk Density (RHOB), and Neutron
Porosity (NPHI).

divided into two major zones, the First Porosity and the Main
Pay (Figure 5). The top of the First Porosity occurs approximately 400 feet below the top of the San Andres Formation.
The two productive zones were further subdivided in Shell's
Denver Unit into the First Porosity (F1-F5) and Main Pay
(M1-M8) using pressure test data. Ballard (1984) used Shell's
zonation to subdivide the reservoir on the Mahoney lease. The
best reservoir interval is the Main Pay, zones M1-M4, which
lie underneath the First Porosity at an average depth of 5100
feet.

Core from wells #13A, 46, and 48 (Figure 6), were visually examined in slabs and thin sections to determine the lithologic, depositional and reservoir characteristics of the First Porosity and Main Pay on the Mahoney lease. Dolomitization has destroyed most of the original fabric and depositional features; however, two dominant facies are still distinguishable.

The Main Pay consists predominately of a subtidal facies characterized by dolomitized peloidal packstones with lesser amounts of wackestone and grainstone. Molds of bivalves, crinoids, ostracods, sponge spicules, and bryozoans occur throughout the Main Pay, but are most heavily concentrated in

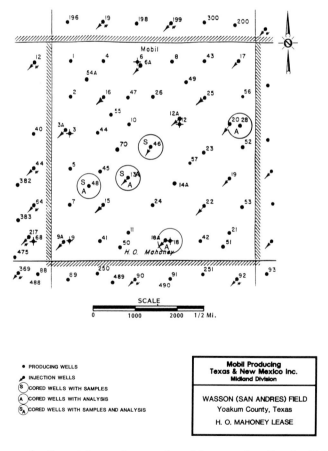

Figure 6. Location of cored wells on the H. O. Mahoney lease.

the M1 zone at the top of the section. Burrows filled with
dolomitized mud or anhydrite are also present, and the
percentage filled with anhydrite increases upsection.
Anhydrite occurs as nodular fill that is either replacive or
displacive, pore-filling in fossil molds, the replacement
product of dolomite crystals, and cement. The percentage of
anhydrite present in the Main Pay is low, as are the occurrences of algal laminations and dessication features. Four
subtidal facies: biogenic banks, shoals, shallow marine,
and lagoonal were identified and described in detail by
Ballard (1984).

Subtidal intervals similar in lithology to the Main Pay
comprise the majority of the First Porosity; however, intertidal zones are also common. A characteristic feature of
the intertidal zones are the 6 inch to 4 foot thick
algal-laminated beds. Lithologically, these intervals are
comprised of dolomitized peloidal packstones and wackestones
with trace amounts of burrows and fossils. Nodular,
pore-filling and replacement anhydrite are also abundant.
Dessication cracks and chicken-wire anhydrite, charactistic of
supratidal environments, are present in a few thin dolomitized
mudstone beds overlying algal laminations. Three major
differences distinguish the subtidal facies in the First
Porosity from the subtidal facies in the Main Pay: a higher
percentage of anhydrite, lower percentage of fossil molds, and
higher percentage of wispy clay laminations.

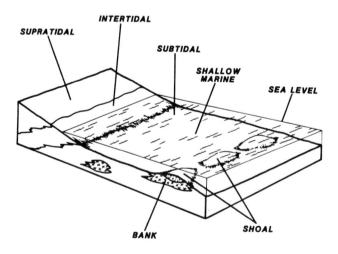

Figure 7. Generalized depositional model for the H. O.
Mahoney lease (from Ballard, 1984).

The San Andres Formation was deposited in an arid, tidal
flat environment similar to the Trucial Coast of the Persian
Gulf (Purser 1973, and Scholle and Kinsman, 1974). A
depositional model for the H. O. Mahoney lease is shown in
Figure 7. Here, the San Andres Formation is a shoaling upward
sequence of subtidal (Main Pay), subtidal and intertidal
(First Porosity), and supratidal deposits (post First
Porosity). Minor repeated transgressions produced the
intertidal-subtidal cycles within the First Porosity (Figure
8).

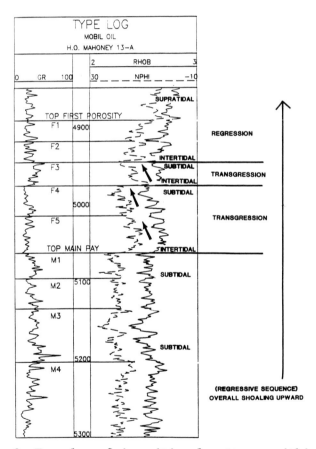

Figure 8. Type log of depositional patterns within the
First Porosity and Main Pay intervals. Arrows indicate
increasing porosity within transgressive intervals.

All porosity formed in the facies described is secondary;
the product of diagenetic processes. Five porosity types are
recognized: intercrystalline-intragranular, intercrystalline,
moldic, leached-vuggy, and fracture (Ballard 1984). In
general, porosity was formed by dolomitization, and the
dissolution of allochems, dolomite, and evaporites during
changes in relative sealevel. Meteoric water percolating into
the subaerially exposed carbonates acted as an agent of
dissolution and dolomitization (Ramondetta, 1982). Porosity
was destroyed when hypersaline fluids, associated with
prograding sabkha environments, precipitated pore-filling
anhydrite into the underlying permeable zones, and when
dolomite was replaced by anhydrite. Several cycles of dissolution, dolomitization, precipitation, and replacement, as
well as burial diagenesis, created the porosity and
permeability in the San Andres Formation on the H. O. Mahoney
lease.

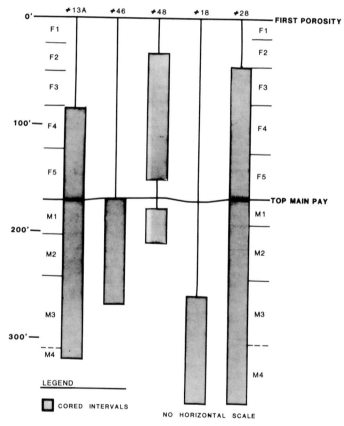

Figure 9. Intervals cored in H. O. Mahoney wells.

IV. METHODOLOGY

Listed below are the four major steps followed to derive the permeability-porosity transforms.

1.) Data Preparation
2.) Data Correlation and Grouping
3.) Derivation of Permeability Cutoff
4.) Derivation of Permeability-Porosity Transforms

A. Data Preparation

All core samples and core data available for the H. O. Mahoney lease were used in this study. Five wells have been cored through various intervals of the productive San Andres section: 13A, 18, 28, 46, and 48 (Figure 9). Visual

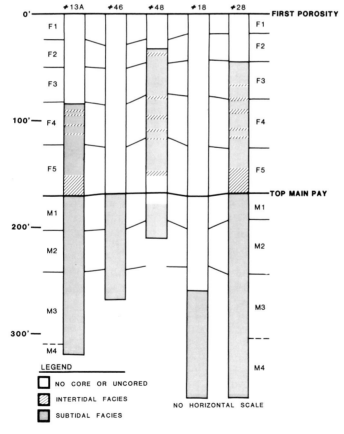

Figure 10. Stratigraphic position of facies in core.

descriptions of facies are based on 512 ft of core samples
from wells 13A, 46 and 48; whereas, all cutoffs and transforms
are based on 771 ft of core analysis data from wells 13A, 18,
28, and 48. Visual descriptions were not done on the core from
wells 18 and 28 because chips are all that has been preserved.
Also, core analysis data from well 46 was not used to derive
cutoffs and transforms because the core from this well was
never analyzed.

Core analysis data from wells 13A, 18, 28, and 48 were
input from digital tape into a Mobil VAX-driven log analysis
program. The porosity and permeability information from these
data were printed and plotted to compare with the original
analysis from the core contractor. The output data duplicated
the contractors data, confirming that the data had been input
into the VAX program correctly.

As part of the quality control procedure, 35 datapoints
below 0.01 md permeability and 4 datapoints above 150 md were
removed from the dataset, leaving 732 datapoints. The low
permeability datapoints were removed because 0.01 md is too
low to be measured accurately. The high values probably
resulted from incorrect measurements or measurement across
natural or induced hairline fractures. Fractures are present
and do contribute to the permeability of the reservoir, but
only in thin localized intervals (Ballard, 1984).
Consequently, the permeability contribution from fractures is
not representative of the reservoir intervals for which
transforms will be calculated. The permeability and porosity
values for the remaining dataset ranges from 0.01 to 100 md
and 0.3 to 28.6%, respectively.

B. Data Correlation and Grouping

Examination of core slabs and thin-sections indicate that
permeability-porosity characteristics vary from facies to
facies; hence, the first grouping of the core analysis data
was by facies. Using this approach three datasets were formed:
Main Pay subtidal (MPS), First Porosity subtidal (FPS), and
First Porosity intertidal (FPI). The stratigraphic position of
these facies in the cored wells is shown in Figure 10. No core
samples were available for wells 18 and 28. Facies identified
in the core samples from wells 13A, 46 and 48 were transferred
to wells 18 and 28 using well log response and porosity
delimitors derived from a permeability-porosity plot of the
facies in well 13A (Figure 11). A porosity delimitor was used
where wireline log reponses could not be correlated from well
to well because of lateral variations in facies. In Figure 11,
the line through the FPI facies intersects the two subtidal
facies between approximately 8.5% and 9.5% porosity.
Therefore, though the data sets overlap, intervals below 9%
porosity are assumed to be largely intertidal and those above
9% are largely subtidal. Using this cutoff and the wireline
log responses of general depositional patterns, e. g., the
occurrence of subtidal transgressions in the First Porosity,
facies were marked on logs for wells 18 and 28.

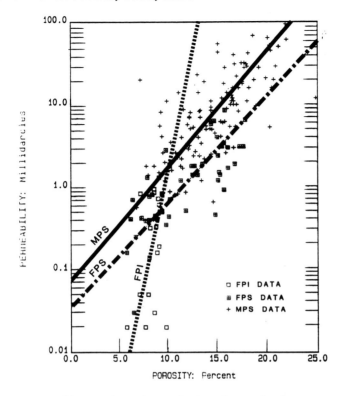

Figure 11. K-phi plot of the three facies dependent
datasets in H. O. Mahoney 13A. Datasets are First
Porosity intertidal (FPI), First Porosity subtidal
(FPS), and Main Pay subtidal (MPS). Transforms show
the relationship between permeability and porosity
for each dataset.

Permeability-porosity characteristics for the MPS, FPS,
and FPI are shown in Figures 11 and 12. The Main Pay subtidal
(MPS) has an average porosity of 14.0% and an average permeability of 9.2 millidarcies. This facies is dominated by
moldic, vuggy, and intercrystalline porosity. In general,
estimates from thin sections place the ratio of moldic and
vuggy porosity to intercrystalline porosity at approximately
3:1. Isolated molds and vugs contribute significantly to
reservoir volume, but have low permeability. As total porosity
increases, the percentage of intercrystalline porosity
increases more rapidly than moldic and vuggy porosity. At
approximately 15% total porosity, molds and vugs become interconnected by intercrystalline porosity, and permeability
increases substantially. Similar increases in permeability for

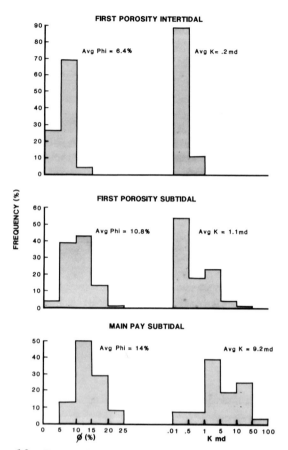

Figure 12. Permeability and porosity distribution for the three facies dependent datasets. Data are from wells 13A, 18, 28, and 48.

intervals with greater than 17% total porosity have been documented in the Littlefield Northeast (San Andres) Field, Lamb County, Texas (Chuber and Pusey, 1967). Ramondetta (1982) attributes this increase in permeability, which occurs in intervals of seemingly uniform lithology, to a change in porosity type above 17% porosity.

The First Porosity subtidal (FPS) is dominated by moldic and vuggy porosity with minor amounts of intercrystalline porosity in the most porous zones. The average porosity and permeability for this facies are 10.8% and 1.1 millidarcies, respectively. This facies has a lower average permeability than the MPS (Figure 12), because molds and vugs are seldom connected. Furthermore, the occurrence of moldic and vuggy porosity has decreased, causing the average porosity of this facies to be 3.2% lower than that of the MPS. The differences in permeability and porosity between the two facies are due to the higher percentage of anhydrite and lower percentage of fossil molds in the FPS. Most of the intercrystalline porosity and some of the moldic and vuggy porosity in the FPS have been filled by anhydrite. The anhydrite may have been created by the localized replacement of dolomite (Ballard, 1984), or it may have precipitated from hypersaline brines percolating in from above (Ramondetta, 1982). Eighty feet of downward percolation by hypersaline fluids has been documented in the Palo Duro Basin in Texas (Barone, 1976). The amount of anhydrite and fossil molds in the FPS suggests this facies was deposited in a shallower portion of the subtidal environment than the MPS. This is further supported by the permeability and porosity distribution of this facies, which lies between the MPS and FPI (Figure 12).

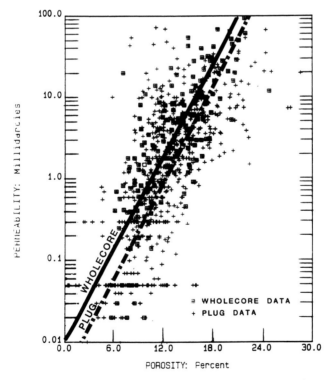

Figure 13. K-phi plot of all whole core and plug analysis data. Whole core data are from well 13A; plug data are from wells 18, 28, and 48.

The First Porosity intertidal (FPI) has moldic and vuggy porosity only; essentially all intercrystalline porosity has been occluded by anhydrite. Consequently, the average permeability of this facies is only 0.2 millidarcies. The average porosity is only 6.4% due to the decreased amounts of moldic and vuggy porosity. Anhydrite comprises 30% or more of some of the lowest porosity intervals.

Core data were grouped into the three facies described, and by type of core analysis, whole core versus plug analysis. Plug core analysis was performed on wells 18, 28, and 48; whereas, whole core analysis was performed on well 13A. The permeability and porosity data for these two datasets are shown in Figure 13. Best-fit regression lines, calculated by the method of reduced major axis (Middleton, 1963), show the permeabilities of the whole core data to exceed those of the plug data

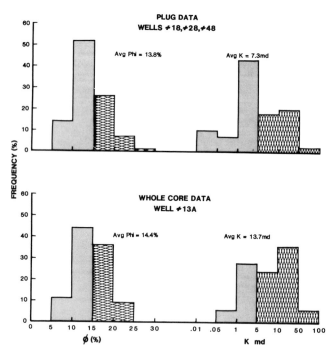

Figure 14. Permeability and porosity distribution for whole core and plug analysis data from the Main Pay only.

throughout the entire porosity range. A similar pattern is
shown in Figure 14, where histograms are used to compare data
from only the Main Pay subtidal facies. Sixty-six percent of
the whole core samples measured exceed 5 millidarcies permeability, compared with only 40% of the plug samples. Similarly, 45% of the whole core samples exceed 15% porosity,
compared with 34% of the plug samples. Variations between the
permeability and porosity characteristics of the plug analysis
data and the whole core data are attributed to sampling techniques (bias) and sample size versus reservoir heterogeneities.
Ideally, in whole core analysis 100% of the 12 inch interval
to be analyzed is sampled; whereas, in plug analysis a 1 inch
X 1.5 inch sample is selected to represent a segment of core 4
inches X 12 inches in size. The San Andres Formation on the
Mahoney lease is very heterogenous; consequently, plug samples
are not as representative of the reservoir as whole core
samples. For this reason, only the whole core data from well
13A will be used to derive the final permeability cutoffs and
permeability-porosity transforms. Plug data values will be
presented in the following sections for comparison purposes
only.

C. Derivation of Permeability Cutoffs

   Core porosity and permeability data from wells 13A, 18,
28, and 48 were subdivided into four datasets: all facies
combined, First Porosity intertidal (FPI), First Porosity
subtidal (FPS), and Main Pay Subtidal (MPS). Each of the datasets was input into a Mobil pc-driven software program that
calculates the series of parameters listed below (some definitions are included):

   Cumulative Capacity: Permeability-feet
   Cumulative Thickness
   Cumulative Pore Volume: Porosity-feet
   50% Cumulative Thickness Value of Permeability
   84.1% Cumulative Thickness Value of Permeability

   Permeability Variation: a coefficient introduced by
      Dykstra and Parsons (1950) defined as

      $V = (km-ko)/km$ where km = median permeability
                              ko = permeability at 84.1%
                                 of the cumulative
                                 sample

Total Vertical Sweep: percent of the vertical interval
swept when fluid has broken through in the most
permeable layer.

Coefficient of Heterogeneity: calculated by plotting the
percent cumulative capacity as a function of the
percent cumulative thickness. Capacity is the
permeability-thickness product for a given layer.

The following guidelines are used by the Midland Division
of Mobil Exploration and Producing U.S. Inc to derive permeability (k) cutoffs: reservoirs under primary depletion k
cutoff = k value at 95% kh; reservoirs under secondary depletion k cutoff = k value at 90% kh; and reservoirs under tertiary depletion k cutoff = k value at 85% kh. These kh percent-
ages are based on computer studies and theoretical
experiments. They represent a "best estimate" of the percent
of kh that will transmit fluid under a given depletion
mechanism.

The H. O. Mahoney lease has undergone 46 years of primary
and secondary depletion and only 3 years of tertiary depletion
so 90% kh was selected for the permeability cutoffs. Using
this guideline and the output from the facies-biased whole
core datasets in well 13A, the following cutoffs were selected:
FPI 0.22 md, FPS 0.92 md, and MPS 6.2 md. These cutoffs apply
to reservoirs where fluid is selectively injected into each
zone to assure that the lowest permeability reservoir rock
receives fluid. However, selective injectivity is not used in
Mahoney wellbores; all zones receive fluid at the same
injection pressure and fluid preferentially flows into zones
with the highest permeability. Under these conditions, only
one permeability cutoff; representative of the least permeable
reservoir rock that can effectively be flooded, should be
used.

Fluid injection profiles show that fluid does enter both
the FPS and MPS. The FPI is not perforated in most injection
wells, therefore its ability to receive and transmit fluid is
largely unknown. However, the permeability of reservoir quality
rock in this facies ranges from only 0.22 to 0.94 md;
consequently, if it were perforated fluid would probably
bypass this facies and enter the more permeable subtidal
facies. In addition, the output summarized in Table 1 shows
the First Porosity intertidal contributes only 0.36%
(6.37/1767.27) of the cumulative capacity and 6.2%
(1.53/24.73) of the cumulative pore volume in well 13A.
Therefore, even if this facies could transmit fluid at maximum
efficiency, its contribution would be relatively
insignificant. For this reason, 1.0 md, an approximation of
the First Porosity subtidal cutoff, will be used as the
permeability cutoff for all reservoir rock. This cutoff

Table 1. Summary of output from pc-driven Mobil software.

INTERVAL (FACIES)	WELL NO.	CUMULATIVE CAPACITY Kh (MD/FT)	CUMULATIVE THICKNESS h (FT)	CUMULATIVE PORE VOLUME (Phi-h)	K VALUE @ 90% Kh (MD)	% PORE VOL. @ 90% Kh	K VALUE @ 84.1% h	K VALUE @ 50% h	K VARIATION	COEFF. OF HETERO.	TOTAL VERTICAL SWEEP (%)
FIRST POROSITY (INTERTIDAL)	13A	6.37	19.00	1.53	.22	55	.03	.25	.880	.503	35.67
	28&48	7.88	56.00	3.25	.05	61	.04	.05	.238	.556	15.63
	28	5.90	35.00	1.66	.05	66	.05	.05	.000	.511	18.73
	48	1.98	21.00	1.59	.04	56	.02	.04	.500	.593	16.84
FIRST POROSITY (SUBTIDAL)	13A	82.45	43.00	5.13	.92	72	.44	1.40	.686	.478	22.56
	28&48	176.37	185.00	19.37	.47	48	.05	.30	.833	.724	4.15
	28	94.50	88.00	8.83	.60	44	.05	.30	.833	.766	4.67
	48	81.87	97.00	10.61	.45	54	.07	.32	.781	.669	11.56
MAIN PAY (SUBTIDAL)	13A	1676.46	122.00	17.60	6.20	62	2.30	8.00	.713	.557	14.62
	18,28&48	2220.54	304.50	42.16	3.60	57	.80	3.70	.784	.610	10.13
	18	1014.18	137.00	17.19	3.70	56	.80	3.80	.789	.610	10.28
	28	603.40	106.00	15.17	2.70	58	.60	2.70	.778	.612	8.34
	48	578.59	61.50	10.07	4.10	60	1.30	4.15	.687	.575	17.44
ENTIRE CORE UNEDITED (ALL FACIES COMBINED)	13A	1767.27	188.50	24.73	5.10	53	.61	3.40	.821	.661	9.97
	18	1408.57	171.50	19.35	4.04	49	.31	3.00	.896	.716	8.21
	28	703.80	231.00	25.66	1.80	45	.05	.70	.929	.752	4.55
	48	662.44	179.50	23.28	2.70	40	.06	.58	.897	.770	6.84

describes the least permeable reservoir rock that can
effectively be flooded under current operating conditions.
Using this cutoff, the First Porosity intertidal facies has no
reservoir quality rock because the maximum permeability of
this facies is only 0.94 md.

In the previous section information was given to validate
the grouping of data by facies and type of core analyses, plug
versus whole core. Data summarized in Table 1 further support
this grouping. Generally, the coefficient of heterogeneity and
permeability variation are higher for datasets where all
facies are combined than they are for the separate facies
datasets in each well. This implies the proper criteria were
used to distinguish between facies because the resulting
datasets are more homogenous, i. e., have more similar
permeability-porosity distributions than the entire core datasets where all facies are combined. Also, the coefficient and
variation values are lower for most of the wholecore data in
well 13A than the plug data in wells 18, 28, and 48. These
values suggest whole core data and plug data do differ for
each facies and as such should not be combined. Furthermore,
because the whole core values are lower, these data have less
scatter and are more reliable.

D. Derivation of Permeability-Porosity Transforms

Permeability and porosity data for all datasets used to
calculate permeability cutoffs and make facies comparisons
were input into "SELECT". SELECT is Core Laboratories pc-
driven software that computes transforms for data.

Transforms derived for the Main Pay subtidal (MPS) and
First Porosity subtidal (FPS) facies, from whole core data in
well 13A, are shown in Figure 15. These transforms represent
the best linear relationship between permeability and porosity
in the reservoir quality rock of the San Andres Formation on
the H. O. Mahoney lease. The formulas for these transforms
are:

FPS Log (k) = (0.0928 * Phi) - 0.9127
MPS Log (k) = (0.1281 * Phi) - 0.9500

where k = permeability in millidarcies
       Phi = porosity in percent

Reservoir quality rock is defined here as any rock which
exceeds 1.0 md in permeability. As previously mentioned, the
maximum permeability measured in the First Porosity intertidal

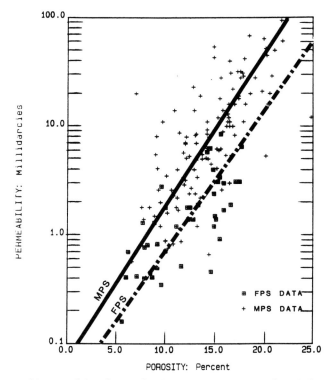

Figure 15. K-phi plot of First Porosity subtidal and
Main Pay subtidal whole core data from well 13A.

(FPI) was 0.94 md; hence, this facies has no reservoir quality
rock. Using the formulas, the porosity at 1.0 md permeability
is 9.84% for the FPS and 7.42% for the MPS, or approximately
10% and 7.5%, respectively. These porosity cutoffs were used
during later phases of the reservoir description to distinguish
between reservoir and non-reservoir quailty rock on wireline
porosity logs.

Differences in the transforms and porosity cutoffs for the
two facies discussed above substantiate the grouping of data by
facies. Furthermore, transforms for whole core data and plug
data from the MPS differ (Figure 16), suggesting that these
two data types should not be combined. Whole core data are
considered to be more representative of the reservoir for
reasons previously discussed.

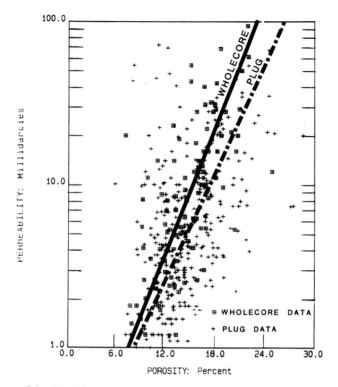

Figure 16. K-phi plot of all reservoir quality whole
core and plug analysis data for the Main Pay subtidal.
Reservoir quality is any rock which exceeds 1.0 md in
permeability. Whole core data are from well 13A; plug
data are from wells 18, 28, and 48.

V.   SUMMARY AND CONCLUSIONS

The derivation of permeability-porosity transforms for the
H. O. Mahoney lease was a five step procedure involving the
identification of facies in core, data preparation, data
grouping, the calculation of permeability cutoffs, and derivation of transforms. Each part of the procedure had an impact
on the transforms, but the grouping of data by facies and type
of analysis (whole core versus plug), and the calculation of
permeability cutoffs had the most significant impact.
   The grouping of core data into three facies-dependent
datasets is supported by several findings which show the data-

sets have unique permeability and porosity characteristics. First, the amount and type of porosity varies for each facies in core. Second, the average permeability and porosity for each dataset are different: FPI 0.2 md and 6.4%; FPS 1.1 md and 10.8%;and MPS 9.2 md and 14.0%, respectively. Third, the co-efficient of heterogeneity and permeability variation are lower for individual facies datasets in each well than they are for datasets where all facies are combined. Fourth, the regression lines "transforms" for the three facies datasets in well 13A are vastly different.

The grouping of data by type of analysis, whole core versus plug, is also supported by several findings. First, whole core permeabilities exceed plug permeabilities over the entire porosity range measured. Second, 45% of the whole core samples exceed 15% porosity compared with only 34% of the plug samples. Third, the coefficient of heterogeneity and permeability varia-tion are lower for most whole core datasets than for plug data-sets. Whole core data were used to derive the final cutoffs and transforms because plug samples are too small to adequately sample the heterogeneities present in the San Andres Formation.

A single permeability cutoff of 1.0 md was selected for the two reservoir quality facies, the FPS and MPS. The maximum permeability measured for the FPI was 0.94 md; consequently it is not a reservoir quality facies. The cutoff of 1.0 md is the k value at 90% kh of the lowest permeable reservoir rock that can effectively be flooded.

The permeability-porosity transforms derived for the two reservoir quality facies are:

FPS   Log (K) = (0.0928 * Phi) - 0.9127
MPS   Log (K) = (0.1281 * Phi) - 0.9500

Using these formulas, the porosity at 1.0 md is 9.84% for the FPS and 7.42% for the MPS, or approximately 10% and 7.5%, res-pectively.

These porosity cutoffs were used in later phases of the reservoir characterization to map porosity-feet, permeability-feet and other reservoir parameters.

Recently, dissolution of anhydrite by waterflooding has been documented in the Wasson(San Andres)Field and Vacuum Field (W. Hermance, personal communication). The effects of dissolution on permeability and porosity, and on the transforms derived in this report are currently being studied.

ACKNOWLEDGEMENTS

I thank the management of Mobil Exploration & Producing U.S. Inc and Mobil Oil Corporation for permission to publish this paper. Bill Hermance and Patricia Ballard provided helpful suggestions while reviewing the manuscript.

REFERENCES

Ballard, P. D., 1984, Reservoir description of the H. O. Mahoney lease, Wasson (San Andres) Field: Mobil company report, 85 p.

Barone, W. E., 1976, Depositional environments and diagenesis of the lower San Andres Formation: Texas Tech University, Masters thesis, 93 p.

Chuber, S., and Pusey, W. C., 1967, San Andres facies and their relationship to diagenesis, porosity, and permeability in the Reeves Field, Yoakum County, Texas, in Elam, J. B. and Chuber, S. eds.: Symposium on cyclic sedimentation Midland, West Texas Geological Society, 232 p.

Dykstra, H. and Parsons, R. L., 1950, The prediction of waterfloo performance with variation in permeability profile, Production Monthly, p. 345.

Galley, J. E., 1958, Oil and geology in the Permian Basin of Texas and New Mexico, in Weeks, L. G. ed., Habitat of oil, a symposium: American Association of Petroleum Geologists, p. 395-446.

Middleton, G. V., 1963, Statistical inference in geochemistry: in Studies in Analytical Geochemistry, Shaw, D. M. ed., University of Toronto Press, Toronto, Canada, p. 124-139.

Purser, B. H., 1973, The Persian Gulf - Holocene carbonate sedimentation and diagenesis in a shallow epicontinental sea, Springer, Berlin, 471 p.

Ramondetta, P. J., 1982, Facies and stratigraphy of the San Andres Formation, Northern and Northwestern Shelves of the Midland Basin, Texas and New Mexico: The University of Texas at Austin, Bureau of Economic Geology Report of Investigations no. 128, 56 p.

Scholle, P. A., and Kinsman, D. J. J., 1974, Aragonite and high-Mg calcite caliche from the Persian Gulf - a modern analog for the Permian of Texas and New Mexico: Journal of Sedimentary Petrology, v.44, no. 3, p. 904-916.

Schmalz, J. P. and Rahme, H. D., The variation of waterflood performance with variation in permeability profile, Production Monthly, p. 9-12.

A Discussion of Douglas E. Craig's
"The Derivation of Permeability-Porosity Transforms for the H.O. Mahoney
Lease, Wasson Field, Yoakum County, Texas"

by Jerry L. Jensen

In this paper, the author uses porosity (ϕ) versus permeability (k)
plots and reduced major axis (rma) lines to support contentions that different
facies and different sampling techniques give distinctly different ϕ-k
relationships. Furthermore, Craig uses the rma lines derived to estimate k
from ϕ measurements.

The use of rma lines to compare the ϕ-k relationships of different
facies or sampling methods agrees with statistical practice (e.g. Troutman and
Williams, 1987). Each line is meant to approximate the covariation of ϕ and
log(k) for a given rock type without regard to either variable's role in the
relationship. However, straight lines may not be the most appropriate form to
express the ϕ-log(k) relationship. An examination of Figures 11, 13, and
15, along with the marginal distributions shown in Figures 12 and 14
suggests that curvilinear relationships may exist for some of the data sets. In
particular, the clouds of MPS data appear to have a distinctly concave
downwards appearance. The PFI data also appear to have no ϕ-k
relationship whatever. These behavior patterns could have been recognized
and incorporated into the analysis from residual plots (e.g., Mann, 1987).

While the use of rma lines is common, the least-squares line is more
appropriate here to establish ϕ-k "transforms" since, for each value of ϕ, the
line is intended to give an estimate for log (k). In contrast with the rma line,
the least-squares line method assumes a distinct difference in the roles of ϕ
(predictor variable) and log(k) (response variable). When the coefficient of
determination (R2) is large, the two lines do not give significantly different
predictions. For these data, however, the ϕ-log(k) relationships are weak to
only moderately strong and the rma line will substantially underestimate
log(k) for low ϕ and overestimate log(k) for high ϕ

Another approach in the analysis of the ϕ-k relationships would be to
consider transforming either k, ϕ, or both to make their joint distribution
more like the bivariate normal distribution (e.g., Hald, 1952). In this case,
both the rma and regression curves are straight lines so there is no question of
curvilinear relationships. Other advantages also obtain with this approach.
First, it would be possible to make tests of significance on the lines (e.g.,

RESERVOIR CHARACTERIZATION II
Copyright © 1991 by Academic Press, Inc.
All rights of reproduction in any form reserved.

Kermack and Haldane, 1950) to determine if sampling variation accounted for the different lines of different facies or sampling methods. This would have been particularly useful in comparisons involving the FPS line because it is based on relatively few data and the ϕ-k relationship is rather weak. Second, an easily calculated correction factor can be included when permeability predictions are being made (e.g., Jensen and Lake, 1985). This would give a more reliable cutoff indicator. Third, the sensitivity of the calculated lines to the very low and very high permeability values may be substantially reduced. Thus, the very low and high data which were excluded from the analyses could have been included.

Most of Craig's claims are not changed by the above observations because he has not argued the points solely on the basis of statistical analysis. For example, in the representativeness of the whole core and core plug data, practical and sedimentological considerations would be sufficient without resorting to any statistical arguments. However, the reader should be aware that limitations exist concerning the application and interpretation of lines calculated by the rma procedure.

Jerry L. Jensen
Heriot-Watt University
Edinburgh, Scotland

References

Hald, A., 1952, Statistical Theory with Engineering Applications, J. Wiley and Sons.
Jensen, J.L., and Lake, L.W., 1985, "Optimization of Regression-Based Porosity-Permeability Predictions," Transactions paper R of The Tenth Formation Evaluation Symposium of the Canadian Well Logging Society, Calgary, Sept. 29 - Oct. 2.
Kermack, K.A., and Haldane, J.B.S., 1950, "Organic Correlation and Allometry," Biometrika, Vol. 37, parts 1 and 2, 30-41.
Mann, C.J., 1987, "Misuses of Linear Regression in Earth Sciences," in Use and Abuse of Statistical Methods in the Earth Sciences. W.B. Size, ed., Oxford University Press.
Troutman, B.M., and Williams, G.P., 1987, "Fitting Straight Lines in the Earth Sciences," in Use and Abuse of Statistical Methods in the Earth Sciences. W.B. Size, ed., Oxford University Press.in

PERMEABILITY PATTERNS IN SOME FLUVIAL SANDSTONES.
AN OUTCROP STUDY FROM YORKSHIRE, NORTH EAST ENGLAND.

Torgrim Jacobsen
Hans Rendall

Department of Sedimentology and Stratigraphy
Continental Shelf and Petroleum Technology
Research Institute A/S
Trondheim, Norway

ABSTRACT

Based on minipermeameter data from outcrops the distribution of permeability is studied in some fluvial sandbodies in the Middle Jurassic Scalby Formation. Additionally plugs for laboratory measurements of porosity and permeability were drilled out at some of the sample locations. Thin-section analyses in a scanning electron microscope equipped with an image analysing program were performed on the same plugs to study the level and pattern of diagenesis. Field notes on descriptive sedimentary facies (texture and structures) were recorded at each measurement point, but no close association between permeability and these particular properties is apparent. Instead, there seems to be a significant relationship between genetic sedimentary facies (type of sandbody) and permeability. Permeabilities in the lenticular-bedded sheet sandstone (braided river) are much higher than in the lateral accretion-bedded sandstone (meandering channel), which in turn have higher permeabilities than the crevasse splay sandstone. The variation in permeability between sandbodies seems to be influenced by varying degrees of quartz cementation.

Permeability within this fluvial sequence therefore is significantly influenced by differential cementation which is more related to the genetic sedimentary facies than to the textural aspects of the initially deposited sediment.

RESERVOIR CHARACTERIZATION II

315

1. INTRODUCTION

The description of permeability variation in reservoir
simulators is routinely carried out by averaging values
measured from core plugs, and from evaluating wireline well
logs, and then interpolating these values in a simple manner
between wells to produce a contour map of permeability
variation for each simulation gridblock layer. This method
for representing permeability is criticized as beeing too
simple to represent a realistic level of permeability
variation within reservoirs, and alternative statistical
methods have been proposed (e.g. Matheron et al., 1987). The
use of statistical methods for modelling the permeability
architecture of a reservoir requires that the 2-D and 3-D
distribution of permeability is known in detail from
analogous sequences. In recent years, detailed measurement of
permeability variation in outcrops has provided data
concerning the permeability architecture of a variety of
sandbody types (Goggin et al., 1986; Stalkup, 1986; Tomutsa
et al., 1986.). The aim of this work was to further expand
this database by studying the absolute permeability values,
the internal permeability structure, and their controlling
factors in different types of fluvial sandstones within a
single formation from the Jurassic of N.E. England. Two types
of channel sandstones (lenticular-bedded sheet sandstones,
Fig. 2; lateral accretion-bedded sandstone, Fig. 3), and a
crevasse splay sandstone (Fig. 4) were studied.

2. GEOLOGICAL SETTING

The data collection was carried out on outcrops of the
Middle Jurassic fluvio-deltaic Scalby Formation in Yorkshire,
along the NE coast of England (Fig. 1). The sandstones and
shales crop out as 40-50m high cliffs over a distance of 10km
along the North Sea coastline. The Scalby Formation overlies
the shallow marine Scarborough Formation and is subdivided
into the Moor Grit Member and the Long Nab Member (Leeder and
Nami, 1979) These outcrops are thought to be similar to the
Ness Formation of the oil-bearing Brent Group in the North
Sea (Mjøs and Walderhaug, 1989). Ravenne et al. (1987)
carried out a 2D and 3D geological study of some of the same
outcrops to describe reservoir heterogeneities. Their results
were used as basis for conditional simulation of the geometry
of a fluvio-deltaic reservoir (Matheron et al., 1987).

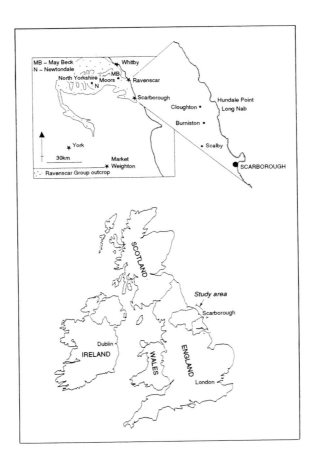

Fig. 1. Map showing location of the field area. The studied outcrops extend from 300m south of Long Nab to Hundale Point.

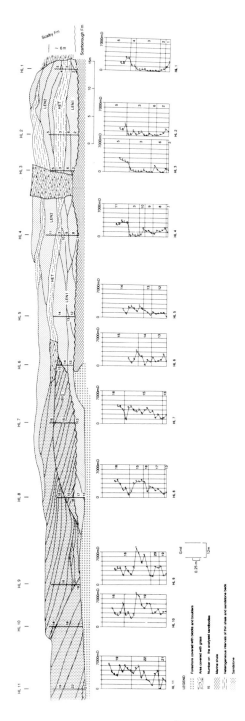

Fig. 2. Sketch and permeability profiles of the lenticular-bedded sheet sandstone (braided river system). The vertical bars mark the measured profiles, and the numbers along the bars and to the right on the permeability profiles refer to the penetrated sandstone bodies. The horizontal space between the bars is constant, but due to the nature of the cliff, is variable on the sketch.

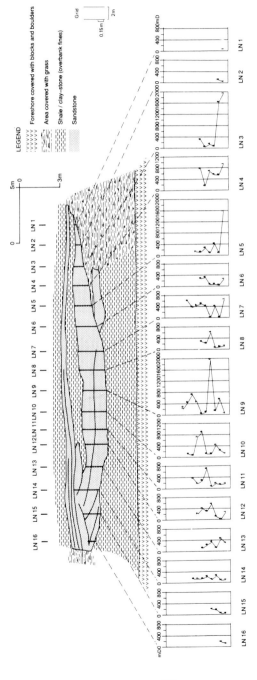

Fig. 3. Sketch and permeability profiles of the
lateral accretion-bedded sandstone (LN; meandering channel).

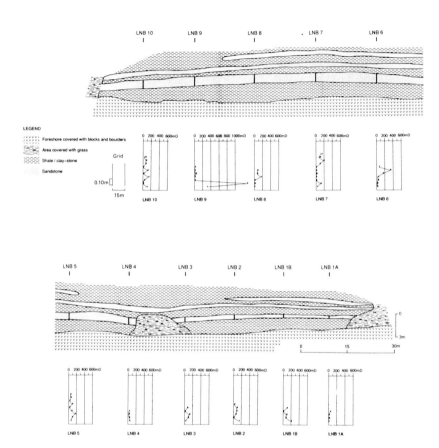

Fig. 4. Sketch and permebility profiles of the crevasse splay sandstone (LNB).

3. FIELDWORK AND ANALYTICAL METHODS

3.1. Data-Collection

Approximately 500 permeability measurements were made in
the field with a minipermeameter, and core plugs were
collected at 10% of the same sample points. The descriptive
sedimentary facies (grain-size, sedimentary structures) was
recorded at all measurement points.The permeability data were
collected in grids specific for each type of sandstone body.
Horizontal spacings of 12m and vertical spacings of 0.25m
were applied to the lenticular-bedded sheet sandstone (Fig.
2) horizontal spacing of 2m and vertical spacing of 0.15m
were applied to the lateral accretion-bedded sandstone (Fig.
3), and horizontal spacing of 15m and vertical spacing of
0.10m for the crevasse splay sandstone (Fig. 4). The
weathered surface crust at each sample locality was removed
with a core drill or a geology hammer before permeability was
measured. If the rock was obviously wet pneumatic air-drying
was done before measurement. We were limited to a height of
about 5.5m (the length of the ladder), and we were able to
map only the lowermost part of the cliffs.

3.2. The Minipermeameter

The minipermeameter is a mechanical, portable device
(Fig. 5) that measures flow rate and injection pressure of
gas into a rock face to determine permeability using Darcy's
law (Goggin et al., 1986) as follows:

$$K_a = \frac{\mu \; Q \; P_i}{a \; G \; \dfrac{(P_i^2 - P_o^2)}{2}}$$

where:

K_a = air permeabilities in darcies
Q = flow rate in ml/sec
P_i = measured flow pressure in atmospheres
P_0 = atmospheric pressure in atmospheres
μ = viscosity of the gas in centipoise
a = internal radius of the tip seal in centimeters
G = geometrical factor (a function of sample geometry and
       tip seal size).

The permeability measurements are strictly localized, because the radius of the invaded area around the injection tip rarely exceeds 1cm (Lake and Kocurek, 1987). The injection tip is fitted with a silicone rubber seal to prevent leakage between the tip and the rock surface. The minipermeameter measurments are rapid, non-destructive and cheap when compared to core plug permeability measurements. Calibration of the instrument was performed by Norsk Hydro by means of open-flow measurements on core plugs of known permeability (Fig. 6). A good correlation between laboratory (Hassler sleeve) and minipermeameter permeabilities was obtained, especially in the 100-7500 mD range.

Fig. 5. Picture demonstrating the minipermeameter in use.

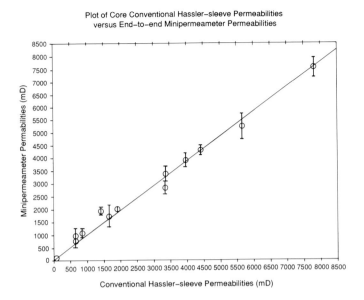

Fig. 6. Plot of core conventional Hassler sleeve permeabilities versus end-to-end minipermeameter permeabilities
(After Dreyer et al., in press).

4. CORE PLUG ANALYSES

4.1. Measurements of Permeability and Porosity

    Porosity and permeability were measured on 50 core plugs.
Permeability was measured using conventional Hassler-sleeve
equipment. Porosity measurements were carried out following
standard routines using helium gas.

4.2. Thin-Section Studies

Thin-sections were made from all the core plugs and
studied qualitatively using a Leitz Laborlux 11 Pol
polarising microscope. Authigenic quartz is the main
diagenetic mineral in all the sandstones and has been
quantified to give a measure of the level of diagenesis.
Analyses on five core plugs were carried out using a Jeol 733
SEM equipped with a solid state backscatter electron detector
(BSE) and a cathodoluminescence detector (CL). Mixed BSE/CL
images (Fig. 7) were analysed with a Kontron SEM-IPS image
analysing system. Magnifications varied from 100 to 150X, and
five random areas were analysed on each thin-section. The
statistical validity of such a few data points can be
questioned, but there was good accordance between these
results and the optical microscope analyses.

4.3. X-ray Diffraction Analyses

Semi-quantitative estimates of whole rock mineralogy were
made using a Phillips PW 1730/10 diffractometer with CuKa-
radiation. Analyses were performed on powdered samples from
the same five core plugs as were subjected to SEM/CL
analyses. The following xrd-peaks and correction factors were
used for quantification of the minerals (Rueslaatten, 1976):

	Angstrom-value	Correction factor
Quartz	4.26	0.7
Potassium feldspar	3.24	0.5
Plagioclase	3.19	0.5
Kaolinite	~7	0.7
Mica	~10	1.0

Fig. 7. Mixed BSE/CL (SEM) image. The light areas are
detrital mineral grains, and the darker outer rim on the
grains is authigenic quartz.

5. ANALYSES AND RESULTS

5.1. Sedimentological Analysis

The lowermost part of the Moor Grit Member (unit A1a in
Ravenne et al., 1987) was subdivided into four units: LEN1,
LEN2, HET, BIG (Fig. 2). This subdivision is based on
geometry, grain size and sedimentary structures. The lenticular sandstones, up to 3m thick and 25m wide, with large-
scale trough cross bedding (LEN1 and LEN2) are interpreted to
represent stacked channels in a multi-storey braided network
of laterally migrating channels. Unit HET, which lies between
LEN1 and LEN2, is composed of laterally quite extensive
interlayers of thin (20-30cm thick) medium/fine-grained sandstones and clayey siltstones. It is interpreted to represent
overbank facies deposited during a period when the main
river system had migrated laterally. The thick,
medium/coarse-grained sandstones (BIG) with giant cross
stratified sets, up to 6m thick and 50m wide, are interpreted

to be the result of migration of large dunes or bars in major distributary channels. The overall system comprising these subunits is thought to be a major, braided river system as proposed by Leeder and Nami (1979), and Ravenne et al (1987). The lateral accretion-bedded sandstone, 32m wide and 2m thick (Fig. 3), is interpreted to represent a pointbar deposit of a relatively small meandering channel. The thin (0.25-0.5m thick) laterally extensive sandstone (Fig. 4), which is at the same stratigraphic level as the lateral accretion-bedded sandstone body, is interpreted to be a crevasse splay deposited by the same meandering river system.

5.2. Analysis of the Permeability Patterns

The permeability profiles (Figs. 2, 3, 4) have no obvious internal trends in the different sandstones. The central parts of some of the sandstones appear to have the highest permeability values, but there is no general pattern. The average absolute permeability values of each unit however, have distinct differences. The lenticular-bedded sheet sandstone, LEN1, has an average permeability of 1110 mD, LEN2 an average permeability of 2152 mD, HET an average permeability of 246 mD, and BIG an average permeability of 3125 mD (Fig. 8). The lateral accretion-bedded sandstone (LN) has an average permeability of 314 mD, and the crevasse splay (LNB) an average permeability of 75 mD. Unfortunately, the chosen spacing between the grid points is too large to provide meaningful semi-variograms and give estimates of correlation lengths. The internal permeability values for each unit are plotted as histograms (Fig. 8). LEN2 and BIG have the highest permeability values and the lowest coefficients of variation, 0.33 and 0.45 respectively. The highest coefficients of variation were estimated for the lateral accretion-bedded sandstone (LN) and the crevasse splay (LNB), 1.18 and 2.11 respectively. These units also have the lowest average permeability.

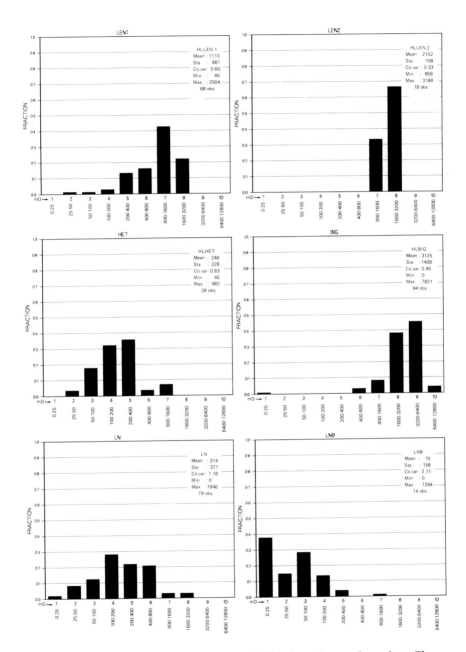

Fig. 8. Histograms of permeabilities in each unit. The ordinate is the fraction of the total number of measurements. The abcissa gives the permeability range on a logarithmic scale. In each diagram the value of mean (arithmetic) permeability, standard devation, coefficient of variation, minimum and maximum values and number of observations are displayed.

5.3. Descriptive sedimentary facies and corresponding
permeabilities.

Descriptive sedimentary facies (texture and structures)
were noted in each point subjected to permeability measurement in order to evaluate the relationship between permeability and the textural aspects of the sedimentary rocks.
Some recent work (Weber, 1982; Stalkup and Ebanks, 1986) has
described a close relationship between descriptive
sedimentary facies and permeability, while others (Hancock,
1978; Tomutsa et al., 1986; Walton et al., 1986) are more
uncertain. In Table I the descriptive sedimentary facies and
corresponding statistics for permeability and porosity are
listed for each unit.

TABLE I. Descriptive sedimentary facies, corresponding
average permeability, and some statistical parameters

LEN1

Facies	Facies description	Mean	Sta	Co.Var	Min	Max	Obs
F2	m gr, mass/ m-l sc cb	1162	923	0,79	324	2152	3
F3	m-f gr, mass/ m-l sc cb	552	433	0,78	188	1510	10
F5	f gr, mass	1251	536	0,43	257	2504	22
F6	f gr, m-l sc cb	1102	648	0,59	45	2319	21
F7	f gr, m-s sc cb/ rippl/par lam	1400	754	0,54	118	2459	12

Average porosity: 21%

LEN2

Facies	Facies description	Mean	Sta	Co.Var	Min	Max	Obs
F1	m-c gr, m-l sc cb	2811	300	0,11	2449	3188	5
F2	m gr, mass/ m-l sc cb	1899	656	0,35	959	2750	13

Average porosity: 19%

HET

Facies	Facies description	Mean	Sta	Co.Var	Min	Max	Obs
F2	m gr, mass/ m-l sc cb	464	470	1,01	64	982	3
F3	m-f gr, mass/ m-l sc cb	235	78	0,33	163	315	4
F4	m-f gr, sm sc cb/ rippl/par lam	116	80	0,69	45	248	5
F5	f gr, mass	55	–	–	–	–	1
F6	f gr, m-l sc cb	391	–	–	–	–	1
F7	f gr, m-s sc cb/ rippl/par lam	302	247	0,82	68	896	10
F8	clayey siltstone, sm sc cb/par lam	125	18	0,14	111	150	4

Average porosity: 21%

BIG

Facies	Facies description	Mean	Sta	Co.Var	Min	Max	Obs
F1	m-c gr, m-l sc cb	3094	1172	0,38	1292	5005	11
F2	m gr, mass/ m-l sc cb	3145	1568	0,50	472	7821	54
F3	m-f gr, mass/ m-l sc cb	3226	388	0,12	2663	3687	6
F6	f gr, m-l sc cb	3284	1318	0,40	650	5142	20
F7	f gr, m-s sc cb/ rippl/par lam	3037	1151	0,38	1718	3835	3
F8	clayey siltstone, sm sc cb/par lam	0	–	–	–	–	1

Average porosity: 22%

LN

Facies	Facies description	Mean	Sta	Co.Var	Min	Max	Obs
F5	f gr, mass	210	195	0,93	0	691	10
F6	f gr, m-l sc cb	374	417	1,15	0	1946	68
F7	f gr, m-s sc cb/ rippl/par lam	122	69	0,57	38	295	15
F8	clayey siltstone, sm sc cb/par lam	266	257	0,97	92	643	4

Average porosity: 21%

LNB

Facies	Facies description	Mean	Sta	Co.Var	Min	Max	Obs
F4	m-f gr, sm sc cb/ rippl/par lam	1294	–	–	–	–	1
F5	f gr, mass	69	57	0,83	0	157	21
F6	f gr, m-l sc cb	76	106	1,40	6	329	10
F7	f gr, m-s sc cb/ rippl/par lam	52	59	1,14	0	299	39
F8	clayey siltstone, sm sc cb/par lam	6	11	1,83	0	19	3

Average porosity: 20%

Legend:

f=fine	l=large	mass=massive
m=medium	sm=small	par lam=paralell lamination
c=coarse	sc=scale	ripp=ripples
gr=grained		cb=crossbedding

From these data there is apparently little relationship between descriptive sedimentary facies and permeability in the different sandstones. The mean permeability values are quite similar, and the coefficients of variation are high for all descriptive sedimentary facies. LEN1 and BIG are the best illustrations of this phenomenon. By comparing the mean permeability values for each descriptive sedimentary facies in the different units, one realizes large variations. Facies F5, F6, and F7 illustrate this phenomenon. These results indicate poor correlation between descriptive sedimentary facies and permeability for the whole fluvial formation in general. The average porosity measured on core plugs was quite similar for all the different units, around 20%.

6.   COMPARISON OF CORE PLUG DATA WITH FIELD MEASUREMENTS

6.1. Permeability (field) versus Permeability (lab)

Field measurements of permeability were cross-plotted versus permeability measured on core plugs taken from the same point (Fig. 9). This gave an average correlation coefficient of 0.7, but the correlation seems to vary with

absolute permeability values. The best correlation is between
500 and 2000 mD. At very high permeabilities the minipermeameter measured too small values, and at low permeability the
minipermeameter displayed too high values. This is in
accordance with Lake and Kocurek (1987), and is caused by
turbulence at high permeabilities and gas leakage at low
permeabilities. One has to be aware that the field and lab
methods are measuring on different rock volumes, so exact
correspondence cannot be expected, especially when dealing
with heterogeneous samples.

6.2. Permeability Versus Porosity

Both field and lab measurements of permeability are
plotted versus porosity and do not display a good
correlation. The porosity has about the same value (about
20%) whatever the measured permeability.

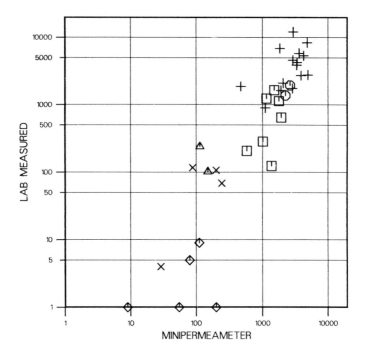

CORRELATIONCOEF.:   0.702752

+ BIG
△ HET
⊙ LEN2
⊡ LEN1
⬦ LNB
✕ LN

Fig. 9.   Cross-plot of field minipermeameter data and core plug data.

7. DIAGENESIS AND MINERALOGY

The qualitative thin-section analyses prove the sandstones to be reasonably well sorted, mineralogically mature quartz arenites. The grain size varies from medium/coarse- (BIG) to fine-grained (LNB). The degree of diagenesis, mainly quartz cementation, seems to be lowest in the braided river facies intermediate in the meandering channel facies, and highest in the crevasse splay facies. The amount of isolated pores also seems to follow the same pattern. One sample from each unit, apart from HET, was picked out for quantitative mineralogical and diagenetic analyses.

7.1. Mineralogy

The semi-quantitative whole rock XRD-analyses (Table II) gave results that are in accordance with the classification of rock types outlined above.

TABLE II. Results of semi-quantitative mineralogical analysis using XRD.

	Quartz	Plagioclase	K-feldspar	Kaolinite
LEN1(HL4N4) :	96	1	–	3
LEN2(HL3N17) :	100	–	–	–
BIG(HL10N17) :	100	–	–	–
LN(LN9N5) :	89	2	4	5
LNB(LNB2N7) :	89	6	1	4

7.2. Diagenesis

Thin-sections from the same five samples were subjected to quantitative SEM (BSE/CL) analyses (see procedure above). The following parameters were quantified (Table III):

- total areal percentage of quartz (TQtz)
- areal percentage of quartz cement (Qtzce)
- areal percentage of quartz cement/areal percentage of total quartz *100 (Qtz/TQtz)*100
- aluminum phase (including plagioclase and clay minerals) (Al- ph)
- potassium feldspar (Kfsp)

TABLE III. Results from the quantitative SEM (BSE/CL)
analysis.

LEN1(HL4N4):

	POR	TQtz	Qtzce	(Qtz/TQtz)*100	AL-ph	Kfsp
Mean:	22	75	17	23	3	–
Sta :	1	2	2	2	2	–
Min :	21	72	15	20	1	–
Max :	24	77	19	25	6	–

LEN2(HL3N17):

	POR	TQtz	Qtzce	(Qtz/TQtz)*100	AL-ph	Kfsp
Mean:	18	81	19	24	–	–
Sta :	4	4	3	5	–	–
Min :	13	76	17	20	–	–
Max :	24	86	25	32	–	–

BIG(HL10N17):

	POR	TQtz	Qtzce	(Qtz/TQtz)*100	AL-ph	Kfsp
Mean:	24	75	11	14	–	–
Sta :	6	7	3	3	–	–
Min :	18	66	8	12	–	–
Max :	35	82	14	18	–	–

LN(LN9N5):

	POR	TQtz	Qtzce	(Qtz/TQtz)*100	AL-ph	Kfsp
Mean:	19	71	23	33	7	1
Sta :	3	4	3	4	1	–
Min :	15	64	20	20	6	–
Max :	22	75	27	27	9	–

LNB(LNB2N7):

	POR	TQtz	Qtzce	(Qtz/TQtz)*100	AL-ph	Kfsp
Mean:	18	68	30	43	10	5
Sta :	6	12	7	7	5	8
Min :	11	51	22	32	4	1
Max :	27	82	37	49	15	20

The porosity estimated from the thin-sections is in accordance with porosity measured on core plugs, and the total mineralogy is in accordance with the XRD-quantifications. The amount of authigenic quartz growth is lowest in BIG(HL10N17), intermediate in LEN1(HL4N4) and LEN2(HL3N17), and highest in LN(LN9N5) and LNB(LNB2N7). This fits well with the qualitative optical microscope analyses and the differences in average permeability in the different units.

The lowest amount of authigenic quartz was found in the sample from the unit with highest average permeability (BIG), and highest amount of authigenic quartz growth was found in the sample from the unit with lowest average permeability (LNB). The SEM analyses also show that the authigenic quartz growth is a result of a pure chemical reaction, and to a small degree caused by pressure solution (Fig. 7). This is probably quite crucial regarding the diagenetic pattern which again exhibits a major control on the permeability pattern. We think that the most important source for the authigenic quartz is silica-saturated porewater migrating from the underlying shale during compaction. Acid porewater generated in the fluvial plain deposits under humid, warm climate has probably dissolved feldspars, and precipitated quartz and kaolinite. Some authigenic kaolinite booklets observed in all the samples provide evidence of such a reaction. This diagenetic process explains, to a large extent, the differences in cementation noticed in the different sandstone types.

LEN1 and BIG directly overlie shallow marine, littoral fine-grained deposits of the Scarborough Formation, and will be drained by less acidic and silica-saturated porewater compared to LEN2, LN and LNB, which directly overlie fluvial plain deposits. Lower degree of quartz cementation would then be expected in LEN1 and BIG than in the three other units, and this fits well with our data.

Units LEN2, LN and LNB overlie similar fluvial deposits, and their thicknesses are probably crucial to their degree of quartz cementation. The thicker the sandstones, the less cemented they are. Certainly this is in agreement with our results. Unit LEN2 is the least cemented, LN has intermediate cementation, and LNB has the highest degree of cementation. The differences in average permeability in the sandstones also seem to be in agreement with the estimated differences in quartz cementation. The only exception being the higher permeability measured in LEN2 than LEN1 despite the areal percentage of authigenic quartz cement being highest in LEN2. This may be caused by the coarser sandstones in LEN2 having higher original permeability than the LEN1 sandstones.

8. SUMMARY AND DISCUSSION

1) The highest permeability values were measured in the
lenticular-bedded sheet sandstone (braided river system).
This sheet sandstone could further be subdivided into four
different units, with different values of average
permeability [3125 mD (BIG), 2152 mD (LEN2), 1110 mD (LEN1),
246 mD (HET)].
Intermediate permeability values were measured in the
lateral accretion-bedded sandstone (meandering channel)
[314 mD (LN)].
The lowest permeability values were measured in the
laterally extensive, thin sandstone (crevasse splay)
[75 mD (LNB)].

2) No obvious internal vertical or horizontal permeability trends are observed in the different sandstone
bodies.

3) There is no clear positive correlation between
descriptive sedimentary facies (texture and structures) and
permeability. Therefore we suggest that the observed permeability pattern is controlled mainly by diagenesis.

4) The quantitative diagenetic analyses indicate that
the diagenesis (and thereby permeability) in this fluvial
formation as a whole is controlled by genetic sedimentary
facies and to a lesser degree by descriptive sedimentary
facies. In other words, diagenesis is controlled by the type
of sandstone, its size, geometry and bounding deposits, in
addition to grain size, sedimentary structures etc.

We cannot assess the controls on the internal diagenetic
structure in the different types of sandstones, but there is
no obvious association in our data set between the
descriptive sedimentary facies and diagenetic pattern as
claimed by Weber (1982) and Stalkup and Ebanks (1986). Had
such an association existed, the permeability would be more
clearly a function of the descriptive sedimentary facies,
which is generally agreed to be the main factor controlling
permeability in non-cemented sediments.
Our data set is limited, however, and more data from this
type of diagenetically altered fluvial environment are
required to verify our conclusions.

ACKNOWLEDGEMENTS

This work is part of a larger project concerning reservoir heterogeneities that is sponsored partly by SPOR (a Programme for Enhanced Oil Recovery and Reservoir Technology funded by the Norwegian state) and partly by Continental Shelf and Petroleum Technology Research Institute A/S (IKU). We are grateful to Norsk Hydro a.s for lending us their field permeameter, and to Tom Dreyer and Åse Scheie for advice and instruction on its use. Alister MacDonald and Vidar Fjerdingstad are thanked for fruitful discussions and proofreading of the manuscript. The authors also wish to express thanks to Mr. Tony Boassen for assistance with the scanning electron microscope, and to Stig Bakke, Jan H. Johansen, Øistein Rossing, Bjørn Ardø and Arnt Stavseth for technical assistance during the completion of this work.

REFERENCES

Dreyer, T., Scheie, Å. and Walderhaug, O. (in press). "A Minipermeameter-Based Study of Permeability Trends in Channel Sandbodies", AAPG Bull.

Goggin, D.J., Chandler, M.A., Kocurek, G.A. and Lake, L.W. (1986). "Patterns of Permeability in Eolian Deposits", SPE/U.S. Dept. of Energy, paper 14893.

Hancock, N.J. (1978). "Possible causes for Rotliegend sandstone diagenesis in northern West Germany", Journ. Geol. Soc. of London, Vol. 135, 35-40.

Lake, L.W. and Kocurek, G. (1987). "Geometrical factors for unconfined core plugs using the field permeameter", In: 4th Annual Center for Enhanced Oil and Gas Recovery Research report, Category A Research, University of Texas at Austin, 3-24.

Leeder, M.R. and Nami, M. (1979). "Sedimentary Models for The Non-marine Scalby Formation (Middle Jurassic), and Evidence for Late Bajocian/Bathonian Uplift of The Yorkshire Basin", Proc. of The Yorkshire Geol. Soc., Vol. 42, part 3, no. 26, 461-482.

Matheron, G., De Fouguet, Ch., Beucher, H., Galli, A., Guerillot, D., and Ravenne, C. (1987) "Conditional Simulation of the Geometry of Fluvio-Deltaic Reservoirs", SPE paper presented at the 1987 SPE Annual Technical Conference and Exhibition, Dallas, September 27-30.

Mjøs, R. and Walderhoug, O. (1989). "Sandstone Geometry in Fluvio-Deltaic Sediments of The Ravenscar Group, Yorkshire" SPOR/RF-report.

Ravenne, C., Eschard, R., Galli, A., Mathieu, Y., Montadert,
L. and Rudkiewicz, J-L. (1987). "Heterogeneities and Geome-
try of Sedimentary Bodies in a Fluvio-Deltaic Reservoir",
SPE, paper 16752.
Rueslaatten, H. (1976). "En kvartærgeologisk kartlegging av
Dagaliområdet med en mineralogisk undersøkelse av podzol-
forvitring". Cand.real. thesis, Univ. of Oslo.
Stalkup, F.I. (1986). "Permeability variations observed at
the faces of crossbedded sanstone outcrops". In: L.W. Lake
and H.B. Carroll (eds.), Reservoir Charachterization, 141-
179, Academic Press.
Stalkup, F.I. and Ebanks, W.J.Jr. (1986). "Permeability Vari-
ation in a Sandstone Barrier Island-Tidal Channel-Tidal
Delta Complex, Ferron Sandstone", SPE, paper 15532.
Tomutsa, L., Jackson, S.R. and Szpakiewicz, M. (1986).
"Geostatistical charachterization and comparison of outcrop
and subsurface facies: Shannon Shelf Ridges", SPE, 15127.
Walton, A.W., Bouquet, D.J., Evenson, R.A., Rofheart, D.H.
and Woody, M.D. (1986). "Charachterization of sandstone
reservoirs in the Cherokee Group (Pennsylvanian,
Desmoinesian) of south-eastern Kansas". In: L.W. Lake and
H.B. Carroll jr. (eds.): Reservoir Charachterization,
39-62.
Weber, K.J. (1982). "Influence of Common Sedimentary
Structures on Fluid Flow in Reservoir Models". J.Pet.Tech,
March, 665-672.
Weber, K.J. (1986). "How heterogeneity affects oil recovery".
In: L.W. Lake and H.B. Carroll jr. (eds.), Reservoir
Characterization, 487-544, Academic Press.

Session 3
Macroscopic

The scale of interwell spacing, usually inferred
from well tests or correlation

RESERVOIR MANAGEMENT USING 3-D SEISMIC DATA

James D. Robertson

ARCO Oil and Gas Company
Dallas, Texas 75221

I. INTRODUCTION

The geologic detail needed to properly develop most
hydrocarbon reservoirs substantially exceeds the detail required to
find them. This obvious, but compelling, precept has fueled the
steadily increasing application of 3-D seismic analyses to reservoir
management. A measure of the increase is that 3-D surveys now
account for half of the seismic activity in the offshore Gulf of Mexico
and North Sea, and the percentage has been rising steadily year by
year since commercial 3-D surveys were first shot in these areas in
1975 (Figure 1). 3-D seismic surveying in other offshore areas and on
land likewise is growing rapidly. The Fall 1988 Distinguished Lecture
of the Society of Exploration Geophysicists addressed the general
subject of managing reservoirs using 3-D seismic data, and this paper
is a condensed version of that lecture. There are three parts to the
paper: a definition of reservoir management; a discussion of the
various kinds of 3-D seismic analyses that can impact the development
and production of a field; and a synopsis of the past history and future
potential of the 3-D seismic technique.

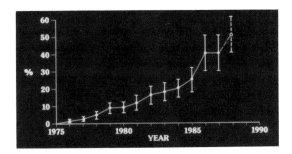

Fig. 1. 3-D seismic surveying as a percentage of total seismic
surveying in the Gulf of Mexico and North Sea.

II. DEFINITION OF RESERVOIR MANAGEMENT

A good working definition of reservoir management is
*maximizing the economic value of a reservoir by optimizing recovery of
hydrocarbons while minimizing capital investments and operating
expenses.*
The first thing to note is that this definition is not geophysical or
geological; in fact, it's not even an engineering definition. Reservoir
management really is the economic process of raising the worth of a
property to the highest possible level. We generally measure
economic value by yardsticks like present worth, investor's rate of
return, payout, and investment efficiency. The task is to maximize
(minimize in the case of payout) these economic descriptors.
Economic value generally increases when more reserves are proved
or when the reservoir's producing rate increases. Of course, capital
investments (drilling, seismic shooting, and lease bonuses) and
operating expenses (lease rentals, staff costs, taxes, etc.) must be
incurred to originally find and subsequently develop and produce
these reserves, and these expenditures by themselves detract from the
economic value. The reservoir manager thus trades off expenditures
that drain present worth against the chance of increasing present
worth by adding reserves and/or increasing production. The process
is a continuous balancing act.
What is the role of seismic surveying, particularly 3-D, in this
balance? Basically, it impacts reservoir management in two different
ways. First, a 3-D seismic analysis can lead to identification of

reserves that will not be produced optimally, or perhaps not produced at all, by the existing reservoir management plan. Second, the analysis can save costs by minimizing dry holes and poor producers, condemning leases that can then be dropped to avoid rental payments, etc. These concepts are summarized in my first major point: *3-D seismic data contribute to reservoir economics by adding reserves and/or by lowering costs.* Either of these impacts can be a sufficient justification for shooting a 3-D seismic survey. Of course, the best situation is when both happen at once, and -- fortunately for geophysicists -- that's generally the case.

One possible model of the reservoir management process is shown in Figure 2 and labeled the linear system. This model consists of the following sequence: a discovery, an evaluation of that discovery, implementation of a development plan leading to production of the field, and final abandonment when the field is no longer economic. In this scheme, a 3-D survey is shot during the evaluation phase and used to assist in the design of the development plan, after which development and production start up.

I suggest that this linear model is not what really happens in reservoir management, except in the very simplest cases (a discovery followed by one or two offsets that fully develop the reservoir). The real world generally is the process shown in Figure 3 and labeled the iterative system. This model also starts with a discovery, but then goes into a loop where data are constantly being evaluated to form the basis for development/production decisions (such as locating production and injection wells, siting and designing platforms, setting flow rates, managing pressure maintenance, performing workovers, planning waterflood and tertiary recovery strategies, etc.) When implemented, the development and production activities in turn generate new information (logs, cores, DSTs, pressure tests, etc.) that change maps, revise structure, alter the reservoir stratigraphic model, and the like.

Fig. 2. Reservoir management - the linear system.

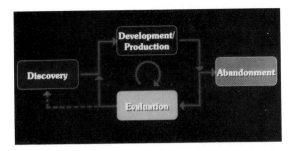

Fig. 3. Reservoir management - the iterative system.

Most of the time spent in managing reservoirs really consists of going around this loop. Occasionally, a deeper pool or offset extension test will spin off from the evaluation, resulting in a new discovery and revitalization of the loop. We continue with this process until the field is finally abandoned.

A 3-D seismic survey is one of the tools in the evaluation tool kit. An initial interpretation of the survey impacts the original development plan. As subsequent events like the drilling of development wells occur, the added information is used to revise and refine the original interpretation. Often, as time passes and the data base builds, elements of the 3-D data that were initially ambiguous begin to make sense, and the interpretation becomes more detailed and sophisticated. This is important enough to be a second major point: *the usefulness of a 3-D seismic survey lasts for the life of a reservoir*. A 3-D survey is not something shot right after the discovery, interpreted once, and then put on the shelf never to be looked at again. It hangs around for years as an active file on one's computer system!

III. TYPES OF 3-D SEISMIC ANALYSES

The interpretations that a geophysicist might perform on a 3-D seismic data volume can conveniently be grouped into those that examine the geometric framework of the hydrocarbon accumulation, those that analyze rock and fluid properties, and those that try to monitor fluid flow and pressure in the reservoir (Figure 4). These analyses impact, and hopefully significantly improve, decisions that

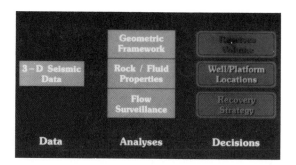

Fig. 4. Applications of 3-D seismic data to reservoir management.

must be made about reserves volume, well/platform locations, and recovery strategy. Thus, the analyses themselves are not the end products, but rather are management tools.

A. Geometric Framework

I am using geometric framework as a collective term for spatial elements like the attitudes of the beds that form the trap, the fault and fracture patterns that guide or block fluid flow, the shapes of the depositional bodies that make up a field's stratigraphy, and the orientations of any unconformity surfaces that might cut through the reservoir. A 3-D seismic data volume samples the geometric framework on a regular 3-D grid (generally 50 to 100 feet laterally and 10 to 50 feet vertically). By mapping travel times to picked events, displaying seismic amplitude variations across selected horizons, isochroning between events, noting event terminations, slicing through the volume at arbitrary angles, compositing horizontal and vertical sections, optimizing the use of color in displays, and employing the wide variety of other interpretive techniques available on a computer workstation, a geophysicist can synthesize a coherent and quite detailed 3-D picture of a field's geometry.

An example of mapping geometric framework is shown in Figures 5 and 6. Figure 5 is an early structure map of the Prudhoe Bay Field on the northern edge of Alaska. The map, which predates any 3-D seismic shooting over the field, was used in a Distinguished Lecture of the American Association of Petroleum Geologists in the early 1970s and subsequently published by Morgridge and Smith

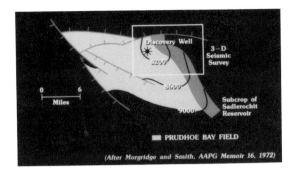

Fig. 5. Prudhoe Bay Field, Alaska - 1971 Sadlerochit structure map. (From Morgridge and Smith, 1972)

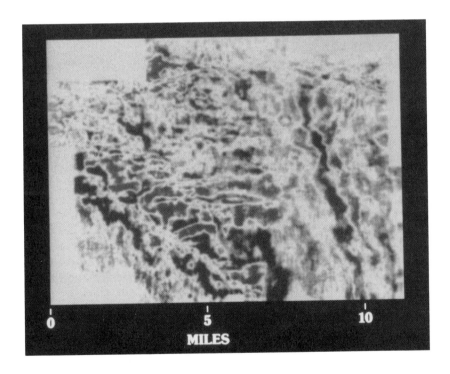

Fig. 6. Prudhoe Bay Field, Alaska - time slice. (Courtesy of D.A. Fisher)

(1972). It shows the basic elements of the trap at Prudhoe Bay: the dip to the south and southwest, the boundary fault to the north, and the erosional truncation of the Sadlerochit reservoir to the east. Figure 6, prepared by David A. Fisher, is a time slice from one of the 3-D seismic volumes that now exist over the field (location shown in Figure 5). The light and dark lineaments are seismic peaks and troughs, and the time slice cuts the volume at about the level of the unconformity truncating the Sadlerochit. One can clearly see the Sadlerochit subcrop and the northern boundary fault on the time slice. More subtle elements of the geometric framework are also evident, such as additional east-west and northwest-southeast faulting. This example illustrates a third major point: *3-D seismic data map the gross, controlling elements of a field.* This type of analysis is the traditional use for the 3-D seismic technique and has contributed very significantly to the geological characterization of many hydrocarbon traps.

An example of imaging stratigraphic shapes with 3-D seismic data has been published by Riese and Winkelman (1989) and is shown in Figures 7 and 8. The data volume comes from the Matagorda area of the offshore Gulf of Mexico. A single seismic section (Figure 7) in the 3-D volume contains a variety of nearly flat events. Some appear and then disappear, others vary laterally in amplitude, and the stratigraphic significance of any particular reflector is not obvious on the 2-D display. Take note of the short black event at about 0.7 second located laterally directly above the zero tick on the scale bar. A time slice through the 3-D volume (Figure 8) reveals that the event is a transverse cut through a meandering stream channel, and the stratigraphic situation becomes clear when the full spatial sampling of the 3-D volume is utilized. (The lineaments crossing the northern part of the slice are fault cuts coming up through the data). The major point here is the following: *minor character changes in 3-D seismic data tend to correlate with real geologic changes.* The variations generally are not noise or acquisition/processing errors, and the challenge is to correctly deduce their geologic significance.

B. Rock and Fluid Properties

The second general grouping of 3-D seismic analyses (Figure 4) encompasses those targeted at the qualitative and quantitative definition of rock and fluid properties. Amplitudes, phase changes, interval travel times between events, frequency variations, and other

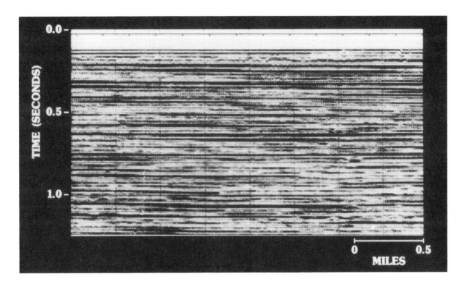

Fig. 7. Matagorda Block 668, Offshore Texas - seismic section.
(From Riese and Winkelman, 1989)

Fig. 8. Matagorda Block 668, Offshore Texas - time slice. (From
Riese and Winkelman, 1989)

characteristics of the seismic data are correlated with porosity, fluid type, lithology, net pay thickness, and other reservoir properties. The correlations usually require borehole control (well logs, cuttings, cores, etc.) both to suggest initial hypotheses and to refine, revise, and test proposed relationships. An interpreter develops a hypothesis by comparing a seismic parameter in the 3-D volume at the location of a well to the well's information, often through the intermediary of a synthetic seismogram match. The hypothesis is then used to predict rock/fluid properties away from the borehole control, and subsequent drilling validates (or invalidates) the concept. For example, 3-D seismic surveys are commonly used in the productive Pleistocene trends of the offshore Gulf of Mexico to directly map gas saturation. One correlates seismic amplitude anomalies with gas-saturated sandstones, and then maps the areal extent (and sometimes net feet of pay) of these bright spots laterally and vertically through the data volume.

A less common type of analysis is illustrated in Figure 9 (personal communication from S. F. Stanulonis and N. Kumar). This example is located on Alaska's North Slope, and the formation of interest is the Lisburne, a carbonate that produces from below the Sadlerochit at Prudhoe Bay. The amplitudes of the Lisburne reflection were determined at points of well control and compared to Lisburne porosity-thicknesses measured in the same wells. The comparison produced a methodology that was used to transform seismic amplitude directly into porosity-thickness values at grid points between wells. Figure 9 is the seismic horizon now scaled to porosity-thickness. Several wells drilled after this analysis have tested the quality of the porosity-thickness predictions, and the tests have matched the predictions to within a few units.

The major point here is the following: *3-D seismic data guide interwell interpolations of reservoir properties.* Given a 3-D survey, one does not have to settle for crude, linear interpolations of reservoir parameters between wells. The reservoir manager can use the seismic volume to pinpoint and understand non-linear lateral changes, an approach that nearly always results in lower costs, fewer surprises during development, and better production.

C. Flow Surveillance

The third, and last, general grouping of 3-D seismic analyses

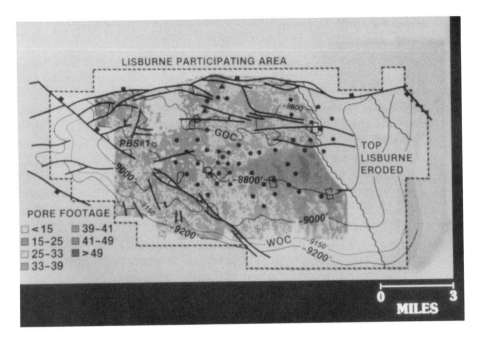

Fig. 9. Prudhoe Bay field, Alaska - Lisburne porosity-thickness from 3-D seismic data. (Courtesy of S.F. Stanulonis and N. Kumar)

(Figure 4) consists of those designed to look at the actual flow of the fluids in a reservoir. Such flow surveillance is possible if one (1) acquires a baseline 3-D data volume at a point in calendar time, (2) allows fluid flow to occur through production and/or injection with attendant pressure/temperature changes, (3) acquires a second 3-D data volume a few weeks or months after the baseline, (4) observes differences between the seismic character of the two volumes at the reservoir horizon, and (5) demonstrates that the differences are the result of fluid flow and pressure/temperature changes. Of course, one must be careful not to vary seismic acquisition and processing parameters drastically between surveys and thereby introduce differences that can be mistaken for fluid flow effects. One expects that the seismic character of horizons above the reservoir would be virtually identical between the volumes (geology generally changing over much a longer time than fluid flow!). Hence, an interpreted flow-induced difference can be indirectly validated by verifying that the difference occurs at the reservoir event, but not elsewhere in the 3-D volume. Of course, one can acquire a third, a fourth, etc. survey and continue the surveillance by computing additional 3-D difference volumes.

Flow surveillance with multiple 3-D seismic surveys is at a very early stage of research and development, but its potential impact on reservoir management is enormous. Most current practice of the technique has been directed toward monitoring enhanced oil recovery (EOR) processes. An example from Greaves and Fulp (1987) is shown in Figure 10. An experimental, oxygen-driven, thermal EOR pilot was performed on a depleted oil field called the Holt Sand Unit located in north Texas. The top section in Figure 10 is a seismic line through a 3-D data volume acquired prior to the start of the pilot. The Holt Sand is the event identified by the white triangles, and its seismic amplitude is low. The underlying bright event is a limestone several hundred feet below the Holt and not associated with the reservoir. The middle section in Figure 10 lies in the same spatial position in the 3-D data volume as the top section, but was acquired a few months after the start of oxygen injection/thermal combustion. Likewise, the bottom section is the same line reshot about a year after startup. One can observe that the oxygen injection/thermal combustion process has produced a dramatic increase in the strength of the Holt Sand reflection, and that more and more of the formation is affected as calendar time passes. The combination of oxygen injection and creation of combustion gases increases gas saturation in the reservoir (in effect, the experiment is creating an artificial bright spot), the thermal process is altering the state of the reservoir, and the multiple seismic datasets are monitoring the changes.

The lesson here is the following: *3-D seismic snapshots can map changes in fluid/pressure/flow regimes.* Although not yet as commercialized as mapping geometric frameworks or estimating rock/fluid properties between wells, this application of 3-D seismic surveying may eventually become just as important as the other two. It has the potential to directly measure reservoir performance, provide timely feedback that can change development/production plans, and be far more spatially specific than pressure tests. In the future, the technique might be used to monitor gas cap movements, control production and injection rates for optimum recovery, map pressure/temperature distributions, and even decipher stress patterns, particularly in tectonically active areas.

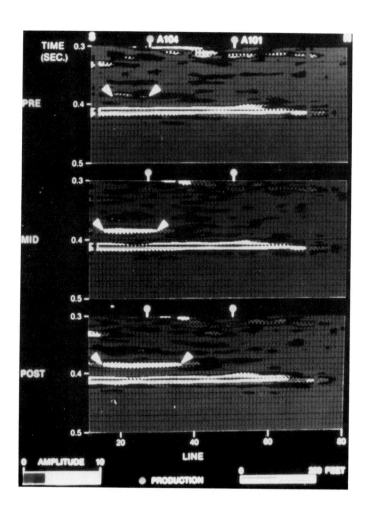

Fig. 10. Holt Sand Unit EOR pilot - seismic sections. (From Greaves and Fulp, 1987)

IV. HISTORY AND FUTURE OF 3-D SEISMIC SURVEYING

The first 3-D seismic survey ever shot appears to have been an experimental survey acquired by Exxon Production Research

Company in 1967 at Friendswood Field near Houston, Texas. A description of the survey and some of the data were published by Walton (1972). Various petroleum companies carried out other experimental surveys in the 1967 to 1972 time period, including some performed with transducers in water tanks by William S. French and co-workers at Gulf Oil Corporation about 1970. It was not until 1973, however, that the first commercial 3-D seismic program was conducted - a land survey in Lea County, New Mexico, shot by Geophysical Service Inc. for a group consisting of Amoco, ARCO, Chevron, Mobil, Phillips, and Texaco. The first commercial marine survey followed two years later in 1975 - one acquired in the High Island area of the Gulf of Mexico by Geophysical Service Inc. for Sun Oil Company. Usage of the technology grew steadily from the mid-1970s onward, particularly in the marine environment, as acquisition and processing improved. The best available evidence is that more than a hundred 3-D seismic surveys were shot worldwide by 1980. Steady innovations in acquisition and processing techniques (streamer tracking, real-time binning, 3-D migration, 3-D velocity analysis, etc.), the advent of supercomputers, and the explosive growth in computer graphics workstations for interpretation have continued to fuel the usage of 3-D seismology. It is virtually certain that more than a thousand 3-D surveys have now been acquired worldwide by the petroleum industry.

A very safe prediction is that steady growth in the application and sophistication of 3-D seismic technology is going to continue for the foreseeable future. The following are some specific areas in which progress is occurring.

(1) *Acquisition/processing/interpretation methods.* At sea, operators are shooting surveys with various combinations of multiple streamers, multiple source arrays, and multiple boats; are experimenting with towing streamers in circles around targets like salt domes to improve the imaging of radial faults; and are shooting into receivers fixed on the ocean bottom. On land, where one is free from the constraints of towing a streamer, operators are deploying many innovative acquisition geometries that make full use of the multichannel capabilities of modern seismic systems. The geometries account for terrain and cultural obstacles while optimizing subsurface coverage and mixes of offsets and azimuths, all at the lowest possible cost. Some experimental work is underway to acquire 3-D three-component surveys, thus adding shear and converted wave data volumes to the standard compressional wave volume. Advances in supercomputing will continue to speed up processing and permit the

inclusion of more sophisticated algorithms in processing schemes. Infusion of analytical techniques from remote sensing and other image processing disciplines is beginning to affect 3-D seismic interpretation. Automated information extraction (for example, algorithms that pick events after a few control points are specified) is becoming a routine part of interpretation, and many facets of 3-D seismic analysis are amenable to being impacted eventually by artificial intelligence technology.

(2) *Routine use in exploration.* A recent innovation in 3-D seismic surveying has been acquisition along lines spaced widely apart followed by filling in of the data volume by numerical interpolation prior to performing 3-D migration. This 3-D scheme (known variously as reconnaissance, exploration, or wide-line 3-D) depends on a good interpolation algorithm to be successful, and, even then, some steep-dip information is lost. However, the technique has the potential to lower acquisition costs to a point where it is feasible to shoot 3-D for exploration, and these types of surveys are now penetrating the seismic market.

(3) *3-D seismic with downhole sources/receivers.* The standard 3-D seismic data volume is acquired with sources and receivers at the earth's surface. It is logistically possible to put sources and/or receivers in boreholes and to record part or all of the 3-D data volume with this downhole hardware. This approach is an active area of research. Depending upon the acquisition configuration, one records various kinds and amounts of reflected and transmitted seismic energy, which can then be sorted out to provide information on geometric framework, rock/fluid properties, and flow surveillance just like surface surveys. Advantages of downhole placement are that higher seismic frequencies generally can be recorded, thereby improving resolution, and surface-associated seismic noise and statics problems are lessened or avoided. The main disadvantages are that one's source/receiver plants are constrained by the physical locations of available boreholes; borehole seismology can be affected by tube waves and the like so is not noise-free; a borehole source cannot be so strong as to damage the well; and the logistics and economics of operating in boreholes are complex, though not necessarily always worse, compared to operating on the surface. One can imagine a time when borehole seismic sources and receivers might be standard components of the hardware run into wells and accepted as routine and valuable devices for reservoir characterization and flow surveillance.

V. Summary.

The petroleum industry's twenty-year experience with 3-D seismic surveying is an example of a technological and economic success. Today, the investment in a 3-D survey typically results in fewer development dry holes, improved placement of drilling locations to maximize recovery, recognition of new drilling opportunities, and more accurate estimates of hydrocarbon volume and recovery rate. These outcomes improve the economics of development/production plans and make the surveys cost-effective. More skillful reservoir management will be a theme of the 1990s, and 3-D seismic technology will be part of the advancement.

ACKNOWLEDGMENTS

I thank the many individuals whose suggestions and data contributed to the Fall 1988 SEG Distinguished Lecture; the local SEG sections and universities for their invitations to speak to them; and the SEG and ARCO for their support of the Lecture tour.

REFERENCES

Greaves, R. J., and T. J. Fulp, 1987. Three-dimensional seismic monitoring of an enhanced oil recovery process. GEOPHYSICS, Vol. 52, No. 9, pp. 1175-1187.

Morgridge, D. L., and W. B. Smith, Jr., 1972. Geology and discovery of Prudhoe Bay Field, Eastern Arctic Slope, Alaska. Memoir 16, American Association of Petroleum Geologists, pp. 489-501.

Riese, W. C., and B. E. Winkelman, 1989. Atlas of Seismic Stratigraphy, Volume 3. American Association of Petroleum Geologists.

Walton, G. G., 1972. Three-dimensional seismic method. GEOPHYSICS, Vol. 37, No. 3, pp. 417-430.

STOCHASTIC SIMULATION OF INTERWELL-SCALE HETEROGENEITY FOR IMPROVED PREDICTION OF SWEEP EFFICIENCY IN A CARBONATE RESERVOIR

Graham E. Fogg[1]
F. Jerry Lucia
R. K. Senger

Bureau of Economic Geology
The University of Texas at Austin
Austin, Texas

I. INTRODUCTION

Petroleum reservoirs typically yield only a fraction of the oil initially in place. A major reason for poor recovery efficiencies is the presence of geologic heterogeneity, which causes highly nonuniform fluid flow patterns and incomplete drainage of oil. Two scales of heterogeneity are most relevant to the problem: (1) a megascopic scale, in which untapped oil-bearing facies occur as either a small number of discrete, elongate, semi-continuous bodies that are easily missed by the majority of wells drilled on a grid or as tabular bodies occurring at deeper, less explored horizons; and (2) a macroscopic scale, in which pervasive heterogeneities that are small in size compared to the well spacing cause local compartmentalization and bypassing of oil during both primary and secondary production. An example of these two scales of heterogeneity is illustrated in figures 1a and b for a 10- or 20- acre well spacing. The well spacing is already small enough to tap the relatively productive grainstone-dominated facies, which comprise the megascopic heterogeneity (fig. 1a). The macroscopic interwell heterogeneities, however, continue to hamper oil recovery efficiency (fig. 1b).

This study was aimed at numerically simulating flow of water and oil in geologically realistic, heterogeneous media in order to better understand effects of macroscopic heterogeneity on oil recovery and on the outcome of infill drilling programs. The simulations are based on data from Section 15 of the Dune Field reservoir, which is composed of dolomitic carbonate rocks deposited in a subtidal to supratidal environment on the Central Basin Platform of the Permian Basin (fig. 2). Recovery efficiency in Section 15 as of 1981 was only 25 percent (Bebout et al., 1987), which is typical for reservoirs of the Permian Basin.

---

[1]Now at University of California, Davis, Department of Land, Air, and Water Resources.

RESERVOIR CHARACTERIZATION II

355

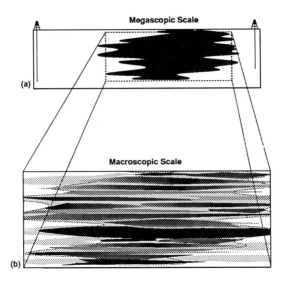

Figure 1. Schematic showing styles of megascopic and macroscopic heterogeneity that are common to San Andres and Grayburg reservoirs. Megascopic scale heterogeneity typically occurs as elongate (into the cross section) grainstones that are relatively high in permeability. The grainstones are easily missed by wells drilled on a regular grid. Even when tapped by wells, however, recovery efficiency commonly remains poor due to channeling of injected fluids through high-permeability stringers that comprise a macroscopic scale of heterogeneity.

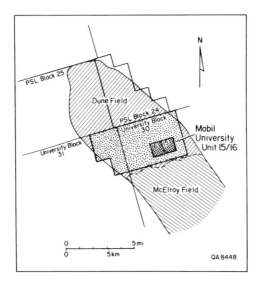

Figure 2. Location of study area.

The approach was to simulate waterflooding in cross sections in which the interwell permeability (k) distribution was estimated on the basis of detailed vertical profiles of k at well locations. Interwell k values were estimated by stochastic methods wherein several possible geologic interpretations of the k patterns were made with the geostatistical technique known as conditional simulation. Conditionally simulated interwell k distributions preserved the high degree of complexity and discontinuity observed in real-world rocks. Further, various possible interwell distributions generated by the technique made it possible to systematically estimate the maximum and minimum expected recovery efficiencies as a function of well spacing.

In contrast to most petroleum simulation studies, the objective here was not to construct a highly accurate predictive model that is calibrated by history-matching procedures and used to forecast future pressures, saturations, and production rates. Rather, the objective was to conduct numerical experiments that elucidate the physical processes affecting recovery efficiency. We avoided the more traditional reservoir modeling approach for two reasons. First, in the Dune Field, as in most other Permian Basin fields, the fluid pressure history data needed for history-matching are virtually nonexistent. Second, a field-scale predictive model must generally be three-dimensional and encompass a large area, requiring the use of simulator grid blocks that average across many scales of heterogeneity. Thus, such a model excludes the details of heterogeneity needed to address issues of recovery efficiency and infill drilling.

For a more detailed description of the Dune Field history and the numerical modeling experiments that we conducted, refer to Bebout et al.(1987) and Fogg and Lucia (1990).

II. SKETCH OF DUNE FIELD RESERVOIR GEOLOGY

The southwest-northeast cross section in figure 3 shows the distribution of lithofacies in Dune Field Unit 15/16. Bebout et al. (1987) subdivided the Grayburg into a non-productive lower unit, a middle unit consisting of the "MA" zone that is topped by the "A" silt, and an upper unit topped by the Queen Formation (fig. 3). In Section 15, the upper unit yielded an estimated 58 percent of total production (as of 1981), with the remainder coming from the MA zone, which contained 59 percent of the original oil in place. Because only 42 percent of total production has come from the oil-rich MA zone, which in 1981 contained an estimated 68 percent of the remaining mobile oil, it appears that the MA zone has been under-exploited.

Grainstone and packstone facies such as the crinoid packstone/grainstone and the pellet grainstone (fig. 3) tend to have the highest permeability (k) values. As demonstrated by Bebout et al. (1987), however, abundant k data measured on core plugs show that k values within facies generally vary by four to five orders of magnitude, rendering the facies ineffective as predictors of local k. Each facies contains a variety of pore types (intergranular, intercrystalline, and mixed), causing the high variability of k within facies. Vugs are not abundant in the Grayburg reservoir of the Dune Field and are seldom numerous enough to form interconnected conduits of extreme permeability. Natural, open fractures are rare in cores taken from 17 wells drilled in and around Unit 15/16 (Bebout et al., 1987).

Our modeling experiments focus on the MA zone because it has the greatest potential for increased production. However, the geologic architecture and patterns of k used in our models are similar to those expected in the upper unit and in many other unfractured, non-vuggy carbonates of the Permian Basin.

358   *Graham E. Fogg et al.*

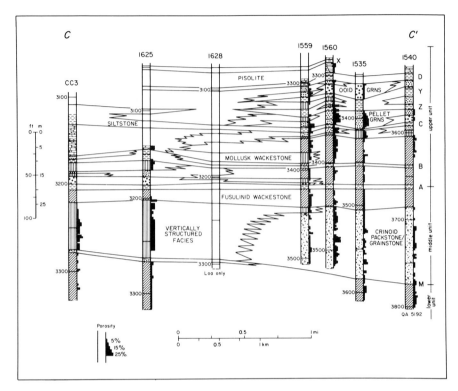

Figure 3. Lithofacies dip section from southwest (left) to northeast (right) across Sections 15 and 16 of the Dune field (from Bebout and others, 1987). Letters A, B, C, D, Y, and Z indicate quartz silts used for correlation. "MA" zone is between the "A" silt and "M" marker.

III. METHODS

A. Estimating Absolute Permeability

Direct measurements of absolute permeability (k) are available for core plugs from five wells drilled in Section 15; however, only the data from two of these wells are trustworthy, the remainder being unreliable due to core damage caused by destruction of gypsum during high-temperature core analysis (Lucia et al., 1987). To provide adequate data control for mapping detailed heterogeneity, k was therefore estimated by a method based on the wireline geophysical log data and petrographic description of the carbonate pore types. This method, which applies only to non-vuggy carbonate rocks like those in the Dune Field, is briefly described below; details can be found in Lucia et al. (1987).

Permeability of non-vuggy dolomites in Section 15 was estimated from sonic and resistivity logs based on a technique that relates permeability to porosity and carbonate pore type (i.e., intergranular, intercrystalline, and mixed). The different pore types can be recognized on thin sections and are each characterized by a different porosity versus permeability (ϕ-k) line. The pore types can also be estimated based on a relationship between pore size, porosity, and oil saturation, which is obtained from the resistivity log. Thus, permeability was estimated from geophysical logs by reading porosity on the sonic log, estimating the pore type, and then reading permeability from the appropriate empirically determined ϕ-k line. Using this approach, permeability was estimated at 1-foot intervals in 33 boreholes spaced approximately 300 m apart on a grid. The estimated permeability values show very close agreement (within less than one-half a $\log_{10}$ unit) with permeabilities measured in cores (Bebout et al., 1987).

Figure 4 is a kh (permeability × thickness) map of the MA zone in Section 15. Thickness of the MA zone here is approximately 100 ft. Note the northwest trending region of high-k values. This region has yielded most of the oil recovered from the MA zone and, interestingly, still contains most of the remaining mobile oil in the MA zone (Bebout et al., 1987).

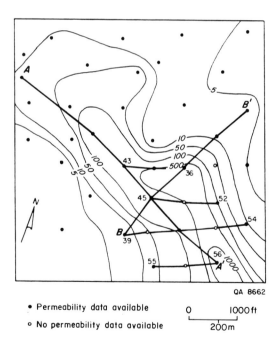

QA 8662

• Permeability data available   0   1000 ft

○ No permeability data available   200 m

Figure 4. Map showing contours on kh (permeability X thickness; md-ft) for the MA zone. Contour interval is variable.

B. Geostatistics

1. Variography

Variography is the use of variograms to statistically characterize spatial variability of a property. Geostatistical estimation and conditional simulation of a property are based on the data and on the estimated variogram.

A common misconception about variography is that it can be routinely used to detect spatial correlation without prior knowledge of the cause or expected style of spatial variability. To the contrary, the variogram can yield misleading results if applied blindly to data having a complex spatial pattern (Fogg, 1989). The trained human eye remains the better tool for **detecting** in a qualitative sense the spatial correlation or pattern. The variogram is mainly a tool for **quantifying** spatial correlation that is either explicitly exhibited in the data or inferred from the data using geologic interpretation. With prior information on the style of spatial variability expected in a given property, the scientist or engineer can often make adjustments in the variogram computations such that the essence of a particular spatial pattern is indeed quantified.

An experimental variogram value (γ^*) can be calculated as a function of distance and direction by computing the mean of the squared differences between pairs of data values (Journel and Huijbregts, 1978; Clark, 1979):

$$\gamma^*(s) = \frac{1}{2N(s)}[Y(x_i) - Y(x_i+s)]^2 \tag{1}$$

where N(s) is the number of data pairs in distance/direction class corresponding to the vector s, $Y(x_i)$ is the value of Y at location x_i and $Y(x_i+s)$ is the value of Y at another location offset from x_i by the vector s. Uneven spacing of data is compensated for by treating the length and orientation of each vector s as variables within specified ± tolerances.

The modifier "experimental" indicates the variogram is based on actual data and therefore only approximates the theoretical variogram, which is written:

$$\gamma(x,x+s) = \frac{1}{2}E\left\{[Y(x+s) - Y(x)]^2\right\} \tag{2}$$

where E indicates "expected value of".

2. Conditional simulation

Conditional simulation is a method for reproducing or simulating the "true" variability of a spatial variable such that the simulated values (1) vary stochastically between data points as a function of the variogram and data distribution, and (2) honor the data points. This is performed via a Monte Carlo procedure wherein many (e.g., 100's) of different patterns or realizations of a key variable such as permeability are generated. Each realization is different but is conditional to the same data and variogram model. The multiple realizations can be used to make predictions under uncertainty. For example, conditional simulation of permeability will produce realizations having relatively poor spatial continuity and realizations having relatively good spatial continuity. Such contrasting realizations can be used in the reservoir simulator to estimate minimum and maximum recovery efficiencies, respectively.

Conditional simulation was performed with the LU-matrix technique (Wilson, 1979; Clifton and Neuman, 1982; Alabert, 1987) implemented in the computer program SIMPAN (Fogg, 1986). Each conditionally simulated value $\tilde{Y}_n$ at grid block n consists of an underlying trend $\hat{Y}$ that is estimated by point kriging and a stochastic component $\tilde{e}_n$ that is semi-random:

$$\tilde{Y}_n = \hat{Y}_n + \tilde{e}_n \tag{3}$$

Two important properties of $\tilde{e}$ are that it preserves the spatial correlation structure represented in the variogram and it shrinks to zero at locations where Y is known. Thus, because $\hat{Y}_n$ honors the data, the conditionally simulated values $\tilde{Y}$ also honor (or are conditioned by) the data.

Kriged values $\hat{Y}_n$ are calculated with the equation

$$\hat{Y}_n = \sum_{i=1}^{I_n} \lambda_{ni} Y^*(x_{ni}) \tag{4}$$

where subscript n denotes the point at which Y is being estimated, $Y^*(x_{ni})$ is a neighboring measurement of Y located at x_{ni}, I_n is the number of neighboring measurements, and λ_{ni} are "kriging weights". The kriging weights are obtained through minimization of the variance of the estimation error e_n subject to the unbias condition that λ_{ni} sum to unity. The value of e_n is simply the difference between the true and kriged values of Y:

$$e_n = Y_n - \hat{Y}_n \tag{5}$$

In practice, the true values Y_n are never known except at points of measurement, but the variance and covariance of e_n can be calculated with the variogram γ. We conditionally simulated permeability using an LU-matrix approach based on the variance and covariance of e_n. The appendix provides a brief description of the LU-matrix algorithm as well as other methods of performing conditional simulation.

Conditional simulation of k was performed for two 1,400-ft cross sections through the MA zone: cross section 55-56, running between wells 1555 and 1556, and cross section 43-36, running between wells 1543 and 1536 (fig. 4). Each simulation generated 200 realizations over a grid having horizontal and vertical spacings of 100 and 4 ft, respectively. Additional simulations were run for section 43-36 with a vertical spacing of 1 ft in order to test the effects of submacroscopic-scale heterogeneity on recovery. This report presents results for cross section 43-36 only.

C. Simulation of Two-Phase Fluid Flow

Numerical simulation of flow of water and oil in response to production and waterflooding was accomplished with the program BOAST, a finite-difference, IMPES (implicit pressure, explicit saturation) code written by Fanchi et al. (1982). Direct factorization by Crout's method with D4 ordering (Price and Coats, 1974) was used to solve for pressures.

With one exception, horizontal and vertical finite-difference grid spacings were 50 and 4 ft, respectively. Although most of the k simulations were generated on a 100- by 4-ft spacing, horizontal lengths in the flow simulator were reduced to 50 ft by doubling the number of columns in the grid in order to allow more freedom in the placement of infill wells. To handle the k values that were simulated at a 100- by 1-ft spacing, the vertical node spacing was reduced first to 1 ft and then increased to 2 ft. We could not achieve mass balance convergence with the 1-ft vertical spacing. Effective k values were scaled from the 1-ft to the 2-ft grid by calculating k_h and k_v with the arithmetic and harmonic means, respectively.

Production and injection were modeled with Dirichlet boundary conditions by assuming that fluid levels were just above the uppermost producing interval in production wells and approximately at land surface in injectors. Most injectors have been operating with little to no injection pressure. No hard data on pressure in production wells were available.

Two sets of relative permeability curves used in the flow simulations are shown in figure 5. The "set 1" were used in most of the simulations. Owing to the general lack of data on relative permeability in the Dune Field, the "set 1" curves were assumed based on data from the literature on similar rocks (Ader and Stein, 1982). The "set 2" curves are based on relative permeability measurements that became available midway through our study. Mobility ratios are 1.5 and 0.5 for sets 1 and 2, respectively. The "set 1" curves are probably most representative of the high-k rocks. The set "2" curves were obtained from a moderately permeable (15 md) core sample.

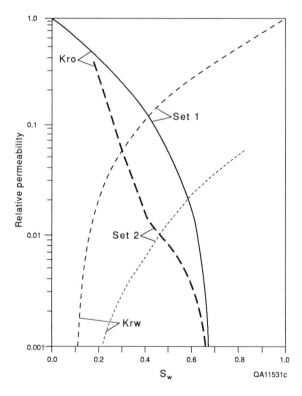

Figure 5. Relative permeability curves used in the flow simulations. Set 1 was used in all but two simulations, in which set 2 was employed. Mobility ratios for sets 1 and 2 are 1.5 and 0.5, respectively.

IV. STOCHASTIC MODELING OF HETEROGENEITY AND ITS EFFECTS ON IN-FILL DRILLING PERFORMANCE

This phase of the study consisted of three parts. First, spatial variability of k was quantified by variogram analysis. Second, the spatial distributions of k in cross section 43-36 were modeled stochastically with the method of conditional simulation (Journel and Huijbregts, 1978). The geologic architecture and continuity of k were imposed with the variogram, and the resulting stochastic "simulations" of k provided hundreds of different possible spatial distributions, or realizations, that were conditioned on (agree with) the well data. Each set of realizations was evaluated statistically and geologically. Third, simulation of production and waterflooding for two infill drilling schemes was performed with selected realizations of k.

A. Variogram Analysis (Variography)

1. Vertical variography

Experimental variograms of log permeability were computed for the MA zone of Section 15. Figure 6 shows vertical variograms representing both grainstone-dominated facies and non-grainstone facies in the MA zone. The grainstone facies include all wells located in the northwest-trending high-permeability zone delineated by the 100 md-ft contour of the kh map (fig. 4), while the non-grainstone facies include the remaining wells.

Although experimental variograms are often highly variable and difficult to interpret, these vertical variograms clearly summarize the vertical continuity characteristics of both facies. The non-grainstone facies reach a much lower sill, or variance, than the grainstone facies because permeabilities of so many of the non-grainstone rocks fall below the lower detection limit of both laboratory and log techniques (0.01 to 0.1 md). In other words, permeability of the non-grainstone rocks may range from 0.000001 to 10 md, but only the upper half or so of this distribution could be sampled, leading to an artificially low variance. Both vertical variograms in figure 6 nevertheless show the same ranges of correlation and evidence of nested structures. Two ranges of correlation are evident, one at 4 to 5 ft and another at 12 to 13 ft. Close examination of the permeability curves for each well in Section 15 (for example, see fig. 7), shows that these variogram ranges reflect two different thickness scales. The short 4- to 5-ft range corresponds to the average thickness of the smallest measurable beds ("type A"). The 12- to 13-ft range corresponds to the average thickness of thicker units ("type B') that are either massive or composed of the thinner beds.

For conditionally simulating k in cross section 43-36, a local vertical experimental variogram was generated using data from just wells 1543 and 1536. Figure 8 shows the resulting experimental variogram and fitted spherical model. The spherical model with a range of 10 ft and sill of 1.2 provides a reasonably good fit, especially for the first 10 feet. Beyond 10 ft, the experimental variogram fluctuates noticeably about the model. Experience has shown, however, that fluctuations of this magnitude generally do not seriously affect the kriging or conditional simulation results.

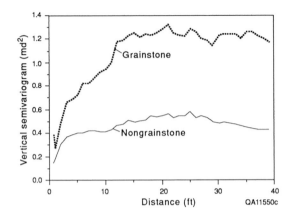

Figure 6. Vertical experimental variograms (γ*) showing vertical correlation of permeability in grainstone- and nongrainstone-dominated facies of the MA zone. Ranges of correlation at approximately 4 and 12 ft indicate nested structures caused by two scales of bed thickness.

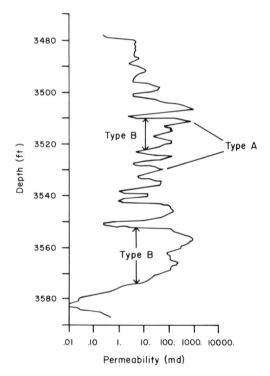

Figure 7. Permeability profile for well 1556. Permeability values were estimated at 1-ft intervals using the relationship between porosity, particle size (Lucia et al., 1987). In figure 6 the ~4-ft variogram range reflects the thin, type A beds, while the ~12-ft range reflects the relatively thick, type B beds.

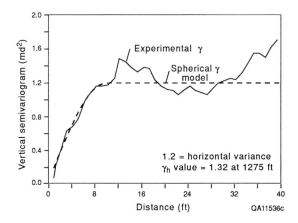

Figure 8. Vertical experimental variogram (γ^*) showing vertical correlation of permeability in the MA zone of Mobil University wells 1543 and 1536. A spherical variogram model with a range of 10 ft and a sill value of 0.7 was fit to this experimental variogram and applied in stochastic simulation of permeability between wells 1543 and 1536.

2. Horizontal variography

Horizontal experimental variograms of permeability in Section 15 are shown in figure 9 for three different azimuths: parallel and perpendicular to the high-permeability grainstone trend (N37°E and N53°W; fig. 4) and East-West. The jagged nature of these variograms as compared with the vertical variograms is due to the relatively sparse data distribution along the horizontal. The irregularities are therefore not necessarily significant statistically or geologically.

To eliminate structural effects in construction of the horizontal variograms, the permeability profile for each well was hung on the base of the "A" silt (fig. 3). Well-to-well variations in thickness of the MA zone are less than 5 percent. The variograms were computed with a search window of ±10° in the horizontal plane and a vertical search band of ±1.1 ft (Fogg and Lucia, 1990).

The experimental variograms in figure 9 hint at an anisotropic correlation structure that is consistent with the high-permeability grainstone trend running diagonally across Section 15 (figure 4). The N53°W variogram, which parallels the grainstone trend, shows fairly good continuity, rising gradually to a range of approximately 2,000 ft and then leveling off. In contrast, the N37°E variogram, which runs perpendicular to the grainstone trend, shows poorer continuity with a range of perhaps 1,000 ft. Thus, the ratio in correlation scales measured parallel and perpendicular to the high-permeability trend appears to be roughly 2 to 1.

We included the East-West experimental variogram because it parallels the orientation of simulated cross section 43-36, which lies midway between the N53°W and N37°E azimuths. Therefore it is not surprising that the resulting variogram lies roughly between the N53°W and N37°E variograms. The low value at approximately 2,000 ft in the E-W variogram is an anomaly caused by the low number of data pairs available for calculating that value.

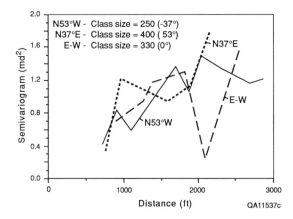

Figure 9. Horizontal experimental variograms (γ*) showing directional correlation of permeability in the horizontal plane, MA zone, Section 15. The N53°W variogram structure, which is oriented parallel to the high-permeability grainstone trend (fig. 4), shows stronger spatial correlation, reaching a range of approximately 2,000 ft as compared to approximately 1,000 ft for the transverse directions (E-W and N37°E). Greater scatter as compared to the vertical experimental variograms (figs. 6 and 8) is due to the coarse distribution of data in the horizontal plane. The low value at about 1,900 ft on the E-W line is probably caused by an insufficient number of data pairs at that spacing.

Because no permeability data occur between wells 1543 and 1536, experimental variograms representing lateral continuity could not be calculated directly. Ideally, we would like to choose a variogram model based on geologic models of facies continuity, but the available carbonate depositional models are not detailed enough to provide guidance at this small scale. Two end-member models were therefore assumed in order to gain a feel for sensitivity of the results to errors in the variogram: a poor continuity case with a variogram range of 400 ft and a good continuity case with a range of 1,300 ft. The sills are 1.2, in accordance with the vertical variogram for wells 1543 and 1536 (fig. 8) and with the experimental variogram value of γ*(1,275)=1.3 calculated between the wells using equation (1). Stochastic simulations of permeability using the variogram models having the alternate ranges of 400 and 1,300 ft are hereafter referred to as cases 43-36-A and 43-36-B, respectively. Simulations of permeability with the refined 100- by 1-ft grid spacing will be referred to as F/43-36-A and F/43-36-B.

B. Conditional Simulation

Two realizations from each conditional simulation were selected for waterflood simulation and are shown in figures 10 (case 43-36-A) and 11 (case 43-36-B). Additionally, figure 12 shows one realization based on the refined grid. The selection criteria used are discussed below.

The principal difference between the simulated cross sections and those created by kriging or deterministic methods is the presence in the former of isolated

REALIZATION NO. 192
PERMEABILITY SIMULATION 43-36-A

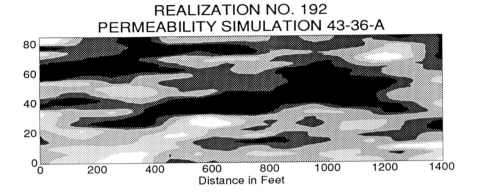

Distance in Feet

REALIZATION NO. 90
PERMEABILITY SIMULATION 43-36-A

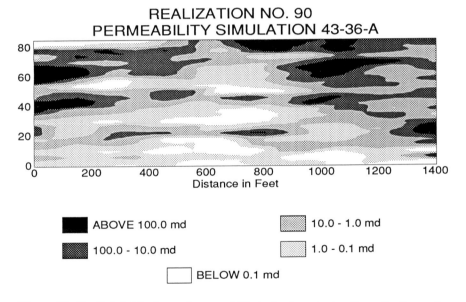

Distance in Feet

ABOVE 100.0 md

10.0 - 1.0 md

100.0 - 10.0 md

1.0 - 0.1 md

BELOW 0.1 md

Figure 10. Spatial distributions of permeability in the two stochastic realizations chosen from Case 43-36-A for use in the waterflood simulations. Realization A192 represents relatively high continuity, lying at the 90 percentile of C_h (see Table 1).
Realization A90 represents relatively low continuity, lying at the 10 percentile of C_h.
Wells 1543 and 1536 are located at the left and right sides of the cross section,
respectively. Well 1508 is located approximately in the middle.

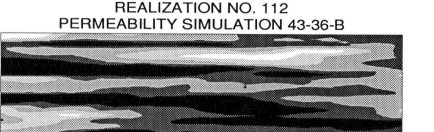

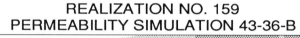

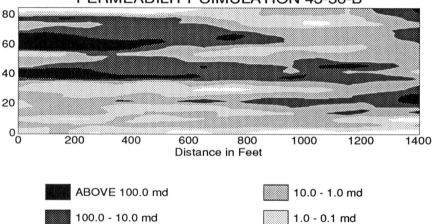

Figure 11. Spatial distributions of permeability in the two stochastic realizations chosen from Case 43-36-B for use in the waterflood simulations. Realization A112 represents relatively high continuity, lying at the 90 percentile of C_h (see Table 1.
Realization A159 represents relatively low continuity, lying at the 10 percentile of C_h.
Wells 1543 and 1536 are located at the left and right sides of the cross section,
respectively. Well 1508 is located approximately in the middle.

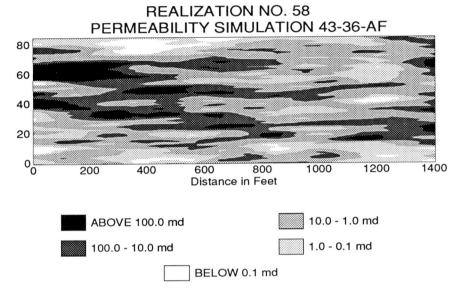

REALIZATION NO. 58
PERMEABILITY SIMULATION 43-36-AF

Distance in Feet

ABOVE 100.0 md

100.0 - 10.0 md

10.0 - 1.0 md

1.0 - 0.1 md

BELOW 0.1 md

Figure 12. Spatial distribution of permeability in stochastic realization A58 obtained from the fine-grid conditional simulations. This realization represents relatively low continuity, lying at the 10 percentile of C_h (see Table 1). Wells 1543 and 1536 are located at the left and right sides of the cross section, respectively. Well 1508 is located approximately in the middle.

stringers of high or low permeability between the wells. Clearly, such discontinuities can strongly influence recovery efficiency and the outcome of an infill drilling program.

1. Choosing realizations for simulation of flow

Each pair of permeability (k) realizations (figs. 10 through 11) chosen from simulations 43-36-A and -B and F/43-36-A and -B consists of one with poor continuity and the other with good continuity. The selection criteria were based on continuity statistics generated with the computer program MCSTAT (Fogg, 1986), which scans each realization to determine vertical and horizontal continuity statistics of the high-k facies, defined as those rocks ≥ 100 md in k. That is, the program measures the total length (C_h) and thickness (C_v) of domains consisting of contiguous blocks with simulated k values ≥ 100 md. Two blocks are considered contiguous when they share a side. The 100-md cutoff was chosen because modeling experiments performed earlier (Fogg and Lucia, 1990) showed that waterflood efficiency tends to be dominated by the distribution and geometry of facies having permeabilities >100 md.

Table 1 shows permeability and continuity statistics for the eight chosen realizations. The choice of poor- and good-continuity realizations was based primarily on C_h, the mean horizontal continuity of high-k facies. Realizations with poor continuity (i.e., A90 and B159; A58 and B 68) have C_h values at approximately the 10 percentile

Table 1. Continuity of simulated facies having k≥100 md (units are ft).

	Mean log-k	Std. Dev. log-k	Volume fraction	No. of domains[a]	Horizontal Continuity (Ch)[b]			Vertical Continuity (Cv)[b]		
					Max.	Min.	Mean ($\overline{C_h}$)	Max.	Min.	Mean ($\overline{C_v}$)
Case 43-36-A: γs range = 400 ft (100- by 4-ft grid)										
Realization 90	0.35	1.41	0.092	12	400	100	**150**	12	4	6.7
Realization 192	1.11	1.69	0.241	8	900	100	**338**	36	4	11.0
Ensemble avg.	0.80	1.40	0.156	11	638	100	**227**	20	4	7.6
Case 43-36-B: γs range = 1,300 ft (100- by 4-ft grid)										
Realization 159	0.61	1.03	0.095	8	600	100	**225**	12	4	6.0
Realization 112	1.11	1.38	0.241	7	1,400	100	**571**	24	4	8.6
Ensemble avg.	0.79	1.24	0.140	6	931	107	**372**	15	4	7.5
Case F/43-36-A: γh range = 400 ft (100- by 1-ft grid)										
Realization 58	0.62	1.01	0.082	25	400	100	**144**	13	1	2.4
Realization 84	1.05	1.15	0.200	18	900	100	**261**	18	1	5.6
Ensemble avg.	0.77	1.21	0.153	23	805	100	**197**	19	1	3.6
Case F/43-36-B: γh range = 1,300 (100- by 1-ft grid)										
Realization 68	0.98	1.15	0.171	26	1,300	100	**188**	15	1	2.2
Realization 21	0.92	1.30	0.205	14	1,300	100	**357**	16	1	3.8
Ensemble avg.	0.77	1.11	0.131	15	1,029	100	**265**	14	1	2.9

a A domain is a group of contiguous blocks in which k≥100 md.
b "Continuity" refers to the total length of a domain having k≥100 md.

(i.e., 10 percent of the realizations have lower $\overline{C}_h$ values); and realizations with good continuity (i.e., A192 and B112; A84 and B21) have $\overline{C}_h$ values at approximately the 90 percentile (i.e., 90 percent of the realizations have lower $\overline{C}_h$ values.

Note in Table 1 that, with the exception of "number of domains", statistics from the fine-grid simulations fluctuate much less between realizations than do the same statistics from the coarse-grid simulations. The greater number of simulation points in the fine grid causes better preservation of the mean characteristics.

C. Simulation of Waterflooding and Infill Drilling

To account for correlations between porosity (ϕ) and permeability, values of ϕ were calculated from the corresponding stochastic k value for each block based on the general regression equation (Bebout et al., 1987):

$$k = \left(3.5 \times 10^4\right)\phi^{4.25} \tag{10}$$

Each realization of k values therefore has a unique realization of ϕ values, making ϕ a stochastic variable by virtue of its correlation to k.

Flow through each k realization was simulated in two steps: a 7-yr simulation with no infill wells followed by a 6-yr simulation with two infill wells. Final conditions from the 7-yr simulations were used as initial conditions in the 6-yr simulations. These time frames were chosen to recover most of the oil producible with the particular well spacing and k architecture. Thus, by the end of each simulation, production had reached fairly low levels.

In each infill case, the end wells (1543 and 1536) were maintained as producers, while the interior wells were injectors and producers. During the first 7-yr simulation, the interior well 1508 was an injector, and during the subsequent 6-yr simulation well 1508 was switched to production and the two infill wells were injectors.

D. Results

1. Coarse grid

Figures 13 and 14 show cumulative oil production (expressed as percentage of the mobile original oil in place [OMOIP]), and producing water/oil ratio (WOR) versus time for k realizations.A90, A192, B159, and B112 as computed in the flow simulations with the coarse finite-difference grid (50- by 4-ft spacing). The infill wells in these simulations did not always increase production significantly. Permeability realization B112, which is the high-k case obtained with variogram model B (range=1,300 ft), yielded nearly 80 percent of the OMOIP during the first seven years of simulation with the original 10-acre well spacing. The high recovery efficiency is due to the presence of laterally extensive high-permeability zones.

If realization B112 were representative of field conditions, it would indicate that infill wells are not needed. Results of this model, however, are considered unrepresentative of field conditions because of the unrealistically high recovery efficiencies, initial production rates, and WOR's that it yields both with and without infill wells. Wells in the Dune Field have historically yielded initial production rates of 50 to 300 BPD, which, when scaled down to the volume of the two-dimensional cross section, would become 0.04 to 0.22 BPD. In contrast, simulated production from well 1543 for

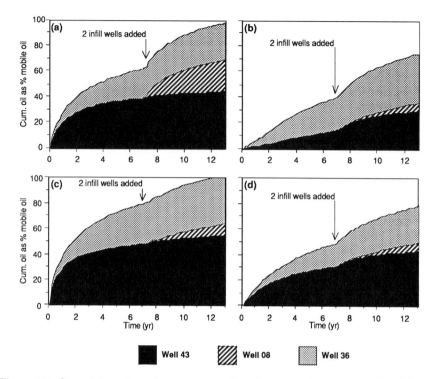

Figure 13. Cumulative oil production versus time (expressed as percent of mobile oil originally in place) computed for cross section 43-36 in waterflood simulations with 2 infill wells in permeability realizations (a) A192, (b) A90, (c) B112, and (d) B159.

case B112 at early time is 1.25 BPD, which is equivalent to approximately 1,700 BPD when scaled up to three dimensions. Similarly, simulated WOR for well 1543 for case B112 reaches nearly 500, compared to a measured value for the well of 113 in November 1987. As discussed in a later section, the two-dimensionality of the model is partly to blame for the high production rates and WOR's, but we do not believe it can account for the large numbers associated with realization B112.

The other three realizations A192, A90, and B159 yield recovery efficiencies and WOR's (figs. 13 and 14) that appear more reasonable, although well 1543 of case A192 yields rather high values of initial production and maximum WOR of 0.57 BPD and 300, respectively. These three realizations also show the most significant responses to infill wells, giving increases in recovery efficiency of 29 to 36 percentage points after addition of the two infill wells.

The infill wells in cross section 43-36 resulted in some high WOR values in every case except A90 (fig. 14b), which had the lowest permeability and continuity. Both A90 and B159, representing the low-permeability and -continuity extremes, gave significantly lower values of WOR. In A159, only well 43 had an excessively high WOR after addition of the infill wells; however well 43 also yielded a significant portion of the additional oil production for the same time period (7 to 13 yr). High WOR values for well 1508 immediately after the seven-year mark result from the well being switched from an injector to a producer.

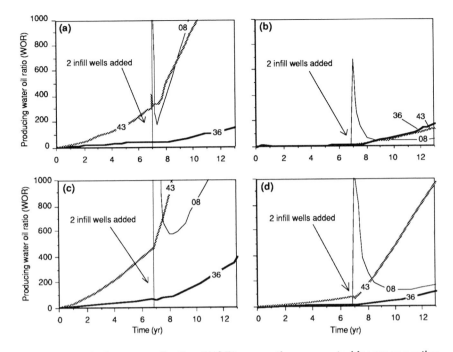

Figure 14. Producing water oil ratios (WOR) versus time computed for cross section 43-36 in waterflood simulations with 2 infill wells in permeability realizations (a) A192, (b) A90, (c) B112, and (d) B159.

2. Fine grid

The fine-grid (100- by 2-ft spacing) simulations are still underway; thus, only re-sults from realization A58 have been generated (figs. 15 and 16). These results nev-ertheless provide noteworthy insights. Realization A58 is most similar to the coarse-grid realizations A90 and B159 in terms of general appearance of the permeability fields (figs. 10b, 11b, 12) and statistics (Table 1). Yet the recovery efficiencies for A58 are lower by 10 to 30 percentage points (figs. 13b and d, 15). This indicates that the smaller, more discontinuous heterogeneities afforded by the fine-grid simulations cause significantly greater compartmentalization and bypass of oil.

3. Alternate relative permeability curves

Flow simulations with the set 2 relative permeability curves having the low mobil-ity ratio of 0.5 computed sharply lower recovery efficiencies and WOR's as compared to the values generated with the set 1 curves (Fogg and Lucia, 1990). In fact, the effect of switching to the set 2 relative k's was quite similar to the effect of refining the grid. Addition of the infill wells still improved recovery efficiency by 20 to 30 percent-age points, however.

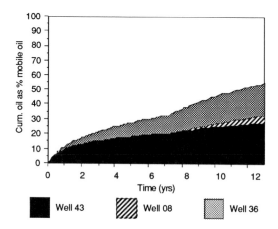

Figure 15. Cumulative oil production versus time (expressed as percent of mobile oil originally in place) computed for cross section 43-36 in the fine-grid waterflood simulation with 2 infill wells in permeability realization A58.

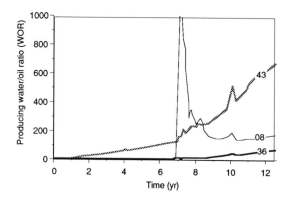

Figure 16. Producing water oil ratios (WOR) versus time computed for cross section 43-36 in the fine-grid waterflood simulation with 2 infill wells in permeability realization A58.

V. DISCUSSION

A. Effects of Grid Refinement

A comprehensive discussion of the effects of the grid refinement cannot be offered until more of the fine-grid k realizations are tested in the flow simulator. Nevertheless, the results obtained thus far strongly suggest that the finer-scale heterogeneities (<4 ft) are important in this case. From the standpoint of geostatistics, the finer-scale conditional simulations are more proper because they more accurately reflect the vertical variogram structure. Range of the vertical variogram for cross section 43-36 is 8 ft; thus, much of the vertical correlation is lost or averaged out when using the 4-ft spacing. Consequently, realization A58 contains thinner permeability beds than realizations A90 or B159, and the difference between the high- and low-k extremes for these thin beds is greater. This leads to greater variations from layer to layer in horizontal fluid displacements and, in turn, lower sweep efficiency.

Recent studies of spatial variability of k in carbonate rocks suggest that, at horizontal distances <10 ft, point values of k are distributed more or less randomly (Kitteridge, 1988). Thus, horizontal correlation of k in these kinds of rocks may occur as thicker beds within which k is semi-random. If this conceptual model is correct, then the coarse-grid simulations with their greater vertical averaging of k may actually be more realistic.

Another important effect of the finer conditional simulation grid is to partially homogenize the realizations from a given conditional simulation. That is, the high- and low-continuity cases from the fine-grid simulations do not differ from one another as much as their counterparts from the coarse-grid simulations. Thus the highs and lows in production, recovery efficiency, and WOR's computed by the coarse-grid models may be too extreme. Indeed, the fact that the two sets of relative k curves never gave reasonable results for both the continuous and discontinuous realizations suggests that the range in conditions portrayed by the coarse-grid realizations is too broad. Given the uncertainties inherent to the simulation models, however, one might desire the wide margin of error provided by the different coarse-grid realizations.

B. Influence of Variogram Model

The two horizontal variogram models applied with ranges of 400 and 1,300 might be considered radically different by some; yet a comparison of results based on these two models indicates that even though the differing ranges caused significantly different looking k patterns (figs. 10 and 11), the flow modeling results look strikingly similar. In figs. 13 and 14 the cumulative recoveries and WOR's for A192 are practically identical to those for B112, which was created with a substantially larger variogram range. The same can be said in comparing A90 and B159. In fact, there is more variation between realizations within a single conditional simulation than between realizations created with different variogram models. Admittedly, much of the variation within conditional simulations is caused by the coarse vertical grid spacing.

Similar results were generated by Fogg (1989) in stochastically modeling single-phase flow through multi-sand-body systems. Although different variograms may create significantly different spatial patterns in a porous media property, the integral response of fluid flow to such patterns will not necessarily vary much between realizations created with those variograms.

C. Effects of Neglecting the Third Dimension

A limitation of the cross-sectional models is their two-dimensionality, which forces all fluid flow into one plane aligned with the string of wells. At best, this plane only represents the flow occurring along the most direct streamline connecting injectors and producers. Due to radial flow to the wells, additional streamlines in the third dimension represent longer flow paths. Thus, producing wells in the cross-sectional models tend to water-out quickly, while in three dimensions primary and secondary production from the other streamlines would tend to increase simulated oil production and decrease simulated WOR's. Consequently the cross-sectional models discussed above reach quasi-steady states in only six to seven years. Clearly, any attempt to history match the results of these models would be fruitless and potentially misleading.

Three-dimensional heterogeneity adds another degree of complexity that is at least as important as the radial flow issues discussed above. It is well known that the effective permeability and connectivity of a porous medium increase as the dimensionality of the system increases from one to three. This trend occurs simply because in three dimensions the number of permeable pathways along which fluids can bypass the low-permeability facies in a reservoir or aquifer is maximized. Thus, cross-sectional models containing media that are discontinuous between wells may tend to overestimate sweep efficiencies because the injected fluids typically contact a greater percentage of the reservoir pay than they would in three dimensions. On the other hand, one might argue that the longer, more circuitous three-dimensional flow path would contact a greater reservoir volume. These issues can only be resolved with three-dimensional modeling experiments.

Summarizing, the cross-sectional models neglect the three-dimensional aspects of radial flow and heterogeneity. Neglect of radial flow leads to abrupt peaks and steep declines in the rate of oil production as well as premature increases in WOR's. The effect of neglecting heterogeneity in the third dimension is problematic. If the three-dimensional model allows greater connectivity between wells, which is typically the case, the two-dimensional model might tend to overestimate sweep efficiency and, perhaps, underestimate WOR. This effect could be counteracted by the neglect of radial flow in the two-dimensional model. If in the two-dimensional model, the reservoir architecture and well spacing are such that connectivity of high-permeability zones between injectors and producers is already good and these zones are laterally extensive normal to the section, results might be similar to those from a three-dimensional model.

D. Implications for Infill Drilling

Except for realization 112, all the simulation results show that infill drilling to reduce the current 10-acre spacing to a 2.5-acre spacing in Dune Field Section 15 causes a significant increase in oil production and at least a 20-percentage-point increase in recovery efficiency. Realization 112 does not respond very well to infill drilling because it contains high-permeability facies that interconnect laterally across the entire cross section, essentially leaving no untapped reservoir compartments. Thus, if permeability between wells is significantly discontinuous, infill drilling will most likely lead to substantial improvement in production.

The results also show that addition of infill wells can lead to excessively high rates of water production, depending on the well spacing and degree of lateral discontinuity of permeability facies. The k realizations that gave the most realistic producing water/oil ratios (1 < WOR < 100) during the first seven years of simulation (pre-infill) generally gave low to moderate values of WOR after two infill wells are added. These realizations are A90 and B159 (fig. 14) with the set 1 relative permeability curves and realization A112 with the set 2 relative permeability curves.

The realizations having high permeability and continuity (A192 and B112) give unrealistically high WOR values for both pre- and post- infill phases of simulation when the set 1 relative permeability curves are used (figs. 14a and b). Realization A90 with the set 2 relative permeability curves gives unrealistically low WOR values both with and without infill wells.

E. Other Methods of Estimating Recoverable Oil

Analytical and semi-empirical methods for estimating waterflood sweep efficiencies are generally based either on the concept of pay continuity or on the influence of vertical stratification on oil recovery. George and Stiles (1987) assume the percentage of the formation that can be flooded is approximately equal to the pay continuity, defined as the fraction of total pay (k>0.1 md) in a well connected to another well. This approach proves to be inappropriate in Dune Section 15, where MA strata between wells 1543, 1508, and 1536 would indicate a pay continuity of 95 percent (ϕ>0.6) but the simulated recovery efficiencies (with no infill wells) are generally less than 60 percent. Even if one defines pay more conservatively as any zone with k > 1.0 md, the pay continuity remains high at approximately 80 percent.

The Dykstra and Parsons (1950) method estimates fractional waterflood coverage (fraction of the reservoir contacted by the waterflood) in a vertically stratified reservoir with no lateral discontinuities. Application of the method in Section 15 of the Dune Field results in predicted recoveries that are lower than the coarse-grid simulated recovery efficiencies by a factor of up to 2 (Fogg and Lucia, 1990). The Dykstra Parsons method underestimates recovery efficiency in the coarse-grid cases because it assumes the layers of different permeability are continuous and the flow is linear. The lateral discontinuities in permeability and the potential for vertical cross flow create greater tortuosity in the flow paths. Consequently, the waterflood contacts a greater percentage of the formation than would be predicted by the Dykstra-Parsons method.

Recovery efficiencies predicted with the Dykstra-Parsons method agree more closely with the fine-grid simulation results. This is presumably because the thinly bedded high- and low-k in the fine-grid facies are more prominent and hence more laterally continuous.

VI. CONCLUSIONS

Stochastic analysis of spatial variability of k in cross sections and black-oil simulation of production and water injection in those cross sections provide several important insights into the effects of realistically complex heterogeneity on fluid flow and infill drilling performance. The variogram analysis (variography) elucidated three-dimensional anisotropic correlation features in Section 15 that are consistent with the geology. Vertical correlation ranges of 4 to 5 ft and 12 to 13 ft in the vertical variograms reflect average thicknesses of alternating grainstone-rich and mud-rich strata as well as average thicknesses of packages of these strata. Anisotropic horizontal variograms suggest an approximately 2,000-ft range of correlation parallel to the elongate, grainstone-rich zone of high permeability that trends northwest across Section 15. Perpendicular to this trend, the variograms suggest a shorter range of roughly 1,000 ft. Horizontal variogram ranges of 400 and 1,300 ft were used in different conditional simulation trials to account for uncertainty in the horizontal variogram structure.

Conditional simulation of k for a number of different scenarios yielded several hundred different stochastic/geologic interpretations of interwell heterogeneity. The diversity of k patterns exhibited in these interpretations, or realizations, is consistent with the fact that estimation of interwell heterogeneity is inherently uncertain. The fine-grid (100- by 1-ft spacing) simulations show less fluctuation in mean k and continuity statistics among the realizations than the coarse-grid (100- by 4-ft spacing) simulations owing to the greater number of simulation points in the former.

Two-phase, black-oil simulations of waterflooding in the stochastic permeability distributions for cross section 43-36 of the MA zone were implemented with finite-difference grid spacings of 50 to 100 ft horizontally and 4 ft to 2 ft vertically. This relatively fine-scale discretization captures nearly all of the permeability variation, including much of the vertical anisotropy caused by horizontal stratification of high- and low-permeability facies.

The one fine-grid simulation (2-ft vertical spacing) conducted thus far strongly suggests that heterogeneities at this fine scale can have a significant negative impact on recovery efficiency. The k realization on which this simulations is based contains a greater number of thin beds of extreme high and low k values, which leads to larger differential displacements of oil and water and hence more bypassing. However, this effect could be offset if local spatial correlation in the horizontal plane were characterized by greater randomness (nugget effect) than was included in the stochastic realizations.

Long-term simulations of waterflooding in stochastically generated permeability distributions indicate that, with the existing well spacing, recovery efficiency (expressed as percent of OMOIP) will be no greater than 40 to 55 percent. Addition of two infill wells to achieve an average well spacing of approximately 2.5 acres generally increases the simulated recovery efficiency of OMOIP by at least 27 to 32 percentage points. Cases giving computed WOR values consistent with field WOR values also yielded moderate WOR values after addition of two infill wells. Cases with either poor lateral continuity of permeability facies or low relative permeability to water (set 2 relative permeability curves; fig. 5) gave the most realistic pre-infill values of WOR, and, in turn, moderate post-infill values of WOR. Thus, the results suggest that infill drilling to 2.5 acres could significantly improve recovery efficiencies without leading to exorbitant WOR values.

Other methods may either over- or underestimate recovery efficiency because they neglect effects of lateral discontinuities in permeability, effects of vertical cross flow between layers, or effects of complex fluid hydrodynamics occurring in heterogeneous geologic media.

APPENDIX: STOCHASTIC SIMULATION ALGORITHM

The stochastic components $\tilde{e}$ of equation (3) are derived in part from the variance-covariance matrix of e. The covariance of e evaluated between points n and m is

$$V_{nm} = E\{e_n \cdot e_m\} \tag{6}$$

By substituting equations (4) and (5) into (6) and expanding, we get an equation expressing V_{nm} in terms the kriging weights and the variogram (Clifton and Neuman, 1982; Wilson, 1979):

$$V_{nm} = -\sum_{i=1}^{l_n} \sum_{j=1}^{l_m} \lambda_{ni}\lambda_{mj}\gamma(x_{ni},x_{mj}) - \gamma(R_n,R_m) + \sum_{i=1}^{l_n} \lambda_{ni}\gamma(x_{ni},R_m) + \sum_{j=1}^{l_m} \lambda_{mj}\gamma(x_{mj},R_n). \tag{7}$$

R_n and R_m represent grid points and x_{ni} and x_{mj} are locations of data in the neighborhoods associated with R_n and R_m, respectively. The quantity $\gamma(R_n,R_m)$ is the variogram value evaluated between grid points R_n and R_m. Likewise, $\gamma(x_{ni},R_m)$ is the variogram value evaluated between data point x_{ni} and grid point R_m. When n=m in equation (7), one gets an expression for the variance of e, V_{nn}, or the kriging variance. V_{nm} and V_{nn} form the elements of the variance-covariance matrix **V**.

The output of a conditional simulation typically consists of at least hundreds of generated realizations $\widetilde{Y}$, each one based on different values of the stochastic component $\widetilde{e}$. In the LU-matrix technique employed here, each $\widetilde{e}$ vector is obtained from the equation

$$\widetilde{e} = \mathbf{M} \cdot \Theta \qquad (8)$$

in which **M** is a lower triangular matrix obtained from the **V** matrix by Cholesky decomposition, and Θ is a vector of normal random deviates (mean = 0; variance = 1). While **M** remains fixed, a new Θ vector is generated for each realization, thereby making each realization different. The Cholesky decomposition and generation of normal random deviates were performed with IMSL (1979) routines LUDECP and GGNML, respectively. For an explanation of why equation (8) gives the appropriate values of $\widetilde{e}$, refer to Alabert (1987) or Fogg (1989).

Vectorization of this LU-matrix algorithm on The University of Texas Cray XMP/24 computer led to an approximately 100-fold reduction in execution times. Most of the reduction came through faster computation of the summations in equation (7).

Another way of calculating $\widetilde{e}$ is as follows (Journel and Huijbregts, 1978):

$$\widetilde{e} = \widetilde{Y}^u - \widehat{Y}^u \qquad (9)$$

where $\widetilde{Y}^u$ represents an unconditional simulation (honors the specified variogram but not measured values of Y) and $\widehat{Y}^u$ is a kriged field obtained by kriging values of $\widetilde{Y}^u$ taken at the points of measurement of Y_n. That is, $\widetilde{Y}_n^u$ values would be calculated from equation (4) by replacing $Y^*(x_{ni})$ with $\widetilde{Y}^u(x_{ni})$. The unconditional simulation $\widetilde{Y}^u$ can be generated with the LU-matrix method described above (where **V** would represent the variance and covariance of Y rather than of e), the turning-bands method (Journel and Huijbregts, 1978; Mantoglou and Wilson, 1982), the Fast-Fourier Transform (FFT) method (Borgman et al., 1983), or using fractal techniques (Hewett, 1986). The main advantage of the turning-bands and FFT methods is they do not require storage and manipulation of large matrices.

Computer algorithms for both the LU-matrix and turning bands methods in the solution of equation (9) were tested on The University of Texas Cray XMP/24 computer. The results show that, for a moderate number of simulation nodes (<1,000) on a vector processor, there is no significant advantage of one method over the other. When dealing with much more than 1,000 nodes, however, turning bands and FFT are the methods of choice because they consume relatively little central memory.

ACKNOWLEDGEMENTS

Funding for this work was provided by The University of Texas System, Texas Higher Education Coordinating Board, and Cray Research, Inc. Mobil Producing Texas and New Mexico provided most of the data. We are also grateful to Rick Edson, who created the computer images of heterogeneity and to Richard Dillon and his carto-graphic staff for drafting the figures.

REFERENCES

Ader, J. C., and Stein, M. H., 1982, Slaughter Estate Unit CO2 pilot reservoir descrip-tion via a black oil model waterflood history match: Society of Petroleum Engineers/U. S. Department of Energy 10727, p. 817-823 plus 19 figures.

Alabert, F., 1987, The practice of fast contitional simulations through the LU decom-position of the covariance matrix: Mathematical Geology, vol. 19, no. 5, p. 369-386.

Bebout, D. G., Lucia, F. J., Hocott, C. R., Fogg, G. E., and Vander Stoep, G. W., 1987, Characterization of the Grayburg Reservoir, University Lands Dune Field, Crane County, Texas: The University of Texas at Austin, Bureau of Economic Geology Report of Investigations No. 168., 98 p.

Borgman, L., Taheri, M., and Hagan, R., 1983, Three-dimensional, frequency-domain simulations of geological variables: in Verly, G., David, M., Journel, A. G., and Marechal, A., Geostatistics for Natural Resources Characterization, Part 1, Dordrecht, D. Reidel Publishing Co., 517-541.

Clark, I., 1979, Practical Geostatistics: London, Applied Science Publishers Ltd., 129 p.

Dykstra, H., and Parsons, R. L., 1950, The prediction of oil recovery by water flood: in Secondary Recovery of Oil in the United States, American Petroleum Institute, New York, p. 160-174.

Fanchi, J. R., Harpole, K. J., and Bujnowski, S. W., 1982, BOAST: A three-dimen-sional, three-phase black oil applied simulation tool (Version 1.1), Volume I: Technical Description and Fortran code: Work performed for the U. S. Depart-ment of Energy under contract no. DE-AC19-80BC10033, 180 p.

Fogg, G. E., 1986, Stochastic analysis of aquifer interconnectedness, with a test case in the Wilcox Group, East Texas: The University of Texas at Austin, Ph.D. dis-sertation, 216 p.

Fogg, G. E., 1989, Stochastic analysis of aquifer interconnectedness, with a test case in the Wilcox Group, East Texas: The University of Texas at Austin, Bureau of Economic Geology Report of Investigations No. 189, 68 p.

Fogg, G. E., and Lucia, F. J., 1990, Reservoir modeling of restricted platform carbon-ates: geologic/geostatistical characterization of interwell-scale reservoir hetero-geneity, Dune Field, Crane County, Texas: The University of Texas at Austin, Bureau of Economic Geology Report of Investigations No. 190, 66 p.

George, C. J., and Stiles, L. H., 1987, Improved techniques for evaluating carbonate waterfloods in West Texas: Journal of Petroleum Technology (November), p. 1547-1554.

Hewett, T. A., 1986, Fractal distributions of reservoir heterogeneity and their influ-ence on fluid transport, Society of Petroleum Engineers 15386, paper pre-sented at 61st Annual Tech. Conf. and Exhibition, New Orleans, 16 p.

IMSL Library Reference Manual, Edition 7, 1979.

Journel, A. G., and Huijbregts, Ch. J., 1978, Mining Geostatistics: New York, Aca-demic Press Inc., 600 p.

Kitteridge, M. G., 1988, Areal permeability variations - San Andres Formation (Guadalupian): Algerita Escarpment, Otero County, New Mexico: The University of Texas at Austin, Masters Thesis, 361 p.

Lucia, F. J., Hocott, C. R., and Vander Stoep, G. W., 1987, Distribution of remaining mobile oil: in Bebout, D. G., Lucia, F. J., Hocott, C. R., Fogg, G. E., and Vander Stoep, G. W., Characterization of the Grayburg Reservoir, University Lands Dune Field, Crane County, Texas: The University of Texas at Austin, Bureau of Economic Geology Report of Investigations No. 168, p. 70-94.

Price, H. S. and Coats, K. H., 1973, Direct methods in reservoir simulation, Transac-tions of Society of Petroleum Engineers of A.I.M.E., 257, p. 295-308.

Wilson, J. L., 1979, The synthetic generation of areal averages of random fields: paper presented at Socorro Workshop on Stochastic Methods in Subsurface Hydrology, New Mex. Tech., Socorro, New Mex., April 26-27.

MAXIMUM ENTROPY LITHO-POROSITY VOLUME
FRACTION PREDICTIONS
FROM Vp/Vs RATIO MEASUREMENTS

Joseph G. Gallagher Jr.

Phillips Petroleum Company
Bartlesville, Oklahoma 74004

I. INTRODUCTION

Numerous studies in the literature (Pickett, 1963;
Gregory, 1976; Hamilton, 1979; and Domenico, 1984) have
shown that the compressional (P) to shear (S) wave velocity
ratio, Vp/Vs, can be used in a qualitative sense to predict
changes in lithology and porosity in a sedimentary
interval. For example, modeled binary mixture studies by
Eastwood and Castagna (1983) have revealed that P and S
wave velocities and Vp/Vs ratio exhibit the greatest
sensitivity to variations of concentration for binary
mixtures containing sands, i.e. sand-shale and sand-
limestone mixtures. Typically, the Vp/Vs ratio for these
mixtures range from 1.5 for 100 percent sand to 1.9 and
higher for 100 percent limestone and shale, respectively.
The largest P and S wave velocity variations with concentration are also observed for sand-shale mixtures, with the
P wave and S wave velocities increasing by factors of 1.35
and almost 2, respectively, as the mixture approaches 100
percent sand. Consequently, this enhanced P and S wave
velocity discrimination with concentration found with sand-
shale mixtures make the Vp/Vs ratio measurement a potential
tool for detecting, especially, reservoir sands in field
development.
    In the real layered earth, sand is usually not found in
binary mixtures but, for example, in porous multicomponent
lithological mixtures of shale, limestone, and dolomite.
The Vp/Vs for such an interval is usually determined from

RESERVOIR CHARACTERIZATION II

382

measurements of the S and P wave transit times across a
pair of equivalent P and S wave seismic reflection events
that form the top and bottom boundaries of the interval
(McCormack et al., 1984).

Incorporating well log data from a nearby well as a
calibration point, a maximum entropy method has been used
to obtain quantitative estimates of the volume fraction of
sand in a porous multicomponent, lithological interval,
laterally away from the calibration well, from Vp/Vs ratio
measurements made on P and S wave surface seismic data.

The P and S wave borehole and surface seismic data used
in this study were acquired in West Texas. The borehole
data included (1) P and S wave vertical seismic profiles
(VSP) for obtaining correlations of P and S wave
reflections bordering the interval, (2) a full waveform
sonic log for obtaining in-situ measurements of the P and S
wave velocities and their Vp/Vs ratios, and (3) a standard
suite of well logs for the determination of the
corresponding lithology and porosity in the interval. The
in-situ Vp/Vs ratio-lithology-porosity trends derived from
the well log data were used in the form of constraint
equations in the maximum entropy method to predict the
variational dependence of the Vp/Vs ratio on lithology and
porosity for the interval. Measurements of the S and P
wave transit times and their ratio were obtained from
collateral P and SH (horizontally polarized shear) wave
surface seismic data acquired over the calibration well.

The target in the area was the Canyon sands, which are
part of a fluvial deltaic fan system that formed in the
late Pennsylvanian time. The sands are distributed over a
500 ft interval between 6100 and 6600 ft. The lithological
environment in the interval, in increasing order of
predominance, is dolomite, sandstone, limestone and shale,
with the amount of sand varying in these fan systems from 5
to 30 percent.

II. BOREHOLE AND SURFACE SEISMIC DATA ANALYSIS

A. VSP Data

Correlation of P and SH wave reflection events
bordering the Canyon sands were obtained from vertical
seismic profiles (VSP) acquired in well "A". The complete
P and SH wave fields propagating across the borehole were

recorded with a triaxial orthogonal geophone system containing three geophones inclined at an angle of 54.7 degrees to the longitudinal axis of a VSP tool. Surface vibrator sources (Mertz Model 18) placed approximately 1500 ft from the wellhead were used to generate P and SH wave energies. The VSP data were recorded from geophone depths 6726 ft to 2075 ft at a geophone spacing of 75 ft. Using the amplitude of either the direct P or the direct SH wave arrival wavelets, tool rotation effects were removed from the data with an orthogonal coordinate transformation from the triaxial geophone response to a standard (x,y,z) cartesian geophone response. After the coordinate rotation at each depth, it was assumed the radial geophone pointed in the x-y plane in the direction of the source. The transverse geophone pointed in the direction perpendicular to the source, with the vertical geophone aligned along the borehole axis. Frequency-wavenumber (F-K) filtering was used to extract the P and SH wave upgoing events from the vertical geophone and transverse geophone signals, respectively.

Figure 1 shows the correlation of the P and SH wave reflection events obtained from the VSP data. The SH wave data have been rescaled using a data processing method that transforms an SH wave time section into an equivalent P wave time section. The mapping function for accomplishing this nonlinear rescaling of the SH wave time section is provided by the normal moveout-corrected (NMO-corrected) P and SH wave direct arrival times and depths from both VSP data sets. The rescaling of the SH wave events allows for the exact correlation of all P and SH wave reflection events to total depth (TD) in the well and an estimate of the correlation below TD from regression analysis on the P and SH time-depth data above TD.

From these data, two reflection events, the Breckenridge (depth 5950 ft) and the Palo Pinto (depth 6590 ft), were found that could be used as reference reflectors for the measurement of the S and P wave transit times (their ratio is equal to an average Vp/Vs ratio) across the interval of the Canyon sands (depth 6190 - 6590 ft). The two-way times to these reflection events at well "A" are approximately 0.97 sec and 1.07 sec on the P wave VSP data and 1.91 and 2.08 sec on the SH wave VSP data, respectively (see Figure 3).

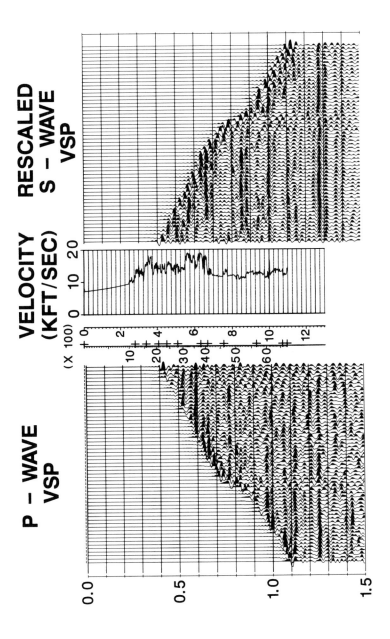

Figure 1. Correlation of P and SH wave reflection events from vertical seismic profile (VSP) data.

B. Surface Seismic Data

The measurements of the S and P wave transit times
across the Canyon sand interval away from the calibration
well were obtained from collateral P and SH wave surface
seismic data acquired over well "A". These data were
acquired with nonlinear vibrator sweeps to offset the
attenuation that occurs, especially in the SH wave
propagation, in the near surface. Using a high frequency
emphasis +7db/octave (8-48 Hz) nonlinear sweep, the
bandwidth of the S wave data was increased to two octaves
(8-32 Hz) at the target depth (5950-6590 ft); an 8-80 Hz
bandwidth was achieved in the P wave data using the same
amount of emphasis in an 8-80 Hz nonlinear sweep. Three
Mertz Model 18 land vibrators were used to acquire the P
and SH wave data sets. Twelve 20-sec sweeps were recorded
for the P wave data; sixteen 24-sec sweeps were recorded
for the SH wave data. The shot point spacing was 220 ft
and the geophone array was 12 phones over 220 ft for both
wave modes; 8 Hz vertical and horizontal geophones were
used to record the P and SH wave data, respectively. A 96-
trace straddle spread was used; the near trace distance was
1100 ft and the far trace distance was 10,340 ft.
     In addition to the standard 2-D processing, a source-
receiver deconvolution and a post-residual statics
processing step were performed on both the P and SH wave
data set. The former process was used to remove source-
receiver coupling variations and to improve the bandwidth
of the final stacked sections, while the latter process was
used to increase the multitrace coherence of the reflection
events for a more precise S to P transit time ratio (ΔTs/
ΔTp) measurement in the target area.
     Figures 2 and 3 show the correlation obtained between
the P and SH wave reflection events from the VSP data
acquired in well "A" and the surface seismic data over the
well (arrows on the figures mark the locations of the P and
SH wave Breckenridge and Palo Pinto reflection events).
Static shifts have been applied to the P and SH wave VSP
data to account for the datum shifts in the surface seismic
data. In the vicinity of well "A", the VSP data and
surface seismic data agree with respect to the amplitude of
the reference reflection events. For example, at the well
the SH wave reflection from the Breckenridge (t=1.785 sec)
is weak while the P wave reflection event (t=0.915 sec) is
strong. The P wave reflection continuity is good from the
Palo Pinto (t=1.010 sec), while the SH wave event is
discontinuous from the well (t=1.970 sec) out to common

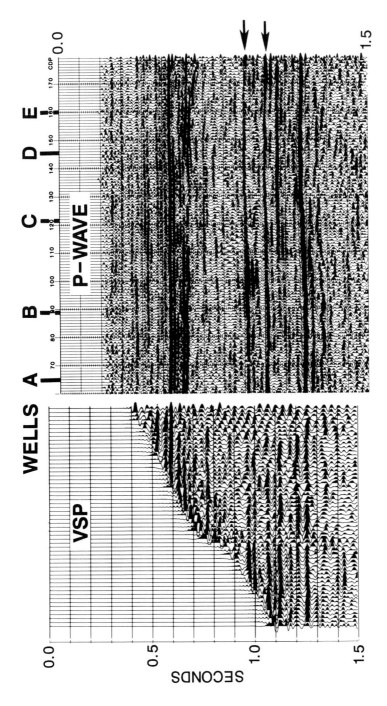

Figure 2. Correlation of P wave reflection events on VSP data and surface seismic data.

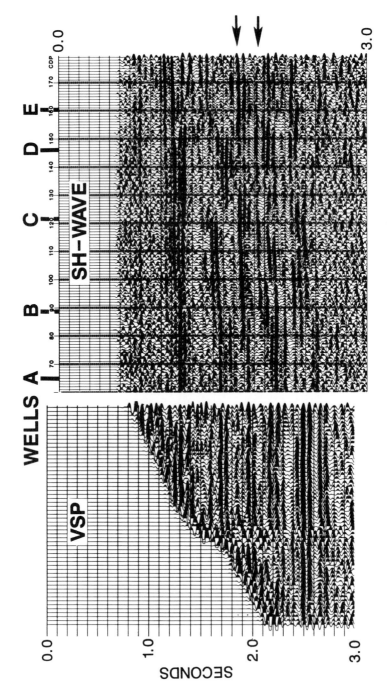

Figure 3. Correlation of SH wave reflection events on VSP data and surface seismic data.

depth point location 80. The variation in the volume
fraction of sand in the Canyon interval away from well "A"
was then obtained from the P and SH wave transit times
measured across these two reference reflection events on
the surface seismic sections.

C. Well Log Data and Maximum Entropy Estimation

  The well log data acquired in the calibration well were
used to obtain both qualitative and quantitative estimates
of how the Vp/Vs ratio is expected to vary in this interval
with respect to changes in the volume fraction of sand.
In-situ measurements of the P and S wave velocities and
their Vp/Vs ratios were obtained every 0.5 ft in this
interval from the P and S wave onset times on a full
waveform sonic log. The lithology in the interval, for
correlation with the Vp/Vs ratios, was estimated from the
gamma ray, the neutron porosity, density and the P wave
sonic logs. A three-rock-matrix solution, consisting of
two minerals and percent volume shale, with percent volume
porosity were derived at every 0.5 ft interval. The two-
mineral solutions allowed from bulk density-neutron
porosity cross plots were either percent volume limestone
and sand for the density range 2.65 to 2.7 gm/cc or percent
volume limestone and dolomite for the density range 2.7 to
2.87 gm/cc. The rock matrix was estimated to be anhydrite
for densities greater than 2.87 gm/cc.
  Figures 4 and 5 depict the correlation found between
measured Vp/Vs ratios and the lithology and porosity
estimates for the interval between the two reference
seismic reflection events. Figure 4 shows the measured
values from the well log data obtained every 0.5 ft, while
Figure 5 shows an interval average value obtained at each
depth within the interval. The latter average is calculated from the well log measured values every 0.5 ft, from
5950 ft to each depth in the interval. It is used to
simulate the averaging process that occurs in the seismic
measurements as the waves propagate across this interval.
The interval average Vp/Vs ratio calculated across this
interval is obtained at 6590 ft and is approximately 1.77.
The well log estimated lithology corresponding to this
Vp/Vs ratio is 62% shale, 20% limestone, 12% sand and 4%
dolomite; the average porosity is 2% over the interval.
  A qualitative indication of how the interval average
Vp/Vs ratio will vary in this multicomponent lithological
interval with respect to percent volume of sand can be

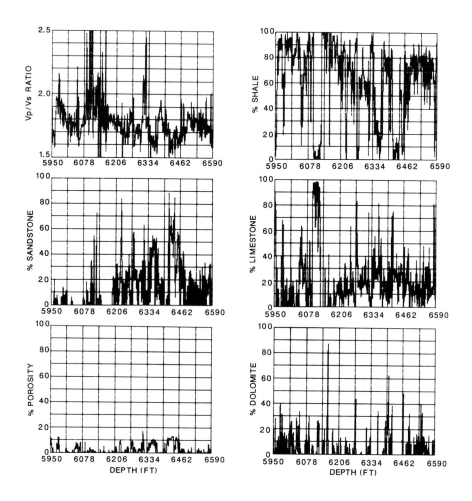

Figure 4. Well log Vp/Vs ratios and percent volume lithology and porosity estimates between Breckenridge and Palo
Pinto horizons.

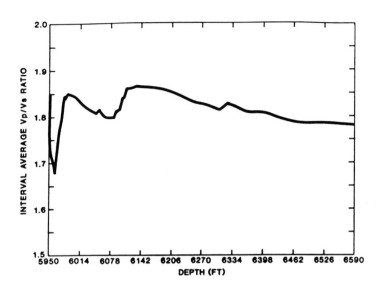

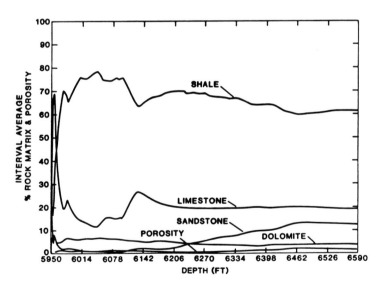

Figure 5.   Depth correlations of interval average Vp/Vs
ratios and interval average percent volume lithology and
porosity.

inferred from Figures 4 and 5. For example, below 6206 ft, the decrease in the interval average Vp/Vs ratio from 1.85 to 1.77 at 6590 ft is caused by a decrease in the percent volume of shale and an increase in the percentage of sand, with the percent volume of limestone constant over this depth interval. This suggests that further increases in the percent volume of shale and/or percent volume of limestone and decrease in the percent volume of sand will cause the Vp/Vs ratio to become greater than 1.77, while decreases in the percent volume of shale and/or limestone and increases in the percentage of sand will cause the Vp/Vs ratio to become less than 1.77 in this interval.

A quantitative estimate of how the Vp/Vs ratio in this interval will vary with the percent volume of sand has also been derived from the well log data using a maximum entropy method. The well log data are entered into the maximum entropy formulation in the form of constraint equations calculated from multiple regressions on the discrete Vp/Vs ratio, lithology and porosity data measured in the interval. A brief description of the method follows.

Consider the volume fractions (f_i) of sand, limestone, shale, dolomite and porosity, which must sum to unity in the interval

$$\sum_{i=1}^{N} f_i = 1 \qquad (1)$$

as probabilities of occurrence of each component. It is assumed that the Vp/Vs ratio trend in the interval can be expressed as a linear function of the volume fractions of these components

$$Vp/Vs = a_0 + \sum_{i=1}^{N} a_i f_i \qquad (2)$$

where $a_0 \ldots a_N$ are a set of regression coefficients. Furthermore, assume that only sand, limestone and dolomite can have porosity and that the porosity trend in the interval can also be expressed as a linear function of the volume fraction of these three components

$$f_\phi = \sum_{i=1}^{N-2} b_i f_i \qquad (3a)$$

or

$$0 = \sum_{i=1}^{N} b_i f_i \qquad (3b)$$

where b_i is a second set of regression coefficients with $b_{sh} = 0$ for the shale volume fraction and $b_\phi = -1$ for the porosity volume fraction (f_ϕ). Since the number of equations is less than the number of unknowns, a unique solution for the volume fractions characterizing a given Vp/Vs ratio of the interval cannot be obtained by standard methods. Using an entropy measure introduced by Shannon (1948) in information theory, a least-bias set of volume fractions that are consistent with the constraint equations (1) - (3) can however be estimated from the quantity

$$H = -\sum_{i=1}^{N} f_i \ln f_i \qquad (4)$$

which is a measure of the entropy (Jaynes, 1957, 1985), i.e. the degree of disorder or mixing in the system. For the system considered in equation (1), it can be shown that the entropy will be a maximum when the volume fractions are all equal, i.e. $f_i = 1/N$. Maximizing the entropy for a given Vp/Vs ratio using the additional constraint equations (2) and (3) provides a least-bias estimate of the volume fractions of five components in the interval consistent with trends calculated from the two regression equations. An alternate entropy expression that contains additional information on prior probabilities (p_i) of occurrence of each component in the interval, which will be used here, is given by the Kullback-Leibler norm of estimation (Frieden, 1985)

$$H_{KL} = -\sum_{i=1}^{N} f_i \ln(f_i/p_i) \qquad (5)$$

This relative entropy expression has its maximum when $f_i = p_i$. In this formulation, the prior probabilities are related to the interval average volume fractions of each component in the interval and are obtained from Figure 5 at the depth 6590 ft.

The expression to maximize is then given by

$$G = - \sum_{i=1}^{N} f_i \ln(f_i/p_i) + \lambda_1 (Vp/Vs - a_0 - \sum_{i=1}^{N} a_i f_i)$$

$$+ \lambda_2 (- \sum_{i=1}^{N} b_i f_i) + (\lambda_3 - 1)(1 - \sum_{i=1}^{N} f_i) \qquad (6)$$

where λ_i represent Lagrangian multipliers. Maximizing equation (6) with respect to the volume fraction yields the following expression:

$$\delta G = - \sum_{i=1}^{N} \delta f_i \ln(f_i/p_i) + \lambda_1 a_i + \lambda_2 b_i + \lambda_3 \cong 0 \qquad (7)$$

which, upon using equation (1), reduces to

$$f_i = \frac{p_i e^{-\lambda_1 a_i - \lambda_2 b_i}}{\sum_{i=1}^{N} p_i e^{-\lambda_1 a_i - \lambda_2 b_i}} \qquad (8)$$

Substituting equation (8) into equations (2) and (3b), a solution to the two Lagrangian multipliers can be obtained from these two equations for a given Vp/Vs ratio using the Newton-Raphson method. The volume fraction corresponding to the Vp/Vs ratio then follows from equation (8) and the values of the Lagrangian multipliers and the prior probabilities.

The results from the maximum entropy method are illustrated in Figure 6 and provide the least-bias estimate of how the Vp/Vs ratio is expected to vary quantitatively in the interval, 5950 ft to 6590 ft, with the volume fractions of the five components. They indicated that changes in the volume fractions of sand, shale and limestone provide the principal influence for changes in the value of the Vp/Vs ratio. As the volume fractions of shale and limestone are replaced with sand, the Vp/Vs ratio decreases from 1.80 to 1.61 in this interval in good agreement with literature studies. The average porosity in the interval is also seen to increase with the increase in volume fraction of sand in quantitative agreement with the well log data (see Figure 4).

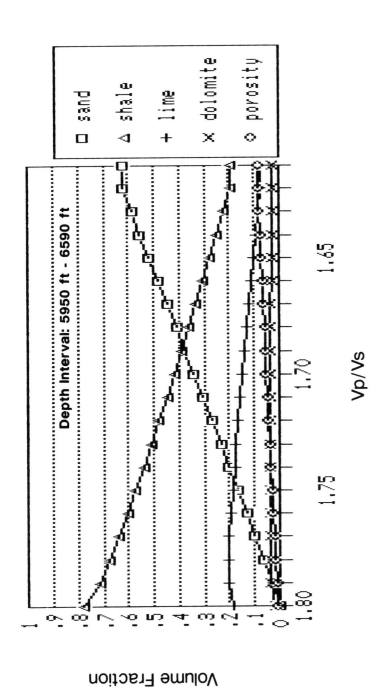

Figure 6. Maximum entropy estimation of the dependence of Vp/Vs ratio on lithology and porosity.

III. Δ Ts/Δ Tp RATIO MEASUREMENTS

Measurements of the P and SH wave transit times and the ΔTs/ΔTp ratio across the two reference reflection events not only reflect changes in the volume fraction of the rock matrix but are also affected by changes in the character of the seismic signal. For example, the presence of residual NMO errors, bandwidth and phase variations in the propagating seismic wavelet, wave interference effects from neighboring reflection events, and signal-to-noise ratio variations on a common depth point (CDP) gather prior to stacking can all affect the transit time measurements made on a stacked section. To minimize the effects of these influences on the data, source-receiver deconvolution and post-residual static correction steps, mentioned previously, have been included in the processing of both the P and SH wave surface seismic data. Moreover, since the seismic signal from a reflection event boundary arises from an area in the subsurface (Fresnel zone) larger than the common depth point spacing (110 ft in these data), it was assumed that abrupt isolated changes in the P and SH wave transit times were not associated with changes in the lithology but rather changes in the character of the seismic signal that act as noise spikes. Consequently, the analysis of the lateral variations in the Vp/Vs ratio has been performed on smoothed versions of the original stacked sections. The type of smoothing operator used was a three-point, trace-weighted average.

Figure 7 shows the lateral variations of the measured P and SH wave seismic transit times (open circles) and their corresponding Vp/Vs (= Δ Ts/Δ Tp) ratios (solid line) measured across the interval containing the Canyon sands. The total distance covered by this measurement is approximately 2 miles. The times are measured from the Breckenridge to the Palo Pinto reflection events. The windowed sections of the smoothed P and SH wave traces in the target area are also included in the figure. Since these reflection events are discontinuous on the SH wave data to the left of CDP 78 (well "A" is located at CDP 65), only data to the right of CDP 78 were used in this analysis. The transit times were calculated from the difference in the peak maximum times, determined from a three-point quadratic curve fit, on the two reference reflectors. It is seen in this figure that the lateral variations measured in the Vp/Vs ratio closely follow the SH wave transit time variations along the line. This suggests that the S wave velocity is more sensitive to

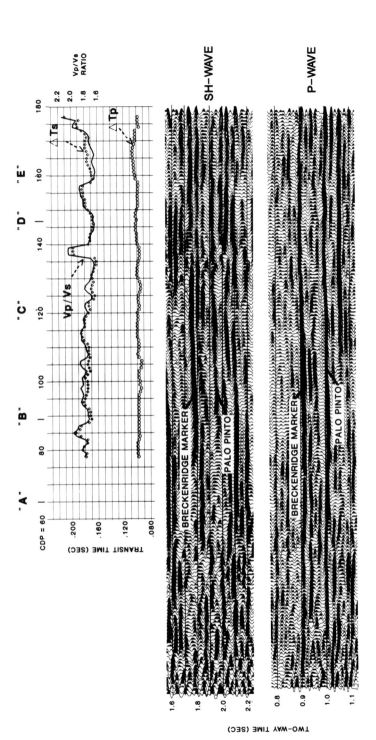

Figure 7. Measurement of SH and P wave transit times and their Vp/Vs ratios ($=\Delta Ts/\Delta Tp$) across the Breckenridge and Palo Pinto reflection events.

changes in lithology than the P wave velocity, in agreement
with binary model studies of Eastwood and Castagna (1983).
Except for the sudden increases in the SH wave transit
times and the V_p/V_s between CDP's 137 and 139, which can be
attributed to changes in the SH wave reflection character
at the Breckenridge, there appears to be a gradual
decreasing trend in the V_p/V_s ratios to the right of CDP
120 from an average value of approximately 1.80 to a value
below 1.70. This drop in the V_p/V_s ratio is observed
between CDP 140 and 150 and again between CDP 158 and 168.
After CDP 168, the V_p/V_s ratio increases again. These two
locations are coincident with two gas-producing wells that
have well log estimated sand-shale ratios of 28% (well "D")
and 18% (well "E"). It is seen that the SH wave transit
time decreases around these two wells consistently with the
faster rate of increase of the S wave velocity in sands
(Eastwood and Castagna, 1983). In addition, the slight
increase in the P wave transit times at these two locations
suggests a lower average interval P wave velocity, which is
indicative of gas found in these two wells.

Table 1 shows for five wells the values for the volume
fraction of sand in the Canyon interval obtained from
direct well log measurements and from the V_p/V_s ratio
measurements and the maximum entropy estimates (Figure 6).
In the case of the two gas wells ("D" and "E"), the
measured V_p/V_s ratios at these two locations were corrected
for the presence of gas. It is mentioned that, since the
calibration well data did not contain a gas volume fraction
component, a lower V_p/V_s ratio value from a formation
containing gas would yield a higher volume fraction of sand
when directly compared to the V_p/V_s ratio-volume fraction
sand estimates derived from the maximum entropy prediction
(see Figure 6). In making this correction, it was assumed
that the difference between the P wave transit times
measured in the vicinity of these two wells and the P wave
transit time measured between the two wells (CDP locations
150 to 158) is associated with the presence of gas.
Consequently, the gas correction to the V_p/V_s ratio was
obtained by using the latter P wave transit time in
calculating the V_p/V_s ratio in the vicinity of these two
wells. These corrected V_p/V_s ratio values are shown in
Table 1, and the corresponding volume fractions of sand
predicted by the maximum entropy method are 34% (well "D")
and 22% (well "E"), in reasonable agreement with direct
well log measurements.

TABLE I. A Comparison of Percentage Sand Calculated from
Direct Well Log Measurements (w.l.) and from Vp/Vs Ratio
Measurements and Maximum Entropy Estimation (m.e.e.)

Well	Vp/Vs (meas.)	Vp/Vs* (corr.)	% Sand (w.l.)	% Sand (m.e.e.)
A	1.77	–	12	10
B	1.75	–	12	16
C	1.76	–	13	13
D	1.66	1.70	28	34
E	1.65	1.73	18	22

* gas correction

IV. CONCLUSION

Using well log data from a nearby well as a calibration
point and a maximum entropy method, a quantitative estimate
of the volume fraction of sand in a complex multicomponent
lithological interval has been obtained from Vp/Vs ratio
measurements made on P and SH wave surface seismic data.
From Vp/Vs ratios derived from a full waveform sonic log
and corresponding lithology (sand, shale, dolomite and
limestone) and porosity models derived from a standard set
of well log data, a set of regression equations were
obtained that characterize a Vp/Vs ratio-litho-porosity
trend for the interval. Incorporating these equations into
an entropy formulation and maximizing the entropy for a
given Vp/Vs ratio provided a least-bias set of predictions
of the volume fractions of these components consistent with
the interval's Vp/Vs ratio-litho-porosity trend. Good
quantitative agreement with direct well log measurements
was obtained with this prediction method for the volume
fraction of sand encountered in wells nearly two miles from
the calibration well. Since this maximum entropy method
uses log data from a single well, it can be used initially
to determine the potential of using P and S wave data
acquisition in field development for detecting variations
in lithology and porosity from surface seismic derived
Vp/Vs ratio measurements. In development areas where
stratigraphic sands are the target, these results suggest
that Vp/Vs ratio measurements from P and SH wave surface
seismic data can be used to map changes in the percent sand
away from a well on the order of 10% or greater.

ACKNOWLEDGMENTS

The author thanks Phillips Petroleum Company for permission to publish this work. John Hensley calculated the P and S wave velocities from the full waveform sonic log data and the lithology and porosity from the well log data. Discussions with J. E. Smith on maximum entropy methods were invaluable in formulating the method for estimating the lithology and porosity dependence of the Vp/Vs ratio.

REFERENCES

Domenico, S. N., 1984, Rock lithology and porosity determination from shear and compressional wave velocity, Geophysics, Vol. 49, No. 8, p. 1188.

Eastwood, R. and Castagna, J., 1983, Basis for the interpretation of Vp/Vs in complex lithologies, SPWLA 24th Annual Logging Transactions, p. 19.

Gregory, A. R., 1976, Fluid saturation effects on dynamic elastic properties of sedimentary rocks, Geophysics, Vol. 41, p. 895.

Hamilton, E. L., 1979, Vp/Vs and Poisson's ratio in marine sediments and rocks, J. Acoustic Soc. Am., Vol. 66, p. 1093

Jaynes, E. T., 1957, Information theory and statistical mechanics, Phys. Rev., Vol. 106, p. 620.

Jaynes, E. T., 1985, Where do we go from here?, Maximum-Entropy and Bayesian Methods in Inverse Problems, eds. Smith, C. R. and Grandy, Jr., W. T. (D. Reidel Publishing Co.), p. 21.

Frieden, B. R., 1985, Estimating occurrence laws with maximum probability, and transition to entropic estimates, Maximum-Entropy and Bayesian Methods in Inverse Problems, eds. Smith, C. R. and Grandy, Jr., W. T. (D. Reidel Publishing Co.), p. 133.

McCormack, M. D., Dunbar, J. A. and Sharp, W. W., 1984, A case study of stratigraphic interpretation using shear and compression seismic data, Geophysics, Vol. 49, No. 5, p. 509.

Pickett, G. R., 1963, Acoustic character logs and their applications in formulation evaluation, J. Petr. Tech, Vol. 15, p. 659.

Shannon, C. E., 1948, A mathematical theory of communication, Bell Sys. Tech. Jour., Vol. 27, p. 379-423, p. 623-656.

SCALING LAWS IN RESERVOIR SIMULATION AND THEIR USE IN A HYBRID FINITE DIFFERENCE/STREAMTUBE APPROACH TO SIMULATING THE EFFECTS OF PERMEABILITY HETEROGENEITY

T. A. Hewett
R. A. Behrens

Chevron Oil Field Research Company
La Habra, California

I. ABSTRACT

The scaling behavior of solutions to the transport equations for flow in permeable media is reviewed. It is shown that when written in terms of the appropriate variables, one-dimensional solutions for both continuous and slug injections can be scaled for use with quasi-one-dimensional streamtubes of differing geometries and pore volumes. In flows described by a single fractional flow relation, the effects of numerical dispersion only act to thicken sharp fronts, but in flows involving changes in fluid viscosity, numerical dispersion can introduce anomalous waves that alter the nature of the solution. Simulations of displacements in heterogeneous cross sections can be reduced to equivalent single-layer solutions for use in streamtube displacement calculations or for deriving pseudo-relative permeabilities. The limitations and advantages of these two approaches are discussed.

II. INTRODUCTION

The influence of reservoir heterogeneity on the performance of oil recovery processes has become the subject of increased attention in the petroleum literature in recent years.[1-5] Although no single approach to characterizing heterogeneity and including its effects in performance predictions has emerged pre-eminent, important conclusions can be drawn from the work to date.

RESERVOIR CHARACTERIZATION II
402

One is that the use of appropriately averaged values of permeability may be satisfactory for modeling the "flux problem", but it is unsatisfactory for modeling the "transport problem". By the "flux problem" we mean the prediction of the fluid flux in response to an imposed pressure gradient. This entails the calculation of an "effective permeability" as some kind of average of measured permeability values, usually from cores. Several different approaches to deriving effective permeabilities have been proposed. These include the use of harmonic, geometric, or power law averages, depending on the flow configuration and direction relative to anisotropies in the permeability distribution,[6-8] or weighted averages of bimodal distributions of sand and shale permeabilities.[9-10]

By the "transport problem" we mean the prediction of the motion of displacing fluid fronts and the breakthrough and subsequent production of fluids injected into wells or encroaching from aquifers. This requires the calculation of "effective relative permeabilities" and "effective dispersion coefficients" to model the effect of permeability heterogeneity on frontal displacements in coarsely gridded areal models, or the development of alternative scale-up procedures. The calculation of effective flow properties requires a flow simulation on a grid of permeability values that maintains the extremes of the distribution as well as the mean, since the extremes represent the flow conduits and barriers that control the details of fluid movement.[11] In addition, it has been shown that the nature of spatial correlations in the permeability distribution influences the definition of these effective properties and may preclude their unique definition when the range of correlations is large compared to the length of the flow path.[2,12] Proper evaluation of effective flow properties for modeling the transport problem requires simulation of flow processes on permeability fields that not only reproduce the values sampled at wells, but also reproduce the character of permeability variations throughout the interwell region.

Since data is seldom, if ever, available at sufficient resolution to completely describe interwell property variations, the use of probabilistic methods has gained favor for constructing permeability fields with the desired characteristics at the required resolution. These methods rely on statistical measures of spatial correlation observed in measured data and introduce synthetic variations in the permeability values between wells in accordance with the statistical model chosen. While these synthetic fields are not unique and cannot be expected to be accurate on a point-by-point basis, they do honor the available data and mimic the nature of property variations at unsampled locations. In so doing, they preserve the extremes of the permeability distribution and the spatial correlations that control the transport problem.

A variety of approaches for constructing permeability distributions that honor measurements at wells and reproduce different statistical measures of spatial continuity and variability have been introduced. These range in complexity from simple Monte Carlo methods, which only reproduce the univariate statistics of the desired distribution with no spatial correlations,[6] to Sequential Indicator Simulation, which can accommodate different structures of spatial correlation for each of an arbitrary number of ranked classes of data values.[11] In between are stochastic facies models, which reproduce the size distribution of facies appropriate to the depositional environment with no regard for correlations between associated facies,[13] and conditional simulations derived by geostatistical methods, which reproduce a single structure of spatial correlations in a distribution.[4,14]

These detailed simulations of geologic property variations can be used for a variety of purposes. One of the most straightforward is the determination of the statistical continuity of permeable units or impermeable flow barriers.[13] A second use of these geological simulations is for numerical simulations of reservoir flow processes to determine the range of recoveries expected, given the amount of uncertainty remaining in the model.[4] Perhaps the most important use of detailed simulations of geologic property distributions is the derivation of methods for scaling up the flow performance observed in detailed simulation models to field-scale performance predictions. Previous reference has been made to the results of this procedure for the flux problem. For the transport problem in multiphase flows, this requires the calculation of pseudorelative permeabilities for use in numerical simulation models, or the development of scalable one-dimensional solutions to the transport equations, analogous to the Buckley-Leverett solution, for use in streamtube calculations.

Previous studies of the use of pseudofunctions have shown that they can effectively reduce the results of waterflood simulations in detailed vertical cross-section models to equivalent coarsely gridded one-dimensional models with pseudo-relative permeabilities. These same studies showed that the resulting pseudofunctions can have a flow-rate dependence.[15] When used in areal models to make predictions for a three-dimensional flow problem, flow rates can vary substantially between near wellbore regions and regions away from wells. In addition, when used in the coarsely-gridded models required by constraints on computer resources, additional errors due to grid orientation effects can affect the predicted performance.

In this paper, we review the use of a hybrid finite difference/streamtube approach for making field-scale predictions of oil recovery processes.[16,17] This method was introduced to overcome the difficulties of modeling the performance of a field-scale surfactant/polymer project with sufficient grid resolution to reduce the level of numerical dispersion to acceptable levels. The method is readily generalizable to many kinds of displacement processes. The idea behind the method is to run a simulation of a process in a vertical cross section and develop the vertically averaged performance of the displacement. With a proper understanding of the scaling behavior of this solution, it can be mapped onto quasi-one-dimensional streamtubes of differing length in the streamtube model. The production from these streamtubes is then summed to predict the performance of the total project.

The key to practical implementation of this procedure is an understanding of the scaling relations for the one-dimensional solution. This need arises because each streamtube in the model may have a different pore volume, and each streamtube receives a different fractional pore volume of any fluids we inject. In the following sections we develop the appropriate scaling laws for one-dimensional displacement processes with an eye to their use in the hybrid method. The use of streamtube models with injection rate updating to reflect the influence of non-unit mobility ratios is reviewed. We then discuss the derivation of pseudo-relative permeabilities and compare the two approaches to scale-up of the transport problem.

III. SCALING LAWS IN MULTIPHASE DISPLACEMENT PROCESSES

The one-dimensional transport equation for multicomponent, multiphase flow in a porous medium is:

$$\frac{\partial C_i}{\partial t} + \frac{q(t)}{\phi(x)A(x)} \frac{\partial F_i}{\partial x} = 0$$

where

$$C_i = \sum_{j=1}^{N_p} c_{ij} S_j$$

$$F_i = \sum_{j=1}^{N_p} c_{ij} f_j \tag{1}$$

$$\sum_{j=1}^{N_p} S_j = 1$$

In this equation, $q(t)$ is the instantaneous volumetric flow rate, and $\phi(x)$ and $A(x)$ are the average porosity and cross-sectional area, which may be functions of streamwise distance. F_i is the fractional flow of component i and depends only on the local composition of the fluids, C_i. S_j is the saturation of phase j, f_j is the fractional flux of phase j, c_{ij} is the concentration of component i in phase j, and N_p is the total number of phases.

This equation includes the simplifying assumptions that the fluids and rock are incompressible and isothermal, the fluids flowing through the medium are in local thermodynamic equilibrium and do not undergo chemical reactions, and the displacement is dissipation-free.[18] By dissipation-free, we mean that transport caused by flowing phase capillary pressure gradients or by dispersion within a phase is negligible compared to advective transport.

The neglect of the dissipative terms in the transport equations is justified by the observation that transport by the terms remaining in Equation (1) scales linearly with the streamwise distance from the injection point, while transport by the dissipative terms scales with the square root of the streamwise distance. This means that for field-scale displacements the effects of the dissipative terms will be small compared to the advective transport terms and will only act to modify the shape of displacement fronts. A quantitative measure of the relative importance of advective transport to dispersion is the macroscopic Peclet number defined as:

$$Pe_x = \frac{Vx}{D} = \frac{x}{\alpha}$$

where $D = \alpha V$ is the macroscopic dispersion coefficient, α is the dispersivity, a microscopic mixing length, and x is the streamwise distance from the injection point. Laboratory measurements of the value of α from outcrop sandstones show values typically less than 0.5 cm,[19] while measurements of α in carbonates are more variable and typically in the 0.1-2 cm range.[20] For typical interwell distances (> 200 m), the Peclet number is greater than 10,000. In simulations with finite difference models, the effects of truncation error introduce a numerical dispersion that, for small timesteps, scales as $Pe \approx 2N$, where N is the number of of gridblocks along the flow direction.[21] This means that in most flow simulations, the effects of numerical dispersion will be larger than properly scaled physical dispersion. Since we are interested in deriving the scalable solutions to Equation (1) from finite difference simulations, it will be important to understand the effects of numerical dispersion on our results.

Similar arguments show that the effects of ignoring capillary pressure gradients will be negligible except in the vicinity of discontinuities in fluid saturations.[22] For this case, however, the spreading effects of capillary forces are balanced by a tendency for saturation waves to sharpen, and a wave of permanent form develops. This is demonstrated in the simulation results presented in a later section.

Although Equation (1) is restricted by the assumptions listed above, it should be understood that it is quite general with regard to the types of displacement processes that can be modeled. By allowing components to partition between the phases (and including a stationary rock phase) according to their equilibrium phase relationships, such complex processes as ion exchange, surfactant adsorption, interphase mass transfer, and multiple-contact miscibility can be modeled along with first-contact miscible and immiscible displacements. An understanding of the scaling behavior of solutions to Equation (1) provides a means of using streamtube methods for the scale-up of most non-thermal recovery processes.

Since our primary interest is in applying solutions to Equation (1) to streamtubes that may have variable width with position along the streamtube and for displacements that may have variable flow rates, it is desirable to introduce a change of variables that makes the same solution applicable to any streamtube. Defining

$$Q(t) = \int_0^t q(t')dt'$$

(2)

$$V_p(x) = \int_0^x \phi(x')A(x')dx'$$

and changing variables, Equation (1) becomes:

$$\frac{\partial C_i}{\partial Q} + \frac{\partial F_i}{\partial V_p} = 0$$

(3)

or

$$\frac{\partial C_i}{\partial Q} + \frac{dF_i}{dC_i}\frac{\partial C_i}{\partial V_p} = 0$$

(4)

This change of variables shows that the relevant variables for describing a displacement are $Q(t)$, the cumulative fluid volume injected, and $V_p(x)$, the "local pore volume", or the cumulative pore volume between the injection point and the local streamwise position.

The form of Equation (4) is that which describes a class of flow phenomena called "kinematic waves".[23] The name comes from the fact that the wave properties of these flows arise from the equation of continuity alone, with no reference to the pressure equation. In general, solution of Equation (4) relies on the Method of Characteristics. A characteristic is defined as a line in the Q-V_p plane along which the solution may be described by an ordinary, or total, differential equation. The total derivative of C_i is:

$$\frac{dC_i}{dQ} = \frac{\partial C_i}{\partial Q}\bigg|_{V_p} + \frac{\partial C_i}{\partial V_p}\bigg|_Q \frac{\partial V_p}{\partial Q}\bigg|_{C_i}$$

(5)

Comparison of Equation (5) with (4) shows that the characteristic lines for Equation (4) must satisfy the condition that:

$$\frac{\partial V_p}{\partial Q}\bigg|_{C_i} = \frac{dF_i}{dC_i} = v_i^*$$

(6)

where v_i^* is a dimensionless velocity normalized by the local fluid velocity. Along these lines the total derivative of C_i is

$$\frac{dC_i}{dQ} = 0$$

This pair of ordinary differential equations is equivalent to Equation (4) and is much easier to solve. An alternative way of stating this result is that constant values of C_i propagate through the flow along wavelets having different velocities corresponding to the slope of the fractional flux curve. Whenever the fractional flux relation is nonlinear, waves carrying different values of C_i move at different velocities producing a waveform which changes shape as it propagates. This can result in the formation of composition discontinuities, or "shocks", which travel at a dimensionless velocity[23]

$$v_s^* = \frac{\Delta F_i}{\Delta C_i} \tag{7}$$

where Δ denotes the jump in value across the discontinuity.

When the initial conditions are restricted to uniform distributions of composition, and changes in boundary conditions are restricted to continuous injection following a step change in injected fluid composition, further simplifications in the solution procedure are possible. Under these circumstances the condition of wave coherence is observed.[18,24] This simply means that all of the nonlinear wave interactions occur instantaneously at the time of the step change in boundary conditions and all of the waves introduced by the change in boundary conditions are sorted into the order of their propagation velocities, with faster waves traveling ahead of slower waves. This means that no further wave interactions can occur. All components that have the same velocity travel together as "coherent", or unchanging, compositions.

When conditions for wave coherence apply, simplified solutions based on graphical constructions using the fractional flow curves can be used.[18,25-27] These solution techniques are referred to as "fractional flow theories". Fractional flow theories provide considerable insight into the behavior of kinematic waves and are suitable for most problems involving continuous injection of a displacing fluid. Readers unfamiliar with the behavior of nonlinear waves should consult these references.

When the inlet boundary condition is changed gradually or more than one step change in flow conditions is introduced, as occurs in the injection of a finite slug, faster waves can be introduced upstream of slower moving waves. This results in a region of nonlinear wave interactions, or noncoherence, as the faster waves overtake the slower waves. This lasts until the waves sort themselves into the order of their propagation velocities and the coherence condition again applies.

   Solutions to Equation (4) in noncoherent regions generally
require a numerical integration along characteristics, although
some analytical results can be obtained for the case of a finite
slug injected as a step change of displacing fluid followed by a
step change to chase fluid. Since we will be deriving the scalable
solutions by finite difference simulation, we need not concern
ourselves with the details of the solution procedure for these
cases, although the analytical results for a finite slug are shown
for comparison with the finite difference results in a later section.
   We will restrict our attention to either continuous injections or
finite slug injections into uniform initial conditions. The scaling
relations for these two types of flows are different. The results
of fractional flow theory for the case of first-contact miscible
displacements have recently been published.[18] Since, in the
absence of dispersion, these results are indistinguishable from
those for multiple-contact miscible displacements and exhibit all
of the features of a flow that affect their scaling, the results for
first-contact miscible and immiscible displacements will be used
to elucidate the scaling relations for any process described by
Equation (4). It should be mentioned in passing that the break-
down of the assumption of dissipation-free flow has more drastic
consequences for displacements with generated miscibility than
for the first-contact miscible displacements considered here.
   We will begin by considering the scaling behavior of displacements by the continuous injection of a displacing fluid into uniform initial conditions that correspond to either "secondary recovery conditions" (initial saturations corresponding to connate water
and mobile oil), or "tertiary recovery conditions" (initial saturations corresponding to residual oil saturation and mobile water).

A. Continuous Injections

   Mathematically speaking, Equation (4) is a first-order, nonlinear, hyperbolic partial differential equation. This means that its
solutions have one system of characteristics, all of which propagate in the downstream direction. There is no upstream influence
from any downstream condition. The solution then develops
independently of the overall length, or volume, of the system
considered. Since the solution is independent of the overall
length of the system, there is no characteristic length or volume
appropriate for scaling.
   In the absence of a characteristic length or volume scale, the
solution can be reduced to a function of a single nondimensional
variable, the ratio of the two primary variables, viz., $Q(t)/V_p(x)$.
This variable is just the "local pore volumes injected" since it
measures the amount of fluid injected relative to the cumulative
pore volume up to the local position considered. For some purposes, such as viewing the scaled saturation profiles, it is more
convenient to consider the solution as a function of $V_p(x)/Q(t)$,
which corresponds to the wave velocities of the individual compositions.

This reduction of variables can be seen most easily by considering the solution to the familiar Buckley-Leverett problem where water is continuously injected into uniform secondary recovery initial conditions. For this example and those that follow, the relative permeabilities used are shown later in Figure 28. The oil/water viscosity ratio is 2. The solvent/water viscosity ratio (to be referred to later) is 0.1. The resulting fractional flow curves are shown in Figure 1. A characteristics diagram for this problem is shown in Figure 2, where the units for the axes are arbitrary, and the saturation contour intervals behind the shock are $\Delta S_w = 0.01$. A characteristics diagram is simply a contour plot of composition as a function of $Q(t)$ and $V_p(x)$, with contours of constant composition corresponding to the characteristic lines of the solution. Since this problem satisfies the conditions for coherence, all of the characteristics are straight lines emanating from the origin.

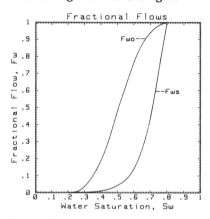

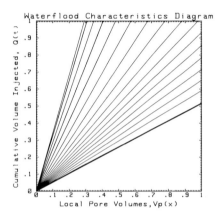

Fig. 1. Fractional flow relations for oil and solvent.

Fig. 2. Waterflood characteristics, analytical solution.

The nonlinear wave interactions introduced by the step change at the inlet result in a shock at the leading edge of the displacement followed by a spreading wave fan. The height and velocity of the shock are determined by the Welge construction. Since each characteristic is a straight line from the origin, its entire trajectory can be described by its slope, i.e., the value of $Q(t)/V_p(x)$, which is constant along it. The characteristics diagram is unchanged if both axes are scaled with the same constant; only the units change. Although it is common practice to normalize the axes of characteristics diagrams with the pore volume corresponding to the total system length, $V_p(L)$, this places an unwarranted emphasis on the influence of the system length on the solution.

The solution to this same problem was also obtained by finite difference simulation (fully implicit, upwind differences with a 1/4 power blending law for solvent and oil viscosities) with 200 gridblocks and a constant timestep equal to the injection of 1/5 gridblock pore volumes per timestep. This produces a numerical dispersion corresponding to a Peclet number of a little more than 300.[21] A contour plot of the resulting solution for water saturation, with a contour interval of $\Delta S_w = 0.05$, is shown in Figure 3, where the axes have been normalized by the pore volume of a single gridblock, the characteristic volume introduced by the discretization of Equation (4). The use of this normalization allows the results of simulations with different numbers of gridblocks to be presented on the same diagram. Any other normalization produces slightly different looking diagrams, since the effects of numerical dispersion scale with the number of upstream gridblocks.

It can be seen that the finite difference solution reproduces the behavior of the characteristics solutions, with the effects of discretization error only acting to thicken the shock front. The staircase appearance of the front is more a consequence of sampling the solution at only a limited number of time steps than any shortcoming of the finite difference solution. Also shown in Figure 3 is a surface plot of the same solution. Presentation of the solution in this form allows for an easy interpretation of the characteristics diagram as a series of saturation profiles observed at successive times.

The behavior of these solutions as a function of the reduced variable $V_p(x)/Q(t)$ is shown in Figure 4, along with the method of characteristics solution. In Figure 4a, the profiles at 51 different timesteps are shown. At early timesteps the front is quite diffuse and numerical dispersion produces an asymmetrical spreading around the position of the shock. At later times, the self-sharpening tendency of the waves sharpens the profiles into a constant shape where the spreading effects of numerical dispersion are balanced by the sharpening tendency of the nonlinear waves. This can be seen in Figure 4b, where only the scaled profiles from injection volumes greater than 50 gridblock pore volumes are shown.

Similar results are obtained when oil, or solvent with a viscosity equal to the oil viscosity, is injected into tertiary recovery conditions. When the viscosity of the solvent equals that of the oil, the flow may still be described by a single fractional flow curve. The scaled late profiles of water, oil and solvent saturation are shown in Figure 5 along with the method of characteristics solution. The shape and velocity of the oil bank fronts are the same as those expected at the leading edge of the oil bank in the miscible displacements from tertiary recovery conditions described later. If we look at the contact line between the oil and solvent, we can see the effects of numerical dispersion on the propagation of miscible fronts, or contact lines. This is also shown in the characteristics diagram of solvent saturation in Figure 6, where the axes are again normalized with the pore volume of a single gridblock. The contours of constant composition are parabolas centered on the contact line moving with the velocity shown in Figure 5c. The parabolic nature of the

contours results from the square-root dependence of dispersion
with travel distance or time.

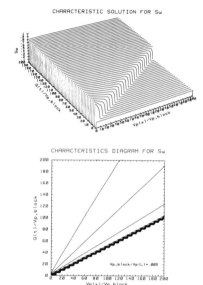

Fig. 3. Waterflood water saturations
(time vs. distance).

Fig. 4. Waterflood scaled profiles,
analytical and numerical, (a) all profiles, (b) later profiles.

From the results of these simulations and their comparison
with analytical solutions, we can conclude that the primary effect
of numerical dispersion, in cases where the path of the displacement follows a single fractional flow curve, is simply a thickening
of immiscible shock fronts and a spreading of miscible contact
lines. The velocity of propagation and the magnitude of saturation changes across these waves is not affected appreciably by
numerical dispersion. These conclusions are altered when the
injected fluids are described by fractional flow curves different
from the oil-water fractional flows.

This can be seen by considering the continuous injection of
the less viscous solvent into tertiary recovery initial conditions.
This produces a displacement that combines the features observed above and introduces new effects resulting from the
change in viscosity of the solvent relative to the oil. The injection
of solvent raises the oleic-phase saturation and sets off a mobile
oil shock and fan, followed by an indifferent wave at the contact
between the injected solvent and mobilized oil. At this point the
path of the displacement switches from the oil-water fractional
flow curve to the solvent-water fractional flow curve. This
produces a zone of constant water and solvent saturations followed by a spreading wave that reduces the water saturation to
its irreducible value. The graphical construction technique of

fractional flow theory produces the characteristics diagram shown in Figure 7.[18] This can be compared to the saturation contour plots from a simulation with 200 gridblocks shown in Figures 8-10, which show the water saturation, oil saturation, and solvent saturation contours. The oil saturation contours show the mobilized oil bank traveling with a sharp immiscible front, and a diffuse rear due to mixing with the solvent by numerical dispersion.

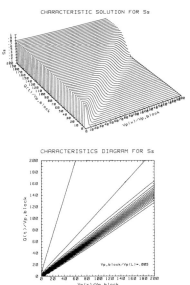

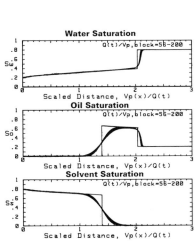

Fig. 5. Equal viscosity continuous solvent flood scaled late profiles, analytical, numerical.

Fig. 6. Equal viscosity continuous solvent flood solvent saturation (time vs. distance).

In addition to the expected smearing of the miscible contact line, we see an even more significant effect of numerical dispersion in the water saturation profiles, where an immiscible wave not predicted by the characteristics solution is observed moving at a slow velocity ahead of the anticipated spreading wave which reduces the water saturation to immobile conditions. In the method of characteristics, it is assumed that the switch from the water-oil fractional flow curve to the water-solvent curve is instantaneous. In the finite difference solution, there is a finite mixing zone in which the flow traverses a whole family of fractional flow curves corresponding to viscosities between the solvent and oil viscosities.

Near the origin, the mixing zone between the oil and solvent overlaps the immiscible waves that produce the oil bank and the trailing fan of waves that reduce the water saturation to its immobile value. These intermediate fractional flow curves introduce waves with higher velocities and saturations than the waves from the method of characteristics solution. These waves move back onto the oil-water and solvent-water fractional flow curves as they leave the mixing zone. The differences in the resulting scaled late-saturation profiles can be seen in Figure 11, where an anomalous water saturation wave is seen in front of the trailing fan (at $Q/V_p ≈ 0.3$) and the speed and saturation of the oil bank are higher than predicted by the method of characteristics. Even though the mixing zone only overlaps these immiscible waves near the origin, the effects introduced by numerical dispersion near the origin are felt far downstream. The introduction of anomalous waves by the action of dispersion shown here is similar to the "dispersion-induced ion exchange" previously observed in simulations of ion exchange in chemical flooding[28] and the strong dependence of recovery on the level of numerical dispersion in compositional simulations.[29]

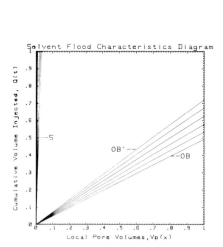

Fig. 7. Continuous solvent flood saturation characteristics, analytical.

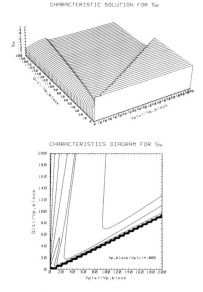

Fig. 8. Continuous solvent flood water saturations (time vs. distance).

Further grid refinement can reduce these effects, but does not entirely eliminate them. This can be seen in Figure 12 where the scaled late profiles from a simulation with 1000 gridblocks are shown. As finer grids are used the impact of numerical dispersion on the propagation of the oil bank is nearly eliminated, but the anomalous wave introduced ahead of the trailing fan is still present. For our purposes, the propagation of the oil bank and solvent front are the most important features of the displacement, and the slow moving anomalous water saturation wave will not have much impact on the oil recovery.

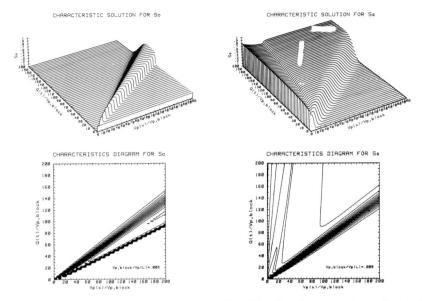

Fig. 9. Continuous solvent flood solvent saturations (time vs. distance).

Fig. 10. Continuous solvent flood oil saturations (time vs. distance).

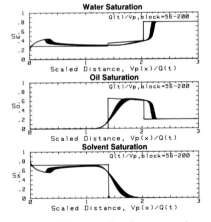

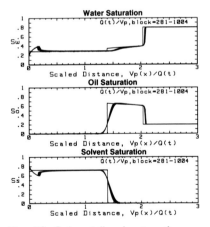

Fig. 11. Solvent flood saturation scaled late profiles, analytical and numerical.

Fig. 12. Solvent flood saturation scaled late profiles, analytical and numerical (1000 cell model).

However, the influence of numerical dispersion on the propagation velocity of oil and solvent displacement fronts raises serious questions about the use of coarsely gridded models for simulating processes that entail a change in fractional flow curves. As an example, the scaled saturation profiles after the injection of 8 gridblock pore volumes of solvent are shown in Figure 13a, along with the scaled analytical solution. Since the shock at the front of the oil bank moves with a velocity about twice that of the piston displacement velocity, the profile shown corresponds to a shock position about 16 gridblocks from the injector. Comparison of the two solutions shows that the finite difference solution is a poor approximation to the analytical solution, both in terms of the propagation velocities and saturation changes. The finite difference solution shows the tertiary oil bank moving with a velocity nearly 50% higher than that predicted analytically. Clearly, simulations with fewer than 20 cells between injector-producer pairs will bear little resemblance to the analytical results. The comparison between the simulated and analytical profiles improves with increasing numbers of gridblocks from the injector, but the rate of improvement is slow as the profiles shown in Figure 13b, with the oil shock 40 gridblocks downstream, indicate.

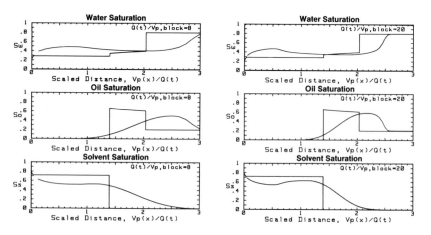

Fig. 13. Solvent flood saturation scaled profiles, analytical and numerical, (a) Q(t)/Vp, block = 8, (b) Q(t)/Vp, block = 20.

B. Single-Slug Injections

When the displacing fluid is injected as a finite slug a characteristic volume, the slug volume, is introduced. The dimensionality of the problem can no longer be reduced and the solution is now a function of two dimensionless variables, $Q(t)/Q_s$ and $V_p(x)/Q_s$, where Q_s is the slug volume. This is referred to as "volumetric linear scaling."[30] With this non-dimensionalization, the same characteristics diagram will apply to displacements with any slug size. The use of any other volume to normalize the variables (e.g., $V_p(L)$) would result in a different characteristics diagram for each slug size. The same scaling applies in multiple slug processes, provided the ratio of successive slug sizes is held constant.

The method of characteristics solution for the injection of a finite slug of solvent with a viscosity equal to that of the oil is shown in Figure 14 at two different scales. Because the oil and solvent viscosity are equal, there is no change in the fractional flow curves for oil and solvent. The injected solvent initiates a mobile oil shock identical to the one described above, which persists until it is overtaken by a second shock introduced by the injection of chase water. The trajectory of the second shock is slightly curved due to nonlinear interactions with the fan of spreading waves behind the first shock, and there is a fan of spreading waves behind the second shock. The solution for the trajectory of the trailing shock requires a numerical integration up to the point where the shocks intersect. The trajectory of the merged shocks can be calculated analytically by noting that their trajectory corresponds to intersections of characteristics originating from the uniform region ahead of the slug with characteristics in the fan behind the second shock.[23] As shown in the Appendix, conservation of mass requires the trajectory to be described by the parametric relations,

$$\frac{V_p(x)}{Q_s} = \left[(S_{o2} - S_{orw}) - \frac{f_{o2}}{V_{o2}^*} \right]^{-1}$$

$$\frac{Q(t)}{Q_s} = \frac{1}{V_{o2}^*} \left(\frac{V_p}{Q_s} \right) + 1$$

(10)

where f_{o2}, S_{o2}, and V_{o2}^* are values associated with individual characteristics in the fan behind the second shock that intersect the merged shock. The finite difference solution for the same problem is shown in Figure 15, showing the total oleic phase (oil plus solvent) saturations.

When the solvent viscosity is less than the oil viscosity, the flow switches fractional flow curves and simple analytical results are not available. As shown in Figures 16 and 17, the trailing solvent-water shock interacts with the waves behind the oil bank and overtakes the leading shock less quickly, owing to the lower viscosity of the solvent. The effects of numerical dispersion can be seen here as the trailing shock overtakes the mixing zone and a mixture of oil and solvent is immobilized by the chase water. As the trailing shock moves from the solvent-water fractional flow curve to the oil-water fractional flow curve, new waves are introduced behind the trailing shock. Although the wave interactions for a finite slug process are complex and not easily described analytically, the solution can still be reduced to a function of two variables for any slug size.

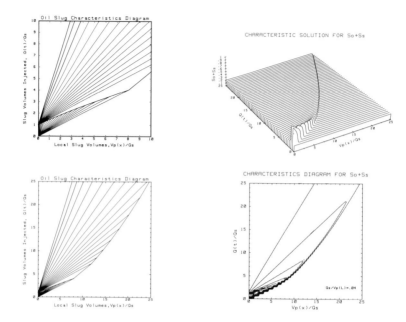

Fig. 14. Equal viscosity solvent slug saturation characteristics, analytical.

Fig. 15. Equal viscosity solvent slug (4% TPVI) oleic saturations (time vs. distance).

C. Alternating Slug Injections

It is common practice to inject alternating slugs of solvent and water in a WAG (water alternating gas) process. The injection of multiple slugs can still be scaled to a single characteristics diagram, provided the ratio of successive slugs is the same for all initial slug sizes, i.e. the WAG ratio and volume injected per cycle are held constant. The finite difference solution for the water saturations for a 1:1 WAG is shown in Figure 18. The size of the solvent slug injected represented 10% of the total pore volume (TPV) and each WAG cycle was 20% of the total pore volume. The nature of the nonlinear wave interactions can clearly be seen. The first WAG cycle behaves like the isolated slug, with the water displacement front starting to catch the initial shock. The shock from the second solvent slug catches up with the water shock and slows it down. The second water shock quickly catches up to this merged shock and remobilizes it. This alternating interaction of the solvent and water shocks continues until the waves have sorted themselves into their coherent order. This is shown better in Figure 19, showing the same process with smaller WAG cycles. The contours shown in Figure 18 are contained within the lower left corner of Figure 19. It can be seen that the region ⌄⸍ nonlinear wave interactions in this process persists for almost 25 slug volumes downstream. The length of this region of wave interactions is highly dependent on the fractional flow curves for the fluids, and no general conclusions about the length of the noncoherent region can be drawn.

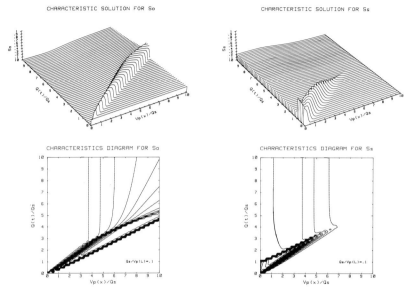

Fig. 16. Solvent slug (10% TPVI) oil saturations (time vs. distance).

Fig. 17. Solvent slug (10% TPVI) solvent saturations (time vs. distance).

The results of a finite difference simulation for simultaneous injection of solvent and water in the same proportions is shown in Figure 20, where the coordinates are normalized by the gridblock pore volume since there is no characteristic slug volume. The slopes of the mobile oil shock and the solvent-water shock are the same as those in the region of coherence in Figure 19. These results validate the use of simultaneous injections to model WAG processes[18] and suggest a way of determining the limitations on slug size for which this approximation is valid. Since the mixture of water and solvent is injected continuously, the characteristics solution can be obtained from fractional flow theory.[18] Comparison of the simulated late scaled profiles for simultaneous injection with those obtained from the method of characteristics in Figure 21 show that the effects of numerical dispersion near the origin, where the mixing zone overlaps the immiscible waves, alter the solution far downstream by introducing higher velocity waves than predicted by the analytical solution. Refinement of the grid to 1000 cells results in smaller discrepancies between the wave velocities in the simulated solution and the analytical solution.

With the ability to represent WAG processes as simultaneous injections, finite WAG processes can be scaled like a single-slug process, with Q_s now corresponding to the total WAG fluid injection. For a fixed WAG ratio, the results for any size total WAG injection can be reduced to a single characteristic solution. An example for a finite 1:1 WAG is shown in Figures 22 and 23, where the oil and solvent saturations are plotted.

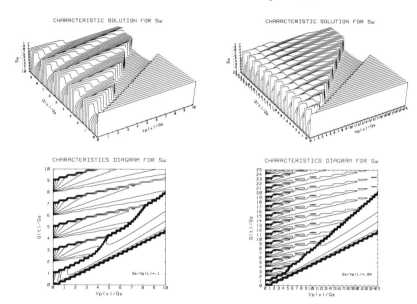

Fig. 18. 1:1 WAG slug (10% + 10% TPVI) water saturations (time vs. distance).

Fig. 19. 1:1 WAG slug (4% + 4% TPVI) water saturations (time vs. distance).

D.   Review of Scaling Laws in 1-D Displacements

This review of the scaling behavior of the results of simulated one-dimensional displacement processes, and comparison with analytical solutions where they are available, has illustrated several important results. (1) For continuous injection of a displacing fluid, the solution can be reduced to a function of a single dimensionless variable that can be stored as a one-dimensional table for mapping onto streamtubes of any length and geometry. (2) For slug injections, the solution can be reduced to a function of two dimensionless variables that can be stored as a two-dimensional table for mapping onto streamtubes. These solutions are valid for arbitrary slug sizes and can be mapped onto streamtubes of any length and geometry. (3) WAG processes can be modeled as a simultaneous injection when the WAG slug size is small enough for the region of noncoherent wave interactions to sort the waves into coherent order. This slug size depends on the fractional flow curves of the fluids involved in the displacement. For the example shown, the region of noncoherent interactions persisted for 25 slug volumes downstream. (4) When the flow is described by a single fractional flow curve, the effects of numerical dispersion only act to spread immiscible shocks and miscible contact lines. The propagation velocities and magnitudes of changes are relatively unaffected by dispersion for this case. (5) When the flow is described by more than one fractional flow curve, the effects of numerical dispersion can also introduce anomalous waves in the region of overlap of mixing zones and immiscible waves, and these waves can persist far downstream and alter both the propagation velocities and magnitudes of displacement fronts. Elimination of the effects of numerical dispersion for this case requires very finely resolved grids.

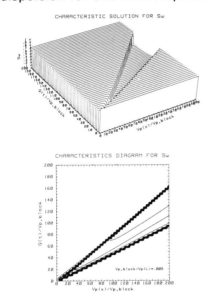

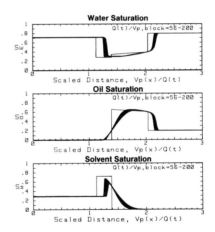

Fig. 20. 1:1 simultaneous injection water saturations (time vs. distance).

Fig 21. 1:1 simultaneous injection saturations scaled late profiles, analytical and numerical.

IV. THE IMPACT OF PERMEABILITY HETEROGENEITY ON SCALING IN MULTIPHASE FLOWS

To understand the influence of permeability heterogeneity on displacement processes, we must consider flow in more than one dimension. For two-dimensional flows in homogeneous permeability fields with uniform injection along a line source, the same frontal advance relations will apply along each parallel flow path and the results for one-dimensional flows will be preserved. When permeability heterogeneity is introduced the velocity along each flow path can be different and the displacing fronts will move with a wave velocity that is proportional to the local flow velocity. These differential velocities distort the shape of the displacing fronts, which can become highly irregular.

For constant-rate miscible displacements with a unit mobility ratio, the velocity field is steady. The spreading of a miscible front for this case can be analyzed theoretically by integrating the velocity along individual flow paths to determine the displacement distances of different parts of the front. Since detailed descriptions of permeability fields are seldom available, statistical theories based on measures of the spatial correlations of permeability fields have been used to derive the statistics of displacement fronts.[2,7,12] These theories show that the spreading of miscible fronts can be modeled by adding an effective dispersion coefficient to the macroscopic dispersion coefficient in the transport equation. The solution of the resulting equation gives the expected value of the concentration of injected fluid.

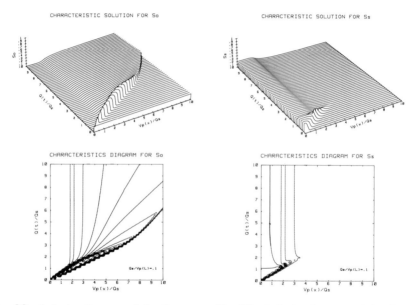

Fig. 22. 1:1 simultaneous injection slug (10% TPVI) oil saturations (time vs. distance).

Fig. 23. 1:1 simultaneous injection slug (10% TPVI) solvent saturations (time vs. distance).

These same theories also show that the expected value of the concentration is a good predictor of true local values of concentration only when the flow path from the injection point is large compared to the range of permeability correlations (several tens of integral scales).[12] Until that condition is reached, the fluid is not truly mixed, but merely spread around, and the solution of the transport equations with an augmented dispersion coefficient to account for the effects of heterogeneity is a bad predictor of local compositions. These same theories show that, if the condition of small correlation length compared to flow path length is not satisfied, the calculated effective dispersion coefficient has a scale dependence that has also been observed in field measurements of tracer dispersion over distances greater than several kilometers.[2,3] For typical oil field problems, where correlations between wells are commonly observed, these solutions will be of little value in describing interwell flow.

A similar theoretical treatment is not available for non-unit mobility ratio miscible displacements, or immiscible displacement processes, since the velocity field continuously changes in response to the changing mobilities as the displacement proceeds. We expect, however, that the use of numerical dispersion, or an augmented physical dispersion, will not provide accurate predictions of the effects of permeability heterogeneity when correlations are present at the scale of the interwell spacing. For these cases, it is better to simulate the displacement process at a scale where laboratory measured values of permeability are appropriate, and let the flow do a dynamic average of the frontal displacements along different flow paths. Based on the results for single-phase flow discussed above, we expect this averaged frontal displacement to exhibit a spreading of the front with a streamwise dependence when there are correlations in the permeability field over distances comparable to the well spacing.

In natural porous media, the nature of permeability variations is highly anisotropic, owing to the stratified nature of sedimentary deposits. In these systems, the most rapid variations occur in the direction normal to the bedding planes, usually the vertical direction. In most approaches to scaling flow processes, the predominant effects of vertical property variations are considered separately from areal property variations, and the scale-up process is accomplished in a two-step procedure.

In the hybrid finite difference/streamtube approach, this is done by first constructing a conditional simulation of interwell permeability variations in a representative vertical cross section between two wells and running a simulation of the process of interest in this cross section. The results of this simulation are then reduced to an equivalent single-layer solution that depends only on streamwise distance and time. If the statistics of vertical permeability variations are similar throughout the area of interest, a single characteristic single-layer solution can be used in the streamtube simulation of areal fluid flow. If several distinct regions of vertical stratification are present, more than one cross-section and finite difference solution may be required.

As an example of this procedure, consider the conditional simulation of permeability variations in a vertical cross-section between two wells shown in Figure 24b. A finite difference simulation of a waterflood was performed on this cross-section. A representative saturation distribution part way through the flood is shown in Figure 24a. The characteristics diagram for the vertically averaged saturations (a porosity-weighted average) is shown in Figure 25. The effects of heterogeneity on the solution

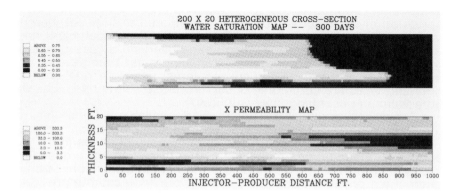

Fig. 24. Cross section (a) horizontal permeability, (b) waterflood water saturation at 32%.

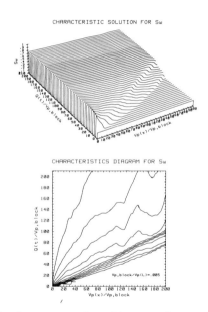

Fig. 25. 2-D waterflood water satations (time vs. distance).

result in a spreading of the sharp front and earlier breakthrough of the injected water than seen in the homogeneous solutions. This is due to preferential flow channels breaking through ahead of the mean flow. The characteristic line at the leading edge of the front is quite linear, but streamwise effects can be seen in the contours behind the front. Although the injection of water is continuous, the presence of correlated heterogeneity introduces length scales that preclude the reduction of the solution to a function of a single dimensionless variable.

The injection of a finite 1:1 WAG of solvent and water was also simulated. The presence of length scales associated with the heterogeneity as well as a length scale associated with the slug means that the solution should now be a function of three dimensionless variables, the two for volumetric linear scaling and a third for the ratio of slug length to the scales of heterogeneity. The characteristics diagram for the oil saturations for a single slug size is shown in Figure 26 as a function of the volumetric linear scaling variables. It can be seen that the effects of heterogeneity cause a spreading and earlier breakthrough of the fronts that were sharp in the homogeneous simulations and introduce further streamwise variations in the later contours.

The saturation contours for a displacement with a slug one-third the size of the original are shown in Figure 27. Comparison of the lower left corner of this diagram with the previous diagram shows that the propagation velocity and fluid saturations of the oil bank are similar, but the streamwise dependence of the effects of permeability are not captured by the linear scaling variables alone. Displacement calculations using a two-dimensional solution based on a particular slug size will only approximate the streamwise effects of heterogeneity on slug processes. Considering the added complexity of constructing and using a three-dimensional solution, this level of approximation for slug processes will be acceptable. We can now proceed to examine the effects of areal permeability variations and the geometry of well placement on the production of fluids.

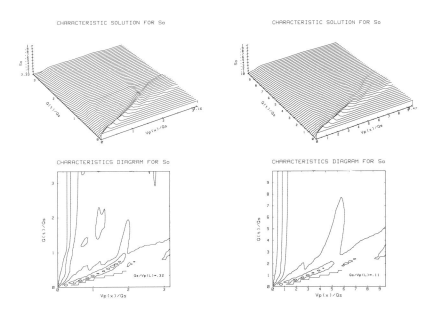

Fig. 26. 2-D 1:1 simultaneous injection slug (32% TPVI) oil saturations (time vs. distance).

Fig. 27. 2-D 1:1 simultaneous injection slug (11% TPVI) oil saturations (time vs. distance).

V. STREAMTUBE DISPLACEMENT CALCULATIONS

When areal variations in permeability are correlated over distances comparable to the well spacing, the effect on frontal displacements and the breakthrough of injected fluids is very similar to the effects seen in layered systems. This has been shown by the results of statistical transport theories[2] and by the simulation of interwell tracer flows in heterogeneous permeability fields.[31] This occurs because the permeability variations are dominated by a few large-scale features that produce preferential flow paths, which channel the fluids in much the same way a stratified system does.

The approach to areal simulation used here is based on the use of streamtubes. The solution for the stream function is first obtained by solving Laplace's equation with a distribution of sources and sinks with strengths corresponding to the average historical well rates. The streamtube geometries are found by calculating contours at equal increments in the stream function. These contours of constant stream function represent the boundaries of the streamtubes and are lines which are parallel to the local flow velocities. For a unit mobility ratio miscible displacement at a constant injection rate, the volumetric flow rate down each of the streamtubes is equal.

For mobility ratios other than unity, the velocity field continuously changes. One approach to accounting for the changing velocity field is to recalculate the streamtubes as the displacement proceeds. It has been shown, however, that the effect of non-unit mobility ratios can also be accounted for to a good approximation by relaxing the requirement of equal volume flow down each tube and allocating the fluid injected into each streamtube according to the total mobility of the fluids along the tube.[32] When this is done, the streamtube boundaries no longer represent equal increments in stream function, but rather act as a curvilinear coordinate system for performing the displacement calculations. This approximation is valid for mobility ratios from 0.1 to 100. Mobility ratios outside this range require a periodic updating of the streamtube geometries.[32]

Before the results of a fine-scale vertical cross-section simulation can be used in streamtube calculations, they must first be reduced to an equivalent single-layer solution. This is done by integrating over the vertical dimension. The total volume flow through the cross section is calculated as

$$q(t) = W(x) \int_0^{H(x)} k(x,z) \sum_{j=1}^{N_p} \frac{k_{rj}(x,z,t)}{\mu_j} \frac{\partial P_j(x,z,t)}{\partial x} \, dz \qquad (9)$$

where $P_j(x,z,t) = p_j(x,z,t) + \rho_j gz$

with z measured from an arbitrary datum. W(x) and H(x) are the local width and height of the cross section, respectively. When the gradients of capillary pressure in the direction of flow are small,

$$\frac{\partial P_j(x,z,t)}{\partial x} \simeq \frac{\partial P(x,z,t)}{\partial x}$$

for all j, and Equation (9) can be rewritten as

$$q(t) = W(x) \int_0^{H(x)} k(x,z) M_T(x,z,t) \frac{\partial P(x,z,t)}{\partial x} dz \qquad (10)$$

with

$$M_T(x,z,t) = \sum_{j=1}^{N_p} \frac{k_{rj}(x,z,t)}{\mu_j}$$

The fractional flow of individual components is calculated as

$$\bar{F}_i(x,t) = \frac{W(x) \int_0^{H(x)} \left[\sum_{j=1}^{N_p} c_{ij}(x,z,t) \frac{k_{rj}(x,z,t)}{\mu_j} \right] \frac{\partial P(x,z,t)}{\partial x} dz}{q(t)} \qquad (11)$$

In order to properly allocate the total fluid volume to each of the streamtubes, we must derive the effective total mobility from the cross section simulation. If the conditions for vertical equilibrium[33] are satisfied, the pressure gradient in Equation (10) is independent of z, and the equation may be rewritten as

$$q(t) = W(x) \frac{\partial P(x,t)}{\partial x}\bigg|_{VE} \int_0^{H(x)} k(x,z) M_T(x,z,t) dz \qquad (12)$$

where the VE subscript denotes vertical equilibrium. For this case we can define an effective total mobility as

$$\bar{M}_T(x,t) = \frac{\int_0^{H(x)} k(x,z) M_T(x,z,t) dz}{\int_0^{H(x)} k(x,z) dz} = \frac{\int_0^{H(x)} k(x,z) M_T(x,z,t) dz}{H(x)\bar{k}(x)} \qquad (13)$$

where $\bar{k}(x)$ is the local arithmetic average of permeability over the cross section. The relation between flow and pressure gradient in the resulting equivalent single-layer solution is then

$$q(t) = W(x)H(x)\bar{k}(x)\bar{M}_T(x,t) \frac{\partial P(x,t)}{\partial x}\bigg|_{VE} \qquad (14)$$

If the effective mobility defined in Equation (13) is used when the conditions for vertical equilibrium are not satisfied, the appropriate flow rate expression is

$$q(t) = W(x)H(x)\overline{k}(x)\overline{M}_T(x,t) \frac{\overline{\partial P}(x,t)}{\partial x} \qquad (15)$$

where the pressure gradient is now a pseudo-pressure gradient defined as the conductance (mobility times permeability) weighted average of the local pressure gradient, i.e.,

$$\frac{\overline{\partial P}(x,t)}{\partial x} = \frac{\displaystyle\int_0^{H(x)} k(x,z)M_T(x,z,t) \frac{\partial P(x,z,t)}{\partial x} \, dz}{\displaystyle\int_0^{H(x)} k(x,z)M_T(x,z,t)dz} \qquad (16)$$

This pseudo-pressure gradient can be different from the average pressure gradient, where the average pressure is defined as

$$\overline{P}(x,t) = \frac{1}{H(x)} \int_0^{H(x)} P(x,z,t)dz \qquad (17)$$

The pseudo-pressure gradient defined in Equation (16) accounts for the effects of the two-dimensional heterogeneous permeability field on the pressure field when vertical equilibrium is not satisfied.

Approaches to defining effective properties based on dividing the flow rate by a pseudo-pressure gradient[15,34] recover the effective total mobility defined in Equation (13). The intended application for these effective properties is in single-layer simulations where the boundary conditions are specified in terms of average pressures. Since a single-layer simulation cannot reproduce the modifications of the pressure field caused by a two-dimensional heterogeneous permeability field, and correct calculation of the volumetric flow rate using the mobility defined by Equation (13) requires a knowledge of the pseudo-pressure gradient, we prefer an alternative definition of effective mobility which captures all of the effects of heterogeneity on both the mobility and pressure fields. The preferred definition of effective mobility for use in single-layer simulations is

$$\overline{M}_T(x,t) = \frac{\displaystyle\int_0^{H(x)} k(x,z)M_T(x,z,t) \frac{\partial P(x,z,t)}{\partial x}dz}{H(x)\overline{k}(x) \frac{\partial \overline{P}(x,t)}{\partial x}} \qquad (18)$$

with the resulting expression for flow-rate being

$$q(t) = W(x)H(x)\bar{k}(x)\bar{M}_T(x,t) \frac{\partial \bar{P}}{\partial x}(x,t) \tag{19}$$

When conditions of vertical equilibrium apply, Equations (18) and (13) are equivalent. Comparison of mobilities calculated from Equations (13) and (18) for a variety of cases shows that the biggest differences between them occur in the vicinity of rapid changes in mobility and that these differences can be large.

When deriving effective mobilities from the results of numerical simulations, the quantities in Equation (18) must be approximated from their finite difference representations. The flow across the boundary between cells in two adjacent columns is calculated using the pressure difference between the two cells, the harmonic average of their permeabilities, and the total mobility of the upstream cell. The average permeability, $k(x)$, is the arithmetic average of the harmonic averages of the cells used for calculating the pressure gradient at each level. For use with streamtubes of varying geometries and overall lengths, the solutions for the fractional flows and effective total mobility are stored as functions of $V_p(x)$ and $Q(t)$, or $V_p(x)/Q_s$ and $Q(t)/Q_s$ when a slug process is modeled.

For each of the streamtubes connecting an injector-producer pair, we define a streamwise coordinate, s, running along the centerline of the streamtube, beginning at the injector and ending with a value S at the producer. Knowing the areal distribution of average flow properties and the streamtube geometries, we tabulate the static properties $A(s) = H(s)W(s)$, $\phi(s)$, $\bar{k}(s)$, and $V_p(s)$ for all of the streamtubes. The displacement calculation is done in discrete steps, keeping track of the cumulative volume injected into each streamtube. With a knowledge of the static properties, the cumulative volume injected, and the scalable solution for total effective mobility, we also have $\bar{M}_T(s,t)$ and can calculate the pressure drop along the streamtube as

$$\Delta P(t) = \int_0^S \frac{\partial \bar{P}}{\partial s} ds = q(t) \int_0^S \left[\bar{k}(s)A(s)\bar{M}_T(s,t)\right]^{-1} ds \tag{20}$$

or

$$\Delta \bar{P}(t) = q(t) R_\psi(t)$$

where

$$R_\psi(t) = \int_0^S \left[\bar{k}(s)A(s)\bar{M}_T(s,t)\right]^{-1} ds$$

is the total resistance to flow in streamtube ψ. When $\Delta P(t)$ is specified, Equation (20) can be used to calculate the instantaneous flow rate in each tube. When q(t) is specified, the fraction allocated to each streamtube should be in proportion to the ratio

$$\frac{R_T(t)}{R_\psi(t)} \text{ , where } \frac{1}{R_T(t)} = \sum_\psi \frac{1}{R_\psi(t)} \tag{21}$$

This assumes that all of the wells in communication with the injector considered have been pumped off to the same level. If this is not the case, the total flow can be allocated in proportion to the pressure drop/resistance ratios.

The displacement calculations are based on the assumption of incompressible flow, so the volume flow-rate of a component produced from a streamtube in any timestep is just $F_i(V_p(S),Q(t),Q_s)$ times the instantaneous flow-rate in that tube. When the only effect of compressibility on a displacement is the presence of a compliance in the fluid, the effect on production is simply a delay in the production response.

VI. PSEUDOFUNCTION SCALE-UP

With the effective mobility and fractional flow solutions defined above, it is also possible to define local pseudo-relative permeabilities. The pseudo-relative permeabilities of the individual phases are defined as

$$\bar{k}_{rj}(x,t) = \mu_j \bar{F}_j(x,t) \bar{M}_T(x,t) \tag{22}$$

It should be understood that these pseudofunctions depend on the mobility ratio of the fluids involved and cannot be regarded as rock properties that can be used with other fluids.[35] When the porosity-weighted average saturations are monotonically increasing or decreasing functions of time at a particular location, there will be a one-to-one correspondence between $k_{rj}(x,t)$ and $S_j(x,t)$ and we can calculate $k_{rj}(S_j(x,t))$.

These pseudofunctions account for the dynamic averaging of the flow by permeability variations but do not contain corrections for numerical dispersion if a coarsening of the grid in the single-layer simulation is anticipated. This can be accomplished using procedures similar to those given by Kyte and Berry,[15] with their pseudopressures replaced by the average pressures.

Pseudofunctions derived by Equation (22) were obtained for each of the 200 columns of a simulated waterflood in the heterogeneous cross section presented in Figure 24b. Five of these 200, taken at equal spacing along the section, are shown in Figure 28a, along with the original rock curves. These 200 pseudofunctions were used in a one-dimensional simulation with the same streamwise discretization as that used in the cross section simulation. The resulting produced oil fractional flow is shown in Figure 29 (dotted line) along with the results of the cross-section simulation (solid line). Also shown in this figure (dashed line) are the results of scaling up the grid size in the flow direction to five total cells and using a separate pseudofunction that has been corrected for numerical dispersion for each cell. These results show that pseudofunctions can effectively reproduce the results of the cross section simulation from which they were derived.

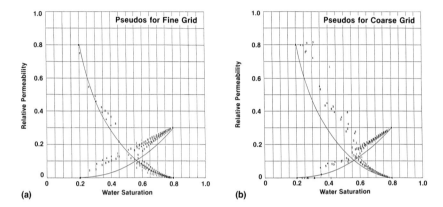

Fig. 28. Rock and pseudo-relative permeabilities (a) 5 of 200 fine grid, (b) 5 of 5 coarse grid.

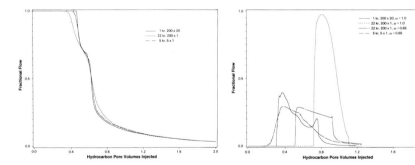

Fig. 29. Waterflood oil fractional flows, 2-D and 1-D using pseudo kr's.

Fig. 30. Solvent slug (32% TPVI) oil fractional flows, 2-D and 1-D using pseudo kr's.

These same pseudofunctions were used in a one-dimensional simulation of a solvent flood. The resulting fractional oil flow is shown in Figure 30 (dotted line), along with the results of the cross section simulation (solid line). Since the flow is one-dimensional, there is no opportunity for the solvent to bypass any oil and the sweep is complete. Provision for modeling the bypassing of oil by solvent can be included by adopting the approach of

Todd and Longstaff,[36] which includes the use of a mixing factor, ω, to control the amount of bypassing that occurs. The results of a simulation with $\omega = 0.65$ are also shown in Figure 30 (broken line) with the improvement in simulated oil response shown. The value of ω was chosen to match the total oil recovery. It can be seen that, although the use of a mixing factor to model the two-dimensional effects of solvent bypassing oil improves the re-covery predictions, they are still far from correct both in terms of the timing of the oil bank and the peak oil flow rate. Simulations using lower values of ω gave better predictions of the timing of the oil bank but lower overall recoveries. Conversely, higher values of ω retarded the movement of the oil bank and overpredicted the total recovery.

A method of accelerating the oil bank velocity without changing the recovery is required to match the cross section results. As noted in the discussion of Figure 13, the simulated velocity of the oil bank was too high when a small number of cells was used to model solvent displacement from tertiary recov-ery conditions. The error in the propagation velocity from using waterflood pseudo-relative permeabilities can be partially offset by using a coarser representation of the grid. An example of this is shown (dashed line) in Figure 30, where the results of a sol-vent displacement with $\omega = 0.65$ are shown for a five-cell model using the five pseudofunctions corrected for numerical dispersion. For this example, the two kinds of errors cancel each other and result in a good prediction of the oil bank arrival. This fortuitous result cannot be expected for other levels of discretization.

VII.   COMPARISON OF METHODS

Both of the approaches to scale-up of multiphase flows des-cribed above have their limitations. Streamtube methods are based on a fixed spatial distribution of sources and sinks repre-senting wells with constant relative flow rates. Any change in the well configuration would require a recalculation of the stream-tubes with the distribution of mobilities present at the time of realignment. In addition, since the scalable solutions are based on displacement from a uniform initial condition, it is not clear how those solutions can be used for displacement calculations in streamtubes with non-uniform initial saturation distributions.

Streamtube methods are not suitable for modeling primary depletion, since they are based on incompressible flow. They also are not suitable for flows where the principal drive mecha-nism acts in the vertical direction, as occurs in a bottomwater drive or gas-cap expansion. The scalable solutions used for the displacement calculations can be affected by flow rate, but this can be accounted for approximately by doing the cross section simulation in a cross section with a variable area approximating the geometry of a typical streamtube. The review of the scaling relations observed in a variety of processes showed that the scalable solutions used in streamtube methods can accommodate a wide variety of processes, including those involving fluids of differing viscosity injected continuously or as discrete slugs.

Pseudofunctions offer the apparent flexibility of allowing well realignments during a simulation. However, since they are derived from simulations of displacements from uniform initial conditions, they will not accurately model flow in other situations, since the one-to-one correspondence of saturations and mobilities used in their derivation is particular to the displacement used in their derivation.

In most cases, both methods are used in a single-layer solution with a coarser resolution than that used in the cross-section simulation. Streamtube methods are free of grid orientation and numerical dispersion effects, while coarsely-gridded finite difference models are sensitive to these. The alterations of the pseudofunctions required to correct for numerical dispersion tend to exacerbate these grid orientation effects.

The displacement calculations of streamtube methods involve a simple mapping of the scalable solutions from the cross-section simulation to obtain the production curves. The procedure for flow-rate updating to accommodate the changing mobilities does involve some simple integrals, but the complete flow equations are only solved once in the cross section. The use of pseudofunctions in a finite difference simulation requires the solution of the complete flow equations in the single-layer model, with the attendant computational overhead.

A comparison of the relative amounts of computation time required by each of the methods can be seen in the following example of a field-scale solvent displacement involving five injectors and seven producers in the configuration shown in Figure 31. The layout of the grid and the resulting streamtube pattern are also shown. The fractional flows and mobilities were

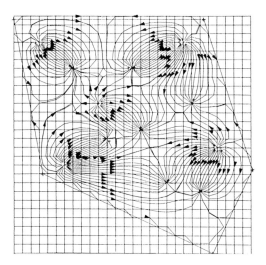

Fig. 31. Streamtubes and areal grid.

calculated from the solvent flood and the pseudofunctions were calculated from the waterflood on the cross section shown in Figure 24b. The simulation of 2.5 pore volumes of injection of a solvent from tertiary recovery conditions using the streamtube method required 30 seconds on a MicroVAX 3200 workstation. The same period of recovery was simulated, using the coarse-grid pseudos with $\omega = 0.65$, on the grid shown (cells outside the no-flow boundaries are inactive) and required 630 seconds on a Cray XMP-48. The resulting fractional flow curves show similar break-throughs, but the overall recovery is significantly higher for the finite difference model. Although no "truth set" is available for validating either method, the close adherence of this problem to the assumptions of the streamtube method favor the validity of the results derived by that method.

VIII.   CONCLUSIONS

1. One-dimensional solutions to the transport equations can be scaled and applied to quasi-one-dimensional streamtubes of differing geometry and total pore volume if the solutions are obtained as functions of local pore volume, $V_p(x)$, and cumulative volume injected, $Q(t)$.

2. One-dimensional displacements involving the continuous injection of a displacing fluid can be reduced to a function of a single dimensionless variable, $Q(t)/V_p(x)$.

3. One-dimensional displacements involving the injection of a slug of displacing fluid can be reduced to a single function of two dimensionless variables, $Q(t)/Q_s$ and $V_p(x)/Q_s$, for all slug volumes, Q_s. These variables are the volumetric linear scaling variables.

4. In flows with a single fractional flow curve the effects of numerical dispersion act to thicken the width of sharp fronts. In flows described by more than one fractional flow curve, numerical dispersion can, in addition, introduce anomalous waves to the solution that alter its behavior far downstream, changing the velocity and height of displacement fronts.

5. Simulations of continuous displacements in heterogeneous vertical cross sections can be reduced to equivalent single-layer solutions as a function of two variables for use in streamtube displacement calculations.

6. Simulations of slug injections in heterogeneous vertical cross sections can be reduced to equivalent single-layer solutions as a function of three variables. Approximate displacement calculations based on the two volumetric linear scaling variables include the effect of stratification on the spreading of displacement fronts, but do not completely reproduce the streamwise dependence of the effects of heterogeneity.

7. Pseudo-relative permeabilities that change along the flow path can reproduce the behavior of equivalent single-layer solutions derived from heterogeneous cross sections when a single fractional flow relation describes the flow, but cannot

reproduce the behavior of processes involving a change in fluid viscosities.

8. Methods based on the use of pseudo-relative permeabilities and streamtubes for scaling the effects of permeability heterogeneity in field-scale problems both have limitations that should be understood before using them. Streamtube methods can accommodate processes that involve a change in fluid viscosity and offer substantial savings in the amount of computation required.

APPENDIX

Calculating Shock Trajectories

Shocks, or discontinuities in properties, arise when characteristics introduced in the flow at different times, and traveling at different velocities, intersect. Since the different compositions traveling along two characteristics cannot simultaneously be present at the same location, the intersection of two characteristics results in a step change in composition from that carried along the downstream characteristic to that carried along the upstream characteristic. The trajectory and height of the resulting shock can be calculated by determining which characteristics intersect along it and the locus of those intersections.[23]

The motion of individual characteristics is given by

$$dx = v_i dt = \frac{q(t)}{\phi(x)A(x)} v_i^* \, dt \qquad \text{(A-1)}$$

where v_i^* is given by Equation (6).

Denoting the time at which a characteristic is introduced at $x = 0$ by $t = T_k$, Equation (A-1) may be rearranged and integrated to give

$$V_p(x) = v_{ik}^* \left[Q(t) - Q_k \right] \qquad \text{(A-2)}$$

where

$$Q_k = \int_0^{T_k} q(t) dt$$

The rate at which component i crosses a wave traveling at velocity v_{ik} is the difference between the rate at which that component flows relative to the stationary porous medium and the rate at which the wave sweeps past that component in the medium. This net rate of flow is just $\{q(t) F_i - v_i \phi(x) A(x) C_i\}$. In traveling for a time interval $t - T_k$, a volume of component i equal to

$$\int_{T_k}^{t} \left[q(t')F_i - v_{ik}A(x)\phi(x)C_i \right] dt' = \int_0^{x} \left[\frac{q(t)F_i}{v_{ik}} - A(x')\phi(x')C_i \right] dx'$$

will cross any characteristic. For two characteristics introduced at times T_1 and T_2 to meet at a given position x, the total amount of component i introduced in the flow between T_1 and T_2 must flow out of the region between the characteristics. Thus,

$$\int_{T_1}^{T_2} q(t')F_i(0,t')dt' = \int_0^x \left\{ \left[\frac{q(t)F_{i1}}{v_{i1}} - A(x')\phi(x')C_{i1} \right] \right.$$

$$\left. - \left[\frac{q(t)F_{i2}}{v_{i2}} - A(x')\phi(x')C_{i2} \right] \right\} dx'$$

where $C_{ik} = C_i(0,T_k)$, $F_{ik} = F_i(0,T_k)$ and the v_{ik} are the velocities of the characteristics introduced at time T_k. In terms of the variables defined in Equation (2), this expression becomes

$$\int_{Q_1}^{Q_2} F_i(0,Q)dQ = V_p \left[\left(\frac{F_{i1}}{v_{i1}^*} - C_{i1} \right) - \left(\frac{F_{i2}}{v_{i2}^*} - C_{i2} \right) \right] \qquad \text{(A-3)}$$

This expression, combined with Equations (A-2) for the trajectories of the characteristics, specifies the position of the discontinuity resulting from their intersection.

When a slug of displacing fluid is injected into a region of uniform composition, all of the characteristics ahead of the slug have the same properties and velocities. Denoting the time at which the slug injection is initiated by $Q = 0$, the total volume of slug by Q_s, the constant fractional flow of component i in the slug by F_{is}, and using Equations (A-2) to eliminate Q_1, the locus of intersections of the shock with characteristics introduced at Q_s is given by

$$\frac{V_p}{Q_s} = \frac{F_{is} - F_{i1}}{\left(\dfrac{F_{i1}}{v_{i2}^*} - C_{i1} \right) - \left(\dfrac{F_{i2}}{v_{i2}^*} - C_{i2} \right)} \qquad \text{(A-4)}$$

and the time of the intersections is given by

$$\frac{Q}{Q_s} = \frac{1}{v_{i2}^*} \left(\frac{V_p}{Q_s} \right) + 1 \qquad \text{(A-5)}$$

For the injection of an oleic slug with the same viscosity as the residual oil at tertiary recovery conditions, $F_{i1} = 0$, $C_{i1} = S_{orw}$, $F_{is} = 1$, $C_{i2} = S_{o2}$, $F_{i2} = f_{o2}$, and

$$\frac{V_p}{Q_s} = \frac{1}{(S_{02} - S_{orw}) - \dfrac{f_{o2}}{v_{o2}^*}}$$ (A-5)

It is interesting to note that these same arguments can be used to rigorously derive the Welge tangent construction for the Buckley-Leverett shock velocity without resort to ad hoc assumptions about equal area integrals of a physically unrealizable saturation distribution as was required in earlier analyses.

ACKNOWLEDGEMENTS

The authors would like to thank the management of Chevron Oil Field Research Company for permission to publish this work.

NOMENCLATURE

$A(x)$	=	Cross-sectional area at downstream position x, m^2
c_{ij}	=	Concentration of component i in phase j
C_i	=	Concentration of component i
D	=	Dispersion coefficient, m^2/s
f_j	=	Volumetric fractional flux of phase j
F_i	=	Fractional flow of component i
g	=	Gravitational force, m/s^2
$H(x)$	=	Height, m
$k(x,z)$	=	Absolute permeability at position (x,z), m^2
$k_{rj}(x,z,t)$	=	Relative permeability of phase j at position (x,z) and time t
L	=	Cross-section length in streamwise direction, m
$M_T(x,z,t)$	=	Total mobility, 1/Pa•s
N_p	=	Number of phases
$p(x,z,t)$	=	Pressure, Pa
$P(x,z,t)$	=	Potential, Pa
Pe	=	Peclet number
$q(t)$	=	Flow rate at time t, m^3/s
$Q(t)$	=	Cumulative injection up to time t, m^3
Q_s	=	Slug volume injected, m^3
$R_T(t)$	=	Total resistance of all streamtubes, Pa•s/m^3
$R_\psi(t)$	=	Resistance of a streamtube, Pa•s/m^3

s	=	Streamwise position along streamtube centerline, m
S	=	Streamwise length of streamtube, m
S_j	=	Saturation of phase j
t	=	Time, s
T_k	=	Time when kth characteristic crosses $x = 0$, s
v_i	=	Wave velocity of component i, m/s
v_i^*	=	Wave velocity of component i normalized by local fluid velocity
v_s^*	=	Shock velocity normalized by local fluid velocity
V	=	Velocity, m/s
$V_p(x)$	=	Cumulative pore volume to downstream position x, m^3
$W(x)$	=	Width, m
x	=	Downstream position, m
z	=	Distance above a datum, m

Greek

α	=	Dispersivity, m
μ	=	Viscosity, Pa•s
ψ	=	Streamtube
ω	=	Mixing parameter

Subscripts

i	=	Component
j	=	Phase
k	=	Characteristic
o	=	Oil
orw	=	Oil residual to water
s	=	Slug
VE	=	Vertical equilibrium
w	=	Water
1	=	Specific value of k
2	=	Specific value of k

REFERENCES

1. Lake, L. W., and Carroll, H. B., Reservoir Characterization, Academic Press, Orlando, 1986.

2. Arya, A., Hewett, T. A., Larson, R. G., and Lake, L. W., "Dispersion and Reservoir Heterogeneity", SPE Res. Eng., p. 139-148, Feb. 1988.

3. Hewett, T. A., "Fractal Distributions of Reservoir Heterogeneity and Their Influence on Fluid Transport", SPE 15386, 61st Ann. Tech. Conf. of SPE, New Orleans, 1986.

4. Hewett, T. A. and Behrens, R. A., "Conditional Simulation of Reservoir Heterogeneity with Fractals" SPE 18326, 63rd Ann. Tech. Conf. of SPE, Houston, 1988.

5. Haldorsen, H. H., and Lake, L. W., "A New Approach to Shale Management in Field-Scale Models", SPEJ, p. 447-457, Aug. 1984.

6. Warren, J. E., and Price, H. S., "Flow in Heterogeneous Porous Media", SPEJ, p. 153-169, Sept. 1961.

7. Dagan, G., "Models of Groundwater Flows in Statistically Homogeneous Porous Formations", Water Resour. Res., 15, 1, p. 47-63, 1979.

8. Desbarats, A. J., "Numerical Estimation of Effective Permeability in Sand-Shale Formations", Water Resour. Res., 23, 2, p. 273-286, Feb. 1987.

9. Begg, S. H., Chang, D. M., and Haldorsen, H. H., "A Simple Statistical Method for Calculating the Effective Vertical Permeability of a Reservoir Containing Discontinuous Shales", SPE 14271, 60th Ann. Tech. Conf. of SPE, Las Vegas, 1985.

10. Begg, S. H., Carter, R. R., and Dranfield, P., "Assigning Effective Values to Simulator Grid-Block Parameters in Heterogeneous Reservoirs, SPE 16754, 62nd Ann. Tech. Conf. of SPE, Dallas, 1987.

11. Journel, A. G. and Alabert, F. "Focusing on the Connectivity of Extreme-Valued Attributes: Stochastic Indicator Models of Reservoir Heterogeneities", SPE 18324, 63rd Ann. Tech. Conf. of SPE, Houston, 1988.

12. Dagan, G., "Solute Transport in Heterogeneous Porous Media", J. Fluid Mech., 145, p. 151-177, 1984.

13. Haldorsen, H. H. and MacDonald, C. J., "Stochastic Modeling of Underground Reservoir Facies", SPE 16751

14.   Matheron, G., De Fouquet, Ch., Beucher, H., Galli, A.,
        Guerillot, D., and Ravenne, C., "Conditional Simulation of
        Fluvio-Deltaic Reservoirs", SPE 16753, 62nd Ann. Tech.
        Conf. of SPE, Dallas, 1987.

15.   Kyte, J. R. and Berry, D. W., "New Pseudo Functions to
        Control Numerical Dispersion", SPEJ, p. 269-276, Aug.
        1975.

16.   Lake, L. W., Johnston, J. R., and Stegemeier, G. L.,
        "Simulation and Performance Prediction of a Large-Scale
        Surfactant/Polymer Project", SPEJ, p. 731-739, Dec.
        1981.

17.   Emanuel, A. S., Alameda, G. K., Behrens, R. A., and
        Hewett, T. A., "Reservoir Performance Prediction
        Methods Based on Fractal Geostatistics", SPE Res. Eng.,
        p. 311-318, Aug. 1989.

18.   Walsh, M. P., and Lake, L. W., "Applying Fractional Flow
        Theory to Solvent Flooding", J. Pet. Sci. and Eng., 2,
        p. 281-303, 1989.

19.   Perkins, T. K., and Johnston, O. C., "A Review of
        Diffusion and Dispersion in Porous Media", SPEJ, p. 70-
        80, March 1963.

20.   Spence, A. P., and Watkins, R. W., "The Effect of Micro-
        scopic Core Heterogeneity on Miscible Flood Residual Oil
        Saturation", SPE 9229, 55th Ann. Tech. Conf. of SPE,
        Dallas, 1980.

21.   Lantz, R. B., "Quantitative Evaluation of Numerical Dif-
        fusion (Truncation Error)", SPEJ, p. 315-320, Sept. 1971.

22.   Bentsen, R. G., "Conditions Under Which the Capillary
        Term May be Neglected", J. Can. Pet. Tech., 17, 4,
        p. 25-30, Oct.-Dec. 1978.

23.   Lighthill, M. J., and Whitham, G. B., "On Kinematic
        Waves I. Flood Movement in Long Rivers", Proc. Roy.
        Soc. (London), A229, p. 281-316, 1955.

24.   Helfferich, F. G., "General Theory of Multicomponent,
        Multiphase Displacement in Porous Media", SPE 8372,
        54th Ann. Tech. Conf. of SPE, Las Vegas, 1979.

25.   Pope, G. A., "The Application of Fractional Flow Theory
        to Enhanced Oil Recovery", SPEJ, p. 191-205, June
        1980.

26.   Larson, R. G., Davis, H. T., and Scriven, L. E.,
        "Elementary Mechanisms of Oil Recovery by Chemical
        Methods", SPE 8840, SPE/DOE Symp. Enhanced Oil Rec.,
        Tulsa, 1980.

27. Helfferich, F. G., "Generalized Welge Construction for Two-Phase Flow in Porous Media with Limited Miscibility", SPE 9730, 57th Ann. Tech. Conf. of SPE, New Orleans, 1982.

28. Lake, L. W. and Helfferich, F., "Cation Exchange in Chemical Flooding: Part 2 - The Effect of Dispersion, Cation Exchange, and Polymer/Surfactant Adsorption on Chemical Flood Environment", SPEJ, p. 435-444, Dec. 1978.

29. Stalkup, F. I., "Effect of Gas Enrichment and Numerical Dispersion on Compositional Simulator Predictions of Oil Recovery in Reservoir Condensing and Condensing/Vaporizing Gas Drives", SPE 18060, 63rd Ann. Tech. Conf. of SPE, Houston, 1988.

30. Parsons, R. W., and Jones, S. C., "Linear Scaling in Slug-Type Processes - Application to Micellar Flooding", SPEJ, p. 11-26, Feb. 1977.

31. Mishra, S., Brigham, W. E., and Orr, F. M., "Analysis of Tracer and Pressure Data For Characterization of Areally Heterogeneous Reservoirs, SPE 17365, SPE/DOE 6th Symp. Enhanced Oil Rec., Tulsa, 1988.

32. Martin, J. C., and Wegner, R. E., "Numerical Solution of Multiphase, Two-Dimensional Incompressible Flow Using Stream-Tube Relationships", SPEJ, p. 313-323, Oct. 1979.

33. Lake, L. W., Enhanced Oil Recovery, Prentice-Hall, Englewood Cliffs, N.J., 1989.

34. Lasseter, T. J., Waggoner, J. R., and Lake, L. W., "Reservoir Heterogeneities and Their Influence on Ultimate Recovery", Reservoir Characterization, ed. Lake and Carroll, Academic Press, Orlando, 1986.

35. Pande, K. K., Ramey, H. J., Brigham, W. E., and Orr, F. M., "Frontal Advance Theory for Flow in Heterogeneous Porous Media", SPE 16344, SPE Calif. Reg. Mtg., Ventura, April 8-10, 1987.

36. Todd, M. R., and Longstaff, W. J., "The Development, Testing, and Application of a Numerical Simulator for Predicting Miscible Flood Performance", J. Pet. Tech., p. 341-379, July 1972.

INTERWELL GEOLOGY FROM GEOPHYSICAL DATA[1]

Douglas A. Lawson[2]

Department of Applied Earth Sciences
Stanford University
Stanford, California

Abstract

Reflection seismology can be used to constrain reservoir models by providing two- or three-dimensional lithologic information. A multilithologic frequency distribution is obtained by compositional log analysis for a well on or near the seismic line. Associated with this distribution is a sonic velocity distribution. In additional to these distributions a lithologic probability matrix is developed from the log analysis. This matrix includes the probability of vertical transitions among lithologies and the distribution of lithology thicknesses. This information in employed along with the instantaneous velocity inversion of the seismic line. The outcome is not only the most probable subsurface pattern of velocity, density, and constituent lithologies but also an indication of the reliability of the velocity inversion. Also, as a consequence of the calculations, the subsurface pattern of probability associated with each depth point is generated, and all of the subsurface sections can readily converted from time to depth. In this way, a measure of the interwell geology can be provided for reservoir simulation.

---

[1]Supported by the Stanford Center for Reservoir Forecasting
[2]Present address: ARCO Oil and Gas Company, Plano, Texas 75075

442

I. INTRODUCTION

To develop an intelligent field infill strategy, a detailed flow model must be generated that represents any significant heterogeneity. At present surface reflection seismic data are overwhelmingly the most common source for deriving interwell geologic information. The two basic limiting factors for geologic interpretation are vertical resolution and the underdetermined nature of the problem.

Considering the common depth and thickness of the reservoir rocks and standard surface seismic data with a frequency of 50 Hz, inversion interpretations would be pushed beyond their limit if the resolution sought is greater than about 100 to 50 feet. Methods for carrying out a velocity inversion of a seismic reflection amplitude section are discussed by Lindseth (1979) and Gelfand & Larner (1984). Normally with moderately high resolution seismic data, it would be difficult to extract information about the nature of 25- to 12- foot thick beds at a depth of 5000 feet. However, if this resolution is sufficient, petrophysically conditioned interpretive processing of the inverted seismic data can provide information about the interior of the reservoir. Furthermore, if this resolution is not sufficient, the same conditioning information can be used to generate the most probable internal stratigraphic architecture of the resevoir.

Since velocity is the single most readily obtained bulk property from seismic data, and there are generally more than two major constituents in sedimentary rocks of distinct velocity, the possible number of mixtures of these constituents that would have the same bulk velocity is infinite. If, however, sedimentary rocks fall into only a limited number of these possible mixtures and their distribution in space is in some way systematic, then there may be but one or, at most, a few more probable sedimentary circumstances associated with a particular velocity. The conditional processing described here is one way of compiling and applying this circumstantial knowledge to the interpretation of a single bulk property measurement.

II. COMPOSITIONAL PETROPHYSICAL ANALYSIS

To constrain the interpretive processing of the seismic velocity a
large body of petrological information must be built up. This
information will describe the most likely geological states that the
rocks of the study are found in. Because large samples are necessary
to describe the compositional distribution of rock types, core-derived
petrology is inadequate. Compositional analysis of wireline logs
must be the source of this large sample.

Given the bulk properties of major constituents of the subsurface
rocks to be analyzed, the volume of each constituent (e.g. dolomite,
quartz, water, oil, clay) at depth can be determined if the number of
bulk properties measured by a suite of wireline logs is nearly equal to
the number of major constituents. If the response of the instrument
is linearly proportional to the volume of some material in the
sample, then the bulk measurement is the sum of the effect of each
material in proportion to its volume (Figure 1) e.g.:

$$\rho_b = \rho_1 v_1 + \rho_2 v_2 + \cdots + \rho_n v_n$$

where

ρ_b = the bulk density
ρ_i = the density of the *ith* material
v_i = the volume of the *ith* material

The volume of materials can be solved for with the
following system:

$$l_1 = c_{11} v_1 + c_{12} v_2 + \cdots + c_{1n} v_n$$
$$l_2 = c_{11} v_1 + c_{12} v_2 + \cdots + c_{2n} v_n$$
$$\vdots \qquad \vdots$$
$$l_m = c_{m1} v_1 + c_{m2} v_2 + \cdots + c_{mn} v_n$$

or represented more compactly with matrix notation as:

$$[L] = [C] [V]. \tag{1}$$

The solution is obtained by dividing the instrument responses by the coefficients:

$$[V] = [C]^{-1} [L]$$. (2)

A large enough sample from each significant lithologic type can be obtained by doing such calculations for log suites from wells penetrating the reservoir and surrounding rocks near the seismic line.

Quite often, a grouping of the analyzed depth points into classes such as quartzose, carbonate, and shale sedimentary rocks is adequate for reservoir description. This can reduce the size of the matrix describing the lithologic transition distribution but would hide the fine variation in the relation between lithologic composition and velocity.

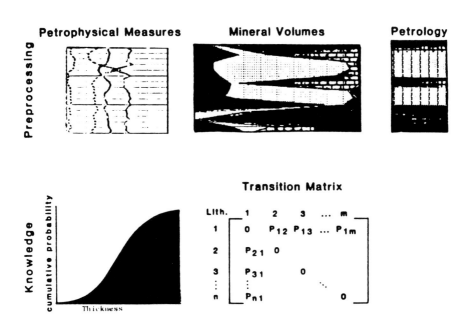

Figure 1. Log derived lithologic information.

III. LITHOLOGY VELOCITIES

Within the space defined by the variations in lithologic composition, there is a velocity trend. The degree of overlap in the velocity range of the lithologies determines the contribution of the petrophysical conditioning information in the interpretive processing of the seismic velocities.

A. F and t Tests of Velocity Populations

It is useful to study the celerial distinctiveness of the lithologies to gain some insight into the potential interplay between geologic knowledge and velocity in the final interpretive processing. If the populations have little overlap in the velocity ranges, then the conditional information will have little influence in the processing. However, if the populations are nearly indistinguishable on the basis of velocity, then the conditional knowledge will be the determining factor in the processing. Standard inferential F and students-t tests can be used to indicate the predominance of either geologic knowledge or velocity by testing the hypotheses of equal variance among the population and then equal means among the populations.

In the case discussed here, the variances of the velocity for the carbonate and shales were indistinguishable. Therefore the t-test could be used to test whether the the shale and carbonate velocities were the same. This hypothesis was rejected with $t=5.39$, alpha $=.01$ and $N=948$. So geologic information will be important in the interpretation of shales and carbonates.

IV. FREQUENCY DISTRIBUTIONS FOR CONDITIONAL PROCESSING

The next step is to derive the frequency distribution of a variety of lithological dependent properties used in the geological interpretation of the seismogram. Each of the following distributions can be developed independently or as joint distributions. The choice

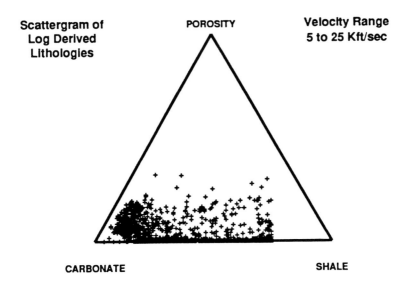

Figure 2. Frequency distribution of lithologies for 5 to 25 Kft/sec.

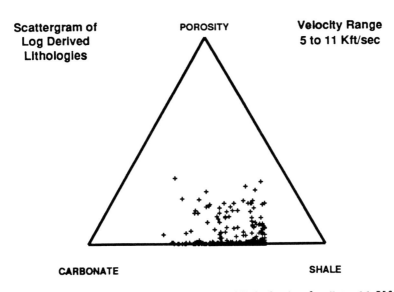

Figure 3. Frequency dirstribution of lithologies for 5 to 11 Kft/sec.

is determined by usefulness of the distributions in defining a single most probable sedimentary state for an observed velocity. The usefulness of a particular distribution in conditioning the processing is demonstrated by comparing the lithologic interpretation based on conditional processing and some conventional petrographic method or to the original log analysis.

In this report only the multilithologic frequency distribution is used to condition the interpretation of the velocity inversion.

A. Multilithologic Distribution

Having determined the composition of a large number of depth point samples, the frequency distribution of these points among the major lithofacies can be derived (Figure 2). Each interpreted depth point can be placed into a percentage class cell (e.g. 0-5 % clay, 0-5 % dolomite, 70-75 % quartzose, 30-15 % porosity) with the sum of the number of points placed in each cell representing the frequency of that composition. These class sizes (lithologies) can be constructed as coarsely as is practical for defining the internal lithological composition of the reservoir. Because there was no significant sand in the section studied, the lithologic axes were shale and carbonate with 100 percent porosity at the origin.

This distribution is essential to the geological conditioning of the processing. Examination of Figures 3 and 4 demonstrates the fundamental importance of the relation that can exist between lithology and velocity. If the lithologic distribution of depth points is generated for points lying within a particular velocity band, it can be seen that certain lithologies are more likely to exhibit a certain velocity. In this example, the bound-water- bearing shales are more likely to have low velocities (Figure 3), and carbonates to have high velocities (Figure 4).

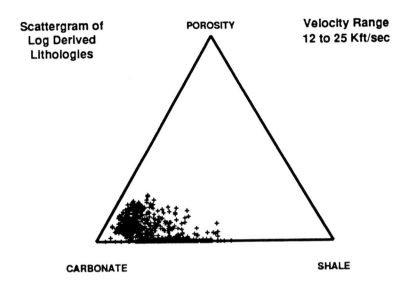

Figure 4. Frequency distribution of lithologies for 11 to 25 Kft/sec.

Table I shows the interpretive results for using only a multilithologic frequency distribution to condition the processing. The sonic log velocities were conditionally processed and compared to the results of the compositional analysis of the entire log suite. The shales and the carbonates are classified correctly 96 % of the time. Only 2 percent of the shales were misclassified as carbonates whereas less than 2 percent of the carbonates were misclassified as shales. Short of developing a synthetic seismogram from the logs and generating a time-depth curve, the likely degree of misclassification is indicated by a comparison of the sonic log velocity distribution (Figure 5) and the same distribution for the trace close to the well (Figure 6). The seismic velocity distribution not only reproduces the two major peaks in the sonic log distribution but also the second-order peaks. Therefore, the misclassification of lithologies should be quite low. Any greater misclassification frequency when using the seismic velocity will be due to the resolution of the seismic data and noise in the seismic signal. Considering the average thickness of the widespread shale beds in the reservoir is 77 feet and the 40Hz frequency of the seismic signal, most misclassification should arise from lack of resolution.

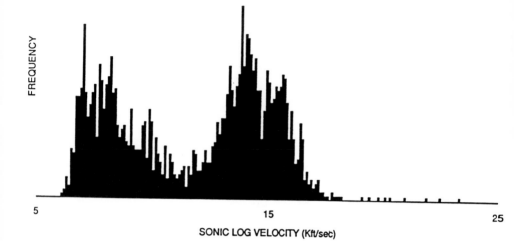

Figure 5. Frequency distribution for sonic log velocity.

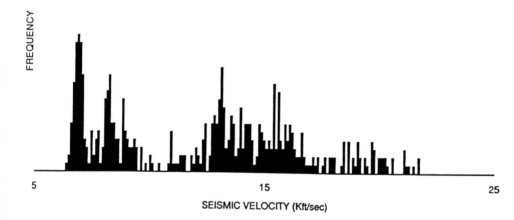

Figure 6. Frequency distribution for seismic velocity from trace near well.

TABLE I. Comparison of classification of common depth points by compositional log analysis and interpretive seismic processing.

	Shale	Carbonate
Log Shale	48.36%	2.19%
Log Carbonate	1.42%	48.03%

B. Lithologic Transition Matrix

The frequency of transitions between lithologies is obtained by summing the number of adjacent depth point sample pairs that are from different lithologies (Figure 1). Both downward and upward distributions can be obtained. This information in conjunction with the lithologic thickness distribution can improve the interpretive processing. The contribution of this geologic knowledge is greatest when the stratigraphic relationship between lithologies is highly predictable. In this study there were only two major lithologies; therefore no transition information could be used. Finer subdivision of the lithologies into tight and porous carbonate might be of some use in conditioning but is not investigated here.

C. Lithology Thickness Distributions

The thickness distribution for each lithology is determined by summing the number of runs that have the same number of consecutive depth point samples from the same lithology (Figure 1). The sum of the points in this distribution for each lithology would be equivalent to the diagonal of the transition matrix if the latter is formed by recording the transitions observed at a regular depth interval. Since the seismic data are in the time domain, thicknesses must be converted to two-way time durations. This kind of conditioning information can be extremely important if there is a significant overlap in the velocity ranges of the lithologies. However, because there was no significant overlap, this information was not used in the processing.

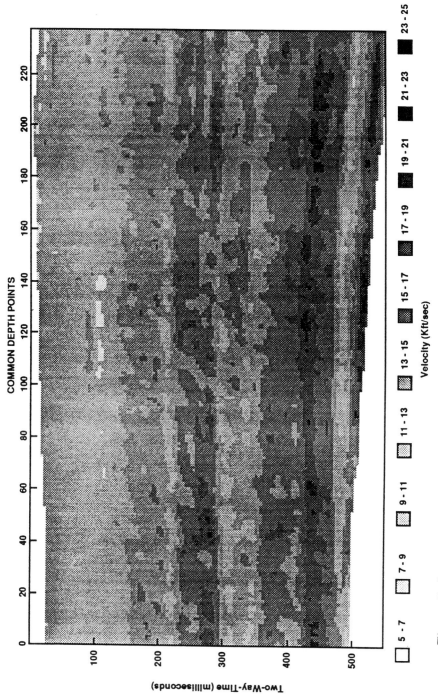

Figure 7. Instantaneous velocity section.

V. MULTILITHOLOGIC CONDITIONED SEISMIC VELOCITY INTERPRETATION

Approximately 2700 compositional log interpretations and sonic log readings were used to produce the three-dimensional discrete frequency distribution. Two zones, one for each lithology, were established by defining contraints on the feasible space for the optimization. Remember that 100% fluid is at the origin of the space. The constraints (bounds) for each zone were:

- no lithologic fraction below zero

- primary lithology of the zone, no less than 50 % and no more than 100 %

- neither secondary lithology of the zone greater than 50 %

- sum of the lithologic fractions, no less than 50 % and no greater than 100 %.

The optimization procedure can either be carried out on a continuous differentiable function (Beightler, Phillips, and Wilde, 1979), which could be a least-squares polynomial fit through the three-dimensional space or a velocity-band-limited direct search of the discretized n-dimensional space. The disadvantages of the first method are:

- high-order polynomials are necessary to represent significant maxima in the distribution

- the global maximum may not be found

- object functions for high-dimensional spaces are hard to obtain.

The disadvantages of the second method are:

- large samples are necessary to finely discretize the space

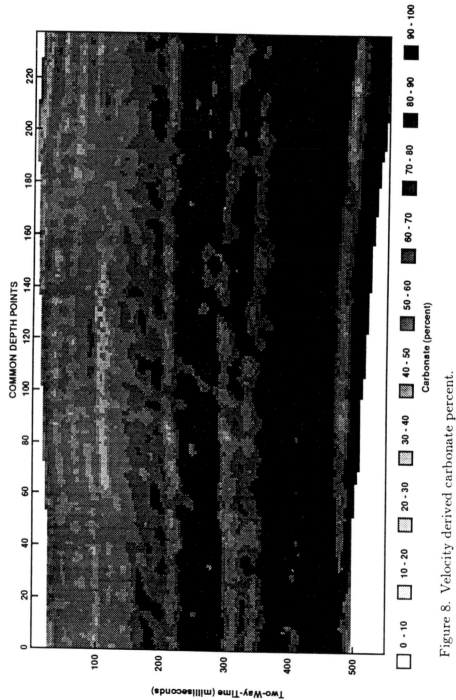

Figure 8. Velocity derived carbonate percent.

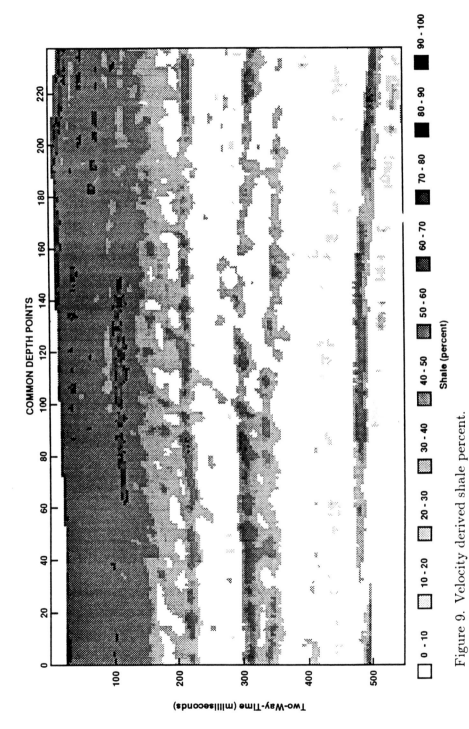

Figure 9. Velocity derived shale percent.

455

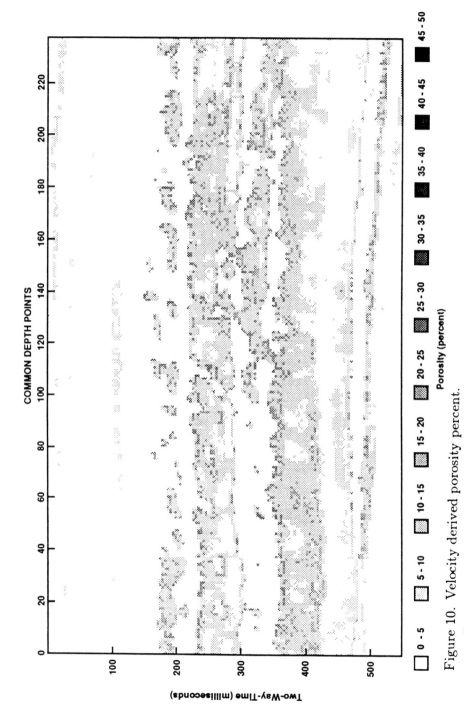

Figure 10. Velocity derived porosity percent.

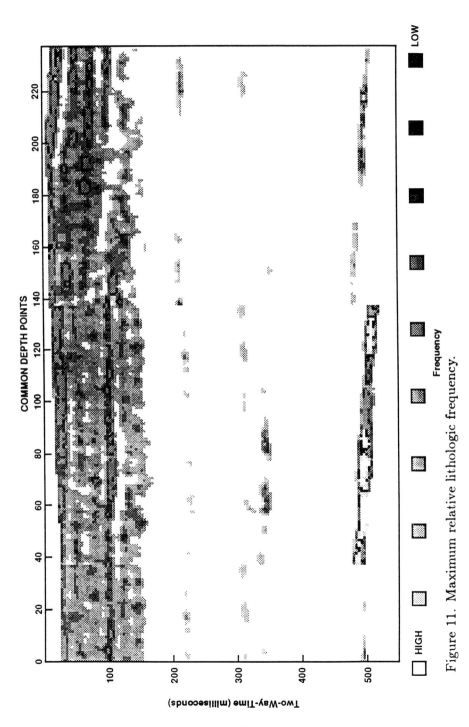

Figure 11. Maximum relative lithologic frequency.

457

- a velocity band as wide as the discrete units must be used

- computer memory use and time increase geometrically with the dimension of the space.

Every point in the velocity inverted seismic line defines the optimization surface for the indirect method or band for the direct method. Figure 7 shows the instantaneous velocity inversion of the ungained seismic section. The section can be divided into a slow zone in the upper 150 milliseconds, which includes the trapping shales over the reservoir, and an underlying fast zone, which includes the reservoir. In the example here, using the first method,the maximum discovered during the optimization is the most likely (frequent) lithologic composition for the given velocity. The result of the interpretation is a set of percentage composition cross sections.

The reservoir is composed of carbonates and lenticular shales, overlain by trapping shale. Figure 8 shows the most likely carbonate concentration over a 550-millisecond two-way time interval. Carbonate makes up less than 15 % of the rock in the upper 100 milliseconds but more than 60 % of the lower 300 milliseconds with some stringers of low concentration. Figure 9 shows that shale makes up over 70 % of the rock in the upper 100 milliseconds, and forms the seal over the reservoir. It can also be seen that most of the velocity drops within the carbonate reservoir are more likely to be shale strings than increases in effective porosity. The effective porosity pattern in the reservoir is shown in Figure 10. The reservoir is definitely stratified with lower connectivity in the lower 100 milliseconds of the reservoir.

The maximum relative frequency value associated with each velocity point is shown in Figure 11. This information can be used to assess the reliability of each interpreted point in the section. Considering the frequency distribution of each major lithology in the log compositional analysis, there are no obvious zones of extremely improbable interpretation.

VI. CONCLUSIONS

Having generated a geological cross section from the seismic data, this section must be converted to a flow properties section to form the flow simulation model. Finally, if the seismic survey is shot over the development area, the lithologic cross section resulting from the processing could be readily converted into a grid of effective flow properties based on core analysis.

Flow properties could be generated for each grid node using a Monte Carlo Simulation. Just as there is a velocity distribution in the multilithologic space, there are flow property distributions associated with it. Many of the cells in this space would actually have a frequency distribution for these properties (e.g. horizontal permeability) because more than one measurement may have been made on core samples with a lithologic composition identified with a particular cell. Therefore, a value for the flow property could be randomly sampled from the distribution associated with the lithologic composition at each point in the interpreted geologic section. Repeating this procedure for every flow property for each point in the section produces the flow simulation model.

Additional flow simulation models could be generated for prescribed flow property values (e.g. median). This could be done by recording the value of the flow property at the desired point in the distribution associated with the lithologic composition at each point in the interpreted geologic section.

REFERENCES

Beightler, C. S., Phillips, D. T., and Wilde, D. J., 1979, Foundations of Optimization, Prentice-Hall, Englewood Cliffs, New Jersey.

Gelfand, V. A., and Larner, K. L., 1984, Seismic lithologic modeling:Geophysics: Leading Edge, Nov. p. 30-34.

Lindseth, R. O., 1979, Synthetic sonic logs - a process for stratigraphic interpretation: Geophysics, v. 44, no. 1, p. 3-26.

CROSS-WELL SEISMOLOGY -
A TOOL FOR RESERVOIR GEOPHYSICS

Björn N.P. Paulsson

Chevron Oil Field Research Company
La Habra, California

I. BACKGROUND

Today it is common that 60 to 70 percent of the mobile oil is left in the ground when an oil reservoir is considered economically depleted (DOE, 1986). The large percentage of mobile oil left in the ground is due, in part, to macroscopic inhomogeneities in the oil reservoirs. This is illustrated in Figure 1. An oil well taps only a small fraction of the reservoir due to impermeable layers, which effectively transform a large oil pool into a number of noninteracting pockets of oil. This fact is known in the oil industry, but little action has been taken so far because few tools are available to define the precise location of the untapped mobile oil. Simple infield drilling with increasingly smaller spacing between the oil wells will eventually drain all the small oil pockets in an oil field. This is a common route taken today. However, it is a very expensive way to drain an oil reservoir because of the limited amount of information available to determine the location of the new wells. This also results in many unnecessary and misplaced wells. Given the cost for an oil well of $0.3M to $10M, better reservoir definition currently receives high priority in the oil industry.

RESERVOIR CHARACTERIZATION II

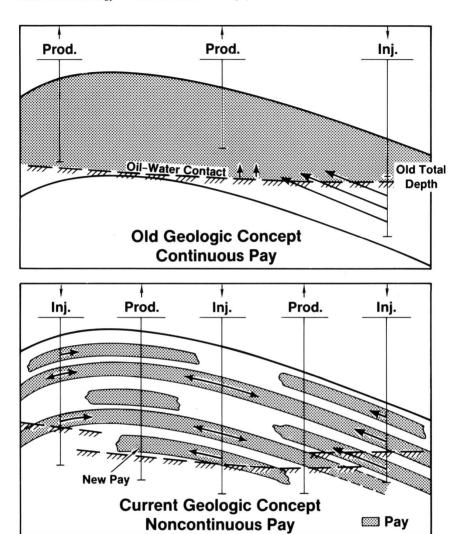

Figure 1. Old and Current Concepts of the Continuity of Oil Reservoirs.

II. INTRODUCTION

Seismic techniques in the oil industry are commonly deployed on two different scales: one, a 3-D, seismic scale that covers square miles; and a 1-D, well logging technique that covers a few inches around the well bore. The 3-D seismic survey covers more terrain but commonly lacks resolution to be effective on the reservoir scale. The seismic- or acoustic-logging technique has tremendous resolution but covers a limited amount of terrain, so that anomalies away from the well bore are not sampled.

Surface reflection and refraction techniques use the surface of the earth for both the seismic source and the receivers. The drawback with this configuration is that the sensors are far from the targets and seismic energy has to penetrate the slow and highly attenuating near-surface layer, both going down and coming up. Furthermore, the available energy in surface seismic sources is mainly converted into undesirable surface waves (Miller and Pursey, 1955).

The surface seismic techniques for monitoring and characterizing oil or gas reservoirs are shown in Figure 2. The common frequency range for surface-recorded reflected events, 10 to 50 Hz, makes imaging of thin beds and other thin features difficult or impossible. Surface seismic techniques also have to contend with a highly attenuating weathered layer, which decreases the signal/noise ratio. Surface noise compounds this signal/noise ratio problem, especially when surface seismic data are collected in oil fields.

The seismic well logging technique samples the geology surrounding the well. In the near-well zone one can find both a borehole-generated anomalous stress field as well as mechanical property and porefluid changes, generated by the process of drilling. In most cases the near-well zone is a poor representation of stress conditions, geology and saturation conditions of oil reservoirs.

Cross-hole seismology, shown in Figure 3, is emerging as a promising technique to evaluate and delineate oil reservoirs. This technique has several advantages because downhole seismic sources and multilevel receiver strings are used for reservoir characterization. One of the advantages is the potential for using order-of-magnitude higher seismic frequencies than surface techniques due to lower attenuation in the sub-weathered layer formation. This, together with the relative closeness of the cross-well transducers to the target, indicates that the potential of an order-of-magnitude improvement or better in the resolution of the reflected events in well-to-well data as compared with conventional surface seismic data. In cross-well seismology it is also possible to use trans-mitted seismic arrivals, which make it possible to perform P- and S-wave cross-well seismic transmission tomography,

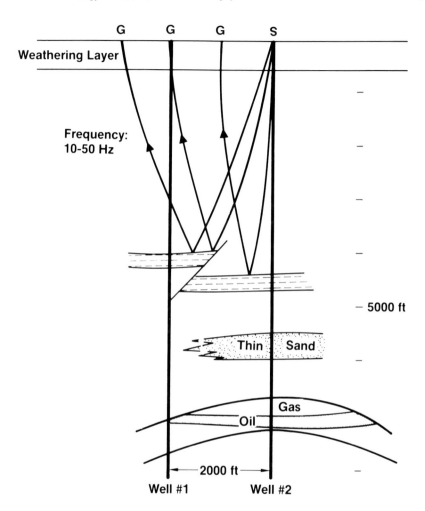

Figure 2. Reservoir Characterization Using Surface Seismic Techniques.

which provides data for an accurate reconstruction of the velocity field.

The 2-D cross-well seismic survey fills the void between the 3-D surface seismic and the 1-D logging techniques both in terms of spatial coverage and the seismic frequency band width. The cross-well seismic technique can thus be seen as a complementary tool to existing techniques for the exploration and development geophysicist. The cross-well configuration allows the

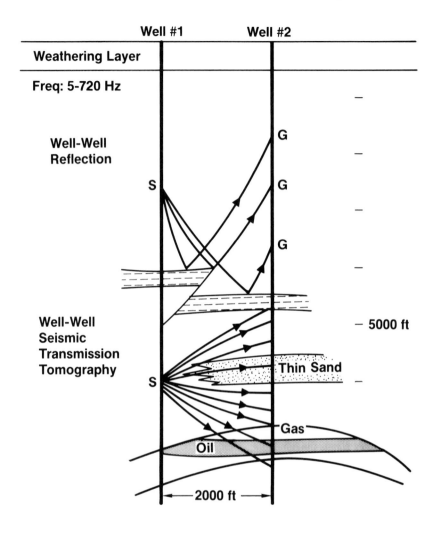

Figure 3. Reservoir Characterization Using Cross-Well
Seismic Techniques.

use of transmitted together with reflected waves. This
allows the construction of an image using a combination of
tomographic velocity analysis using the transmitted seismic
energy and migration of cross-well reflected events.
Figure 3 shows how both these techniques can be used when
seismic transducers are placed in wells. The downhole

source can also be used in a reverse VSP mode. This is an attractive alternative when only one well is available. If a reverse VSP survey is done in a noisy environment, it may be possible to place the geophones below the weathered layer. This allows the use of much higher seismic frequencies in the reverse VSP mode than for a conventional VSP using a surface source. One of the most significant advantages gained by a reverse VSP operation is that it is possible to perform multi-offset VSP's with only one run in the well with the downhole source. These data allow imaging between the well and the surface in as many cross sections as is desired without additional expense except for positioning the surface geophones. Naturally, the cross-well survey can be combined with a reverse VSP to generate both a detailed 2-D image from the cross-well data and a 3-D image from the VSP data. If the cross-well reflected energy is used, it is possible to extend the imaged zone below the wells and still maintain order-of-magnitude higher frequencies than for surface seismic techniques.

In some cases, wells are drilled with little a priori information about the geology of the drill site because severe surface noise or weathered layer problems have prevented adequate surface seismic surveys. If nearby wells are available with a spacing less than 5,000 feet, it is possible in some cases to replace the drilling of an investigation well with a cross-well seismic survey. Most oil wells in the world, 80% reported in 1986, are drilled as development wells in existing fields. In these cases nearby wells exist and the cross-well seismic survey is an option to investigative drilling. In other cases, such as areas with good surface seismic information, a cross-well seismic survey will give more detailed information of the cross-well geology for evaluating in-fill drilling locations for optimizing field development.

One of the primary applications for cross-hole tomography is the pre-EOR site evaluation for bed and shale continuity and for spatial and temporal monitoring of the process of Enhanced Oil Recovery (EOR). This can be done before the steam or gas has reached the production or observation wells and thus make it possible to take corrective steps early in the EOR process. Another important application for cross-well seismology is the evaluation of pilot EOR projects in new areas.

III. MODELING OF CROSS-WELL TRAVEL TIMES

Computer tomography modeling experiments have been performed using complex 2-D velocity sections constructed from real cross-well seismic data and well logs from the Kern River oil field. The simulated, cross-well travel-time data were obtained by raytracing through sections with a well spacing of 200 and 400 feet and a well depth of 1,000 feet. The raytraced travel times were checked using elastic, finite difference modeling through the same section. Travel times for the finite difference and the raytracing modeling for the same source-receiver pair were generally found to be within the sample rate of 1/2 millisecond.

Figure 4 shows the flow of processing cross-well travel time data from both field and modeling experiments. A raytracing algorithm described by Cerveny (1985) was used to obtain both the geometric raypaths through the cross-well velocity fields as well as the total travel time along these paths. The velocity section was reconstructed using an Algebraic Reconstruction Tomography (ART) algorithm as described by Lytle and Dines (1980) and Peterson et al. (1985). The velocity imaging process is iterative with two loops: the inner one is the ART and the outer, the raytracing loop. The ART algorithm used in this paper is using the difference between the observed travel times, real or model, and the travel times through the current model along specific ray paths. This algorithm converges to the minimum-norm, least-squares solution (Ivansson, 1983). The start model might be a constant-velocity field, and the initial ray paths in that case would be straight lines. The outer loop is an iterative raytracing loop, which traces rays through an improved estimate of the velocity field after each ART reconstruction. In each step the estimate of the ray path improves, so velocity corrections which minimize the difference between observed and computed travel times are distributed along better estimates of the raypaths. An important feature in the processing is the smoothing of the reconstructed velocity field prior to any raytracing.

The result of modeling a section between wells separated by 200 feet is shown in Figure 5. This figure shows the input model (Earth), the starting model (Log Model) obtained from two-dimensional extrapolation of a velocity log, and the reconstructed section (Result). To the right of the Result section are velocity profiles taken at three locations from the Earth section, shown by heavy black lines, and velocity profiles from the reconstructed

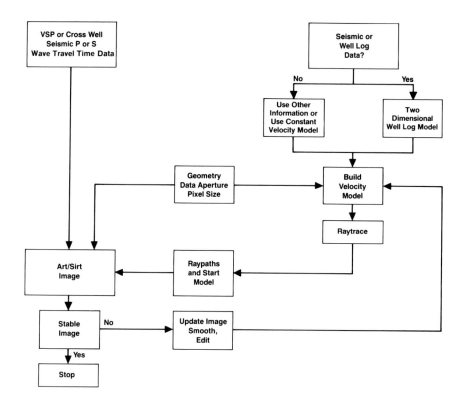

Figure 4.   Flow Diagram for Processing of Tomographic
VSP or Cross-Well Seismic Data.

section, shown by thin lines. The three velocity profiles
were taken from pixel columns next to each well and in the
middle of the section, respectively. To the right of the
Vp Log are five common source point gathers for raytraced
travel times through the Earth and the Result. The small
difference between the two travel-time data sets gives an
indication that, with limited a priori information about
the structure between boreholes, a very good approximation
of the true velocity distribution can be obtained using
travel times, ART tomography, and accurate raytracing.
    In the work of tomographic velocity reconstruction the
raytracing portion represents the vast bulk of the computing time. Without the raytracing the velocity image can
be reconstructed fairly easily on a small computer in the

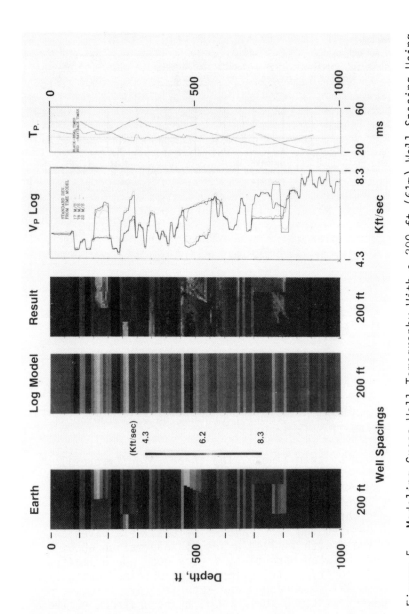

Figure 5. Modeling Cross-Well Tomography With a 200-ft (61m) Well Spacing Using Raytracing. Transducer Spacings and Pixel Size Dimensions are 5 feet (1.5m).

field. This would have advantages in terms of quality
control of the data and for making an informed decision on
the spacing of the transducers in the wells.

In the discussion of the above model, shown in Figure 5,
the vertical spacing between the transducer locations is 5
feet (1.5 m), resulting in 200 source points and 200
receiver points in the 1,000-ft wells. A data aperture of
±45° has been found empirically to be sufficient to
produce good images and is used for all presented tomographic results. A data aperture of ±90° has been found to
be very time-consuming because the raytracing is through
many more pixels. It was also found to introduce noise in
the reconstructed section due to long rays, which are not
necessarily the minimum time paths despite a successfully
traced ray. The ±45° aperture generated over 12,000 rays
for the 200-ft (61-m) model, which was over 98% of all
possible rays for this aperture. The commonly found
distance between wells in the Kern River oil field is
between 200 and 400 feet (61 and 122 m), so the size of
this model is realistic.

It is clear from comparing the two sections Earth and
Result in Figure 5 that Earth was successfully reconstructed using realistic model data. Even the thin truncated
bed at 580 feet was successfully found. This figure shows
the result of four consecutive smoothings, re-raytracing
and reconstructions.

In a study of the influence of noise it was found
random timing errors as high as the maximum travel time in
one pixel, approximately 1 millisecond, could be added to
the travel times without serious degradation of the images.
Each pixel is intersected by many ray segments, so any
random errors tend to be cancelled. A large, systematic
shift in the picked travel times, due to incorrect identification of the arrivals, was, however, found to be more
serious.

IV. TOMOGRAPHIC IMAGING OF AN EOR
    APPLICATION FOR HEAVY OIL

A field experiment was performed in January 1985
between pairs of three wells, which penetrated a
steam-flooded sequence of oil sands in the Kern River Oil
field in California. A plan of the field site is shown in
Figure 6. The reason for conducting the experiment at this
site was the expressed need for a detailed image of the
steam- and water-flooded sequence of sands and the

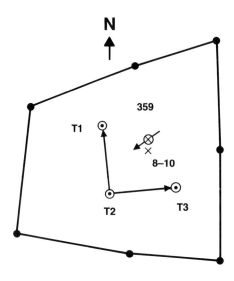

N

⊘ = Steam Injection Well

✕ = Water Injection Well

• = Oil Production Well

⊙ = Seismic and Temperature Monitoring Well

T2: Seismic Source Hole

T1, T3: Seismic Receiver Holes

Figure 6.   Plan view of Seismic Cross-Well Tomography Experiment in Kern River.

anticipated large velocity changes from both the heating of the formation with heavy oil and the increase of gas saturation. Both these effects have been investigated by Tosaya et al. (1984) and Dunn (1986). Their results show that the P-wave velocity decreases sharply both with increasing the temperature and increasing the gas satura- tion in the core sample. By measuring the temperature in the wells, one can obtain an estimate of the gas saturation from the cross-well velocity images. The interest in imaging a thermal EOR situation is derived from the fact that, if the movement of steam can be predicted or moni- tored, methods exist to prevent steam breakthroughs and to guide the steam to unheated parts of the reservoir.

The depth of the wells where the experiment was per- formed is 1,000 feet (305 m) and they were separated by 100 feet (30.5 m). One well was used as a seismic source well (Well T2) and two wells were used as receiver wells (T1 and T3). The source used was a 40 cubic-inch downhole airgun from Bolt Technology, Inc., and the receivers were SSC clamped 3-component K-tools. In an early test during the experiment it was found that single pops with the airgun were sufficient to obtain good quality first arrivals.

The field data were of good quality in the beginning of the experiment but, as the data acquisition proceeded, the signal/noise ratio decreased substantially as a result of a combination of aeration of the well fluid (the well released air for several hours after the airgun operation stopped) and the airgun-induced damaged cement-casing bond in the source well. In a repeat of one receiver position at the end of the experiment, the amplitude of the arrived, horizontally traveling P-wave decreased from 0.2 units to 0.05 units, a decrease in the amplitude of 75%.

The source well was drilled between the two receiver wells so a section 1,000 feet (305 m) deep and 2x100 feet (2x30.5m) wide could be imaged, as shown in Figure 7. In this figure the field Velocity Log is shown for Well II (same as T2) in a heavy black line, together with three logs through one of the reconstructed sections between Wells T2 and T3. The 100-ft (30.5-m) wide images are divided into 10 pixel columns. The three logs are through pixel columns 2, 5, and 9, respectively. To the right of the velocity logs are shown four sets of raypaths for common source points between Wells I (T1) and II (T2). As can be seen in the figure, significant ray bending occurred. These are the raypaths along which slowness is distributed as discussed in Peterson et al. (1985). Finally, to the right of the raypaths the travel times are shown for the same four common source points. Both the travel times picked from the field data and the travel times from raytracing the section between Wells I (T1) and II (T2) are shown. The difference between the two travel-time sets is small, which indicates that a velocity reconstruction fitting the field data was achieved.

When travel-time data are used along straight raypaths, significant lateral and vertical smearing of the velocity field occurs because slowness differences between the model and the image are not distributed into the correct pixels. In Figure 8 the cross-well data from the previous figure were used to create a velocity image using only straight rays and no start model. The result shows that the boundary between the high- and the low-velocity zones is not so sharp as when traced rays were used. However, there is much useful information in this image, which could be obtained on a small field computer, for evaluating the survey in the field or for monitoring a rapidly progressing steam or gas zone.

The interpretation of the cross-well velocity image in terms of the status of different sands is shown in Figure 9.

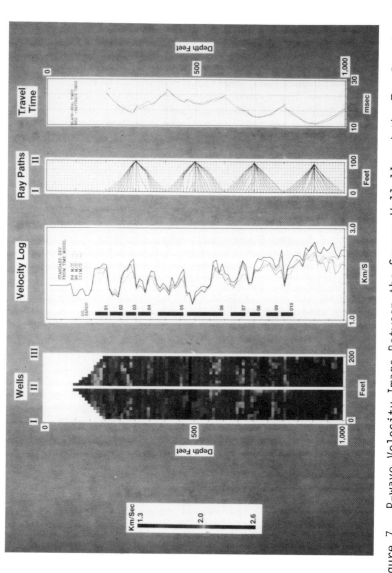

Figure 7. P-wave Velocity Image Between the Source Well II and the Two Receiver Wells I and III Using ART and Several Iterations of Raytracing. Transducer Spacings and Pixel Size Dimensions are 10 feet. A Model Derived From a Well Log Was Used as Start Model.

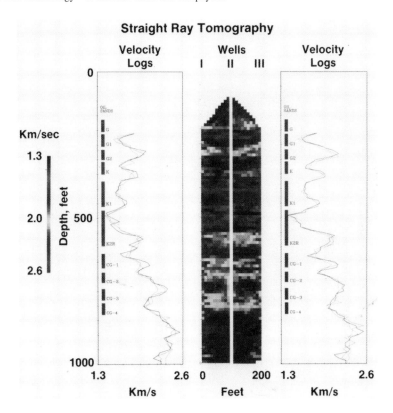

Straight Ray Tomography

Figure 8. P-wave Velocity Image Between the Source
Well II and the Two Receiver Wells I and III Using ART and
Straight Ray Paths. No Start Model is Used.

In this figure the velocity section is interpreted in terms
of fluid and gas saturations. The oil sands, as derived
from the three well logs, are shown as short columns beside
the velocity log and are numbered 1 to 10. The sands in
the bottom of the section, #8 and #9, show up as
high-velocity features due to high water saturation from a
prolonged period of waterflooding. Other sands, #7 and
bottom of #6, show a lateral change of velocity, which may
indicate a lateral change in the fluid saturation. The
reconstructed velocity section shows how sand #6 at a depth
between 480 and 600 feet (146 and 183 m) is gas saturated
in the top (low velocity) and oil saturated near the bottom
(high velocity). Other sands, #4 and #5, are imaged as
low-velocity zones. The low temperature logged in these

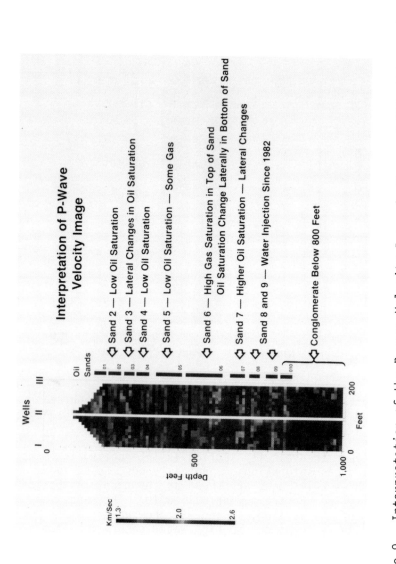

Figure 9. Interpretation of the P-wave Velocity Image in Terms of Oil, Water and Gas Saturations.

two sands, 80°F (27°C) versus 250°F (121°C) in sand #7,
indicates that the low velocity is due to high gas saturation rather than low viscosity from high temperature.

We also recorded an early steam breakthrough, using
monthly temperature logs in Well 359 T1 at the depth of the
recorded low-velocity zone. This confirms that little oil
was left in the sand and that steam permeated quickly
through the sand. Steam broke through in 359 T2 well
before it broke through in 359 T3, where the tomographic
velocity image indicated that the gas zone was smaller
compared with the other two wells. The thickest,
low-velocity zone in the cross-well image in the K2R sand
was found in Well T1, the next largest in Well T2, and the
smallest in Well T3. This is consistent with well log
data, which indicate a thicker gas-saturated zone in Well
T1 compared to Wells T2 and T3.

Sands #6 and #7 were part of a Vertically Expanding
Steam (VES) flood, which started after the conclusion of
the tomographic experiment. The velocity image indicates
that the fluid saturation in the two sands was low.
Production results of the VES were disappointing, indicating that little oil was left in the two sands. This is
also consistent with the velocity image.

V. CONCLUSIONS

Successful tomographic images using realistic model
data provide strong indications that the reconstructed
sections using field data are, indeed, realistic representations of the velocity distributions between pairs of the
three wells.

The most important factor for obtaining a stable
velocity reconstruction is good data. Random travel time
noise of the order of 1 millisecond will not significantly
deteriorate the tomographic image. It is, however,
important that large systematic errors in the picked travel
times be avoided.

The cross-well velocity sections obtained from field
cross-well seismic data obtained in an oil field show that
stable images can be generated with moderately good quality
data. I have been able to correlate the resulting velocity
images with both various well logs obtained before the well
was cased and with monthly temperature logs used to monitor
an advancing steam front. The images are also consistent
with production results following the cross- well
experiment.

These results show that cross-well seismic techniques and seismic tomography in particular potentially are powerful diagnostic and monitoring techniques for thermal EOR situations. These techniques, when fully developed and commercialized, will, because of their spatial resolution, also have a large impact on other aspects of future oil reservoir management.

ACKNOWLEDGEMENT

I would like to thank Chevron Oil Field Research Company for permission to publish this paper. I also would like to acknowledge John MacNider for assistance with the modeling and the data processing presented in this paper.

REFERENCES

Cerveny, V., 1985, "The Application of Ray Tracing to the Numerical Modeling of Seismic Wave Fields in Complex Structures", Seismic Shear Waves, Part A: Theory, pp. 1-124, Geophysical Press, London.

DOE, 1986, "Reserve Growth and Future U.S. Oil Supplies", (Contract DE-AC01-85FE-6063, report prepared for Department of Energy Washington, D.C.).

Dunn, K.J., Personal communication.

Ivansson, S., 1983, "Remark on an Earlier Proposed Iterative Tomographic Algorithm", Geophys. J. Astronom. Soc. 75, p. 855.

Lytle, R.J. & Dines, K.A., 1980, "Interactive Ray Tracing Between Boreholes for Underground Image Reconstruction", IEEE Trans. Geosci. Rem. Sens., GE-18, 234-240.

Miller, G.F., and Pursey, H., 1955, "On the Partition of Energy Between Elastic Waves in a Semi-Infinite Solid", Royal Society of London, Proceedings, Ser. A, Vol. 233, pp. 55-69.

Peterson, J.E., Paulsson, B.N.P., and McEvilly, T.V., 1985, "Applications of Algebraic Reconstruction Techniques to Crosshole Seismic Data", Geophysics, Vol. 50.

Tosaya, C.A., Nur, A.M., and Giovanni, D.P., 1984, "Monitoring of Thermal EOR Fronts by Seismic Methods", SPE 12744, SPE Regional Meeting, Long Beach, California, April 1984.

Session 4
Workshop/Discussions

SECOND INTERNATIONAL RESERVOIR CHARACTERIZATION CONFERENCE
Workshop 6/27/89 Dallas, Texas

Larry W. Lake
Herbert B. Carroll
Thomas Wesson

One of the enduring (and it is to be hoped, endearing) features of the Reservoir Characterization Conference is the opportunity for a much enlarged group participation through the means of workshop sessions. There were five such sessions, each consisting of 40-50 conference participants. Each session was given a study question and about 1-1/2 hours to prepare an informal response to the question. This period was followed by a 1-1/2 hour general session where each group presented and defended their answer to the entire conference.

The following represents a distillation of the results of each group. In each case, the study question is in italics with the committee response following in normal type. We are indebted to the discussion group leaders, each listed at the front of their respective section, for presiding over the sessions and providing the information to create this report.

DISCUSSION GROUP #1

Group leaders: Bob Lemmon, U.S. Department of Energy
                  Aaron Cheng, NIPER

We have said "we've got to get engineers and geologists working together" for so long that it probably isn't true anymore. The statement generally overlooks geophysicists and certainly implies that we aren't working together now.

The challenge of this discussion is to investigate the whole question of collaborative work between disciplines, especially geologists, geophysicists and engineers. Suggest a workable organizational structure where collaboration will take place in a constructive manner. Cite examples of teamwork where each discipline has aided the other in understanding reservoir characterization and simulation.

The consensus of the group is that there already exists a fairly good cooperation and data exchange between geologists, geophysicists and engineers in the area of hydrocarbon exploitation. Communication among these various disciplines is a dynamic process that can be fostered through the team organizational approach. Comments as to how to further the cooperative effort

RESERVOIR CHARACTERIZATION II
Copyright © 1991 by Academic Press, Inc.
All rights of reproduction in any form reserved.

can be grouped into three broad categories; 1) focusing on the objective, 2) team organization and management, and 3) database development and management.

1. The mission of the team should be based on producing an economically successful effort. The objectives should be focused and defined in relation to the realities of dollar/time constraints and personnel allotted to the effort. Common objectives should be defined. The kinds and amounts of data that will be needed to arrive at the results should be identified. The project should be organized into a series of milestones and expected results. There must be built-in flexibility such that the effort is sensitive and adaptable to changing economic and management objectives.

2. The team should be located in close proximity (i.e., clustered in a series of offices that share a common workroom), but should be physically separated from individual team member's management. The project leader should be a working member of the team. The team members should be dedicated to the project for an extended period of time (6 months to years). A core group should be maintained at all times for continuity; however, individuals may be shifted into and out of the team as their speciality is needed. Allowance for creativity (i.e., brainstorming/innovative approaches) should be encouraged if this will contribute to project cost efficiency and technical competitiveness. The team should organize their own efforts so that there is a timely flow of data to maximize group efficiency. Team members should be picked such that individuals are amenable to teach/learn more about each other's discipline and data format requirements. Frequent, informal meetings of team members should occur to keep all members informed of any changes in project emphasis or direction resulting from management decisions. Regular meetings with management should be scheduled and these should include most team members when discussing progress or making recommendations.

3. Database development and management should be computer based and developed for the long-term. Both historical and currently developed information should be incorporated into the database. The data format should be broad in that it incorporates data from all disciplines and at all scales of reservoir heterogeneity. The format should be flexible enough to allow a wide variety of end user products that are a compilation of data from several disciplines/scales. Analog field data for comparison of simulations to actual successful efforts based on geological depositional environmental models should be available.

The group cited three examples illustrating where the team concept has worked:

A. The Belridge Field, CA. Seven cores through the reservoir over a 40 acre area were available for detailed core and environmental description by the geologist. The resulting data allowed a close history match on the second run of a detailed reservoir simulation.

B. A North Sea Field. A 3-D reservoir simulation was successful through a team effort with geologic input. The production team was able to obtain a good history match after only a couple of reservoir simulation runs.

C. Thistle Field, North Sea. A reservoir engineer could not establish a history match without a flow barrier in the field. Working with the geologist a previously unknown fault was discovered.

DISCUSSION GROUP #2

Group leader: Gary M. Hoover, Phillips Petroleum Company
              Mike Stephens, M-I Drilling Fluids

Many exotic oil recovery projects are killed by the perception of being too risky. Yet, many people think "risk" is a board game and confuse subjective judgements for true analysis. Your challenge here is to put the question of risk in perspective.

Give a brief definition of risk (25 words or less) as it applies to oil recovery processes. Experience in exploration may be helpful here, but don't just parrot standard lines about "drilling being a risky venture". You can, of course, explain why some risk is acceptable in wild-cat drilling but not in enhanced oil recovery. In other words, give some idea of how much risk is acceptable and how this matches up with the prevailing economic climate. After this, develop a reasonable procedure for how the techniques being discussed in this conference could be used to quantitatively estimate risk. It is perfectly permissible to conclude that risk can't be estimated.

The group decided that risk could best be defined as "the difference between the value of an acceptable outcome and the value of the possible range of outcomes at any given time." The definition itself gives clues as to the elements involved in estimating risk: there must be an estimate of the acceptable outcome followed by an estimate of the range of possible outcomes. Unfortunately, neither task is especially easy.

An estimate of the acceptable outcome entails a reasonable understanding of the economic climate under which the project is being undertaken. This factor is highly process specific inasmuch as the requirements of acceptable rate of return (short-term benefit) must be balanced against the needs of reserve additions (long-term benefit). Unfortunately, the acceptable outcome estimate is itself uncertain owing to variances in economics, politics and costs.

Estimating the range of possible outcomes is simpler, at least conceptually, and it bears more directly on the themes of this conference. The group thought it important to begin this process by understanding the reservoir and process in question. There are two main parts to this: identify the individual parts of the system, then understand how these parts relate to each other. Since reservoir systems are usually very complex, this understanding must usually come from a variety of sources, hence the need for often-quoted interchange of information and insights among disciplines. Specific technologies involved at this step are production seismic techniques (both three-dimensional and interwell), production data (interpreted perhaps by large simulation models), and a concerted program of technology transfer. Specific factors for estimating the range of uncertainty in enhanced oil recovery then would be original oil in place, recovery efficiency, rate of recovery and the reservoir description. The culmination of this understanding would be a workable model of the reservoir and process.

Estimating the range of uncertainty involves using the model developed above to translate the effects of individual uncertainties into the cumulative uncertainty of the model output. How to do this efficiently is the subject of active research, as evidenced by the number of presentations dealing with mathematical modeling and stochastic assignments at this conference. The cost of many of these techniques is large; however, this cost, like those associated with gathering and analyzing data, must be factored into the notion of a acceptable outcome as discussed above.

As a final point, the group noted that the idea of risk manifests itself in numerous ways in current practice. The focus of this discussion has been on economic risk, but there is now a risk, incompletely manifest in economics, that is associated with environmental factors and a risk associated with spending too much money on data gathering and analyses.

DISCUSSION GROUP #3

Group leader: John Heller, Petroleum Recovery Research Institute

Much of this conference deals with statistics and how they can be applied to reservoir characterization. Yet the use of statistics, by its very nature, implies a lack of precise data; hence, the necessity to speak in terms of likelihoods, expectations, etc.

In reservoir characterization, statistical application must overcome the twin hurdles of sparse data sets and immense natural complexity. The first is rapidly apparent to even a beginning practitioner: a field with as many as twenty wells containing good information is relatively rare. The second hurdle is of such a magnitude that most engineers won't even take a second trip to an outcrop site.

List no more than 10 geostatistical techniques (some suggest the correct phrase should be geoSTATISTICAL) with at least one "pro" and one "con" for each of their applications to reservoir simulation. Suggest ways, both present and future, whereby statistics may become a more useful tool in reservoir characterization. By this, I mean ways to reduce bias and increase precision and to integrate more geological and geophysical insights.

We began this session by asking "what makes a technique distinctively a geostatistical technique". Suggestions ranged from "computing average properties" and "estimating values to use in a simulation", to more detailed descriptions of conditioned simulations. After some more discussion, it was agreed that a reasonably precise description would be "statistical procedures that describe or infer spatial variability".

There were supporters of the idea that "geostatistical techniques", as used in reservoir engineering, should include spatial averaging, indicator methods and the use of empirical correlations for spatial distributions. In fact, it was pointed out that most of the work done by reservoir engineers in "assigning grid block values" was done by the use of empirical correlations. Nevertheless, the consensus seemed to be that the primary geostatistical techniques should be listed as: variogram analysis, kriging, stochastic (conditioned) simulation, and co-kriging.

But there was general recognition that these items were not the complete answer to engineer or even numerical simulation needs. Although they represent a definite advance in recognizing the degree of spatial correlation in a variable field, geostatistical treatments frequently fail to capture a meaningful description of complex systems. A good example of this is a meandering river bed, and from which would be extracted a hypothetical, but not unreasonable, set of data. A geostatistical interpretation of this data could like quite different from reality.

Along these same lines an interpretation was given of Graham Fogg's paper from this morning's session, suggesting that the paper's message was that the range of values of oil recovery, among simulations of different realizations of the same statistics, was so large that doubt was cast on the utility of such geostatistical simulations!

Some time was also devoted to a discussion of whether the computations of "the right pseudos", for a given field and recovery method, could qualify as a geostatistical technique. It did, according to some, and did not, according to others.

One disadvantage of any geostatistical technique that relied upon a large number of realizations is that they would probably not be useable in regular reservoir engineering practice. There's not enough time, or computer facilities available to the working reservoir engineer for such work.

So, the future of developing geostatistical techniques is yet unclear for two basic reasons: results are not yet sufficiently good.and use of the techniques is too costly.

As a final thrust, another participant suggested that the emphasis in the word geostatistics should be more on the first root of the word (GEOstatistics) than on the second.

DISCUSSION GROUP #4

Group leaders: M.L. Fowler, Oxy USA, Inc.
Jim Ebanks, Arco Oil and Gas
Susan Longacre, Texaco

The hypothesis of "reservoir classes" suggests that reservoirs can be grouped on the basis of similarities in depositional environment, diagenesis, structural history, etc. If correct, this means that knowledge based on a few reservoirs in the group can be applied to the development of and production from other reservoirs in the same class. On the other hand, many scientists believe that each reservoir is unique and there are only gross similarities.

Give no more than five examples of how information about one reservoir can be applied to an unstudied but similar reservoir. Give no more than five examples of reservoir characteristics that are unique to each reservoir and therefore not portable.

The pattern of evolution of the group's discussion should provide useful insight into the importance and impact of the topic as well as give some indication of the still critical need for better communication between engineers and geologists.

At the outset of discussion examples were cited of case studies in which reservoirs similar in seemingly all geologic aspects had grossly different reservoir properties. Even different areas within the same reservoir having essentially the same geology were identified as having potentially different properties. Some wondered aloud or flatly stated that if such is the case and no information is really transferable, there is no point in dealing with geological information derived from analogue situations under any circumstances. A comment was also made that the resolution of this matter would answer the question about the value of outcrop work in reservoir characterization as well.

The discussion then settled on identifying what were the common factors influencing recovery performance among all types of reservoirs. The discussion used the partial list presented in the discussion charge as a jumping off point.

After some discussion, it was agreed in general terms that depositional environment was perhaps most influential. At first, examples were cited where similar depositional environments were of no use whatever in prediction of reservoir properties. Diagenesis and tectonic factors were identified as causes and listed as separate categories having influence. Influences on a scale larger than the depositional environment such as source mineralogy and regional tectonics were recognized. A suggestion was made that we enumerate carbonate and clastic depositional environments and discuss the influence of each on the potential analogy of reservoir properties, but this was not followed through, probably because of the formidable size of the task.

As a more general approach it was suggested that we specify characters that potentially might be carried from depositionally analogous reservoirs. We further decided on a list ordered in descending scale, with successively smaller scales having presumably smaller probability of carryover. At the largest scale the resulting list included external geometry of the reservoir, i.e., its shape or proportionate dimensions and its length, width, and thickness. Internal geometry of the reservoir, the next smaller scale, includes lithofacies geometry and distribution and also flow facies geometry and distribution. (Consideration of flow facies - which lead to the definition of flow units - was not accomplished in our initial discussion at this point, but was added later.) Within-facies patterns of variation in reservoir properties constitutes the next smaller scale. Influence of sedimentary structures, bed forms, and bedding units appears at this level. Finally, at the smallest scale, appear absolute values of reservoir properties as they are distributed in three-dimensions.

Diagenetic characteristics were added as a second major category of influences, particularly because their effects on reservoir properties are extremely important, but also because diagenetic characters often are very strongly linked to depositional environment. That is, in many cases an appreciable amount of diagenetic change is predictably related to depositional environment. Burial history (including fluid history) was pointed out as an additional very important factor also.

It was suggested that tectonic factors, as well as having profound influence on reservoir properties, can also be linked to some degree to depositional environment, but the linkage is less strong and less predictable than in the case of diagenesis. More intense fracturing in brittle media was an example cited as a manifestation of this linkage.

At this point the engineers in the group expressed an ardent desire to focus the discussion more on the items of their specific concern (i.e., reservoir properties). We proceeded to list those characteristics they considered to be of most importance to reservoir evaluation and simulation. The list included the following:

Volumetrics - size, shape, pore volume,
Continuity - on well spacing scale,
3-D distribution of porosity,
3-D distribution of permeability (absolute and relative),
3-D distribution of saturations (water, oil, and gas),
Fluid types and distribution, and
Pressure and temperature (function of depth)

It took only a few moments to arrive at the above list.

We then decided that the two lists we had made had to be combined in some way that would allow an evaluation of (1) which geologic characteristics of depositional environment, diagenesis, etc. bear on which of the important reservoir characteristics?, and (2) in which of the above cases can information be reliably transferred from an analogous reservoir A decision tree approach was suggested as a practical implementation, but another suggestion to try a matrix form for organization and presentation quickly supplanted it as being a more possible and straightforward task.

In preparation, our first task was to retrace our path a bit to flesh in some of the details on the diagenetic and tectonic characteristics that bear on reservoir properties. Specific diagenetic characteristics influencing reservoir properties were outlined as : compaction (or lack thereof, i.e., geopressure), cementation, dissolution, and recrystallization. In this process our previous "umbrella" category of burial history was dropped.

Realizing that a consideration of present-day fluid content would be important to retain in some way, we created a new header in our list of influential factors called fluid characteristics. The new header included the following specific categories: hydrocarbon fluids, brines (or perhaps better formation water because high salinity is not a requirement), and injected fluids

Tectonic characteristics were quickly summed up in the terms "folding, faulting, and fracturing." Reflection on this later indicates that these are primarily local structural characteristics. Larger scale considerations, specifically subsidence, was mentioned in our discussion but did not get on our list - time was beginning to run low. The list should probably include a regional tectonic category for both subsidence and uplift effects.

The above information gathered, we constructed a matrix (Fig. 1) using geological characteristics as rows and reservoir parameters as columns. At the intersection of rows and columns "O" was used to indicate that an effect on the reservoir property could be exerted by the particular geological characteristic and that the effect could, under proper circumstances, be predicted by observing or sampling an analogous reservoir. The possibility of assigning numeric or ordinal rank values to indicate relative probability instead of using "O'" marks was considered but rejected to avoid excessive subjective discussion in filling out the chart. To indicate characteristics not likely to carry from analogous reservoirs an "X" was used. (There may be special circumstances where even absolute values of reservoir parameters can be derived from analogous reservoirs!). Blank intersections indicate the geologic characteristic being considered has no generally recognized effect on the reservoir property being addressed.

The group, I think, was somewhat surprised to see the number of "O" marks that appeared on the completed chart. In answer to our charge, the group showed that if circumstances warrant, a large amount of information at various scales can be gleaned from analogous reservoirs, but no information at any scale can categorically be said to always carry or to never carry.

As a concluding side note, it was pointed out that rocks having exactly the same composition and pore structure will exhibit exactly the same fluid performance characteristics (if the fluid used are the same). A catalogue of such microscopic rock characteristics could be used to predict performance quantitatively thus eliminating the need for analogues to draw quantitative data from comparisons.

FIGURE 1 - Matrix of Geological and Reservoir Properties

	VOLUMETRICS	CONTINUITY	POROSITY	ABS PERM	REL PERM	SATURATIONS	FLUID TYPES	PRESSURE TEMP
DEPOSITIONAL CHARACTERISTICS								
External Geometry		O	O					
Internal Geometry			O	O	O	O	O	
Pattern of Variation (Quantitative Res. Properties)			O	O	O	O	O	
Absolute Values (Quantitative Res. Properties)	X	X	X	X	X	X	X	X
DIAGENETIC CHARACTERISTICS								
Compaction (including geopressure)					O	O		O
Cementation				O	O	O	O	O
Dissolution				O	O	O	O	O
Recrystallization				O	O	O	O	O
TECTONIC CHARACTERISTICS								
Folding		O	O	O				
Faulting		O	O					
Fracturing		O	O			O		
FLUID CHARACTERISTICS								
Hydrocarbons								O
Formation Water					O	O		
Injected Fluids					O	O		

O = Property portable   X = Property not portable

DISCUSSION GROUP #5

Group leaders: Edith Allison, U.S. Department of Energy
                  H. Duane Babcock, ResTech, Inc.

This symposium is organized around four scales of reservoir heterogeneity: microscopic, mesoscopic, macroscopic, and megascopic. Each of us considers some of these scales to be more important than others to increased oil production, based on our own experience and the types of problems we are involved in.

Do you really think there are separate, distinguishable scales in reservoirs, or do we make these classifications solely on the basis of convenience (or to organize conferences)? Adopting the point of view that distinct scales do exist, for each of the four scales, give at least two examples in which heterogeneity was the sole or the dominant limitation to efficient reservoir development and production. Give examples of reservoirs in which all four scales of heterogeneity limited production.

The participants had diverse opinions about defining the scales that should be used for description. This diversity narrowed as the subject was examined during the workshop period. There was good agreement that scale is a continuum, from molecular size to the largest scale that could have an effect on reservoir characterization. In some cases, the largest scale might be related to basin size. Even plate tectonics might have application in characterizing a particular reservoir.

During the early discussion period part of the group wanted to classify scales based on the measurement tools. There is a natural tendency to think in terms of data acquisition systems. Later, a consensus seemed to form that numerous "tools" could extend across several scales. For example, core analysis has a several order-of-magnitude scale breadth, detailed imaging techniques are being developed from well logging methods and seismic data is being scaled down in attempting to define fluid flow. This blurs the scale definitions based on measurement tools. Of larger importance is the ability to quantitatively link measurements at one scale to another.

It was agreed that there was a problem in defining any scale boundaries because of the continuum factor, yet artificial scales are primarily necessary for communication purposes. The ability to communicate over a range of many many factors of ten strongly suggests that even artificial or arbitrary scales are better than not using scale descriptions, regardless of what they are called. It was proposed that:

A. We do need some classification primarily for the exchange of communication among the disciplines,

B. The selected scale boundaries need not be exact but sensible to the users,

C. The selected scale boundaries may be process oriented, and

D. Scales relate very often in our ability to measure them.

A group consensus formed that the different disciplines should try to adopt an approximate definition to aid communication during the work on a particular problem. Although the reservoir engineer may never find an application where basin size scale is to be applied, he or she should realize that a scale larger than the limits of an individual reservoir is of definite use to another discipline.

Having discerned that the scale range is large, it was agreed that the scales may change in size depending upon the group of individuals or disciplines working across the scales on a particular problem. For example, it could be agreed that a basin size scale is inappropriate and the largest scale size required would be of "field" size or within the individual reservoir boundaries appropriate to the conditions. The scale size definition might change with the type of reservoir problem under study.

Since we have observed scale definition changing over time, it was speculated that scale size definitions might possibly change sometime in the future. Not too many years ago, prior to such measurement technology as the scanning electron microscopes, the scale microscopic would probably have been defined between pore size and grain size up through core plug size. Today the tendency is to define the smaller end of the micro scale by the pore or grain size. Of course, pores or grain sizes can, in some instances, also be relatively large.

As far as names of scales, the mesoscopic scale was the more controversial. Some thought that it should be between macro and megascale rather than between micro and macroscale. Since "meso" is middle, a valid argument could support either position.

The group then attempted to develop a four-fold classification of heterogeneities. There was some division of opinion about what the underlying scientific basis of the hierarchy should be. There was strong support for the concept that a classification should reflect the reservoir/fluid flow processes that control oil production. There was also support for the concept that the subdivisions should reflect depositional processes that cause the heterogeneities.

There was more agreement reached on the smallest or microscale definition than any other scale. This could be because all disciplines understand that this scale affects their work and the fundamental relationships applicable to their technology. The geophysicist knows that sound waves travel variously through the grains and the fluids in the pores, but yet the end use resolution of geophysics is most often applicable to the larger scales. The descriptive terms used by the group for the smallest scale were:

Pore and/or grain size

Microscopic or less that the naked eye can discern.

Less than 1 mm or only a few millimeters down to the molecular level.

Microscale effects would be related to absorption and residual oil effects through the swept zone for enhanced oil recovery work.

Agreement was also less controversial on the largest scale which is generally referred to as megascopic, but there was more appreciation by the geologist for a larger definition than by the other disciplines. Where an interdisciplinary team is working together on a given project, the definition they should adopt should be the one that is sensible for the problem under study.

In the attempt to define the largest scale, it was generally conceded that the scale should be larger than the discrete reservoir size (boundaries of producing reservoir - vertically and horizontally) and generally of field size. It might occasionally extend to the size of the basin. It should extend vertically beyond the discrete reservoir boundary in some instances since effects from formation stress or over-pressure could readily affect reservoir performance.

Determining the middle scales was done without using the terms of macro or meso. The scale smaller than megascopic scale (including reservoir scale) would be broadly defined as the between well/interwell size. This is the scale where measurement tools are virtually non-existent.

The second middle scale would be one scale larger than microscopic and can be broadly described as the borehole size and near borehole in the lateral dimension and bed related vertically. During the discussion period there was less emphasis throughout the discussion about vertical discrimination and more focus on the lateral dimension.

If the number of divisions had not been arbitrarily limited, additional intermediate subdivisions would have been proposed. The attendees seemed to have a reasonably clear understanding and agreement about the meanings of the classes; however, there was disagreement over the naming of the classes.

As the final phase of the discussion, the group attempted to list specific heterogeneities that exist in all scales of heterogeneity, and to list some heterogeneities that exist in only one scale. Although a list of single-scale heterogeneities was initially drafted, subsequent discussion showed that most of the examples spanned several scale classes. Multiple-scale heterogeneities were easily enumerated. This result is consistent with the initial concept of a continuum of sizes of heterogeneities.

Examples of heterogeneity effects where it was determined that is was the sole or the dominant limitation to efficient reservoir development and production were: tar mats within a reservoir, wettability effect, clay plugging of a pore throat, grain/pore size distribution, sedimentation effects, and bitumen in pores Examples of reservoirs in which all four scales of heterogeneity deterred development were: fracture (displacement), which can be at grain size up through complete fault block size., and permeability effects that can extend from pore size up through discontinuous sand lens size (reservoir).

An informal survey of the group showed that about 75% work in a multi-disciplinary group, dealing with heterogeneities that occur in more than one scale.

Session 5
Megascopic

The scale beginning at interwell spacing and extending up to field dimensions; the domain of geophysical analysis

FIELD-SCALE RESERVOIR CHARACTERIZATION

Timothy A. Cross

Department of Geology and Geological Engineering
Colorado School of Mines
Golden, Colorado

ABSTRACT

Field-scale characterization of reservoirs is the pivotal link between the exploration/discovery process and the development/reservoir management process. Geologic concepts derived from regional studies, and which may have led to discovery of the field, provide the initial understanding of the geologic attributes of the field area. These concepts—which may include trapping mechanism and morphology, fault and fold patterns, fracture densities and orientations, geometries and stratigraphic relationships of seal and reservoir units, and depositional systems and facies architecture of the reservoir—are initially transferred and applied to reservoir development and production. During field development, these concepts will be tested and modified as appropriate. However, the accuracy to which the geologic attributes of the field area were inferred prior to and during development will influence the economic and recovery success of the field.

Linkage with the recovery/reservoir management process arises because field-scale reservoir characterization establishes the spatial framework for describing and predicting smaller scale reservoir geometries and heterogeneities, along with their associated petrophysical and fluid-flow properties. Field-scale description establishes whether the reservoir is compartmentalized into more than one producing zone or unit. If only one compartment is recognized, then smaller scale attributes considered important in controlling fluid flow are studied, evaluated and/or modeled within the spatial framework provided by the field-scale description. If multiple compartments are recognized, the field-scale description may establish that the facies distributions and fluid-flow properties within each are similar, but that the compartments are partially or totally isolated from each other. In this case, the field-scale description indicates that at least two components of the reservoir—the lithohydraulically "homogeneous" compartments and the "heterogeneous" bounding lithologies that disrupt fluid communication through the reservoir unit—must be evaluated and/or simulated in terms of their contributions to fluid flow. Alternatively, the different compartments may have dissimilar facies distributions, heterogeneities and fluid-flow properties. Then the smaller scale attributes that control fluid flow must be evaluated and modeled separately for each compartment as well as between compartments.

To define the numbers and characteristics of compartments and compartment boundaries within reservoirs, and to assess their individual and cumulative contributions in controlling fluid flow, three essential elements must be evaluated in field-scale reservoir characterization. The first is describing or predicting the spatial arrangement of sedimentary facies within the entire reservoir unit. The second element is evaluating the contributions that lithologic heterogeneities of varying scales and characteristics make in dividing the reservoir into compartments. The third is devising a way to convert lithostratigraphic units into lithohydraulic units such that fluid-flow pathways in the reservoir are described.

To understand the arrangement of fluid-flow pathways, along with barriers, filters and retardants to fluid flow, requires an understanding of the geometry of facies and their contributions in controlling fluid flow. Therefore, the first element of field-scale reservoir characterization is the description or prediction of the spatial arrangement—position, geometry, interconnectedness and volume—of sedimentary facies within the entire reservoir unit. The thesis argued in this paper is that depicting the spatial arrangement of sedimentary facies is accomplished most successfully and accurately by placing facies distributions within a high-resolution time framework.

Empirical stratigraphic data of the past century, augmented by numerical stratigraphic models of more recent vintage, indicate that the spatial arrangements of facies change regularly as a function of changes in the space available for sediments to accumulate. One commonly cited example of such changes occurs in fluvial environments. Fluvial channelbelt sandstones deposited during conditions of low subsidence rate may form vertically and laterally interconnected, blanket-like reservoirs. By contrast, the same channelbelt facies of identical depositional systems may occur as isolated, stringer-like reservoirs if deposited during periods of higher subsidence rate.

These observations augur for considerable caution in the application of analog facies and depositional systems models to reservoir simulation. One approach used to characterize reservoir continuity and heterogeneity at bedding- through field-scales is the measurement, cataloging and statistical treatment of sedimentary facies attributes considered important in controlling fluid flow. For example, the dimensions, frequencies of occurrence, and/or interconnectedness ratios of particular facies within particular depositional systems are measured at one locality or at many localities that are considered geologically analogous. From these measurements, the population structures and statistical parameters of these facies attributes may be estimated, with the objective of providing "normal" values for variables used in reservoir simulators. As commonly employed today, this approach is a logical extension of decades of sedimentological studies in which sedimentological attributes have been synthesized into facies and depositional systems models.

Strict application of these models invites the assumption that attributes measured for one facies or facies assemblage in a particular depositional system are applicable to all occurrences of the same facies or facies assemblage in analogous depositional systems. Thus, in the preceding example, it would be regarded as sufficient to measure the geometry, size, frequencies of occurrence, and interconnectedness ratios of fluvial channelbelt sandstones at one locality, and apply those measurements and statistical summaries of those attributes to fluvial depositional systems in reservoirs elsewhere. Similarly, it would be considered sufficient to apply statistical summaries from a catalog of measurements of fluvial channelbelt sandstones collected from analogous depositional systems at multiple localities.

As previously mentioned, empirical observations and numerical models indicate that the arrangements of facies in identical depositional systems vary with changes in first-order controls of stratigraphic architecture: e.g., changes in rates of tectonic movement, base level, sea level and sediment supply. Assessment of these controls requires establishing a high-resolution time framework and placing the distributions of facies within that framework. A catalog of measured sedimentary facies attributes, collected only in the context of similar depositional systems, will amalgamate originally discrete and distinct populations into a heterogeneous, mixed assemblage of attributes that are similar only in name. In summary, if these measured attributes are considered representative of all occurrences of "X" facies within "Y" depositional system, and if they are measured independently of a high-resolution time frame, without assessment of the fundamental controls on the observed stratigraphic architecture, then their indiscriminate application in a reservoir simulator can introduce a false confidence about the accuracy of the geologic framework used in the simulation.

The second essential element that must be evaluated in field-scale reservoir characterization is the role of lithologic heterogeneities in controlling the pathways of fluid flow. Lithologic heterogeneities of different scales and petrophysical characteristics occur within and/or between field-scale reservoir compartments. We conventionally describe scales of reservoir heterogeneity in a somewhat anthropomorphic context: microscopic, mesoscopic, macroscopic and megascopic. For a particular reservoir, which of these is most important, which the least, and do they contribute interdependent effects on fluid flow?

Field-scale reservoir characterization provides the framework for assessing which lithologic heterogeneities of which scales and characteristics are important in controlling the pathways of fluid flow. Depending upon the length of time that flow is considered to operate, and irrespective of spatial scale, fluid type and pressure gradient, various lithologies or facies elements will act as open pipes, baffled pipes, retardants or barriers to fluid flow. Whether these different lithologic units of variable transmissivities act as conduits, retardants or barriers to flow within a particular reservoir will be determined primarily by their size and continuity relative to the scale of the reservoir and the spacing of wells. An *a priori* argument can be made that lithologic heterogeneities of scales similar to size of the reservoir and the well spacing will compartmentalize a reservoir unit. Lithologic heterogeneities of this scale, therefore, are first-order controls on fluid flow for a particular reservoir, and they will vary in absolute size and continuity as a function of well spacing and geologic attributes of the reservoir. Which types and sizes of geologic elements constitute first-order heterogeneities? Because the areal extents, volumes and well spacings of reservoirs are quite variable, the types and scales of geologic features that may be considered first-order also are quite variable. The definition of a first-order heterogeneity should be assigned to individual reservoirs as a function of the size and well spacing of the reservoir, rather than as an arbitrarily chosen dimension.

Because field-scale characterization can potentially describe the distribution of all facies elements within a high-resolution time framework, lithologic elements of any origin can be arranged on a scale relative to the first-order heterogeneities. However, the individual or cumulative contributions of these smaller scale elements in controlling fluid flow must be assessed independently. Whether it will be sufficient to describe reservoirs in terms of bedding, facies architecture, depositional environments or depositional systems will largely depend upon the scale of these attributes relative to the well spacing. This paper presents a conceptual method for relating scales and types of lithologic heterogeneities to scales of reservoirs and well spacing and to the time scales of fluid movement.

These two elements—definition of the spatial arrangement of facies and lithologic heterogeneities—would be sufficient for field-scale characterization of reservoirs if their roles in controlling fluid flow were also established. The third element, conversion of lithostratigraphic units into lithohydraulic units, remains a fundamental weak link in the chain. However, one approach that offers some hope for establishing these relations is to map actual fluid-flow pathways through oil-saturated strata. By observing directly how fluids moved through, around or were retarded by different types and scales of lithologic heterogeneities, the translation of facies units into lithohydraulic units might be accomplished. This paper suggests how sedimentary facies and lithologic heterogeneities might be calibrated to fluid flow using this method.

EXAMPLES OF RESERVOIR SIMULATION STUDIES UTILIZING GEOSTATISTICAL MODELS OF RESERVOIR HETEROGENEITY

Dalian V. Payne
Kelly A. Edwards
Alan S. Emanuel

Chevron Oil Field Research Company
La Habra, California

I. ABSTRACT

The use of geostatistics to represent reservoir properties has been shown to model heterogeneous reservoirs effectively . The technique is useful in predicting overall performance of large pattern floods as well as in modeling well-to-well behavior. Three example studies are presented to show how fractal cross sections are used to model performance. A study of a West Texas carbonate combined a finite-difference simulator with a streamtube model to represent field performance. The model was then used to predict waterflood and CO_2 flood performance. Cross-sectional models were also created for two pinnacle reefs. These models were used to study well responses and fluid saturations. The methodology of incorporating fractal statistics into reservoir simulation is discussed along with the results of the three examples.

II. INTRODUCTION

The use of statistical methods to represent reservoir properties is well known in reservoir engineering and

related fields. The Dykstra-Parsons method, which models
discrete layers based on a statistical permeability
variation, is perhaps the most universally prescribed
technique (Dykstra and Parsons, 1950). While the technique
does focus on properly modeling the average permeability,
it has shortcomings. Notably, the layering scheme produces
a purely hypothetical system that generally bears little
resemblance to spatial permeability distributions.

Other techniques use stochastic modeling to assign
uncorrelated reservoir properties with a predetermined
distribution (Warren and Price, 1961). As originally used,
this technique had the advantage of matching the
range and probability of a property value. However, unless
a provision is made for spatial correlation, the geologic
layering is not modeled.

More recently, geostatistical methods have been employed
to represent the spatial distribution of porosity and
permeability for simulation studies (Tang et al., 1989;
Mathews et al., 1988; Emanuel et al., 1987). The primary
aim is the realization of property variation between
sampling points, i.e., in the interwell region, the process
includes characterization of the variations and extension
of the trend to the interwell region. That is,
characterize the structure of the known data and use the
same structure to interpolate unknown values.

This technique, known as conditional simulation, has
advantages over previous techniques (Hewett and Behrens,
1988). Conditional simulation preserves data values that
are measured by well logs and conventional cores. Previous
techniques relied on hypothetical random realizations that
matched overall average features. The conditional approach
preserves reservoir heterogeneity features that may cause
fluid flow dispersion.

Reservoir phenomena such as channeling and gravity
segregation will undoubtedly show sensitivity to
heterogeneity and its spatial distribution. This study
shows how fine-grid cross sections of reservoir heterogeneity may be incorporated into reservoir simulation
studies. These cross-sectional models are used to
determine the effect of heterogeneity on reservoir fluid
flow. When required, the results are scaled up to the
field level using the hybrid simulation technique, which is
discussed below.

III. METHODOLOGY

The simulation method applied in the first case history is called hybrid simulation. The hybrid simulation methodology makes use of geostatistical analysis, detailed cross-sectional models, and streamtube models to estimate production from reservoirs. The aim of this methodology is to create a model which: (1) has adequate resolution to characterize the effects of heterogeneity; (2) enables the use of primary flow properties; and (3) provides efficiency so that cost does not become prohibitive. The basic steps in creating a hybrid model are (Emanuel et al., 1987):

1. Determine the fractal structure of the porosity and/or permeability data using a geostatistical procedure called the "rescaled range analysis".
2. Generate a finely gridded cross section for a typical injector-producer well pair using a fractal interpolation scheme to condition the data to fit the known well data. A finite-difference simulation model is run under waterflood and/or miscible flood conditions.
3. Develop a streamtube model so that areal conformance can be quantified. Displacement calculations based on the finite-difference simulation results are carried out on each streamtube and summed to estimate field-wide performance.

Each of these steps is detailed below.

A. Fractal Characterization

Fractals are geometric shapes that show variations at all length scales, yet are correlated at any length scale. They can be either exactly self-similar or statistically self-similar. Statistically self-similar fractals represent the heterogeneity found in geological formations better than exactly self-similar fractals (Hewett, 1986).
The fractal distribution of rock properties is quantified using the rescaled range (R/S) procedure. From this procedure, the intermittency exponent (H) is calculated (Mandelbrot and Wallis, 1969; Hewett, 1986). The value of H is a measure of the correlation of the rock property at large length scales . A value of H = 0.5 indicates a property that has no correlation and is, therefore, totally random. A value of H approaching 1.0 indicates that the property is highly correlated and that variations from the

mean value tend to persist. That is, higher-than-average
values of the property tend to be followed by higher-than-
average values, while lower-than-average values tend to be
followed by lower-than-average values.

B. Cross-Sectional Modeling

     Cross-sectional models of 2000 to 4000 cells are
typically used. Simulator cell sizes are generally larger
than the scale at which local properties can be defined,
and the validity of using primary flow properties (such as
absolute permeability or coreflood relative permeability)
rather than effective properties (such as pseudorelative
permeabilities) is a concern. This problem was examined by
Hewett and Behrens (1988). Based on their work, models
with 2000 to 4000 cells have enough resolution so that
primary flow variables can be used for simulator input.
     The interwell cell values of porosity are obtained with
a series of one-dimensional stochastic interpolations
(Emanuel et al., 1987). The stochastic interpolation is
effected by adding a random variance to a linear
interpolation. The initial variance (σ_0^2) is calculated
from the mean square variation of values from one well to
the other on a foot-to-foot basis. The initial variance
is scaled using a power law that is dependent upon H to
obtain the random variance . This process is repeated on
successively finer intervals to the desired resolution.
The final result is a fractal cross section that has a
variance in properties from cell to cell consistent at all
length scales with the initial variance and that is
conditioned to honor the well data.
     Cross sections of horizontal and vertical permeability
can be interpolated in the same manner using the logarithm
of permeability (if the data are available) or can be
calculated from the porosity data using transforms.
     These cross-sectional models are used in a finite
difference simulator to study waterflood and miscible flood
cases. A black oil simulator is used for waterfloods,
while a four-component miscible flood simulator is
generally used to model miscible floods (Todd and
Longstaff, 1972). Output, in the form of average
fractional flow and phase saturation versus pore volumes
injected (PVI), may be incorporated into a streamtube model
to estimate field-wide performance.

C. Streamtube Model and Displacement Calculations

The areal sweep efficiency of a flooding process is estimated using streamtube modeling (Martin and Wegner, 1978). Displacement calculations that combine the cross-sectional results with the streamtube model are carried out. The final result is a field-wide forecast of production and injection.

Fixed streamtubes, as opposed to variable or recalculated streamtubes, are used to reduce computing requirements. The error introduced by the use of fixed streamtubes is usually quite small, especially for the pattern floods with high mobility ratios where this methodology is usually applied. The work of Martin and Wegner showed that the largest error in recovery vs. PVI to be expected between fixed and variable streamtubes was, in most cases, much less than 10%.

The streamtube displacement calculations provide a field-wide forecast of production. Calculations are made along each streamtube using the fractional flow data from the cross-sectional model as a characteristic solution. At each timestep, the fractional flow for each streamtube is based on the pore volume injected into that streamtube. The volume injected into each streamtube depends upon the relative resistance factor and is a function of streamtube geometry, fluid saturations, and fluid mobilities. The production from all the streamtubes is then summed to give a field-wide forecast.

IV. CASE HISTORIES

A. Study of the McElroy Field

The McElroy Field is situated on the eastern edge of the Central Basin Platform of West Texas. Production is from the Grayburg Formation, a 200- to 300-foot thick dolomite-siltstone sequence. The reservoir is subdivided vertically by impervious, sulphate-cemented zones of varying lateral continuity. The complex history of sedimentation and diagenesis has resulted in a heterogeneous reservoir.

The field, with original oil-in-place (OOIP) of 2200 MMSTB, was originally discovered in 1926. A secondary

waterflood project has been in operation since 1960. A
modeling study was undertaken to evaluate a portion of the
field and assess the recovery potential of waterflood
realignment and CO_2 injection.

The 215-acre study area is situated primarily in
Section 194 of McElroy Field (Figure 1). Oil-in-place at
waterflood startup was 15.4 MMSTB. A pattern realignment
to convert from inverted nine-spots to a line-drive was
initiated in 1986 to increase water injection and improve
the pattern sweep efficiency. Initially the producing
water cut was reduced by 2%, but the total fluid production
rate remained at pre-realignment levels.

1. Cross-sectional simulation

   a. Continuous Layer Model. A simulation grid was
constructed from a fractal interpolation of porosity data
from Wells 726 and 1062 (line A-A', Fig. 1). Porosity log
values (Figure 2),

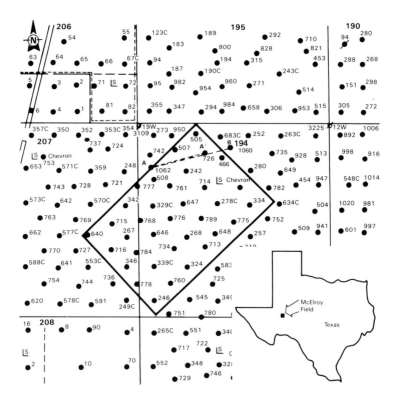

Figure 1. McElroy Section 194 study area.

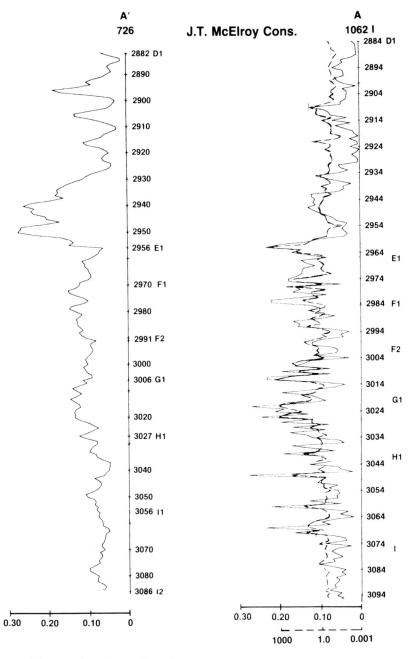

Figure 2. Porosity logs used to construct fractal cross section for McElroy Field. Dashed line shows core permeability values.

which range from 1% to 20%, are an indicator of the degree
of heterogeneity. The wells form an injector-producer
pair, which are completed in zones D1-I1. The porosity
values are highly structured, as indicated by the
intermittency exponent of 0.87. Permeability of the cross
section was calculated with a permeability transform
relationship derived from core data.

The model consisted of 86 layers with a total of 3698
cells. Each two-foot layer of the 700-foot cross section
was divided into 42 equal-width blocks resulting in
dimensions of 17 ft x 2 ft in the x- and z-directions. The
y-dimension was varied from 11 to 110 feet to approximate
the areal profile of a streamtube. (Emanuel et al., 1987).
Water throughput in the model was equivalent to a field
injection rate of 200 reservoir barrels per day (RB/D) per
well.

Finite-difference simulation runs were performed to
calculate waterflood and 1:1 WAG CO_2 flood recovery of the
cross section. Results are shown as fractional flow curves
in Figure 3. The oil curve declines exponentially after
breakthrough until CO_2 injection is initiated at a 98%
water cut. Multiple oil peaks form as the CO_2 sweeps the
cross section. The CO_2 (solvent) curve indicates that
breakthrough occurred after 0.2 PVI.

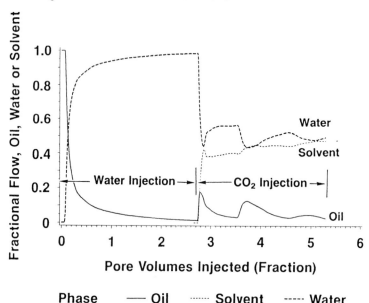

Figure 3.   McElroy Field cross-sectional simulation
results.

Oil saturation maps from solvent injection (Figures 4a and b) show the pronounced channeling of CO_2. The CO_2 channels form multiple oil banks which were observed as the peak oil responses in the fractional flow curves (Figure 3). At the conclusion of the run (Figure 4b), a significant oil saturation remains.

b. Noncontinuous Layer Model. To investigate the effect of horizontal continuity, a second heterogeneous cross section was constructed. Two wells (1060 and 1062) were chosen because the high-permeability channel between them is not continuous (Figure 5c). Although the wells are three well spacings apart and do not conform to the convention of using adjacent pairs, they were chosen to form a hypothetical example. The purpose of this example was to develop a sensitivity to the geologic model and to demonstrate the importance of detailed interwell channel and heterogeneity modeling.

Waterflood simulations of the noncontinuous example were compared to the most likely (continuous layer) case. The resulting oil fractional flow for the noncontinuous example is slightly lower (Figure 6). Oil recovery for the noncontinuous model was 7% OOIP less than the continuous model.

CO_2 flood simulations, each initialized at waterflood residual oil saturation, were run to evaluate the effect of layer continuity. In Figure 5b, the vertical sweep efficiency for the noncontinuous model resulted in an ending oil saturation of 22%. The continuous layer case left a higher oil saturation of 30% (saturation map not shown). Figure 7 shows how the differences in layer continuity impact the oil fractional flow. The continuous-layer case is marked by early oil response and CO_2 break-through. On the other hand, the noncontinuous case is characterized by a delayed but more sustained oil response. These fractional flow characteristics have a controlling impact on the solvent flood performance predictions presented in the following section.

2. Streamtube simulation and performance predictions

The hybrid method was used to combine the cross-sectional simulation fractional flow data with the areal streamtubes to make field-scale performance predictions. Streamtube patterns generated for the inverted nine-spot and line drive patterns are shown in Figure 8.

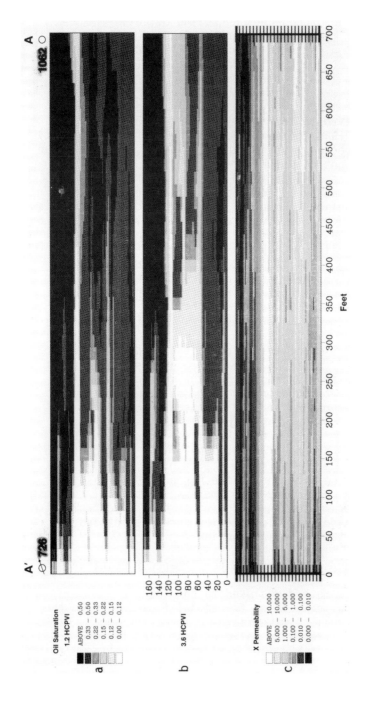

Figure 4. McElroy Field continuous model cross sections showing: (a) oil saturation at 1.2 HCPVI; (b) oil saturation at 3.6 HCPVI; and (c) horizontal permeability of the CO_2 flood run.

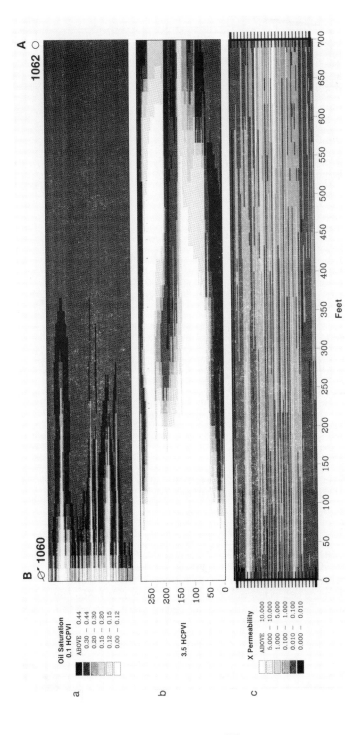

Figure 5. McElroy Field noncontinuous model cross sections showing: (a) oil saturation at 0.1 HCPVI; (b) oil saturation at 3.5 HCPVI; and (c) horizontal permeability of a miscible flood run initialized at $S_o=S_{orw}$.

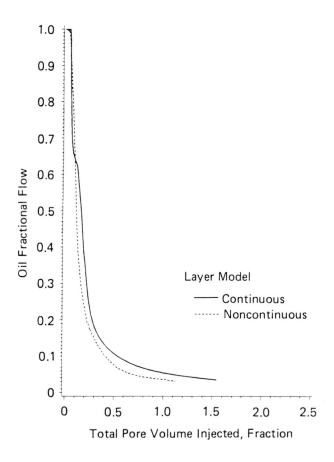

Figure 6. McElroy Field waterflood simulation results showing the effect of layer continuity.

a. <u>Waterflood Verification</u>. Historical waterflood performance was modeled with the continuous-layer model. An injection schedule was developed to equal total fluid withdrawals during the waterflood from 1965 to 1986. The nine injection rates increase from 148 to 1290 Rbl/day and define the gross fluid production rates for the waterflood model.

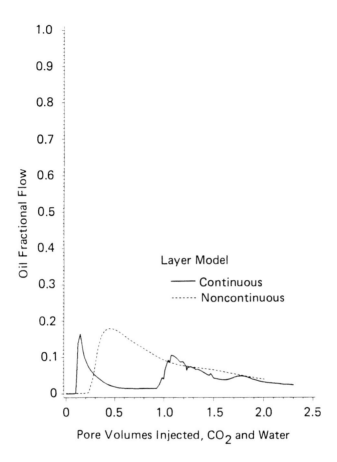

Figure 7.   McElroy Field solvent flood simulation showing the effect of layer continuity.

Streamtube displacement predictions of historical waterflood recovery were made to verify the hybrid simulation model.   The predicted recovery was 19% OOIP to date (0.28 PVI), which comes within 1% OOIP of the actual 21-year performance, as shown in Table I.   This agreement was achieved without any history-matching adjustments to the simulator input data.   Figure 9 compares the actual performance of the waterflood with the streamtube model predictions and shows good agreement.   Ultimate waterflood recovery of 30% (at 97% water cut) was calculated by extending the prediction.

Streamtube Pattern

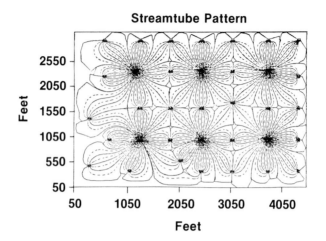

Realigned Streamtube Pattern

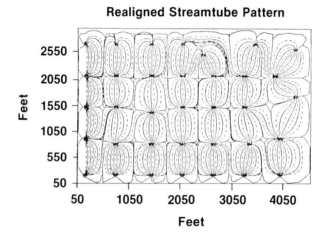

Figure 8.   McElroy Field streamtube models for
(a) inverted nine-spot and (b) line drive patterns.

TABLE I. McElroy Field, Section 194 Waterflood Predictions

Waterflood Case	Injection (PV)	Oil Produced (MSTB)	Recovery (% OOIP)
Historical Data	0.28[a]	2844	18%
Waterflood Match	0.28	2932	19%
Ultimate Waterflood	1.28	4614	30%
Realignment Ultimate	1.28	4710	31%

[a]Gross fluid withdrawals for the period from 1965 to 1986.

b. Waterflood Realignment. The model was converted to a realigned pattern by overlaying a revised streamtube pattern. Saturations were transferred from the old streamtube pattern to the new one, and the numerical recovery calculations were continued. Remaining oil saturation of 60% was carried forward to the realigned waterflood. The model predicted a drop in water cut of 6% from 80% to 74% after realignment, which nets an additional 1% OOIP oil in 14 years. This prediction is based on the fluid injection rate remaining fixed at 1290 RB/D. The resulting oil production rates for the realigned waterflood are shown in Figure 9. Actual water cut decline to date is about 2%.

c. CO_2 Flood Predictions. Miscible flood predictions were completed utilizing the simulation model and streamtube combination verified in the waterflood modeling. The predictive runs used the realigned streamtube pattern. The current waterflood injection rate of 1290 RB/D was assumed unchanged for the CO_2 WAG prediction.

The most likely incremental solvent-flood recovery was determined to be 6% OOIP at 0.92 PV total injection. A 0.20 hydrocarbon pore volume (HCPV) of solvent and a 0.20 HCPV of water were injected at 1:1 WAG, followed by chase water. Solvent injection was begun at a producing water cut of 85%, and the peak oil production response is predicted to occur in six years. Although actual injection rates are expected to increase after realignment, there were insufficient data available to determine the injection increase. Figure 9 shows the solvent-flood oil production prediction.

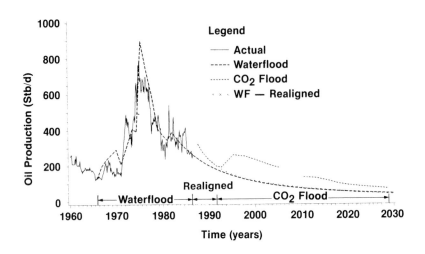

Figure 9.   McElroy Field hybrid simulation results.

    d. Reservoir Continuity. Production predictions were
prepared to isolate the effect of continuity. These
sensitivity runs were based on the cross-sectional oil
fractional flow shown in Figure 7. As shown in Figure 10,
a delayed response to CO_2 injection was observed in the
noncontinuous case. Total recovery for the noncontinuous
case is 5% OOIP higher; however, the bulk of that response
occurs late in the life of the project. Delayed response
to tertiary projects such as that demonstrated here is
detrimental to project economics (when based on rate of
return), thus illustrating the importance of modeling
heterogeneity.

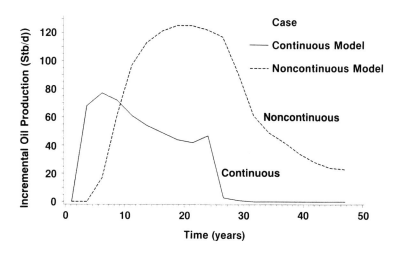

Figure 10. McElroy Field simulation showing the effect
of layer continuity on CO_2 flood performance.

B. Study of Reservoir A

Reservoir A is a pinnacle reef structure with vuggy
porosity dominating. Original oil-in-place is estimated at
13.8 MMSTB. Reservoir A was produced by primary depletion
for approximately one year, after which water injection was
initiated. The waterflood was expected to sweep the pool
in a gravity-stable manner, with injection near the bottom
of the pool and production from the top. However, the pool
was shut in for approximately two years because of early
water breakthrough and injectivity problems.

The early water breakthrough occurred two years after
injection started. A cross-sectional model between the
injector and producer was created to: (1) provide an
explanation for the early water breakthrough; (2) determine
whether fluid saturations at the wellbore could be described
accurately; and (3) estimate sweep performance of a
miscible flood between the two wells.

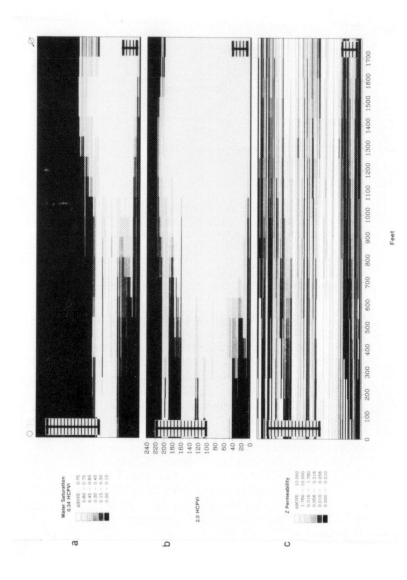

Figure 11. Waterflood model of Reservoir A showing:
(a) water saturation at 0.34 HCPVI; (b) water saturation at
2.0 HCPVI; and (c) vertical permeability distribution.

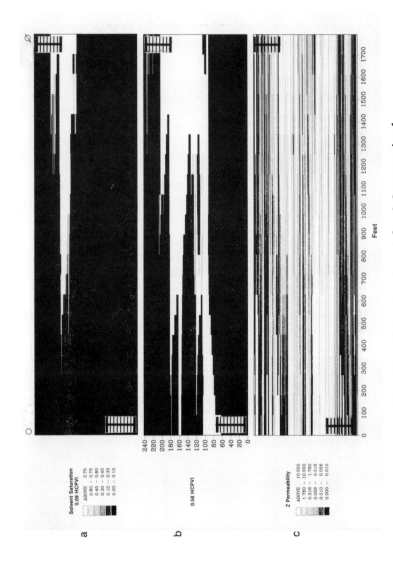

Figure 12. Miscible flood model of Reservoir A showing: (a) solvent saturation at 0.09 HCPVI; (b) solvent saturation at 0.58 HCPVI; and (c) vertical permeability distribution.

515

1. Cross-sectional simulation

A 4000-cell cross-sectional model was constructed using the fractal interpolation routine on the porosity and horizontal permeability data for the two wells. A value of the intermittency exponent (H) equal to 0.95 was obtained for each well. The vertical permeability was calculated using a horizontal to vertical permeability transform. Figure 11 shows the model cross section of vertical permeability.

a. Waterflood Results. Finite-difference simulation was undertaken to study the waterflood performance. The wells in the cross-sectional model were completed as shown in Figure 11. Water injection rates were scaled to approximate field rates on the basis of HCPV.

Water saturation maps of the simulation (Figures 11a and b) show the influence of the reservoir heterogeneity between the two wells. The injected water channeled from the injector to the producer and did not behave as a gravity-stable front.

The fluid movement observed in the model is supported by well log data. A production log run on the producer indicated that the water production was occurring from below the perforated interval. A production log run on the injector showed high water saturations above the perforated interval. Eventually almost the entire wellbore area became water-saturated. Both these anomalies were observed in the model as shown in Figure 11.

b. Miscible Flood Results. The wells were recompleted as shown in Figure 12 to allow injection into the top of the reservoir. The waterflood run was restarted at approximately 1 HCPVI to determine the response to a miscible flood. The injection rates were scaled to be gravity stable in vertical permeability as low as 0.8 md.

Maps of solvent saturation (Figure 12a-b) show that the solvent broke through rapidly at the producer due to the dominance of the horizontal permeability, after which there was very little improvement in sweep efficiency. The reservoir-model heterogeneity played a major role in the performance of the miscible flood, as well as in the waterflood. The miscible flood was expected to act in a gravity-stable manner. Instead, the permeability variations caused the miscible flood to perform in much the same manner as the waterflood, i.e., the dominant flow direction was horizontal instead of vertical.

The cross-sectional model was only used to study fluid movement. Streamtube predictions are not presented for this reservoir because a project forecast using the hybrid method was not completed.

C. Study of the West Pembina Nisku D Pool

The West Pembina Nisku D (WPND) Pool is an isolated pinnacle reef discovered in 1978. Original oil-in-place was estimated at 15.7 MMSTB. It is located approximately 165 miles north-northwest of Calgary, Alberta, Canada. The WPND Pool was put on hydrocarbon miscible flood (HCMF) injection in 1981. After eight years, over 0.34 HCPV of solvent has been injected with no breakthrough.

1. Cross-sectional simulation

A 3999-cell cross-sectional model was created using data from the three wells in the WPND Pool: 6-12-49-13 W5M (6-12), A7-12-49-13 W5M (A7-12), and 10-12-49-13 W5M (10-12). An R/S analysis indicated the intermittency exponent (H) was 0.90. Porosity, k_h, and k_v cross sections were all interpolated independently. The model cross section of vertical permeability is shown in Figure 13c. Note that Well 10-12 was drilled and cored after solvent injection was begun and was not completed in the modeling study. Special core analysis from Well 10-12 was performed to locate the position of the HCMF interface.

2. Miscible flood results

The model was run until 0.31 HCPV of the solvent was injected in Well A7-12 (Figure 13b). At this point there was no production of solvent at Well 6-12, and the solvent-oil interface in the model was stable and level. The position of the solvent-oil interface was compared with special core analysis from Well 10-12 and with a volumetric estimate, and showed agreement within 3% of the total reservoir thickness. Note that the differences in interface depths are consistent with those of the HCPV injected (Table II).

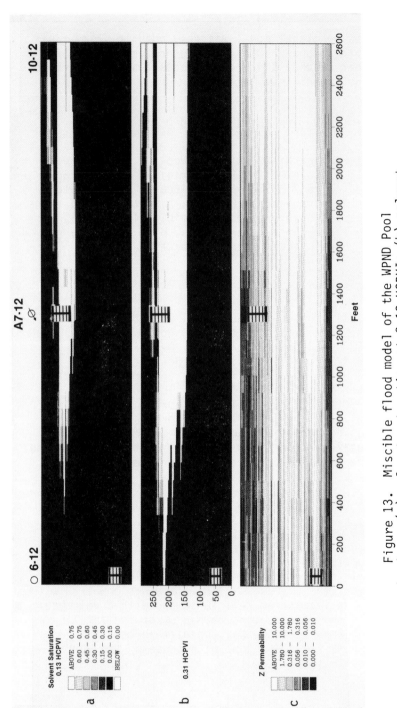

Figure 13. Miscible flood model of the WPND Pool showing: (a) solvent saturation at 0.13 HCPVI; (b) solvent saturation at 0.31 HCPVI; and (c) vertical permeability distribution.

TABLE II. Solvent-oil interface for the West Pembina Study
(DaSie and Guo, 1988).

Source	Interface depth (ft SS)	Solvent injected HCPV
Model prediction	6995	0.31
Special core analysis	6985	0.28
Volumetric estimate	6990	0.29

The transition zone in the model was generally about
two gridblocks deep, or approximately 4.5 feet. The
transition zone in the field was considered small or
nonexistent. Both the model and field data indicate that
the solvent is moving in a gravity-stable manner, with a
sharp interface with the oil.
The size of the WPND Pool and the placement of the
wells precluded the use of a streamtube model. Only a
cross-sectional model was used in this study.

D. Comparison with Conventional Models

Since most reservoir engineers are familiar with the
conventional finite-difference simulation technique, it is
useful to contrast hybrid modeling with the conventional
approach. A modern, full-field simulation model will
generally represent the reservoir with a relatively coarse
grid. A given gridblock may typically be 1 to 10 acres in
areal extent and 30 to 100 feet thick. Block properties
are averaged from detailed log and core data, and then
adjusted to provide as exact a match as possible to
individual well histories. A 10,000-block model may offer
over 40,000 parameters that can be adjusted to match
pressure and fluid history.
The grids used in hybrid modeling are much more
detailed than those used in a conventional model. The
blocks comprising the streamtubes are typically about 0.1
acres. Layers in the cross-sectional model are 1 to 2 feet
thick. The objective is to couple a detailed cross-section
with a detailed areal model to approximate the average
reservoir history.

This detailed modeling of reservoir heterogeneity creates a hybrid model which can be used for predictions without extensive history matching. Our experience to date has been that this approach provides a satisfactory predictive method at a fraction of the computer and personal effort required by an equivalent conventional model. When compared to a conventional simulation with extensive history matching runs, this method is less time consuming and less costly. A single history match iteration may require as much CPU time as a completed cross-sectional run. When multiplied several times for a conventional simulation study, the cost savings are significant.

Hybrid methods are limited to situations in which production principally balances injection, and pressure decline is not an important factor. Streamtube models are not suited for situations in which the flood fronts move vertically. Finally, the wells to be used in the cross section must be chosen carefully. They should represent a typical well pair. When the geology varies significantly, different well pairs may be chosen to represent the different geological features of the formation. The choice of well pairs is subjective, and generally requires the aid of geologists to help determine what properties typify the reservoir.

V.   SOME ASSUMPTIONS AND APPROXIMATIONS

Several points of the technique are based on assumption, rather than proof, and these require some discussion. The heterogeneity of the cross section model is based on H and σ_0. The initial variance, σ_0, is a property of the well pair and depends on the subjective correlation of the well log pairs, that is, how the modeler chooses to line up the stratigraphic markers. The intermittancy coefficient, H, is derived from an analysis of the well log in the vertical direction and then used to create the heterogeneity structure in the horizontal direction. The coefficient H from vertical data represents a time signal of geologic events, such as weather, etc. Properties in the horizontal result from the accumulation of those events, and vary less frequently. The horizontal properties are derived from linear interpolation, with a texture of heterogeneity determined by the H and σ_0 parameters. As a practical outcome, the flow properties of the model are affected more strongly by the well log correlation and the degree of roughness generated by the parameters of the fractal interpolation can be treated as a sensitivity on the underlying geologic structure.

The cross section models are typically 2000-4000 cell blocks based on the practical limits of computer resources. Experience indicates that vertical resolution is more important than horizontal. Consequently cell blocks have a high aspect ratio, typically 50:1.

Ultimately, the justification of the assumptions rests on the applications. The procedures described in Section III have generally provided good descriptions of reservoir performance both in history and prediction. The studies discussed in this paper and in the references will provide the reader with an assessment of how well the assumptions work.

VI. CONCLUSIONS

The following conclusions regarding the hybrid simulation methodology were drawn:

1. Hybrid simulation of McElroy Field accurately reproduced waterflood recovery based on a comparison with field performance. The study utilized the most likely reservoir parameters and did not require extensive history matching.

2. The hybrid method is a cost-efficient way of carrying out a study of an entire field or project area.

The following conclusions regarding geostatistical methods were drawn:

3. Detailed modeling of reservoir heterogeneity was critical for accurate prediction of CO_2 flood response. The initial oil response was accelerated by channeling through continuous layers, but total oil recovery was reduced.

4. Cross-sectional models using fractal geostatistics provide a valuable tool for predicting fluid saturations and movements. The cross-sectional model of Reservoir A showed water saturations that were confirmed by field data, and provided an explanation for the early water breakthrough. The solvent-oil interface predicted by the WPND model compared well to field estimates.

The combination of geostatistical methods with the
hybrid simulation technique considers small-scale heterogeneities, which are generally averaged in conventional
simulation. The presence of such heterogeneity was shown
to have an important influence on simulation results.

ACKNOWLEDGEMENTS

The hybrid reservoir model in the McElroy study was
adapted from a model that was conceived by G. K. Alameda.
Numerous other individuals have contributed their efforts
to this study and are also acknowledged.
The authors thank the management of Chevron Oil Field
Research Company, Chevron USA, Inc. and Chevron Canada
Resources for their permission to publish this paper.

REFERENCES

DaSie, W. and Guo, D. S. (1988). "Assessment of a Vertical
    Hydrocarbon Miscible Flood in the West Pembina Nisku D
    Reef," SPE 17354, SPE/DOE EOR Symposium, Tulsa, April.
Dykstra, H., and Parsons, R. L. (1950). "The Prediction of
    Oil Recovery by Waterflooding," Secondary Recovery of
    Oil in the United States, 2nd ed., API, pp. 160-174.
Emanuel, A. S., Alameda, G. K., Behrens, R. A., and Hewett,
    T. A. (1987). "Reservoir Performance Predictions Based
    on Fractal Geostatistics," SPE 16971, 62nd Annual
    Technical Conference of SPE, Dallas.
Hewett, T. A. (1986). "Fractal Distributions of Reservoir
    Heterogeneity and Their Influence on Fluid Transport,"
    SPE 15386, 61st Annual Technical Conference, New
    Orleans, October.
Hewett, T. A. and Behrens, R. A. (1988). "Conditional
    Simulation of Reservoir Heterogeneity with Fractals,"
    SPE 18326, 63rd Annual Technical Conference of SPE,
    Houston, October, pp. 645-660.
Mandelbrot, B. B., and Wallis, J. R. (1969). "Robustness
    of the Rescaled Range R/S in the Measurement of
    Noncyclic Long Run Statistical Dependence,"
    Water Resources Research 5, October, pp. 967-988.
Martin, J. C., and Wegner, R. E. (1978). "Numerical
    Solution of Multiphase, Two-Dimensional Incompressible
    Flow Using Stream-Tube Relationships," SPE 7140,
    SPE-AIME 48th California Regional Meeting,
    San Francisco, April 12-14, pp. 313-323.

Mathews, J. L., Emanuel, A. S., and Edwards, K. A. (1988). "A Modeling Study of the Mitsue Stage 1 Miscible Flood Using Fractal Geometries," SPE 18327, 63rd Annual Technical Conference of SPE, Houston, October 2-5, pp. 661-674.

Tang, R. W., Behrens, R. A., and Emanuel, A. S. (1989). "Reservoir Studies Using Geostatistics to Forecast Performance," SPE 18432, SPE Symposium on Reservoir Simulation, Houston, February 6-8,pp. 321-337.

Todd, M. R. and Longstaff, W. J. (1972). "The Development, Testing, and Application of a Numerical Flood Simulator for Predicting Miscible Flood Performance," JPT, July, pp. 874-882.

Warren, J. E., and Price, H. S. (1961). "Flow in Heterogeneous Reservoirs," SPEJ, September, pp. 153-169.

CONSTRUCTION OF A RESERVOIR MODEL BY INTEGRATING GEOLOGICAL AND ENGINEERING INFORMATION - BELL CREEK FIELD, A BARRIER/STRANDPLAIN RESERVOIR

S. R. Jackson, L. Tomutsa, M. Szpakiewicz, M. M. Chang, M. M. Honarpour, and R. A. Schatzinger

National Institute for Petroleum and Energy Research (NIPER) Bartlesville, Oklahoma

ABSTRACT

The method for constructing a reservoir model which incorporates, integrates and reconciles various types of geologic and engineering information is presented using an example from Bell Creek field, Montana. The method described is an interactive process with successively more detailed information being incorporated as it becomes available and is relevant for each stage of production. Generally, the trend is increasing complexity and level of detail of the model, as enhanced recovery processes are applied.

The geological model provides the framework for subsequent quantitative engineering models, and incorporates depositional, diagenetic, structural and geochemical information. A simple, permeability layer model is based on the sedimentologically defined units, and provides a first approximation of the degree of layering in the reservoir. This model can be used for calculating reservoir volumetrics and forecasting field and well primary production performance.

A flow unit model incorporates all pertinent geological and engineering information available and provides a reservoir description which retains the complexities of reservoir architecture and variations in reservoir parameters. It is used to predict production performance resulting from secondary and tertiary recovery processes.

The model presented for Bell Creek field is confirmed by comparison with cumulative EOR production data and comparison of the ROS distribution and waterfront advancement rate from simulation results and reservoir data.

I. INTRODUCTION

Development of geologic and engineering models requires the integration and reconciliation of the various types of information. It is an iterative process with successively more detail information being incorporated as it becomes available during each stage of production. Different types and scales of heterogeneities may influence each stage of production, with a general trend of increasing complexity and level of detail and decreasing scale of heterogeneities to be incorporated in the models as enhanced recovery processes are applied.

In the method illustrated in this paper, the geological model provides the framework for subsequent quantitative, engineering models. The geologic model is comprised of four components: depositional , diagenetic, structural and geochemical. A permeability layer model is based on sedimentologically defined units or facies and provides information for calculating reservoir volumetrics and forecasting field and well performance. A flow unit model incorporates more detailed information on the spatial arrangement of fluid flow properties and is used to predict production performance resulting from secondary and tertiary recovery processes.

The examples presented in this paper are from an integrated, geological and engineering study of the Lower Cretaceous Muddy formation in Unit 'A' of Bell Creek field located in southeast Montana. Information from analogous outcrop exposures located 40 miles south near New Haven, Wyoming was also used.. The models presented herein were developed for the Tertiary Incentive Project (TIP) area, which was a pilot micellar-polymer project implemented after 10 years of line-drive waterflooding (Fig. 1).

The Muddy formation in Unit "A", Bell Creek field produces oil primarily from a 30-foot-thick (maximum) barrier-island stratigraphic trap reservoir. It is unconformably overlain by a valley fill complex of poorly productive channel sandstones and marginal to non-productive estuarine siltstones and shales. In places, the valley cuts entirely remove the barrier island sandstone, creating hydraulic barriers within the field and defining production unit boundaries with different oil-water and gas-oil contacts.[1]

II. GEOLOGICAL MODEL

Construction of the geological model incorporated the following information: 1) depositional setting, 2) diagenetic history, 3) structural history, and 4) geochemical (rock and fluid) characteristics. Information resulting from this composite model is the identification of field-scale features such as the reservoir boundaries, laterally extensive shales or cemented zones that act as barriers to fluid flow, erosional features or thinning of the reservoir due to depositional or erosional processes, and major compartments within the reservoir. For each reservoir, the geological features that most effect reservoir quality and production are identified and incorporated into the model.

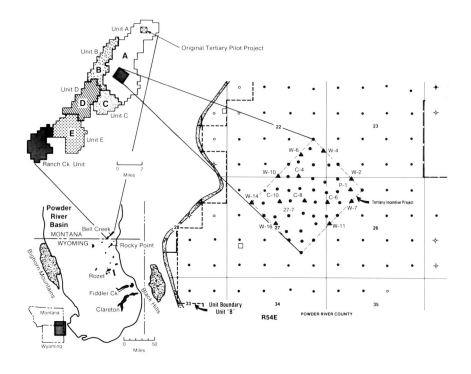

Fig. 1. Location of Bell Creek field and the tertiary incentive project (TIP) area within Unit 'A'.

A. Depositional Setting

The depositional model provides the framework of reservoir architecture and serves as the basis for subsequent fluid flow models. After the depositional processes are identified and the sequence of depositional events are reconstructed, the spatial distribution, geometry and dimensions of facies are determined or predicted.

In the Muddy formation, nine major reservoir sandstone facies were distinguished on the basis of grain size, texture, sedimentary structures, and type and amount of biogenic structures.[2] Information from 26 cores in the northern part of Bell Creek field as well as information from outcrop exposures of the Muddy formation was used to define the facies and reconstruct the depositional and diagenic history.[2] A 3-D conceptual model was developed for Bell Creek field, that includes the interpreted location of Unit 'A' within the deposystem[2-3] (Fig 2).

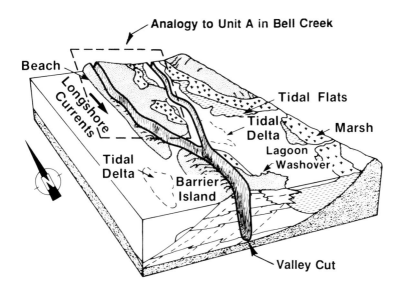

Fig. 2. Conceptual model for the barrier island deposystem of the Muddy formation and the location of Unit 'A', Bell Creek field within the deposystem. Note the non-conformable, valley cuts that represent the younger, deep and narrow style of incision and separate the production units A, B, C, D, and E in Bell Creek field. (Modified from Reinson, 1984)

The foreshore facies and middle, and upper shoreface facies represent deposition by marine processes, and these facies were grouped together into one layer because they contain similar reservoir properties. These facies exhibit the highest quality reservoir rock and comprise the main part of the reservoir. Distinct sedimentological and reservoir properties were noted for the lower shoreface facies. The paralic facies of washover, lagoon, estuarine, tidal channel, and tidal delta exhibit variable reservoir quality characteristics, with the washover facies exhibiting the best reservoir quality of the facies. The distribution of overlying valley fill deposits was controlled primarily by alluvial processes, and these deposits consist of both reservoir sandstone and finer-grained, marginal to non-reservoir sediments. A typical vertical profile of facies with associated petrophysical characteristics is presented in Fig. 3.[4] Figure 4 presents the spatial distribution and thicknesses of these facies in the TIP area. In places, the thickness of the foreshore and the upper shoreface facies is reduced by about 30% (10 feet) by overlying valley fill deposit.

The sequence of depositional events in the Muddy formation in Unit 'A' in the Bell Creek field was the following: 1) deposition of a barrier island sandstone; 2) a transgression and an associated shoreface retreat that

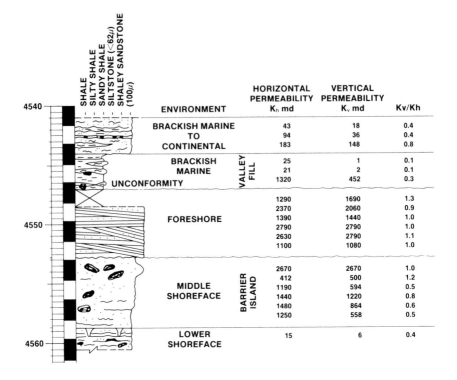

Fig. 3. Vertical sequence of facies in well 27-7, which include regressive barrier island deposits and overlying valley fill deposits. Note petrophysical properties associated with each facies. (After Tillman et al., 1988)

preserved only back-barrier remnants of the first barrier deposit in wells 27-14, W-16, and C-10 (Fig. 4); 3) deposition of a second regressive (progradational) barrier island sandstone that comprises the major reservoir interval in Unit 'A', 4) continued regression that resulted in erosion of the second barrier sand by fluvial channel systems; and 5) transgression that filled the upper parts of the paleovalleys with estuarine deposits, and subsequently deposited the marine Shell Creek and Mowry shales. The regional sea level curves that support this sequence of depositional events have been presented by Szpakiewicz, et al.[2] It has been suggested that two periods of valley cut and fill have incised the barrier island deposits.[2] The earlier valleys are shallower and distributed over a large part of the barrier system while the younger features are narrow and deep.

The initial production potential in the Unit 'A' in Bell Creek field is controlled to a large extent by the distribution of the marine facies and the valley cuts (Fig 5). On the eastern (landward and updip) side of the barrier, primary production rate potential deteriorates very rapidly away from the barrier axis, where reservoir quality sand pinches out and interfingers with back-barrier, lagoonal deposits. On the western (seaward and downdip) side of Unit 'A', barrier island reservoir sandstones are truncated by low-permeability valley fill deposits (Fig. 6), that form hydraulically isolated units and reduce production over the distance of one well spacing, (1,320 feet).

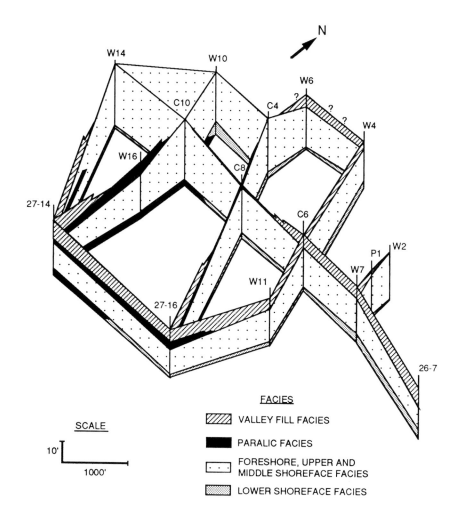

FACIES

▨ VALLEY FILL FACIES

■ PARALIC FACIES

▢ FORESHORE, UPPER AND MIDDLE SHOREFACE FACIES

▨ LOWER SHOREFACE FACIES

SCALE

10'

1000'

Fig. 4. A 3-D diagram showing the spatial distribution and thickness of facies in the TIP area (See Fig. 1 for well locations). Datum is based on the lower Muddy sand, and top is top of the Muddy sand.

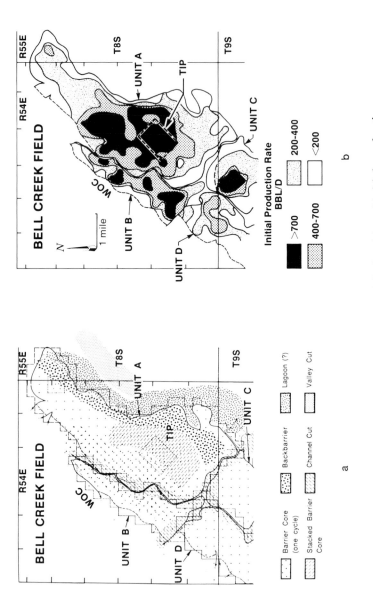

Fig. 5. Comparison of distribution of facies (a) and initial production rate potential (b) (after Szpakiewicz, 1989).

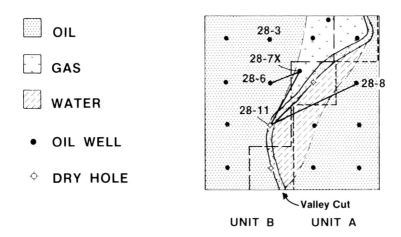

OIL

GAS

WATER

• OIL WELL

◇ DRY HOLE

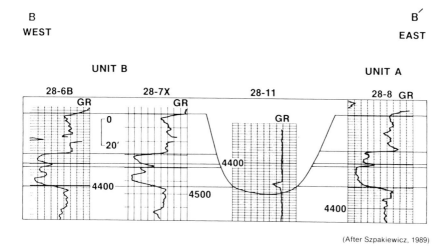

(After Szpakiewicz, 1989)

Fig. 6. Gamma ray log cross section across boundary (and hydraulic barrier) between Unit 'A' and Unit 'B'. Normal thicknesses of barrier sandstone occur on either side of the valley cut but are nearly totally removed in well 28-11. Note the location of oil-water and gas-oil contacts on either side of the valley cut.

B. Petrographic Analysis

The major diagenetic phases established for Unit 'A' were determined from thin-sections of samples primarily from the TIP area.[2] Table 1 outlines the major diagenetic phases identified within the barrier island facies and their potential effect on porosity and permeability. Leaching occurred very early in the paragenetic sequence and significantly increased the pore space in the reservoir by creating intraparticle secondary porosity and oversized pores. Virtually all subsequent diagenetic phases, that include siderite cementation, compaction, silica overgrowths, calcite cementation, later leaching, and clay cementation, reduced the transmissivity of the reservoir rock.

The effect of clay content on permeability can be seen in histograms of permeability from the foreshore facies, where two distinct permeability distributions occur: a relatively sharp-peaked population occurring from 0 to 1,500 md and a broader population from 1,500 to 4,800 md (Fig 7). The samples that comprise the higher permeability population are from wells that contain less than 1% clay cement, whereas those samples in the lower permeability population are from wells that contain greater than 1% clay cement. Statistical tests (Kolmogorov-Smirnoff)[6] indicate that these two populations are distinct and illustrate the diagenetic overprint on the primary depositional permeability fabric.

The spatial variations in the distribution of the clay cement (and therefore permeability) within the TIP area tend to occur laterally rather than vertically and are shown in Fig. 8. Interwell changes in the amount of clay cement tend to correspond with the faults present and crossplots of distance to the nearest fault versus diagenetic clay (Fig. 9) show a positive correlation (correlation coefficient of 0.812). Although the reason for the correlation has not been established at this time, it may be due to the faults providing pathways for fluids which leached the diagenetic clays near the faults.

Different clay assemblages are associated with barrier and valley fill deposits and can be shown to greatly effect chemical EOR production. X-ray diffraction analyses of Muddy formation samples indicate that the barrier island sandstones contain a 2:1 ratio of kaolinite to illite. Total clay comprises less than 15% of the volume of barrier sandstones based on point counts, whereas in valley fill sandstones and mudstones, montmorillonite and kaolinite dominate the clay assemblage.[2]

Sensitivity studies using a three-dimensional chemical flood simulator, UTCHEM[6-7], indicate that the adsorption of surfactant and polymer increases with the amount of montmorillonite reservoir rock (table 2) and that the calcium released due to ion exchange causes precipitation of surfactant . Oil recovery after waterflood varied from 38% in clay-free reservoirs to 35% in low-clay reservoirs and down to 15% recovery in clay-abundant areas of the reservoir.

Table 1. Major Diagenetic Phases Identified Within the Muddy Formation Barrier Island Sandstone Facies. Phases are in Chronological Order From Earliest (top) to Latest (bottom) (after Szpakiewicz, et al., 1989).

Diagenetic phase	Suggested cause	Potential effect on porosity/permeability
Dominant leaching creates secondary porosity, oversize pores, affects chert, feldspars, sedimentary rock fragments; early kaolinization	Meteoric water lens	Major increase
Siderite cement	Mixing of waters at low Eh	Insignificant decrease
Silica overgrowths increase grain eccentricity, grain contact; reduce pore throats	Solution-reprecipitation	Minor decrease Minor k decrease
Calcite cement usually fills all porosity, stops compaction	Deoxygenation, pH and/or temperature changes causing oversaturation	Major decrease Major k decrease
Later leaching corrodes grains and prior cements	Reestablished meteoric water lens	Major or minor increase Major k increase
Clay cement fills or lines pores blocks throats creates microporosity	Changing subsurface water chemistry; new diagenetic fluids along faults	Minor decrease Major k decrease
Hydrocarbon migration	Hydrodynamic forces	Retards or stops diagenesis

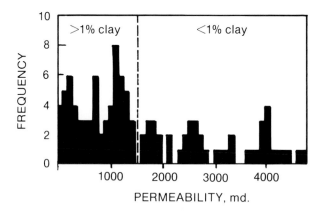

Fig. 7. Frequency histogram of permeability from the foreshore facies indicates two populations that can be related to the diagenetic clay content of the sandstone.

C. Structural Analysis

Recognition of the presence and location of faulting in the TIP area in Bell Creek field resulted from the integration of information from various sources. The first indication of the possibility of faulting was from the inspection of faulted and fractured outcrop exposures. The discrepancy between the actual depth of the Muddy formation in Bell Creek field to calculations of the expected depth based on regional dip, suggested that faults may be present in and around Bell Creek field.[8]

Although not reported in the abundant literature on Bell Creek field and not found in the cores examined, construction of wireline log cross sections indicated the presence of faults with displacements up to 40 feet but generally from 10 to 20 feet.[2] Pressure-pulse and falloff testing as well as the waterfront advancement rate supported the identified fault locations.

The faults identified in Unit 'A' in Bell Creek field are discontinuous, commonly strike 50 degrees northeast and 140 degrees northwest, and are generally parallel to the NW and NE trending lineaments recognized throughout the Powder River Basin. The similarity of azimuths obtained from regional stress orientations,[9] major lineaments identified from Landsat imagery and seismic interpretation[10] suggests that the fault azimuths in Unit 'A' are consistent with the regional trend.

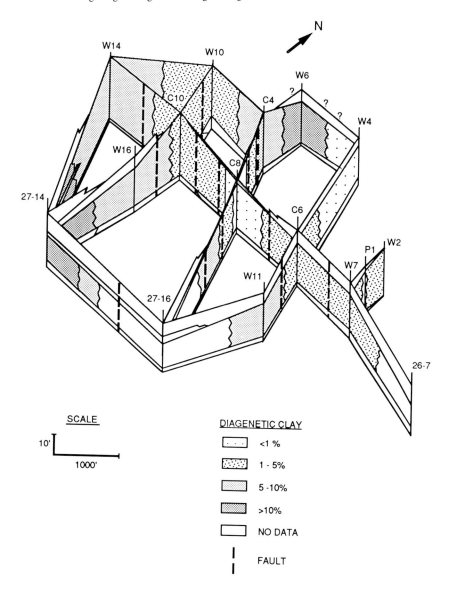

Fig. 8. Spatial distribution of daigenetic clay content in the TIP area. Note that changes in clay content tend to correspond with the presence of faults (See Fig. 1 for well locations).

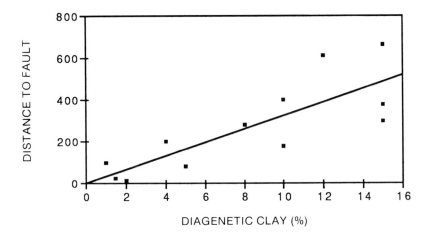

Fig. 9. Crossplot of distance to nearest fault versus amount of diagenetic clay.

Table 2. Simulation Results of Micellar-Polymer Flooding for Various Clay Types and Amounts.

	Clay Content				Adsorption	
Simulation run	Layer 1	Layer 2	Layer 3	Oil recovery, fraction	Surfactant mL/mL PV	Polymer wt %/PV
1	0	0	0	0.376	0	0
2	[1]1% M	[2]0.5% K	0.5% K 0.1% M	0.305	0.0022	0.0051
3	5% M	2.5% K	2.5% K 0.5% M	0.205	0.0049	0.0155
4	10% M	5% K	5% K 1% M	0.168	0.0057	0.0236
5	15% M	7.5% K	7.5% K 1.5% M	0.149	0.0059	0.0294

[1]M = Montmorillonite
[2]K - Kaolinite

Faulting with the reservoir produced a mosaic of small tectonic blocks (Fig. 10). Downthrown tectonic blocks would be expected to produce less oil but high total fluids because of the natural tendency of oil to concentrate in structurally high areas in the reservoir. Based on a comparison of fluid production of well P-14 (structurally low) and well P-11 (structurally high), these expectations are born out.

The effects of structural features on production in Unit 'A' are variable, depending on the stage of production. Studies in the TIP area indicate that the influence of dip and faults on primary production was generally low to negligible; however, greater oil accumulation was found in uplifted tectonic blocks (horsts).[1]

Secondary production was dominantly influenced by structural dip, but not faulting. Wells located updip of the water injection linedrive pattern showed increasingly higher cumulative production eastward, where the oil bank moved updip against a stratigraphic pinchout of the reservoir.[1] Tertiary production was moderately to highly affected by faulting where the disrupted continuity of flow units adversely affected sweep and displacement efficiencies.

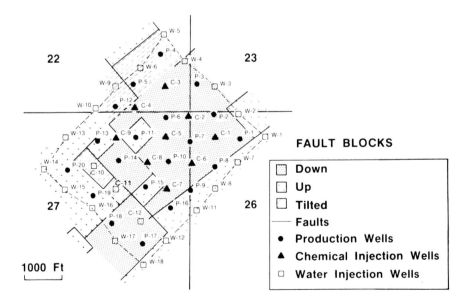

Fig. 10. Separation of the reservoir in the TIP area into small, tectonic blocks as a result of faulting. Note the locations of wells P-11 and P-14 in section 27, which although adjacent to each other, produced different amounts of oil due to the vertical displacement of reservoir blocks. (After Szpakiewicz et al. 1989)

D. Geochemical Analysis

After two decades of reservoir development, including the implementation of two EOR projects, the origin of formation fluids in the productive horizons in Bell Creek Field remains unknown. The enhanced geothermal gradient and a much lower than expected formation water salinity based on a normal hydrogeochemical gradient, strongly indicate a hydraulically dynamic system. However, the variability of chemical and isotopic composition of formation water and oil in the Muddy formation is poorly understood.

Analysis of oil gravity data indicates that, in the northern parts of Bell Creek field (Units 'A', 'B', 'C', and 'D'), oil gravity varies from 28° to 42° API (Fig. 11). There is no obvious relationship between oil gravity and structural dip of the Muddy formation. Except for an isolated area of heavier oil (28° API), 30.5° to 32° API gravity oil exists in the central and central-southern parts of Unit 'A', while in the northern part of Unit 'A', as well as in Units 'B', 'C', and 'D', lighter oil (33°-34° API) predominates. The reason for such a pattern of oil gravity distribution is not clear. There is, however, a possibility that the effects of variable formation temperature, water washing and biodegradation may have influenced the gravity near a network of documented faults.

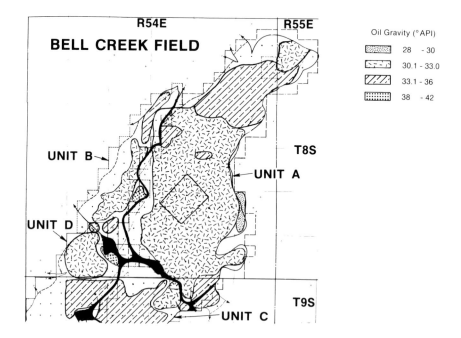

Fig. 11. Areal variations in oil gravity values (° API).

III. PERMEABILITY LAYER MODEL

The geological model presented above provided the framework for the subsequent quantitative permeability layer model and the flow unit model. Permeability layers were based primarily on sedimentological facies divisions and exhibit distinct average reservoir properties such as permeability, porosity, variability of permeability (Dykstra-Parsons coefficient), and the ratio of vertical to horizontal permeability (Table 3). This relatively simple reservoir model provides the framework for calculating reservoir volumetrics as well as forecasting field and well performance, however does not incorporate as much detail as the flow unit model.

Non-parametric (Kolmogorov-Smirnoff) statistical two-sample tests[5] indicated that three distinct permeability distributions occur in the Muddy formation in the area studied. One group includes the higher-energy deposits; a second group includes the lower energy barrier island deposits of lower shoreface; and the third group includes the lowest energy lagoon and marine valley fill deposits. The permeability groups are presented in figure 12, which presents the permeability means and ranges for the barrier facies in 19 wells from the TIP.

Figure 13 presents a 3-D permeability layer model of the TIP area. The datum for the fence diagram is the base of the Muddy sand, and the top of the diagram is the top of the upper sand in the Muddy formation. Layer 1 corresponds to the lower shoreface facies; layer 2 to the foreshore, upper and middle shoreface facies; layer 3 to the lagoonal facies; and layer 4 to the valley fill facies. Although tests indicated that the permeabilities from the lagoonal facies and valley fill facies were from the same distribution, they were distinguished on the diagram because of their very different depositional origins and different clay-type content.

The ratio of vertical permeability to horizontal permeability indicates that layer 2 is essentially isotropic, while layer 4 has the lowest vertical permeability. Dykstra-Parson's coefficients indicate low heterogeneity values for layer 2 and high heterogeneity values for layers 3 and 4 whereas layer 1 exhibits intermediate values.

Application of Outcrop Data

Justification of the lateral continuity of the permeability layers came from outcrop permeability data, where similar permeability averages and vertical profiles extend over 2,000 feet and 1.6 miles (Fig. 14).

Comparison of subsurface and outcrop permeability cumulative distribution functions from the middle shoreface facies in the outcrop and foreshore facies in the subsurface as well as the outcrop and subsurface lower shoreface facies is presented in Fig. 15. Statistical (Kolmogorov-Smirnoff) tests indicate these permeabilities to be from the same population. The comparison of the middle shoreface to the foreshore permeability is valid, based on the similar geologic and petrophysical characteristics observed in both outcrop and subsurface.

Table 3. Permeability Layer Model Characteristics.

Layer	Facies	Grain size (microns)	Number samples[a]	Permeability, md			Porosity, %		k_v/k_h	Dykstra-Parsons coefficient	Cation-exchange capacity
				Arithmetic	Geometric	Range	mean	range			meq/100g
4	Valley fill	100-200	21	193	42	0.6-1320	23	15-33	0.5 (N=16)[b]	0.90	3 (N=4)[b]
3	Paralic facies	75-125	13	256	82	1.3-746	23	14-30	0.87 (N=4)	0.93	24/7[c] (N-4/2)
2	Foreshore, upper and middle shoreface	100-200	233	1662	859	0.01-7400	27	11-34	1 (N=26)	0.59	0.8 (N=7)
1	Lower shoreface	100-150	21	662	276	13-2694	23	9-30	no data	0.78	0.8 (N=1)

[a]for permeability and porosity from conventional core analysis
[b]N = number of samples
[c]from estuarine facies/lagoon facies

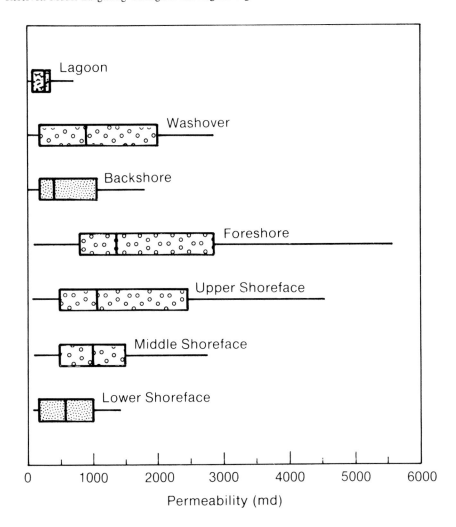

Fig. 12. Mean and range of facies permeability values in the TIP area, Unit 'A', Bell Creek field. The box includes the middle 50% of the permeability values (vertical lines within the box are mean values) while the 'whiskers' extend to 1.5 times the box (or interquartile range). The few values that are greater than this are considered outliers and are excluded from the diagram.

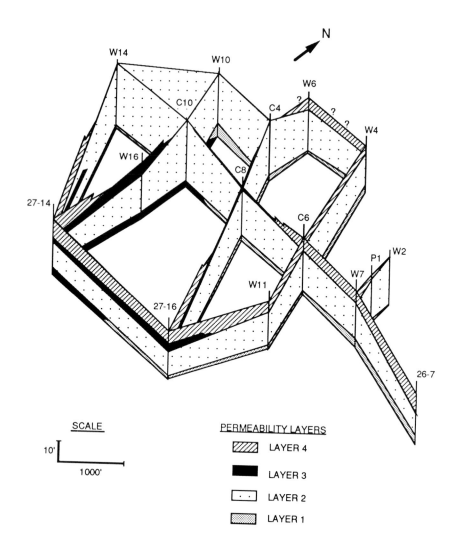

Fig. 13. A 3-D permeability layer model of the TIP area in Unit 'A'.
See Fig. 1 for well locations and Table 3 for layer characteristics.

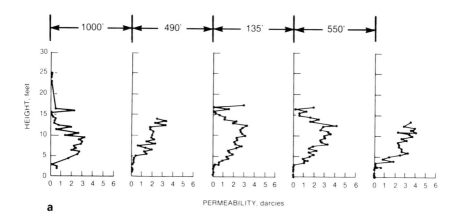

a

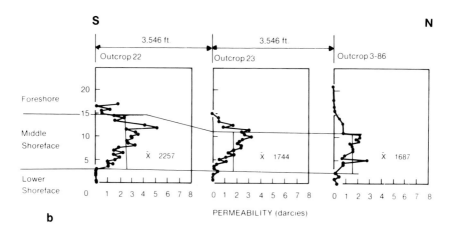

b

Fig. 14. Vertical profiles of permeability across 2000 feet of outcrop exposure of the Muddy formation, Wyoming (a) and from three outcrops over a distance of 1.3 miles (b). Note the similar profiles and average values over distances of 2000 ft. and those greater than a mile.

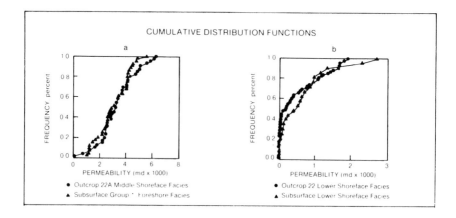

Fig. 15. Comparison of subsurface and outcrop permeability cumulative distribution functions. Similar frequency functions exist for outcrop middle shoreface facies and subsurface, low-diagenetic cement content foreshore facies (a), as well as outcrop and subsurface lower shoreface facies (b).

Other similarities in outcrop and subsurface samples include grain-size frequency distribution and paragenetic diagenetic sequence. Grain size distribution calculated by image analysis of thin-sections indicate similar distributions for outcrop middle shoreface and subsurface upper and middle shoreface facies (Fig. 16). Petrographic studies based on thin-sections indicate a similar paragenetic sequence for outcrop and subsurface barrier island facies. A plot of the natural logarithm of permeability vs. porosity shows similar slopes for outcrop and subsurface data, with outcrop porosity slightly (2%) higher.

Major differences between outcrop and subsurface characteristics documented in this study are the spatial distribution of diagenetic cements. In outcrop, a carbonate-cemented zone in the top of the sandstone sequence (foreshore facies) is present that extended laterally for 1000's of feet, while in the subsurface no laterally continuous, cemented zones were recognized. A second difference is the absence of clay-cemented zones in outcrop which in the reservoir, appear to affect the entire reservoir section and vary over lateral distances of approximately 1,500 feet.

IV. FLOW UNIT MODEL

A flow unit model incorporates all pertinent detailed geologic and petrophysical information available and provides a reservoir description that retains the complexities of reservoir architecture and variations in reservoir parameters. It is most useful in predicting production performance of secondary and tertiary recovery processes.

A flow unit has been defined by Hearn et al.[11-12] as a reservoir zone that is continuous laterally and vertically and has similar averages of those rock properties that affect fluid flow, and has similar bedding characteristics. Ebanks[13] similarly defined a flow unit as a "volume of rock subdivided according to geological and petrophysical properties that influence the flow of fluids through it."

Parameters used by previous workers[12, 14-15] to distinguish flow units include permeability, the product of permeability and thickness (kh), porosity, pore-size distributions determined by mercury-injection and air-brine capillary pressure data, k_v/k_h ratios, oil-saturation, sedimentary structures, lithology, color, grain size and amount of bioturbation.

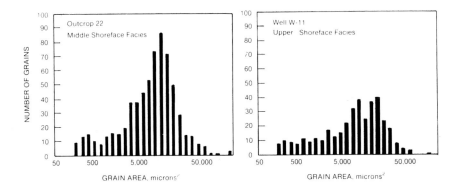

Fig. 16. Frequency distributions of grain sizes calculated by image analysis of thin-sections. Note that the outcrop middle shoreface facies and the subsurface upper shoreface facies have similar distributions.

In Bell Creek field, it was found that the previously constructed permeability layer model based on sedimentologically defined facies provided an acceptable basis for a more detailed flow unit model of the TIP area in Unit 'A'. Permeability, porosity, sedimentologically defined units as well as k_v/k_h ratios, Dykstra-Parsons coefficients, cation exchange capacities, and capillary pressures indicated different rock properties for the layers distinguished (Table 3).

The flow unit model constructed for the TIP area in Unit 'A', Bell Creek field is presented in Fig. 17. Layers were subdivided laterally on the basis of average permeabilities, and porosities, at each well. The resulting model of the study area is one of a mosaic of flow unit blocks where lateral changes in the average permeability values generally correspond to fault locations (Fig. 17) and diagenetic clay content (Fig. 8). Fault locations (shown) and transmissivities (not shown) should also be included in the model.

Variogram analysis of average permeability per well indicates an isotropic, nested pattern consisting of two ranges of correlation lengths: 0.25 and 1.5 to 2.5 miles (Fig. 18a). The shorter range is about the distance between wells and reflects permeability variations within the flow unit.

The longer range is reflected in the permeability layer model and is on the order of the width of the sandstone body in Unit 'A'. This correlation range is consistent with the outcrop permeability variation observed, where similar mean permeability and vertical profiles extend over at least 1.6 miles. This range is significantly larger than the 2,500-ft upper limit observed by Dubrule and Haldorsen[16] for a fluvial braided-stream environment, which forms smaller-scale sandstone units than barrier island-shoreline environments.

The variogram of initial production rate potential also indicates an isotopic nested pattern with ranges in correlation lengths similar to those of average permeability (Fig. 18b). This similarity suggests a dominant control of permeability on initial production.

To determine whether the model developed was an accurate representation of the reservoir in the TIP area, the spatial distribution of flow units and cumulative EOR production were compared. The model was also tested by comparing the spatial distributions of residual oil saturation (ROS) and frontal advancement rate from waterflood simulations of the model with ROS measured from cores taken after waterflooding and the waterfront advancement determined from production rate.

Comparison of cumulative EOR production and the permeability distribution shows similar patterns of distribution and indicates that the model is a reasonable representation of the reservoir (Fig. 19). The similarity illustrates that EOR production is largely controlled by variations in permeability and diagenetic clays. Faulting may have affected the production in well P-12, where production is better than expected based on the permeability and clay content. The lower than expected production in well P-3 may be also attributed to reduced sweep efficiency by low-permeability faults.

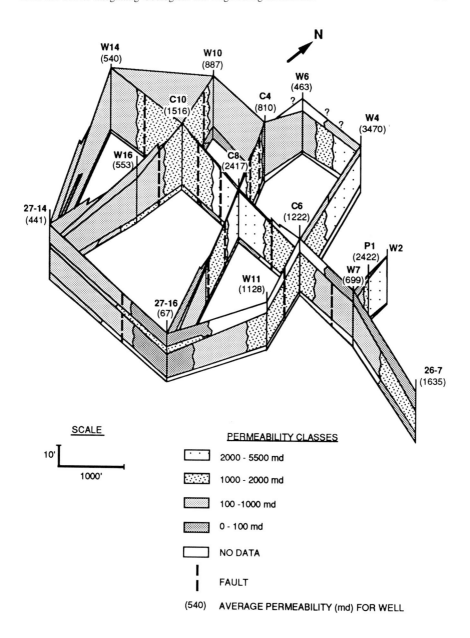

Fig. 17. Flow unit model for the TIP area in Unit 'A'. Note that the lateral changes in permeability correspond with the presence of faults and diagenetic clay content (Fig. 8).

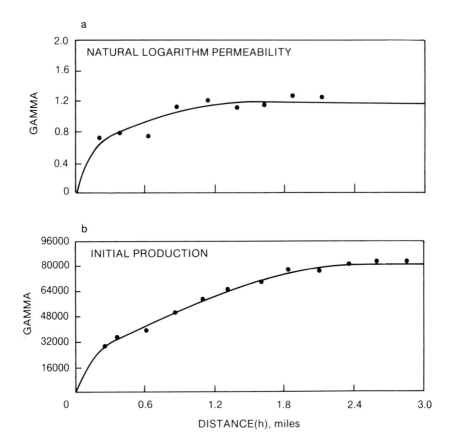

Fig. 18. Variogram for average permeability (a) and initial production
(b) per well. Both variograms indicate two ranges of correlation: 0.25 miles
and 1.5 to 2.5 miles respectively.

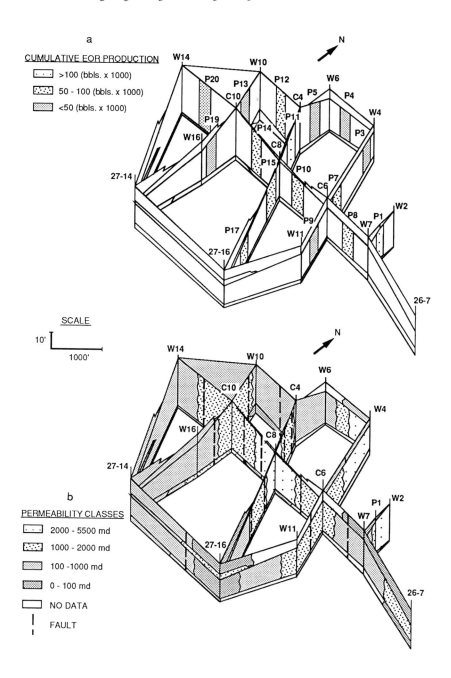

Fig. 19. Comparison of the spatial distribution of cumulative EOR production (a) and permeability (b) in the TIP area.

Model Confirmation

Comparison of the spatial distribution of residual oil saturation obtained from simulation (Fig. 20a) corresponds fairly well to that of core saturations after 10 years of waterflooding (Fig. 20b).[1] In general, greater amounts of oil remained in the southwest part of the TIP, where lower permeabilities and higher diagenetic clay contents prevented good sweep efficiencies. The high ROS values in the southwest part of the pilot may be a result of the presence of nearby faults. Because faults were not included in the simulation model, this area of high ROS is not present in the simulation prediction. Lower residual oil saturations occur in the central portion of the TIP area, where permeabilities are higher and clay contents lower.

Comparison of the field and simulated waterfront advancements of injected water indicate similar frontal movement (Fig. 21). The similar shapes of the fronts suggested that the model developed adequately describes the waterflooding process. The main control mechanism for the frontal advancement is the slope of the reservoir, the location of injectors and major areal permeability variations.

V. MODEL ELEMENTS AND GUIDELINES FOR FIELD DEVELOPMENT

A. Important Model Elements for Each Stage of Production

The heterogeneities important for each stage of production in Unit 'A', were outlined by Honarpour, et al.[1] The relationships between production performance and the various types of heterogeneities found in Bell Creek field may be used as a guide for elements to be included in reservoir models for other barrier island reservoirs.

In Unit 'A', it was found that primary production was dominantly influenced by large-scale depositional heterogeneities and moderately influenced by medium-scale diagenetic heterogeneities. The influence of structural features such as regional dip and faulting was low to negligible. Based on the Bell Creek field example, a sedimentological model including diagenetic information adequately describes the reservoir for prediction of primary production performance.

Secondary production was dominantly influenced by regional dip, moderately to dominantly influenced by medium-scale diagenetic features and moderately influenced by large- to medium-scale depositional features. A simple permeability layer model, as presented in this paper, that includes the dip of the reservoir and additional diagenetic information is necessary to design the waterflood pattern and predict waterflood performance.

Tertiary production was dominantly influenced by depositional features, locally strongly influenced by diagenetic heterogeneities, and moderately to locally strongly by faults. The comparisons of permeability and diagenetic

clay content to EOR production presented in this paper indicate that a
detailed model of these features accompanied by detailed diagenetic and
structural descriptions is necessary to adequately predict EOR production
performance and sweep efficiency.

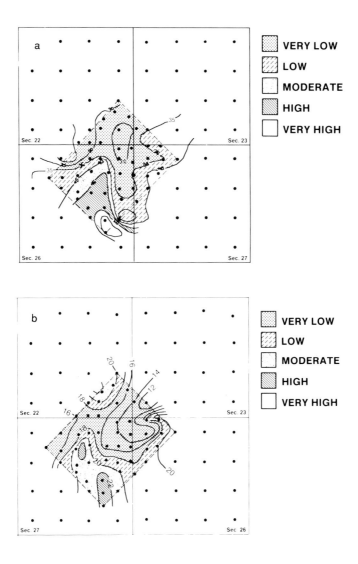

Fig. 20. Comparison of residual oil saturation distribution (a) and that
from measurement of cores drilled after 1980 (b) obtained by full-scale areal
simulation, in percent. (After Honarpour et al., 1988)

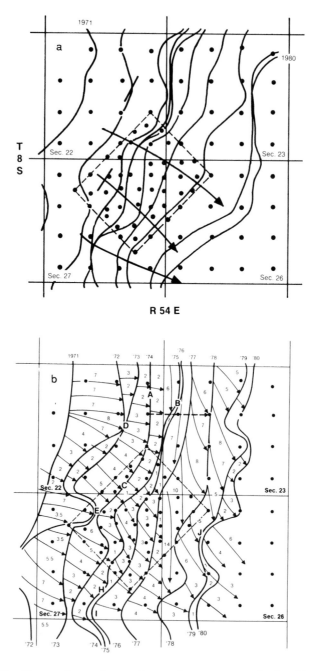

Fig. 21. Comparison of waterfront advancements obtained by full-scale areal simulation (a) and that from the 70% water cut production data (b).

B. Guidelines for Field Development

The following guidelines for field development resulted from NIPER studies of Unit 'A', Bell Creek field and may be useful for developing other barrier-island reservoirs.

1. The best reservoir properties trend along the strike and in the central portions of the sandbody. Well spacings of 40 acres and greater may be applied along the strike of barrier islands, while spacings of 40 acres are adequate perpendicular to the strike for both primary and secondary recovery processes.

The scale of permeability variations from outcrop permeability data supports these spacings and indicates that outcrop permeability data may be useful early in field development to determine the scale of permeability variations.

2. A linedrive waterflooding pattern, with injectors placed down-dip and along the strike of the barrier island sand body and moving the line of injectors updip, is an effective recovery strategy. The saturation of the gas cap and invasion by the oil bank can be prevented by maintaining a high gas-cap pressure.

3. Reservoir characterization for primary and secondary recovery needs to include the definition of external boundaries, lateral variations in reservoir thickness, and the dip and strike of the reservoir. Only major divisions of facies groups with high permeability and kv/kh ratio contrasts are necessary. The importance of permeability contrasts less than two-fold are negligible.

4. Important factors for reservoir characterization for EOR (chemical flooding) are directional permeability (anisotropy), spatial distribution of clay amount and type, and fault locations. A 20-acre spacing and five-spot pattern was adequate for micellar-polymer chemical EOR in the TIP pilot area.

VI. SUMMARY AND CONCLUSIONS

1. A sequence of model development is presented that started with the geological model and was followed by a permeability layer model and a flow unit model. New additional information must be integrated with the previously derived models to enhance the value of the continually changing reservoir model.

2. The geologic model is composed of four components: a) the depositional setting that identifies depositional environment, processes of deposition and erosion, and facies and that contains information on reservoir geometry and dimensions and the internal architecture of facies; b) a diagenetic history, that outlines the paragenetic sequence, documents the stages of reservoir quality enhancement and degradation, and describes the presence of additional heterogeneities developed subsequent to depositional heterogeneities; c) a structural history that identifies the locations, geometries, and dimensions of faults, fractures, folds, and reservoir dip; and d) the geochemical characteristics that contain

information on the origin and type of formation fluids, rock-fluid, and fluid-fluid interactions.

3. The permeability layer model quantifies the sedimentological model by incorporating numerical values of petrophysical properties, which makes it useful for engineering calculations of reservoir volumetrics. In the area studied in Bell Creek field, genetically related groups of facies correspond well with distinct permeability populations.

4. The flow unit model incorporates all available information and provides input for numerical simulation. The model developed for the TIP area in Bell Creek field, illustrated how information from a number of different sources and different scales is combined to form a detailed picture of the reservoir fluid flow properties.

5. A one-layer simulation model that contains lateral permeability variations adequately predicted front movement and ROS distribution in the TIP area because there is little vertical variability within the major part of the reservoir. The greatest variability of permeability on the interwell scale occurred laterally on a scale of 0.25 miles and was controlled by structural and diagenetic processes that, in places, significantly modified the depositionally related permeability pattern. The unmodified depositional pattern and related production characteristics can extend laterally on the order of a few miles.

6. Outcrop data are useful for identification of facies and permeability trends on inter-well scales, as well as important features such as faults and valley fill deposits. Outcrop permeability data compare well with the subsurface data in the TIP area in characteristics such as permeability contrasts, lateral scale of variability grain-size distribution, permeability/porosity relationships and paragenetic sequence. This agreement of properties suggests that outcrop permeability measurements may be used to approximate variations in the subsurface.

7. Two ranges of correlation length from variogram analysis appear to represent features resulting from diagenetic processes (shorter range) and depositional processes of barrier island formation and subsequent erosion by fluvial processes (longer range).

8. The model developed is confirmed by good agreement with cumulative EOR production data and comparison of the ROS distribution and waterfront advancement rate from simulation results and reservoir data.

ACKNOWLEDGMENTS

This work was performed for the U. S. Department of Energy under Cooperative Agreement DE-FC22-83FE60149. The authors thank Herbert B. Carroll, Jr. for his support and encouragement throughout the project. Appreciation is also extended to Edith Allison, of the DOE Bartlesville Project Office and Min K. Tham, Aaron Cheng, Paul Stapp and Bill Linville of NIPER for their valuable suggestions and recommendations. Special thanks are due to Clarence Raible for the CEC measurements, Cindy Robertson for her technical assistance and Edna Hatcher for typing the manuscript.

REFERENCES

1. Honarpour, M., M. J. Szpakiewicz, R. A. Schatzinger, L. Tomutsa, H. B. Carroll, Jr., and R. W. Tillman. Integrated Geological/Engineering Model Barrier Island Deposits in Bell Creek Field, Montana. Pres. at the Sixth SPE/DOE Symposium on Enhanced Oil Recovery, Tulsa, Apr. 14-20. SPE/DOE paper 17366, 1988.

2. Szpakiewicz, M., R. Schatzinger, M. Honarpour, M. Tham, and R. Tillman. Geological/Engineering Evaluation of Heterogeneity, Petrophysical Properties and Productivity of Barrier Island/Valley Fill Lithotypes in the Bell Creek Field; Muddy Sandstone, Powder River Basin, Montana. Rocky Mountain Assoc. Geologists Symposium, Sandstone Reservoirs of the Rocky Mountains, 1989.

3. Reinson, G. E. Barrier Island and Associated Strand Plain System, *in* R. G. Walker, ed., Facies Models: Geoscience Canada, Reprints Series, No. 1., pp. 119-140, 1984.

4. Tillman, R. W., M. Szpakiewicz, M. Honarpour, and S. R. Jackson. Reservoir Description and Production History, Bell Creek Field, Muddy Sandstone, Barrier Island, and Valley Fill Deposits. AAPG, Bull. v 72/2 Abs., 1988.

5. Davis, John C. Statistics and Data Analysis in Geology. John Wiley and Sons, New York, p. 550, 1973.

6. Camilleri, D. Description of an Improved Compositional Micellar/Polymer Simulator. SPE Reservoir Engineering, pp. 427-432, November 1987.

7. Pope, G. A. Mobility Control and Scaleup for Chemical Flooding. Department of Energy Report No. DE85000149, 1986.

8. Szpakiewicz, M., R. Tillman, S. Jackson, and G. deVerges. Sedimentologic Description of Barrier Island and Relate Deposits in the Bell Creek Cores, Montana and Analogous Outcrops Near New Haven, Wyoming. NIPER open file. Available for review at NIPER, 1988.

9. Zoback, M. L. and M. Zoback, State of Stress in the Conterminous United States. Geophy. Res., v. 85, pp. 6113-6156, 1980.

10. Weimer, R. J., C. A. Rebne, and T. L. Davis. "Geologic and Seismic Models, Muddy Sandstone, Lower Cretaceous, Bell Creek--Rocky Point Area, Powder River Basin, Montana and Wyoming." A.A.P.G. Bull. Vol. 72 No. 7, 1988.

11. Hearn, C. L., W. J. Ebanks, R. S. Tye, and V. Ranganathan. Geologic Factors Influencing Reservoir Performance of the Hartzog Draw Field, Wyoming. J. Pet. Tech., pp. 1335-44, 1984.

12. Hearn, C. L., J. P. Hobson, and M. L. Fowler. Reservoir Characterization for Simulation, Hartzog Draw field, Wyoming. L. L. Lake, H. B. Carroll, Jr. eds. Academic Press, 1966.

13. Ebanks, W. J., Jr. Flow Unit Concept - Integrated Approach to Reservoir Description for Engineering Projects. Amer. Assoc. Petrol. Geol. Ann. Mtg., Los Angeles (abs.), 1987.

14. Slatt, R. M. and G. L. Hopkins. Scales of Geological Reservoir
    Description for Engineering Applications: North Sea Oilfield Example.
    Presented at the 63rd Annual Technical Conference and Exhibitor of
    the Society of Petroleum Engineers held in Houston, TX. SPE paper
    18136, October 1988.
15. Rodriguez, A. Facies Modeling and the Flow Unit Concept as a
    Sedimentological Tool in Reservoir Description: A Case Study, Pres.
    at the 63rd Annual Technical Conference and Exhibition of the Society
    of Petroleum Engineers held in Houston, TX. SPE paper 18154,
    October 1988.
16. Dubrule, O. and H. H. Haldorsen. Geostatistics for Permeability
    Estimation in Reservoir Characterization. in L. L. Lake, H. B.
    Carroll, Jr. eds. Proceedings First International Reservoir
    Characterization Technical Conference, Academic Press, 1986.

LITHOLOGY PREDICTION FROM SEISMIC DATA, A MONTE-CARLO APPROACH

P. M. Doyen
T. M. Guidish
M. de Buyl

Western Geophysical,
a Division of Western Atlas International
Houston, Texas

I. SUMMARY

Predicting the areal extent of sand and shale units is essential to delineate the boundaries of clastic reservoirs and understand their flow properties. However, in general, wells are too sparsely distributed to predict lateral lithologic variations accurately. We present a Monte-Carlo technique for simulating sand/shale cross – sectional models from seismic interval velocity profiles. This technique is illustrated with an oil-producing channel-sand reservoir. Alternate sand/shale models are generated that are consistent with lithologic observations at the wells, with the seismic information, and with the spatial autocorrelation and crosscorrelation structure of the data. Compared with lithologic models that rely only on well observations, the simulations that incorporate seismic data are better constrained spatially and, hence, provide more accurate images of the distribution of sands and shales.

II. INTRODUCTION

In reservoirs characterized by alternating shale and sand bodies, prediction of the flow behavior is conditioned by our ability to delineate lithologic variations away from the wells. While shale lenses may act as vertical permeability barriers, they also can control gas coning at the wells. Unfortunately, in areas of rapid lithologic variation, the lateral continuity of sand and shale units can rarely be inferred from sparse gamma-ray log measurements alone. Stratigraphic interpretation of seismically derived impedances or velocities is needed to help differentiate sands from shales and determine their lateral continuity.   We present a Monte-Carlo technique that simulates vertical sand/shale cross sections from seismic interval velocity profiles. This technique is based on the recent work of Alabert (1987) and Journel (1987) in the area of conditional simulation. The simulated models have the following properties:

1. They reproduce the sand/shale vertical sequences interpreted at the wells.
2. The areal extent and thickness of the simulated sand/shale units are consistent with the spatial autocorrelation structure of the binary lithology.
3. The simulations reproduce the spatial crosscorrelation existing between lithology and seismic velocity.

Compared with previous methods for simulating sand/shale sequences (Haldorsen and Lake, 1982; Desbarats, 1987; Matheron et al., 1987), our approach provides models that are more spatially constrained by systematically incorporating densely sampled seismic data indirectly related to lithology.   In the following paragraphs, the Monte-Carlo simulation method is first briefly described. It is then used to predict the lateral variations of lithology in an oil-bearing channel-sand reservoir in the Taber-Turin area of Alberta, Canada.

III. MONTE-CARLO SIMULATION OF SAND/SHALE SEQUENCES

In the geostatistical framework that we adopt here, the unknown vertical sand/shale cross section is interpreted as a particular realization of a two-dimensional, statistically homogeneous binary random field $\{B(\tilde{x})\}$. At each point $\tilde{x} = (x, z)$ of the cross section, the random variable B is defined by

$$B(\tilde{x}) \begin{cases} = 1 & \text{if } \tilde{x} \text{ is in shale} \\ = 0 & \text{if } \tilde{x} \text{ is in sand} \end{cases} \tag{1}$$

with univariate probability distribution given by

$$\text{Prob}\{B(\tilde{x}) = 1\} = \text{shale volume fraction}$$
$$\text{Prob}\{B(\tilde{x}) = 0\} = \text{sandstone volume fraction}. \tag{2}$$

In practice, the sand/shale spatial arrangement must be inferred from a densely and regularly sampled interval velocity profile, $\{V(\tilde{x})\}$, which is indirectly related to lithology, and from observations of the variable B in wells intersecting the velocity cross section (Figure 1). In general, seismic interval velocities are related to lithology only ambiguously. Indeed, the velocity ranges of sand and shale often overlap; i.e., depending on the presence of gas, pore pressure, rock composition, and porosity, reservoir sands can exhibit higher or lower velocities than those of adjacent shales (Mummery, 1988). Here, we do not assume that there is a unique correspondence between B and V. Instead, we model their dependence statistically using a spatial crosscorrelation function.

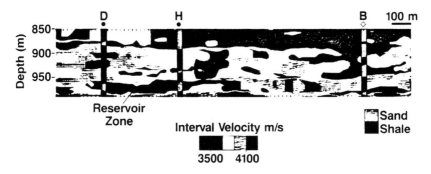

Fig. 1. **Input for Lithologic Simulations**

A critical step in the Monte-Carlo simulation is to obtain a linear-mean-square (LMS) estimate of the following conditional probability at each point $\tilde{x}$ of the cross section:

$$\text{Prob}\{B(\tilde{x}) = 1 \mid B(\tilde{x}_1), ..., B(\tilde{x}_n), V(\tilde{x})\} = \text{Prob}\{B(\tilde{x}) \mid \text{data}\} . \tag{3}$$

In equation (3), $B(\tilde{x}_1), ..., B(\tilde{x}_n)$ are n binary-valued observations of the lithology at spatial locations $\tilde{x}_1, ..., \tilde{x}_n$ and $V(\tilde{x})$ is the seismic interval velocity at $\tilde{x}$. By definition of the binary variable B, this probability also is equal to the conditional expectation of $B(\tilde{x})$, given the data. The LMS estimate, $B^*(\tilde{x})$, is given by

$$\begin{aligned} B^*(\tilde{x}) &= \text{Prob}^* \{B(\tilde{x}) = 1 \,|\, \text{data}\} \\ &= \sum_{i=1}^{n} \omega_i(\tilde{x}) B(\tilde{x}_i) + \alpha(\tilde{x})\, V(\tilde{x}) + c(\tilde{x}), \end{aligned} \qquad (4)$$

where weights $\omega_1, ..., \omega_n$ and α assigned to the data are determined by minimizing the mean square error $E\{[B(\tilde{x}) - B^*(\tilde{x})]^2\}$. This minimization only requires knowledge of the spatial autocorrelation and crosscorrelation structure of the variables B and V. The remaining constant, $c(\tilde{x})$, is determined from the condition that the estimate is globally unbiased; i.e., $E[B^*] = E[B] =$ shale volume fraction. The probability estimate given in equation (3) mixes binary-valued observations with velocity data, which take on continuous values outside the interval [0,1]. In practice, this scaling problem is solved by performing the LMS estimation with the variable V transformed so that its values lie between 0 and 1. At this stage, one limitation of the method is that it assumes that the interval velocity observations are error-free. In reality, the velocities extracted from band-limited and noise-contaminated seismic data are inherently nonunique. However, if the error in the seismic velocity inversion can be quantified, then the Monte-Carlo simulation technique can be constrained by bounds on the velocities instead of single velocity estimates (Journel, 1987).

The sand/shale simulation method involves the sequential estimation of the conditional probability [equation (3)] at all sample points of the seismic velocity profile. The algorithm can be summarized as follows. For all points $\tilde{x}$ in the cross section:

1. Obtain the LMS estimate $\text{Prob}^* \{B(\tilde{x}) = 1 \,|\, \text{data}\} = B^*(\tilde{x})$,
2. At $\tilde{x}$, draw a simulated value B_s that is equal to 1 or 0 with probability $B^*(\tilde{x})$ and $1 - B^*(\tilde{x})$, respectively, and
3. Add the simulated value $B_s(\tilde{x})$ to the conditioning dataset "data."

Note that when the first location $\tilde{x}$ is considered in the simulation, the conditioning dataset contains only the observations of B at the wells. However, when the ith point is selected, the dataset also includes $(i-1)$ previously simulated values.

With the above algorithm, it can be shown that, in theory, the simulated binary-valued field $\{B_s(\tilde{x})\}$ has the same autocorrelation and crosscorrelation structure as the real field $\{B(\tilde{x})\}$ itself. This property directly follows from the fact that, in the LMS estimation, the error $B - B^*$ is orthogonal to the data. Also, it is easy to show that the simulated lithology honors the well information; i.e., $B_s(\tilde{x}) = B(\tilde{x})$ for $\tilde{x}$ in the wells.

IV. A CASE STUDY OF A CHANNEL-SAND RESERVOIR

The simulation technique is applied to predict the spatial arrangement of sands and shales in a section of the Upper Mannville Formation in the Taber-Turin area of Alberta, Canada. Figure 1 shows a seismically derived velocity cross section covering this clastic sequence. This velocity profile was extracted from a 3-D seismic survey using the Seismic Lithologic Modeling (SLIM^{TM})[1] process (de Buyl et al., 1988). At the base of this section is the 15- to 30-m-thick Glauconitic reservoir sand, which produces oil at wells D and H in the northern portion of the field. The velocity profile shows that the reservoir interval exhibits lower velocities than the overlying siltstone. Laterally, to the south, where dry well B was drilled, the channel sand is also surrounded by high-velocity shales and tight siltstones.

Figure 1 also shows the major sand/shale intervals interpreted from log measurements at three well locations intersecting the seismic profile. As only two lithologic classes are modeled in the simulation, the "shale class" includes not only actual shales but also the tight siltstones and shaly sands in the clastic sequence. Comparison at the wells of the binary lithology with the velocity cross section demonstrates that the rocks belonging to the shale class on the average tend to exhibit higher velocities than the sands. Mathematically, this implies that the variables B and V are positively crosscorrelated. In fact, the coefficient of correlation between B and V that was calculated at the wells is equal to 0.7. Note that the sign of this correlation may be reversed, for example, when dealing with low-porosity sands or overpressured shales. Also, the magnitude of the correlation coefficient may be small if the velocity ranges of the two lithologies strongly overlap. In this last case, conditioning the simulations with seismic data will not enhance the sand/shale discrimination.   For the lithologic variable B, a geometrically anisotropic correlation model was selected with major and minor axes oriented in the horizontal and vertical directions, respectively. The autocorrelation function for B is defined by

$$C_{BB}(h_x, h_z) \propto \exp\left\{-[(\frac{h_x}{r_x})^2 + (\frac{h_z}{r_z})^2]^{\frac{1}{2}}\right\} , \qquad (5)$$

which is a function of the horizontal and vertical distances, h_x and h_z in the plane of the cross section. In equation (5), $r_x = 100\,\mathrm{m}$ and $r_z = 10\,\mathrm{m}$ can be considered as one-third of the practical correlation ranges in the horizontal and vertical directions, respectively. In the vertical direction, this exponential correlation model was inferred from observations of B at the wells. In the horizontal direction, the wells are too sparse to allow direct estimation of the correlation structure. Accordingly, an approximate correlation length $3r_x$ was determined from experimental correlations calculated horizontally along the thresholded velocity profile. Indeed, although the velocity intervals of the shale and sand classes overlap, the seismic profile thresholded at $V \approx 3500$ m/s provided a rough image of the spatial distribution of the two lithologies, which was then used to estimate the horizontal correlation structure.

Figure 2 shows two sand/shale simulations performed at the sample points of the SLIM velocity section (Figure 1), which has 90 traces, each containing 71 samples. The spacing between samples in the horizontal and vertical directions is, respectively, 20 and 2 m. The simulated images represent alternate sand/shale distributions, all of which are consistent with both the geophysical and log information at hand. First, the simulated images match the lithology at the wells. Second, they approximately reproduce the autocorrelation model (equation 5) inferred from the reservoir data. Also, the overall shale volume fraction, $E[B] = \mathrm{Prob}\{B = 1\} = 0.55$, which was inferred from the well data, is reproduced by the simulated models. Finally, the simulations are consistent with the spatial crosscorrelation existing between B and V. In particular, the coefficient of correlation between the simulated fields and the velocity profile is approximately equal to that inferred from the data.

---

[1]SLIM is a registered trademark of Western Atlas International, Inc.

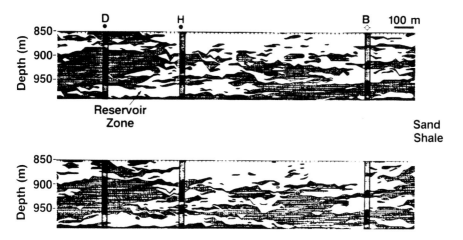

Fig. 2. **Seismically Constrained Simulations**

V. DISCUSSION AND CONCLUSION

We have introduced a Monte-Carlo simulation technique to simulate sand/shale sequences from seismic interval velocity profiles and lithologic observations in wells. This technique yields alternative lithologic models that are conditioned by both the seismic and well data and by their autocorrelation and crosscorrelation structure. Traditionally, lithologic simulations are conditioned by sparse well observations alone. Successive simulations are then widely different in uncontrolled areas, revealing the underconstrained nature of the models. By contrast, the seismically consistent lithologic models displayed in Figure 2 are better constrained spatially and therefore have the same gross layering. In particular, the two models show the truncation of the reservoir sand by shales in the southern region. However, the seismically consistent simulations still differ in the connectivity properties of the sand units. Analyzing the variability among successive simulations is useful for assessing the uncertainty associated with the subsurface models. For example, different simulated models can be selected on the basis of geological expertise and used in a flow simulator to forecast reservoir behavior during production. By comparing the range of forecasts based on the selected models, we can determine the uncertainty in predicting the flow behavior that stems from an imperfect description of lithologic heterogeneities. Moreover, we can assess the need for additional geophysical or geological data to further constrain the reservoir models.

REFERENCES

Alabert, F., 1987, Stochastic imaging of spatial distributions using hard and soft information: M.S. thesis, Stanford Univ.

de Buyl, M., Guidish, T., and Bell, F., 1988, Reservoir description from seismic lithologic parameter estimation: J. Petrol. Technol., **40**, *4*, 475-482.

Desbarats, A. J., 1987, Numerical estimation of effective permeability in sand-shale formations: Water Resources Research, **23**, 273-286.

Haldorsen, H. H., and Lake, L. W., 1982, A new approach to shale management in field scale simulation models: SPE Paper 10976, presented at the 57th Annual Fall Technical Conference and Exhibition of the SPE, New Orleans.

Journel, A. G., 1987, Introduction to geostatistics for reservoir characterization: Class notes, Stanford Univ.

Matheron, G., Beucher, H., de Fouquet, C., Galli, A., Guerillot, D., and Ravenne, C., 1987, Conditional simulation of the geometry of fluvio-deltaic reservoirs: SPE Paper 16753, presented at the 62nd Annual Technical Conference and Exhibition of the SPE, Dallas.

Mummery, R. C., 1988, Discrimination between porous zones and shale intervals using seismic logs: The Leading Edge, **7**.

INFLUENCE OF LITHOLOGY AND GEOLOGIC STRUCTURE ON IN SITU STRESS: EXAMPLES OF STRESS HETEROGENEITY IN RESERVOIRS

Lawrence W. Teufel

Geomechanics Division 6232
Sandia National Laboratories
Albuquerque, New Mexico 87185

ABSTRACT

Knowledge of *in situ* stress is increasingly understood to be an important factor in formulating a multidisciplinary approach to reservoir characterization and in the development and completion of oil and gas reservoirs. Two examples are presented to illustrate the influence of lithology and geologic structure on the distribution and heterogeneity of stress at depth. Stress measurements made in flat-lying, tight gas sandstone reservoirs and bounding mudstone formations of the Mesaverde Group in the Piceance Basin of Colorado show that principal horizontal stress magnitudes do not increase linearly with depth, solely as a function of the overburden, but vary with lithology. Minimum horizontal stresses in sandstone layers are more than 4 MPa less than adjacent mudstones. Stress measurements made in the Ekofisk Field in the North Sea demonstrate that principal horizontal stress directions are not uniform across a large structural dome but are radial, with the azimuth of the maximum horizontal stress oriented perpendicular to the structural contours around the dome. These two examples indicate that *in situ* stress is not always defined by regional boundary conditions of a sedimentary basin, but is also affected by the local geologic environment.

I. INTRODUCTION

In situ stress affects nearly all physical properties of rock and hence the measurement and interpretation of (1) geophysical data, (2) petrophysical properties such as porosity and permeability, (3) rock strength and ductility, and (4) mechanisms of rock deformation and failure. In naturally fractured reservoirs, the influence of stress on reservoir behavior is even more pronounced, particularly with respect to

RESERVOIR CHARACTERIZATION II

565

fluid flow through fractures. Accordingly, knowledge of *in situ* stress is becoming increasingly important in current trends toward a multidisciplinary approach to reservoir characterization and in the development and completion of oil and gas reservoirs. Some important examples include reservoir compaction during fluid withdrawal, behavior of natural fractures, hydraulic fracture growth, interaction between natural and induced hydraulic fractures, fracturing during waterflooding, and wellbore stability.

Although much stress data throughout the world are available in the literature (e.g. Haimson, 1975; Brown and Hoek, 1978; McGarr and Gay, 1978; Zoback and Zoback, 1980), most of these data should be considered isolated and may not give a clear picture of the stress distribution at depth. By isolated, I mean that the results are either: (1) near surface, such as in a quarry or shallow wellbore, (2) have only a few measurements per well and usually a large distance apart, or (3) in a mine where most of the data come from one depth in only one or two lithologies. As such, this information provides little understanding of the local effect of lithology and geologic structure on the measured stress state. Consequently, conflicting observations exist in the literature as to the importance of changes in lithology and structure on *in situ* stress. Some workers indicate that principal horizontal stresses are regionally homogeneous, increasing linearly with depth, solely as a function of the overburden, and principal stress directions are fairly uniform over large areas (e.g. Brown and Hoek, 1978; McGarr and Gay, 1978; Zoback and Zoback, 1980). In sharp contrast, other workers suggest that changes in lithology can significantly affect the local *in situ* stress state (Kry and Gronseth, 1982; Teufel and Warpinski, 1983; Warpinski et al, 1985; and Warpinski and Teufel, 1989a).

The purpose of this paper is to present two examples that illustrate the influence of lithology and geologic structure on the distribution and heterogeneity of stress at depth. The first example is an extensive series of stress measurements made in flat-lying, tight gas sandstone reservoirs and bounding mudstone formations of the Mesaverde Group at the Multiwell Experiment site in the Piceance Basin of Colorado. These measurements show the effect of changes in lithology on the magnitudes of principal horizontal stresses. The second example is stress measurements made in seven wells located at different positions on the elongated structural dome that forms the Ekofisk Field in the North Sea. Principal horizontal stress directions are not uniform across the field and are related to the tectonic evolution of the dome.

II. TECHNIQUES FOR MEASURING IN SITU STRESS
AT DEPTH

Several different techniques have been developed to measure or infer the orientation and magnitude of principal horizontal stresses at depth. These include (1) hydraulic fracture stress tests (Kehle, 1961; Fairhurst, 1967; Haimson, 1981; Warpinski et al, 1985), (2) various core analyses such as anelastic strain recovery (Teufel, 1982; Blanton, 1983; Warpinski and Teufel, 1989b), differential strain curve analysis (Ren and Roegiers, 1983), and differential wave velocity analysis

(Ren and Hudson, 1985), (3) wellbore condition logs (such as televiewers and four-arm calipers) to examine eccentricity and breakouts (Gough and Bell, 1981), and hydraulic fracture diagnostics to infer principal horizontal stress directions (Smith et al, 1986; Lacy, 1987). In this paper we will only present stress data from hydraulic fracture tests and anelastic strain recovery measurements of oriented core. A brief description of the two techniques follows.

A. Hydraulic Fracture Method

Several authors have described the details of stress measurements using the hydraulic fracture technique in open-holes in strong, competent rock (Kehle, 1961; Fairhurst, 1964; Haimson, 1981). The minimum horizontal stress is equal to the instantaneous shut-in pressure (ISIP) for small-volume hydraulic fractures. However, for tests in perforated cased holes, as was done in the Multiwell Experiment, there are additional complications because of the effects of the casing, cement anulus, explosive perforation damage, and random perforation orientation. Warpinski (1983) showed that with a carefully designed perforation schedule, accurate measurements of the ISIP can be made in cased holes. The major uncertainty in determining stress magnitudes with the hydraulic fracture technique, whether the measurement is made in an open-hole or in a cased hole through perforations, is the ability to obtain, measure, and interpret a clear ISIP.

The technique and procedures of hydraulic fracture stress measurements in cased holes through perforations are discussed in detail in Warpinski et al (1985). Briefly they consist of (1) perforating a 0.6 m interval with 8 medium-size perforations (5-15 gm) with 90° or 120° phasing, (2) isolating the interval with straddle packers, (3) fracturing the rock with small volumes (0.04-0.4 m³) of KCl water and recording the pressure with a bottomhole, quartz pressure gage, (4) shutting in with a downhole closure tool, and (5) determining the ISIP. Since the test is conducted in a cased and perforated hole, no information on the maximum horizontal stress or the stress orientations can be obtained.

B. Anelastic Strain Recovery Measurements of Oriented Cores

The Anelastic Strain Recovery (ASR) technique is described in Teufel (1982). Briefly, it consists of mounting clip-on displacement gages (Holcomb and McNamee, 1984) on a piece of sealed, oriented core and recording the time-dependent relaxation of the core. Determination of the orientation of the stress field has been shown to be straight-forward for many sedimentary rocks and is readily calculated by determining the principal strain orientations. If there is no rock fabric to distort the results, the maximum strain direction is found to be coincident with the maximum stress direction as determined by independent methods (Teufel, 1982; Smith et al, 1986; Warpinski and Teufel, 1989a).

The determination of the stress magnitudes is more complicated and requires a constitutive model for the ASR process. Blanton (1983) and Warpinski and Teufel (1989b) have developed different types of viscoelastic models to explain the behavior. Both models will be used in the analysis of data presented below.

Blanton's (1983) solution, which I will refer to as the direct model, is the easiest to apply and yields the direct calculation of the stresses from the principal strains as

$$\sigma_1 = (\sigma_v - \alpha P_p)\frac{(1-\nu)\Delta\epsilon_1 + \nu(\Delta\epsilon_2 + \Delta\epsilon_v)}{(1-\nu)\Delta\epsilon_v + \nu(\Delta\epsilon_1 + \Delta\epsilon_2)} + \alpha P_p \tag{1}$$

$$\sigma_2 = (\sigma_v - \alpha P_p)\frac{(1-\nu)\Delta\epsilon_2 + \nu(\Delta\epsilon_1 + \Delta\epsilon_v)}{(1-\nu)\Delta\epsilon_v + \nu(\Delta\epsilon_1 + \Delta\epsilon_2)} + \alpha P_p \tag{2}$$

where the $\Delta\epsilon$ are the change in the principal strains between any two times, ν is Poisson's ratio, P_p is the pore pressure, α is a poroelastic constant, and the subscripts 1 and 2 refer to the maximum and minimum horizontal directions, respectively, while v refers to the overburden. Important assumptions for the direct model include (1) linear viscoelastic behavior, (2) constant Poisson's ratio throughout the process, (3) step unloading of the *in situ* stresses at the moment of coring, (4) a constant α throughout the process, (5) a vertical overburden stress and wellbore, and (6) isotropic behavior.

The model of Warpinski and Teufel (1989b), which I will refer to as the strain-history model (because it requires fitting a theoretical model to the measured strain history), requires a least-squares fit of the entire strain data set to an expected relaxation behavior of the form

$$\epsilon_r(t) = (2\sigma_1\cos^2\theta + 2\sigma_2\sin^2\theta - \sigma_1\sin^2\theta - \sigma_2\cos^2\theta - \sigma_v)J_1(1 - e^{-t/t_1})$$
$$+ (\sigma_1 + \sigma_2 + \sigma_v - 3P)J_2(1 - e^{-t/t_2}) \tag{3}$$

and

$$\epsilon_v(t) = (2\sigma_v - \sigma_1 - \sigma_2)J_1(1 - e^{-t/t_1})$$
$$+ (\sigma_1 + \sigma_2 + \sigma_v - 3P)J_2(1 - e^{-t/t_2}) \tag{4}$$

where θ is the gage angle orientation with respect to the maximum stress, J_1 and J_2 are the distortional and dilational creep compliance arguments (i.e., equilibrium values of the creep compliance), t is the time, t_1 and t_2 are deviatoric and dilational time constants, respectively, and the subscript r refers to the radial direction in the horizontal plane. Important assumptions for this model are (1) the rock behaves as if it is linearly viscoelastic, (2) the behavior is exponential and can be described using standard models, (3) the overburden stress and wellbore are vertical, (4) the rock is isotropic, (5) the bulk modulus of the grain material is not a viscoelastic parameter (since the process appears to be governed by fracturing), and (6) step unloading of the *in situ* stresses at the moment of coring.

Once the data are fitted by least-squares, estimates of the stresses can be made if J_1 is known. Alternatively, a small-volume hydraulic fracture stress in tandem with the ASR data (so σ_2 is known) allows J_1 to be determined. In this paper hydraulic fracture ISIP data are used to calculate σ_1 and J_1.

The primary problems with ASR are that (1) rock fabric may distort the results and (2) sufficient data must be obtained to use either viscoelastic model to calculate stress magnitudes.

III. DISTRIBUTION OF IN SITU STRESS AT DEPTH

A. Effect of Lithology

One of the best examples showing the effect of changes in lithology on *in situ* stress is an extensive series of stress measurements made at the Multiwell Experiment site in the Piceance Basin, Colorado (Figure 1). The Multiwell Experiment is a U.S. Department of Energy funded experiment to characterize low-permeability reservoirs in the western United States and to test the technology for the production of natural gas from these marginal resources. The facility consists of three closely spaced 2.50 km deep wells that are 30-68 m apart. The Mesaverde Group, the tight sand target, is between 1.25 km and 2.50 km deep at this location, and has been painstakingly characterized by the recovery and analysis of over 1300 m of core, advanced logging programs, detailed well testing, frequent stress testing by several techniques, and highly instrumented stimulation experiments (Northrop, 1988).

The late Cretaceous Mesaverde group at this site consists of (1) a marine-influenced paralic zone from 1250-1340 m, (2) a fluvial meanderbelt interval from 1340-1830 m, (3) an upper deltaic, generally non-coally section from 1830-2010 m, designated the coastal, (4) a deltaic, coal-bearing interval from 2010-2270 m, designated the paludal, and (5) a marine section from 2270-2500 m. These intervals are characterized by their own distinct hydrocarbon reservoirs (Lorenz, 1983). Relatively continuous sandstone units are prevalent in the paralic, wide meanderbelts dominate the fluvial, distributary channels and splay deposits are typical in the paludal and coastal zones, and thick blanket sandstones are found in the marine section. These different depositional characteristics have a strong influence on lithologic varations within each zone.

The burial and tectonic history of the flat-lying formations at the Multiwell site is relatively simple (Lorenz, 1984), consisting of one cycle of subsidence and uplift. Geologic evidence suggests predominantly east-west compression throughout the burial history, as reflected by east-west trending normal faults and extension fractures in the basin. The tight gas sandstone reservoirs at this site are not associated with any tectonic structure.

Determination of *in situ* stress and mapping the distribution of stress as a function of depth and lithology are major objectives of the experiment. Several studies have been completed on the stress measurements at the site in both the marine and nonmarine sections (Teufel and Warpinski, 1984; Teufel et al, 1984; Warpinski et al, 1985; Warpinski and Teufel, 1989a). A representative sample of these stress measurements is a sandstone/mudstone sequence from the coastal zone. Figure 2 shows the magnitudes of the minimum horizontal stress determined from ISIP data from eight hydraulic fracture tests performed in cased holes through perforations. The magnitudes of the maximum and minimum horizontal stresses calculated from ASR core data using the direct and strain history models are shown in Figure 3. The overburden is assumed to be a principal stress and is calculated from gravitational loading using a density of 2.43 gm/cm^3, which was determined from an integrated density log. Pore pressure in the sandstone formations was determined from well testing and is 28.3 ± 1.5 MPa. Lithology is based on core and gamma log data.

The stress measurements clearly show that horizontal stresses are strongly influenced by lithology. Minimum horizontal stress magnitudes in sandstone layers are more than 4 MPa less than in bounding mudstones. The horizontal stresses in the mudstones are close in magnitude to the overburden stress. Using the ASR results, little horizontal stress difference is seen in any of the mudstones. Horizontal stresses in the sandstones are anisotropic. Differences in principal horizontal stress are about 4 - 5 MPa, which is in agreement with results of an open-hole hydraulic fracture stress test conducted in the Rollins marine sandstone at 2304 m depth (Teufel and Warpinski, 1984). The ratios of the minimum and maximum horizontal stresses to the overburden stress range from 0.77 - 0.87 and 0.88 - 0.94, respectively. Magnitudes of the maximum horizontal stress calculated from the direct and strain history ASR models are in close agreement.

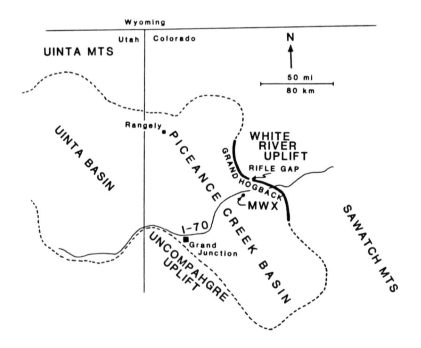

Figure 1. Map showing the location of the Multiwell Experiment (MWX).

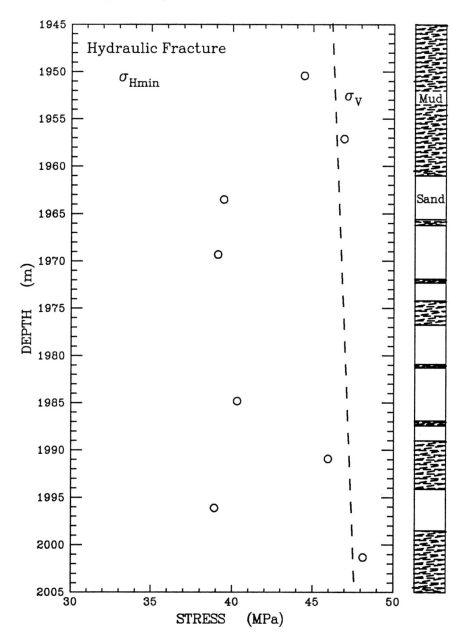

Fig. 2. Plot of minimum horizontal stress magnitude determined from hydraulic fracture tests versus depth and lithology.

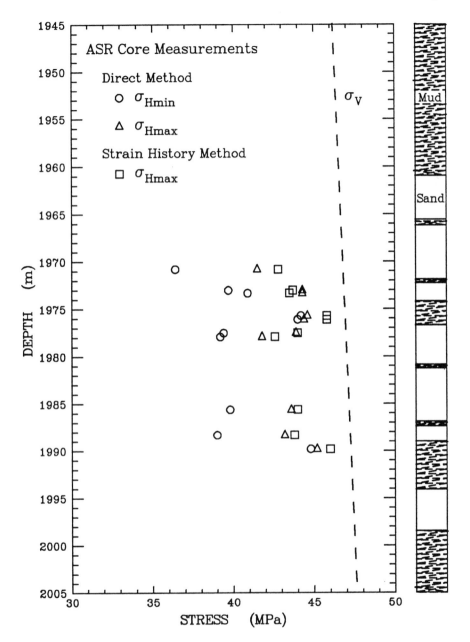

Fig. 3. Plot of maximum and minimum horizontal stress magnitudes determined from anelastic strain recovery measurements versus depth and lithology.

The Multiwell Experiment is unique because of the comprehensive manner in which stress measurements were made, and thus the experiment provides conclusive evidence that lithology significantly influences the *in situ* stress state. The large difference in minimum horizontal stress magntiude between reservoir sandstones and bounding mudstones has important implications for the design of massive hydraulic fractures. These large stress contrasts should result in not only good hydraulic fracture containment, but high treating pressure during the stimulation (Nolte and Smith, 1981). The small difference in principal horizontal stresses in the sandstones may cause problems if any orthogonal or oblique natural fractures are present. These may dilate as treating pressures increase and result in accelerated leakoff (Warpinski and Teufel, 1987).

B. Effect of Geologic Structure

In situ stress measurements using the ASR method have been made in seven wells located at different positions on the elongated structural dome that forms the Ekofisk reservoir. The Ekofisk Field is operated by the Phillips Petroleum Company and lies in the Central Graben region in the southern part of the Norwegian sector of the North Sea (Figure 4). The reservoir is largely chalk and consists of two producing horizons – the Ekofisk Formation (Danian Age,which is approximately 120 m thick, and the 60 m thick Tor Formation (Maastrichtian Age). These two formations are separated by a dense limestone zone which is about 15 - 30 m thick. Except for a few areas of intense fracturing, this zone prevents fluid communication between the producing horizons (Thomas et al, 1987).

Ekofisk is the largest of nine chalk reservoirs in the region. Aerial extent of the field is approximately 9 x 4 km and the thickness of the overlying sediments is about 2.9 km at the crest. The long axis of the elongated dome is north-south.

Porosity of the chalk formations in the producing horizons ranges from about 15 - 40 percent. Matrix permeability is typically about 1 md. However, as result of extensive natural fracturing, the effective permeability of the reservoir is substantially higher, ranging up to 150 md (Thomas et al, 1987).

The natural fracture system at Ekofisk forms the primary conductive path for produced hydrocarbons. Two types of fractures are found: (1) styolite-associated fractures, which form an anastomosing network of nearly vertical extension fractures and (2) tectonic fractures that are conjugate, planar fractures that dip between 65° and 80° (Farrell, 1988). Fracture orientation data suggests that two fracture sets may be present at Ekofisk. One set is associated with a NE-SW regional trend and a second set is associated with a radial pattern around the structural dome (Farrell, 1988; Snow and Hough, 1988).

The ASR method has been used to determine the orientation of principal horizontal stress directions vertically through the reservoir and laterally across the elongated structural dome. ASR measurements were made on oriented cores from 2 to 4 different stratigraphic horizons in each of seven wells located at different positions on the dome. Petrographic observations and petrophysical property measurements were made to eliminate cores with sedimentary or tectonic fabrics that could bias the stress orientations inferred from relaxation data.

Figure 5 shows the orientation of the maximum horizontal stress at seven locations across the reservoir. The stress direction at each well location is given as an average direction. Standard deviations range from 12° to 20° (Table 1). The largest source of error in determining stress directions from ASR measurements is the accuracy of the core orientation survey.

Principal horizontal stress directions are not uniform across the reservoir (Figure 5). On the crest the maximum horizontal stress direction is N8°E ± 15°, which is subparallel to the long axis of the elongated dome. On the flanks, stress directions tend to be radial, with the azimuth of the maximum horizontal stress oriented perpendicular to the structural contours around the dome. At any given well location in the reservoir the principal horizontal stress directions do not show a significant or consistent variation with depth.

Several studies have suggested that the chalk reservoirs in the Ekofisk area are halokinetically induced domes (Ziegler, 1978; Hospers and Holthe, 1980; Watts et al, 1980; Watts, 1983). Principal horizontal stress directions mapped at Ekofisk are in general agreement with tectonic models of elliptical vertical doming described by Withjack and Scheiner (1982). They developed experimental and analytical models for the deformation of a thick homogeneous layer during vertical doming. These models were for circular and elliptical domes, with and without superposed regional horizontal deformation. Analytically derived stress fields for elliptical domes (using a closed-form elastic solution) show that the maximum horizontal stress parallels the long axis of the dome near the crest and becomes radial on the flanks. The theoretical stress pattern is similar to that measured at Ekofisk (Figure 5), and therefore stresses at Ekofisk are thought to be related to the tectonic evolution of the dome.

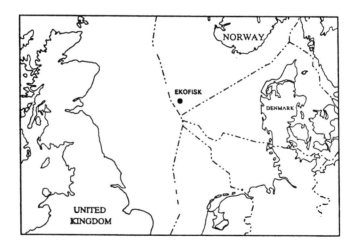

Fig. 4. Map showing location of the Ekofisk Field in the North Sea.

Table 1. Summary of Maximum Horizontal Stress Directions Determined from Partial Anelastic Strain Recovery Measurements Measurements of Oriented Cores from the Ekofisk Field, North Sea

Well	Location	Coring Intervals	Number of Measurements	ϵ_{Hmax} Azimuth
A-4	SE Flank	3	11	N124°E ± 12°
A-5a	S Flank	2	4	N46°E ± 18°
A-15b	SW Flank	3	14	N49°E ± 14°
B-20b	NE Flank	3	12	N27°E ± 16°
C-4	E Flank	4	13	N86°E ± 17°
C-7	W Flank	3	10	N75°E ± 20°
C-11	Crest	3	7	N8°E ± 15°

In addition to tectonic deformation, the stress state at Ekofisk has also been affected recently by reservoir compaction and seafloor subsidence. This deformation is a result of nearly 20 years of petroleum production. Depletion of the reservoir has reduced reservoir pressure by about 20 MPa, which in turn has increased the effective stress and caused compaction of the high-porosity reservoir chalk, and subsidence of the overlying sediments. Seafloor subsidence at the end of 1987 was about 4 m (Sulak and Danielsen, 1988). Numerical models have been used to predict future subsidence at Ekofisk (Boade et al, 1988) and are presently being conducted to determine the effect of subsidence deformation on the *in situ* stress state (R. Boade, personal communication).

Stress measurements in the Ekofisk Field clearly demonstrate that a geologic structure can influence the local *in situ* stress directions. This has important implications to production, development, and waterflood activities, because heterogeneity of *in situ* stresses in a naturally fractured reservoir can affect permeability and permeability anisotropy, hydraulic fracture propagation, and the orientation of fractures induced by waterflooding.

Deformation and conductivity of natural fractures are strongly stress-dependent. For a reservoir with a random population of fractures of similar morphology and roughness, the maximum-permeability direction would be perpendicular to the minimum stress direction. At Ekofisk, there appears to be a well-developed radial fracture pattern around the structure, in addition to a NE-SW regional fracture pattern (Farrell, 1988; Snow and Hough, 1988). The radial fracture pattern is closely aligned with the local maximum horizontal stress direction on the flanks of the structure and probably was created by the structurally-induced stresses during vertical doming. Accordingly, it is suggested that (1) the radial fractures may have the highest permeability, (2) the reservoir permeability should be anisotropic, (3) the maximum flow direction would not be uniform across the field, but would vary radially around the flanks of the structure, and (4) as the effective stress increases due to production, new fractures may be created that are aligned with the local *in situ* stresses.

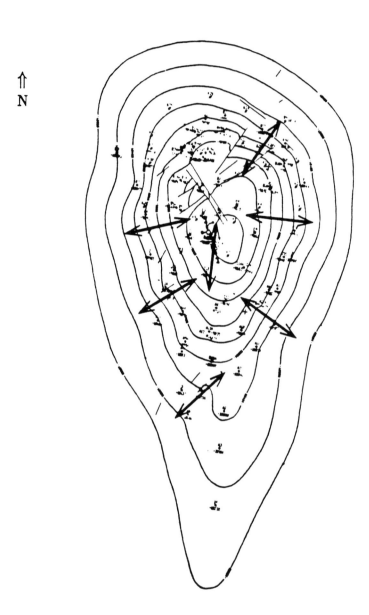

Fig. 5. Structure contour map for top of the Ekofisk Formation (Danian Age) in the Ekofisk Field showing the azimuth of maximum horizontal stress (arrows) determined from anelastic strain recovery measurements of oriented cores from seven wells. The crest of the structure is at a depth of about 9500 ft (2.9 km) and contour intervals are 100 ft (30.48 m).

Limited interference tests and water injection response in a waterflood pilot study conducted in the northern area of the field, near the crest of the structure, indicate that a permeability anisotropy exists (Thomas et al, 1987). The maximum permeability direction was estimated to be in a north-south direction. Additional information on permeability anisotropy at Ekofisk will become available as the waterflood program is expanded.

IV. SUMMARY

Two examples were presented to illustrate the influence of lithology and geologic structure on the distribution and heterogeneity of stress at depth. These examples clearly indicate that *in situ* stress is not always defined by regional boundary conditions of a sedimentary basin, but is also affected by the local geologic environment. Principal horizontal stress magnitudes do not always increase linearly with depth, solely as a function of the overburden, but can be significantly affected by lithologic changes occurring over intervals of a few meters. Principal horizontal stress directions are not always uniform across a field, but can be influenced by geologic structure and how that structure was developed. Therefore, one or two stress measurements within a field may not be sufficient to define the distribution and heterogeneity of stresses within a reservoir and bounding formations necessary for reservoir characterization, design of hydraulic fractures, and optimum location of injection wells for waterflooding.

V. ACKNOWLEDGEMENTS

This work was performed at Sandia National Laboratories, which is operated for the U.S. Department of Energy under contract number DE-AC04-76DP00789. Stress measurements in the Ekofisk Field were conducted through a USDOE/industry cooperative research program with Phillips Petroleum. Discussions with R. Boade, H. Farrell, J. Lorenz, N. Warpinski, and W. Wawersik contributed to the writing of this paper.

VI. REFERENCES

Blanton, T. (1983). presented at SPE/DOE Symp. Low Permeability Gas Reservoirs, Denver, CO.

Boade, R., Chin L., and Siemers, W. (1988). presented at 20th Annual Offshore Technology Conference Houston, TX.

Brown, E. and Hoek, E. (1978). Int. J. Rock Mech. Min. Sci. 15, 211.

Fairhust, C. (1964). Rock Mech. and Eng. Geol. 2, 129.

Farrell, H. (1988). Am. Assoc. Petr. Geol. Bull. 72, 184.

Gough, D. I. and Bell, J. S. (1982). Can. J. Earth Sci. 19, 1358.

Haimson, B. (1975). Reviews of Geophysics and Space Physics 13, 350.

Haimson, B. (1981). Proc. 22nd U. S. Symp. Rock Mechanics (H. Einstein, ed.), 379.

Holcomb, D. and McNamee, M. (1984). Sandia National Laboratories Report SAND84-0651.

Hospers, J. and Holthe, J. (1980). Tectonophysics 68, 257.

Lacy, L. (1987). SPE Prod. Eng. 3, 66.

Lorenz, J. (1983). Sandia National Laboratories Report, Sand83-1078.

Lorenz, J. (1984). Sandia National Laboratories Report, Sand84-2603.

Kehle, R. (1964) J. Geoph. Res. 69, 259.

Kry, R. and Gronseth, M. (1982). presented at 33rd CIM Annual Meeting, Calgary, Alberta.

McGarr, A. and Gay N. (1978). Ann. Rev. Earth Planet. Sci. 6, 405-436.

Nolte, K. and Smith, M. (1981). J. Petr. Tech. 33, 1767.

Northrop, D. (1988). presented at 63rd Annual SPE Meeting, Houston, TX.

Ren, N. and Roegiers, J. (1983). Proc. 5th Int. Soc. Rock Mech. Congress, Melbourne.

Smith, M., Ren, K., Sorells, G., and Teufel, L. (1986). SPE Prod. Eng. 2, 423.

Snow, S. and Hough, E. (1988). presented at 63rd Annual SPE Meeting, Houston, TX.

Sulak, R. and Danielsen, (1988). presented at 20th Annual Offshore Technology Conference Houston, TX.

Teufel, L. (1982). Proc. 23rd U. S. Nat. Mech. Symp. (R. Goodman and F. Heuze, eds.), Berkeley, CA, 238.

Teufel, L. and Warpinski, N. (1983). Proc. 5th Int. Soc. Rock Mech. Congress, Melbourne.

Teufel, L. and Warpinski, N. (1984). Proc. 25th U. S. Rock Mech. Symp. (C. Dowding and M. Singh, eds.), Evanston, IL, 176.

Teufel, L., Hart, C., Sattler, A., Clark, J. (1984). presented at 59th Annual SPE Meeting, Houston, TX.

Thomas, L., Dixon, T., Evans, C., and Vienot, M. (1987). J. Petr. Tech. 39, 221.

Warpinski, N. (1983). Proc. 24th U.S. Rock Mech. Symp. (C. Mathewson, ed.), College Station, TX, 773.

Warpinski, N., Branagan, P., and Wilmer, R. (1985). J. Petr. Tech. 37, 527.

Warpinski, N. and Teufel, L. (1987). J. Petr. Tech. 39, 209.

Warpinski, N. R. and L. W. Teufel (1989a). J. Petr. Tech. 41, 405.

Warpinski, N. R. and L. W. Teufel, (1989b). SPE Res. Eng., in press.

Watts, N. (1983). Am. Assoc. Petr. Geol. Bull. 67, 201.

Watts, N., Larpe, J. van Schijndel-Goester, F., and Ford, A. (1980). Geology 8, 217.

Withjack, M. and Scheiner, C. (1982). Am. Assoc. Petr. Geol. Bull. 66, 302.

Ziegler, P. (1978). Geol. en Mijnbouw 57, 589.

Zoback, M. and Zoback, M. (1980). J. Geophys. Res. 85, 6113.

FRACTAL HETEROGENEITY OF CLASTIC RESERVOIRS

Hans-Henrik Stølum

Norwegian Petroleum Directorate
Stavanger, Norway

ABSTRACT

Significant improvements in the understanding and prediction of reservoir behaviour would result if geological and simulation models could be efficiently integrated. This paper outlines how that aim could be facilitated by using the fractal properties of reservoir rocks.

More precisely, shaly units within a marginal marine sand/shale sequence show a fractal distribution pattern. 'Fractal' meaning that the distribution is self-similar over a wide range of scales. Existence of self-similarity is verified by a number of statistical tests. This scale-invariant pattern is caused by depositional mechanisms, and so the value of its fractal dimension is determined by the depositional environment. The formation studied consists of one complex and rather poor reservoir unit with low connectivity between individual sand lenses and one good reservoir unit. In the core sequences, seen as topologically one-dimensional strings, the lithologic fractal dimension varies between 0.70 and 0.80. Spectral density and R/S-analysis of a digitized core interval showed the sand/shale sequence, recorded as variation in pixel intensity, to be a fractional Gaussian noise. The fractal dimension, paired with a parameter directly related to the shaliness of the formation, forms an excellent tool for quantitatively describing complex heterogeneity.

The fractal pattern observed in one dimension (core) reflects a three-dimensional pattern. Areal photographs/ satellite images of two marginal marine, tide do-

579

minated depositional systems show distinct fractal surface patterns over several orders of scale (10m-km). Such recent analogues give reason to believe that depositional systems can be characterized by multifractal measures in three dimensions.

This possibility opens up some interesting perspectives. Complex heterogeneity in three dimensions could ideally be described quantitatively by a small number of multifractal parameters. If these parameters form an efficient measure of the effect of anisotropy and heterogeneity on permeability, they could readily be utilized in reservoir simulations.

INTRODUCTION

A major task of reservoir geologists is to create geological models of reservoirs. Such models describe the three-dimensional architecture of the reservoir, either for static purposes (volumetrics) or dynamic purposes (fluid flow simulation).

The geological framework of a heterogeneous reservoir influences the motion of fluids during production, but in order to assess how, we are faced with a problem of appropriate description. A good description should take into account the inherent complexity of heterolithic sediments, diagenesis, and also of fracture/fault-patterns, as these forms of complexity are relevant to production. But it would be fair to say that descriptions in general tend to be fairly crude approximations, no doubt leaving behind large amounts of information with significance to reservoir dynamics and management.

A most promising solution to the general problem of describing complexity in nature has emerged with the development of fractal geometry (cf. Mandelbrot 1982, Feder 1988). One avenue to a reservoir model incorporating natural complexity involves the use of fractal techniques of description (Hewett 1986, Thompson et al. 1987, Thomas 1987, Chang and Yortsos 1988).

THE CONCEPT OF FRACTALS

The application of fractal geometry starts with realizing that objects or processes appear in some way similar when observed at different scales. This property, termed self-similarity, or scale invariance, is a definitive characteristic of fractals:

> A fractal is a shape made of parts that are
> similar to, or repeat the whole in some way.
>   (Mandelbrot, quoted in Feder 1988)

In general, self-similarity gives rise to an intrinsic complexity which is amenable to quantitative description in terms of fractal geometry. The basis of fractal description is the relation:

$$MASS \propto LENGTH^D$$

which is a property of fractals. Here MASS is the formal mass of the object, LENGTH is some characteristic length, and D is the similarity dimension of the object. For fractal objects, D is smaller than the topological dimension of the object, which is always an integer. (The topological dimension is the dimension of the linear space in which the object exists.) In these cases D is known as the fractal dimension.

The relation above also holds for familiar geometric shapes like straight lines, squares and cubes, but here the similarity dimension D attains the same value as the topological dimension. For example, the volume (MASS) of a cube is equal to the length of one side (LENGTH) raised to the 3rd power (D=3). Such shapes are termed strictly self-similar, and form a special case in which the whole is exactly identical to the parts. For every integer b, a cube is decomposable into $N = b^3$ sub-cubes of side smaller in ratio $r=1/b$. $D = \log N / \log(1/r) = 3$ for a volume. The same equation holds for fractal objects, but D will attain non-integer values.

STRATIGRAPHIC AND DEPOSITIONAL SETTING

The Middle Jurassic Tilje Formation occurs over most of the Haltenbanken basin, where it forms a significant reservoir unit. Tilje was deposited in a marginal marine setting. It overlies the fluviodeltaic Åre Formation, and is succeeded by the marine shelf deposits of the Ror Formation. Between Tilje and Ror there is a discontinuity, as Tilje terminates with a thin veneer of condensed sediments, showing a phase of non deposition. That Tilje is marginal marine is seen from the diversity of the infauna (burrowing organisms) and the lack of in situ coals that are frequent in the underlying Åre Formation. Tilje also lacks rootlets or root horizons, except for occasional roots in the lowermost part. Thick heterolithic sequences with occasional desiccation cracks, interrupted by sandy channels, indicate a tidal flat environment (Figure 1).

Tilje is comprised by two members, the lowermost (T2) being quite shaly, and the uppermost (T1) quite sandy. There is a sharp transition between them.

MATERIAL AND METHOD

The basis for this study has been the cored intervals of Tilje from five wells in the Haltenbanken basin. A total of 403 m has been cored, comprising 120 m of T1 (30 %) and 283 m of T2 (70 %). These units were treated independently, but sampled according to their proportions. In addition a 1m core interval of T2 was subjected to detailed statistical analyses as a test of self-similarity.

Figure 2 shows the total cored interval, with shaly units down to 1 m in thickness. The figure is based on core descriptions at the scale of 1:200. These descriptions were made from visual inspection of the core material and are a continuous record of lithology (shale/sand proportion), modal grainsize of sands, and dominant structures. This record was mounted with wireline logs and scaled down to 1:500 (Fig. 1). Fig. 3. gives examples of the detailed pattern.

The number of distinct shaly units in Fig. 2 having a thickness of 10 m or more was counted, and their thickness measured. The same was done for the units between 1 and 10 m.

Ten 10-m long intervals, 3 from T1 and 7 from T2, were then picked at random. Shaly units in these intervals between 10 cm and 1 m in thickness were counted, and their thickness measured. The same was done for units between 1 cm and 10 cm. Then a number of 10 cm long intervals were sampled at random from each of the "10 m" samples. Those "10 cm" subsamples were used for counting and measuring of shaly units between 1 mm and 1 cm in thickness. The number of 10 cm intervals picked varies from one "10 m" sample to another, depending on the amount of clean sand sections in the "10 m" sample. For the units between 10 cm and 1 m in thickness, a viewing distance of about 3 m was suitable for appreciating the contrasts in shale density at this scale.

In order to arrive at comparable figures, the number of shaly intervals in the 10 cm - 1 m range of each 10-m-long core sample was averaged arithmetically for the samples of T1 and T2. The mean was then multiplied with a factor of total length of sampled interval / total length of cored interval represented by the samples for T1 and T2. These figures forms inherently non-biased estimators of the number of units in the 10 cm - 1 m range for the total cored interval of T1 and T2, respectively.

The same was done for the intervals in the 1cm - 10 cm range. For the shaly intervals in the 1 mm - 1 cm range, a similar procedure was applied in two steps. The first step gave average figures for each "10 m" sample. Averaging was then done again to obtain estimates for the full core lengths of T1 and T2. To ascertain that the populations did not have properties which would cause bias in the outcome of such averaging, a crude trend analysis was done (Fig. 4).
(For details, see the section "Averaging as a possible source of error".)

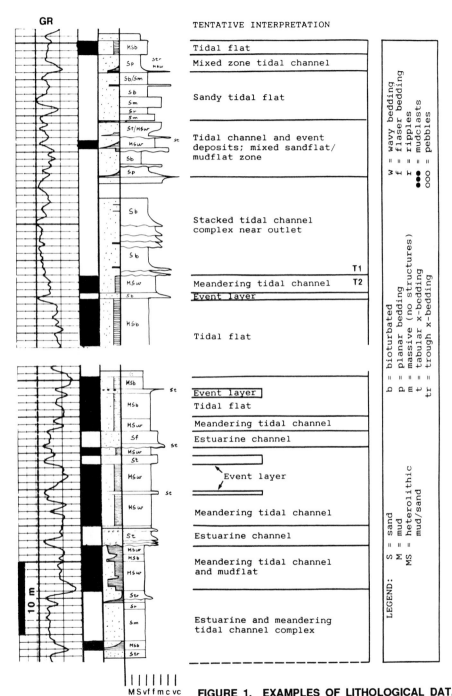

FIGURE 1. EXAMPLES OF LITHOLOGICAL DATA AND A POSSIBLE INTERPRETATION (WELL 2).

VERTICAL DISTRIBUTION OF SHALY UNITS (BLACK) AND SANDY UNITS (WHITE).
TILJE FORMATION, HALTENBANKEN.

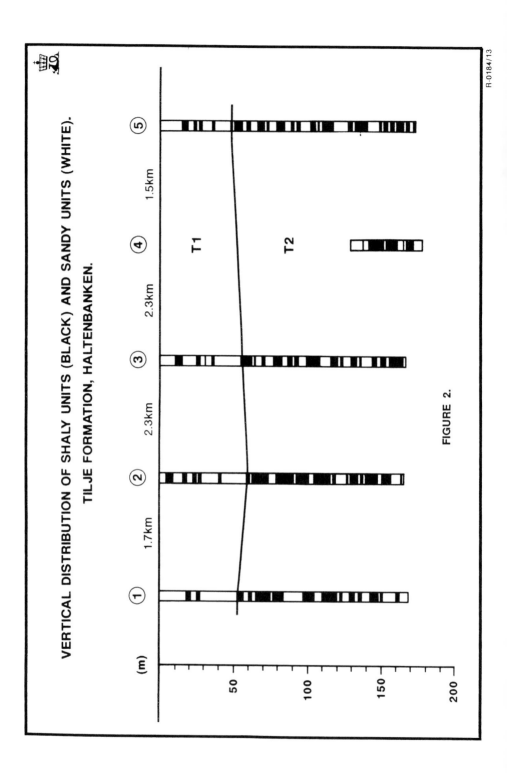

FIGURE 2.

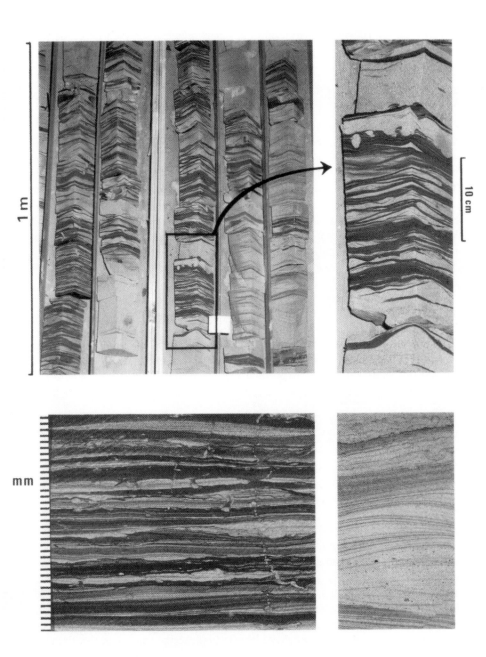

FIGURE 3. CORE PHOTOGRAPHS OF SHALY UNITS WITH HETEROLITHIC WAVY BEDDING AT VARIOUS SCALES. THE UPPER PHOTOS SHOW A SEGMENT OF T2 FROM CORE 1. THE LOWER PHOTOS SHOW THE SMALLEST SCALES OF LAMINATION IN SHALY AND SANDY UNITS.

The thickness measurements were used for a plot of cumulative thickness distribution (Fig. 6). They were also used to calculate medians for each range, as well as histograms showing the actual thickness distributions of shaly units (Fig. 7 and 8).

MEASUREMENT AND THE PROBLEM OF SUBJECTIVITY

What, if anything, is a shaly unit ? Inherent in the described method of data aquisition is a subjective element. A choice has to be made for each scale unit of what to be counted. The problem lies in the fact that the units are 'shaly'. They do not consist of pure shale but of heterolithic alternations of shale and sand that may be more or less shaly. Figure 3 illustrate the concept of heterolithic bedding. The shaly unit must have sharp transitions to the sandy lithologies above and below in order to be visually recognized as a distinct, measurable unit. Fortunately, this has generally been the case in the studied core material.

Having established a method of identification, the next problem is one of measurement. Each identified unit will, seen in isolation, be registered only once. But in addition it is also being seen as part of a nested hierarchy of shaly units, in which larger units are composed of smaller units, which again are composites of even thinner units, and so on until a natural lower limit is reached. Given the method of identification, each shaly unit may be recorded up to three times as a part of larger units at the same scale, in addition to the one registration as an isolated unit.

Some examples may illustrate this: Consider a case of two 20 cm thick shaly intervals separated by a 5 cm thick clean sand. Above and below, there are clean sands. Is the whole interval a 45 cm thick shaly unit, or two minor units separated by a sandy unit ? Both views are of course equally correct. And so the two shaly intervals would be counted three times at the same scale (10cm - 1m). If the units had been closer to the upper limit of the scale, the registration as a composite would take place at the next larger scale (consider, for instance, two 90 cm thick shaly intervals separated by a 20 cm thick clean sand). That the implicit recordings as parts of larger units are sometimes done at the same, and sometimes only at larger scales, is an effect of imposing a rigid and relatively crude frame based on an arbitrarily chosen measuring unit and classification rule. For a large number of recordings, the effect should be the same at all scales, and so will not introduce bias. In the majority of cases, there was none or only one implicit recording of each explicitly identified unit at the same scale as the unit.

A more realistic representation may be obtained by divisive, polythetic cluster analysis, which is a multivariate hierarchical classification technique (Gauch 1982).

Such detailed representation has not been within the scope of this paper. It might be noted that the method used does not differ in principle from the cluster formation procedure of divisive cluster analysis.

Consistency in interpretation can be facilitated by standardization; measurement may be done according to precise definitions. A single criterion might be introduced, for instance, demanding that a shaly unit will only be considered as distinct if the sands above and below are at least half as thick as the shaly unit itself. But this would introduce an arbitrary constraint. Natural variation would be deliberately neglected. In order to take into account natural variation in cleanness of the sandy units according to differences in their visual impression, a well defined range of values was used. For instance, sometimes the intervening sands were clean, and would be considered as separating even when down to 1/5 the thickness of shaly units at either side. Sometimes they too contained shale laminae, and may not have been counted as separating unless at least 1/2 the thickness of shaly units at either side. The minimum thickness of 1/5 for a sand unit to count as separating a shaly unit was applied rigorously as a lower limit. This limit fitted well with intuitive visual impression of the core.

In general, most units were separated from neighbouring shaly units by clean sands or sandy intervals at least 0.3 times as thick as the recognized shaly unit. In nearly all cases, sandy units have a net/gross of 0.90 - 1.0, while the shaly units have net/gross ratios ranging from 0 - 0.70. Within this frame, subjectivity in identifying the units that were counted is not likely to introduce any bias. As for the reality of shaly units, I will encourage the reader to evaluate that for himself by looking at the core photographs shown in Figure 3.

It must be emphasized, however, that in lithologic sequences which come in less discrete packages, subjectivity in measurement may influence the result to an unacceptable extent. A solution may then be to invent a set of rigorous rules and repeat the measurements under those rules in the hope that the figures may show themselves reasonably stable.

AVERAGING AS A POSSIBLE SOURCE OF ERROR

Taking random samples and calculating the number of thin shale layers by averaging over the samples introduces a factor of uncertainty. For this procedure to be valid, only random variation in density of shales must be present. Any systematic vertical trend may cause random sampling to be biased if the number of samples is relatively small.

To check this problem, a shale/sand index was used to gauge shaliness for each 10 m interval of T2 in four wells. The result is shown in Figure 4. As can be seen, the index varies erratically, and so there are no clear trends for the comp-

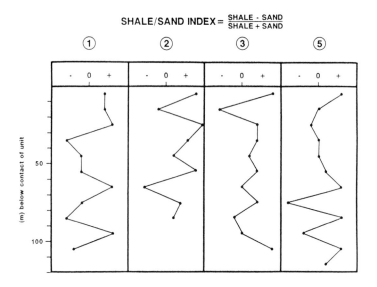

**FIGURE 4. SHALE/SAND INDEX FOR T2 BASED ON FIG. 2. VALUES HAVE BEEN
CALCULATED FOR 10 M INTERVALS DOWNWARDS FROM THE T1/T2
BOUNDARY.**

lete cored intervals. However, there might be minor trends of 20-30 m here and
there. For the present purpose, such variation may be considered random, and
could occur just as well in a random sequence.

Possible trends below the 1 m scale are unlikely, as we are dealing with increasingly more general mechanisms of deposition as the scale becomes smaller.

For T1, trendlines have not been considered. The sand/shale ratio here is much
higher, and so only samples which contained shaly intervals at the 1 m level
were taken into account. By doing so, the samples were focused at the middle
interval and do not record the shale factor of T1 as a whole. The lower part of
T1 is a distinct, good quality reservoir zone, and so was singled out for independent description.

RESULTS

The distribution of number of shaly intervals over four orders of magnitude (mm
to 10-m) follows a power law (Figure 5). This power relation holds over a certain range, bounded by the natural lower limit l_1 and the natural upper limits
T1-l_2 and T2-l_2.

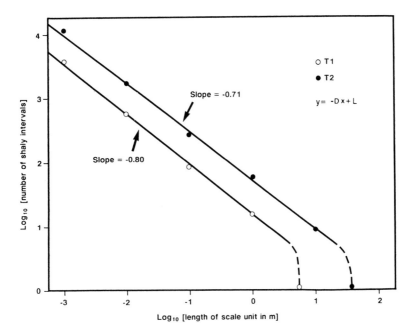

FIGURE 5. THE NUMBER OF SHALY UNITS AS A FUNCTION OF SCALE AT WHICH THE NUMBER HAS BEEN COUNTED. TILJE FORMATION (T1 AND T2).

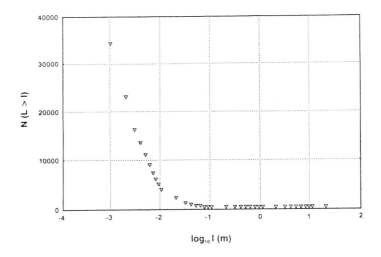

FIGURE 6. THE CUMULATIVE DISTRIBUTION OF THE THICKNESSES OF SHALY UNITS IN T2.

l_1 seems to be in the 0.1 - 1 mm range for both T1 and T2 (fig.3), and is defined by the modal thickness of single pure shale laminae (later termed "unit shales"). The range of 0.1 - 1 mm was not counted, however, due to problems with resolution.

At the other endpoint, the thickest single shaly intervals of T1 and T2 are 7 m and 12 m, respectively. At these extreme thicknesses, the power relationship has broken down, and T1-l_2 is the range of 1 - 10 m, whereas T2-l_2 reaches the range of 10 - 100 m.

The meaning of the power law relationship is that a self-similar or scale-invariant pattern exists within its range, which is also intuitively apparent from Fig. 1 and 2. As noted, this property is a definitive characteristic of fractals.

The similarity dimension D_S can be given as $D_S = D_B =$ absolute value of S, where S is the slope of the log-log function resulting from box-counting, or a homologous procedure like the one applied in this paper (cf. Feder 1988). As self-similarity has been demonstrated in this case, I will refer to D_S only as D. For T1, D = 0.80, while D = 0.71 for T2. This dimension represents the cored sequence considered as a topologically one-dimensional string or set of strings. But it is uncertain what fractal model or process may cause the self-similar pattern observed.

For this reason, the similarity dimension is not immediately translatable into a fractal dimension for the topologically three-dimensional medium, as a multifractal pattern may well be underlying the power-law relationship.

The difference between the values of D for T1 and T2 is not large, and the two lines in Fig. 4 are nearly parallel. This may seem surprising, given the fact that net/gross of T1 is considerably higher than of T1, as may be seen from Fig. 1. But the fractal dimension results from the presence of a self-similar kind of structure or ordering spanning several orders of scale. It is not necessarily dependent upon density. And even if the overall shale fraction is smaller, the same nested hierarchical ordering of shaly intervals is clearly seen in both T1 and T2.

The effect of heterogeneity on permeability will depend significantly on the net/gross difference however. The less shales, the more homogeneous the porous medium, and the less scaling effects due to heterogeneity. There are two ways the net/gross difference is reflected in the log-log plot of Fig.5: the value of the constant term or **amplitude**, L, in the log-linear function, and, more crudely, in the value of l_2. The value of the constant term is also a function of total length of core measured. As the length of core for T2 is more than twice that of T1, it is not surprising that consistently fewer shaly intervals were recorded from T1. Accounting for this effect will reduce the constant term of the T2-line to 1.40, whereas it is in fact 1.77 (this is done by multiplying the actual figure with 0.43=3/7 and then taking the logarithm). The residual difference of 0.25 is truly

T1

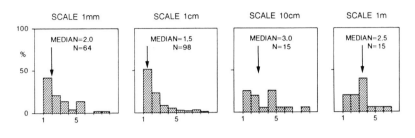

T2

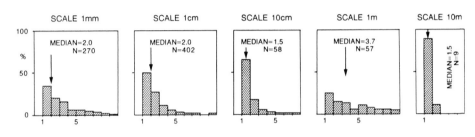

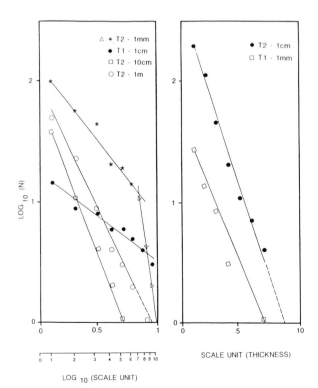

**FIGURE 7.
(ABOVE)
THE THICKNESS
DISTRIBUTIONS
OF SHALY UNITS
AT EACH OF
THE SCALES
IN FIG.5.**

**FIGURE 8.
(LEFT)
LOG-LOG AND
SEMILOG PLOTS
OF THICKNESS
DATA AT
VARIOUS
SCALES.**

reflecting the lower net/gross of T1 relative to T2. The actual value of the constant term, normalized for differences in sample size, is a direct measure of net/gross or "shaliness". We may therefore define as a nondimensional measure of shaliness in fractal sequences the shale factor:

SF = number of shaly units at the 1 m scale / length of sample in m.

By this definition, SF is always a figure between 0 and 1. The number of units at the 1 m scale must be smaller than the total length of the core in m, and so SF will always be smaller than one. If there are less than one unit at that scale, SF will be 0 (fractions of units do not exist, as they would be referred to a smaller scale). Notice that SF by itself does not span the full range from clean sand to shale. A value of 0 means that there are no shaly units at the 1m scale and larger scales. A value of 1 means that all sands are below the 1m scale. In most cases, SF cannot take the value of 1, however, because the median of interval thicknesses in the 1m scale range will be greater than one. For T1 and T2 the normalized SF take the values of 0.13 and 0.20 respectively.

The potential of D and SF as an index of heterogeneity is considered in the discussion.

In Figure 6 is shown the cumulative distribution $N(L>l)$. $N(L>l)$ is the number of shaly units larger than l, and is plotted against $\log_{10} l$. There is a region of one decade of unit thicknesses, where $N(L>l) = - B \log l + A$. This corresponds to a distribution of interval thicknesses $N(l) \propto l^{-1}$, which is typical of many random fractal processes. The shale layer thickness data are summarized in Figure 7 (histograms). As seen from Figure 8, several of these histograms fit quite well with a power law distribution, but some are exponential. Due to few values in the higher end of the range, there are irregularities, and some outliers have been left out of the plots where the lines are punctuated. The power law is to be expected for subsamples of a power-type overall distribution of thicknesses.

FIGURE 9. THE DIGITIZED CORE INTERVAL THAT WAS ANALYZED STATISTICALLY. TOTAL LENGTH: 1.1 M.
THE THIN STRIP IS THE RESULT OF LETTING ALL PIXELS WITH INTENSITY LESS THAN A THRESHOLD ($\xi_T = 128$) BECOME BLACK, AND WHITE IF LARGER. (THE BLACK HALF-CIRCLE IS THE OUTLINE OF A HOLE DRILLED IN THE CORE.)

TESTS OF SELF-SIMILARITY

MATERIAL AND METHOD

In view of the subjectivity of the approach followed so far, a series of rigorous statistical analyses were performed on a 1.1 m long sample of T2 core in order to test the apparent fractal character of the shale and sand distributions (Feder and Jøssang 1989).

The sample was chosen because of its detailed resolution and photographed. The photographs were digitized, and pixel intensities interpreted as increments of an underlying distribution. Methods and results are outlined below. Most of the tests gave consistent results, pointing to a scale-invariant geometry.

The digitized image of the core segment is shown in Fig. 9. Fig. 10 is a profile of the variation in pixel intensities along the core. From this curve, a cumulative function X(z) was derived, being defined as:

$$X(z) = \Sigma \; \{\xi(u) - <\xi>\}$$

where $<\xi>$ is the average intensity of the entire interval. This figure is plotted in Fig. 11. Notice an interesting aspect of this cumulative curve. After an initial region of intensities fluctuating around the average value, there is a large region where the intensities are higher than average, causing the curve to rise. This region

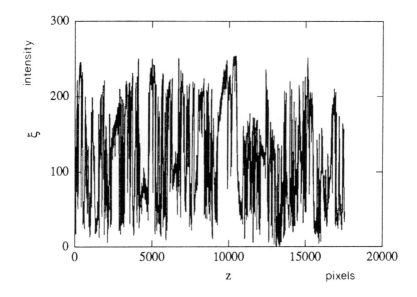

FIGURE 10.   DIGITIZED VALUES OF INTENSITIES AS FUNCTION OF POSITION.

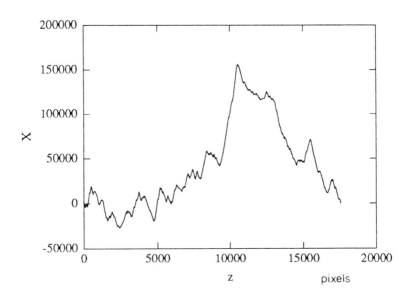

FIGURE 11.   THE CUMULATIVE FUNCTION OF THE RECORD IN FIG. 10.

is followed by a decrease in cumulative function, where the intensities are lower than average. The area of rise must come from a rather large interval containing mostly sand. Now, if one enlarges a short segment of the curve, one will generally see the same overall behaviour, that one large region dominates the cumulative function. This is indicative of scaling behavior and strong persistence in the record.

In conventional terms, unpredictable changes of any quantity varying in time are known as "noise". The noise concept can of course be generalized to curves that only indirectly vary with time, such as those in Fig. 10 and 11. Noise may be characterized by its spectral density, as shown in Fig. 12. The spectral densi-

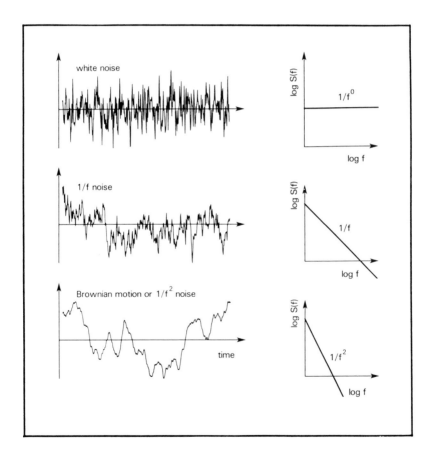

FIGURE 12. SAMPLES OF TYPICAL "NOISES". TO THE RIGHT OF EACH SAMPLE IS A GRAPHICAL REPRESENTATION OF THE SPECTRAL DENSITY, S(f), WHICH CHARACTERIZES THE TIME CORRELATIONS IN THE NOISE. FROM VOSS (1988)

ty $P(f)$ is a measure of the mean size of fluctuations at frequency f. In other words, it is a measure of the variations over a time scale of order 1/f. The three noise samples in Fig. 12 are all fractal curves. White noise is completely uncorrelated from poin to point. Its spectral density is a flat line, representing equal amounts of fluctuation at all frequencies. Brownian motion, or random walk, is the most correlated of the three noise samples. It consists of many more slow (low frequency) than fast (high frequency) fluctuations. Formally, Brownian motion is the integral of the white noise. The intermediate curve is known as 1/f noise, and is ubiquitous in nature.

Comparing Fig. 10 and 11 with the upper and lower curve of Fig. 12 shows a conspicuous resemblance. The curve in Fig. 10 may thus be a form of Gaussian noise, and Fig. 11 a Brownian motion. In the analysis we are testing the hypo thesis that the curve in Fig. 10 represent fractional Gaussian noise (fGn), of which white noise is a special (uncorrelated) case. In the general fGn there is a measure of correlation of the increments that extends to all scales. Further, Fig. 11 may be assumed to represent fractional Brownian motion (fBm), of which the random walk is a special case.

VARIOGRAM AND AUTOCOVARIANCE FUNCTION

The autocovariance function is a measure of the degree of coordinated variation along the core, given as a function of scale (in the sense of resolution). Define the mean value of a (random) function $X(z)$ at a point z as $m(z)=<x(z)>$, where $<...>$ denotes the average over several realizations of the underlying distribution. The autocovariance function is then defined by:

$$C(z_1,z_2) = <[X(z_1) - m(z_1)][X(z_2) - m(z_2)]>$$

If we have a situation of spatial stationarity, then the same statistical distribution will control the random variable X at all points z, and the autocovariance function will depend only on the relative distance $\delta=|z_1-z_2|$. $C(\delta)$ is plotted in Fig. 13. The variance of increments (variogram function, Fig. 14 and 15) is in many cases related to the autocovariance function and is defined as:

$$\gamma (z_1,z_2) = 1/2 <[X(z_1) - X(z_2)]^2>$$

Given stationarity, and existence of the first two moments of the distribution function, the relation between the autocovariance function $C(\delta)$ and the variogram $\gamma (\delta)$ may be written as

$$C(\delta) = Var\{X\} - \gamma (\delta)$$

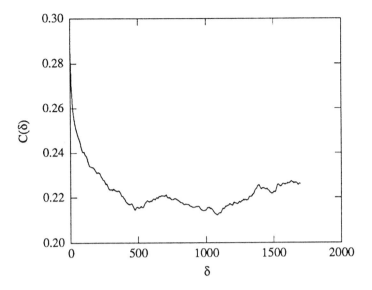

FIGURE 13. THE AUTOCOVARIANCE FUNCTION OF THE PIXEL INTENSITIES CALCULA-TED FROM THE FUNCTION OF INCREMENTS (FIG. 10). THE LAST PART OF THE AUTOCOVARIANCE FUNCTION IS NOISE.

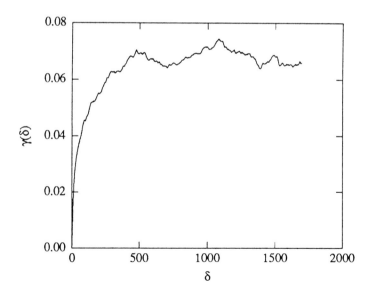

FIGURE 14. THE VARIOGRAM γ CALCULATED FROM THE INCREMENTS

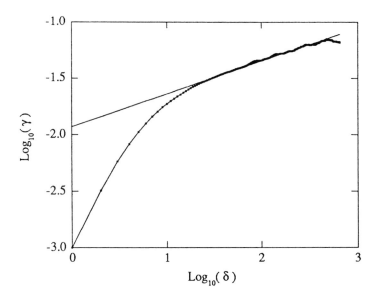

FIGURE 15. THE VARIOGRAM (FIG. 14) PLOTTED IN A LOG-LOG PLOT: THE SLOPE OF THE LINE IS 0.3.

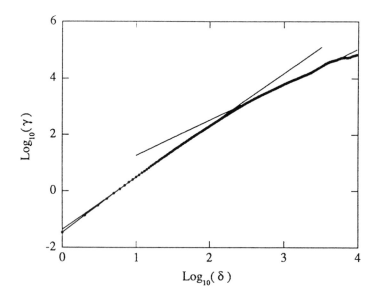

FIGURE 16. THE VARIOGRAM OF THE INTENSITIES CALCULATED FROM THE CUMULATIVE FUNCTION (FIG. 11) AND PLOTTED IN A LOG-LOG PLOT. THE LINES HAVE SLOPES OF 1.8 AND 1.3.

Here Var$\{X\}$ = $<X(z)^2>$ - $<X(z)>^2$ is the variance of the random variable X, and it should be independent of the position z.

As seen from Fig. 13 and 14, the autocovariance and variogram of the pixel intensities are entirely consistent with the equation above. The curves never die down towards a sill or limit value, but continue to move up and down for δ > 500 like a Brownian motion. The form of the variogram and autocovariance function are very similar to the equivalent functions derived from fGn (compare Mandelbrot and Van Ness 1968, Hewett 1986). Notice that when plotted with log scales, the variogram (Fig. 14) does not reach a plateau, but continues to rise steadily. This is characteristic of the variogram of fractal curves, in which variability increase with scale. The logarithm of the variogram of fGn should have a slope of 2H for small lags (Mandelbrot and Van Ness 1968). From Fig. 15 th variogram has a slope of ca. 0.3. This slope is related to the fractal dimension of the intensity curve (Fig. 10) by the equation 4 - 2D = slope, giving D = 1.8. The variogram of the cumulative function X(z) with log scales (Fig. 15) has an initial slope of 1.8, crossing over to 1.3. The cross-over shows that the curve cannot be described by a single power law. It yields H in the range of 0.65 - 0.9.

SPECTRAL DENSITIES

Random functions V(t) are often characterized by their spectral density S(f), as mentioned in the section "Material and method". If V(t) is the input to a narrow bandpass filter at frequency f and bandwidth Δ f, then S(f) is the mean square output V(f) divided by Δ f; S(f) = $|V(f)|^2$ / Δ f. Plotting S(f) against f in a log-log plot will yield straight trend lines if the curves are scaling (self-affine). Since S(f) estimates the number of fluctuations with frequency f, or over a time scale of order 1/f, the S(f) vs. f relationship may be used to define the scaling character of the noise curve as $f^{-\beta}$, where - β is the slope of the trend. The theory of Mandelbrot and Voss (Voss 1988) says that, for a fractional Brownian motion, fBm, there is a relation of the form β - 1 = 2H between the spectral density of a noise curve and the autocorrelation or variogram of the noise. For fGn the relation is β + 1 = 2H.

The spectral density of the intensity variation (Fig. 10) interpreted as fGn gave β = 1.6 and H = 0.3, in agreement with the value of H obtained from the variogram in Fig. 15. The value of β indicates that the pixel intensity variation is between 1/f noise and $1/f^2$ noise.

The spectral density of the cumulative function of intensities (Fig. 11) is shown in Fig. 17 (upper curve). The slope of -2.6 means that S(f) scales with the frequency f as $1/f^{-\beta}$, where β = 2.6. For β = 2.6, H = 0.8, in accordance with the variogram analyses. Thus the cumulative function curve is fractional Brownian motion.

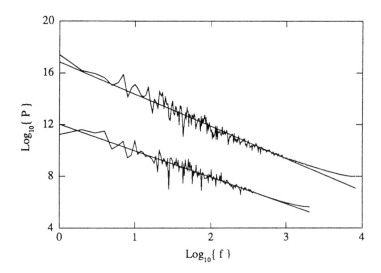

FIGURE 17. THE LOWER CURVE IS THE SPECTRAL DENSITY OF THE NOISE PART OF THE AUTOCOVARIANCE FUNCTION (FIG. 13). THE LINE HAS A SLOPE OF -2. THE UPPER CURVE IS THE SPECTRAL DENSITY OF THE CUMULATIVE FUNCTION (FIG. 11). THE LINE HAS A SLOPE OF -2.6.

The behavior of the autocovariance function in Fig. 13 for values of $\delta > 200$ was analyzed in terms of noise. The spectral density of this curve is shown in Fig. 17 (lower curve). The slope of 2 means that the spectral density S(f) scales with the frequency f as $1/f^2$. This is the characteristic frequency spectrum of Brownian motion (random walk).

R/S-ANALYSIS OF A SEQUENCE SAMPLE

R/S-analysis (cf.Feder 1988) was originally developed by Hurst et al. (1965) as a means of quantifying long term discharge variation of the Nile. Later the method proved to be an efficient technique for analyzing fractal "traces" of variables changing with time (Mandelbrot and Van Ness 1968, Mandelbrot and Wallis 1968, 1969).

In R/S-analysis, the range of variation (R) of a parameter recorded during a certain time interval is measured for longer and longer time intervals. For fractal variables, R will increase as a power law with length of the time interval. In a log-log plot of R versus time, the relation will show up as a straight line. In other words, this kind of fluctuation (Hurst noise) is scaling. In order to facilitate comparison between different kinds of records in time, R is divided by S (standard deviation of mean parameter value) to yield the normalized dimensionless quantity R/S. (For this reason the technique is also known as rescaled range analysis.)

The empirical scaling relation for many records in time is

$$R/S = (\tau/2)^H$$

where τ is the length of the time interval and H is the Hurst exponent. If the fluctuations amounts to white (Gaussian) noise (i.e. each given state is independent of earlier states), then $H = 1/2$ (Feder 1988; pp.154-6). It can be shown that $0 \le H \le 1$ (Mandelbrot 1982), with increasingly positive or negative correlated variability as H goes towards its limits.

When $H > 1/2$, the record is called persistent. Persistence implies that repeating patterns of fluctuation develop in the record, and when broken, the record falls into alternative repeating patterns. These patterns, or the cycling between them, is not periodic, however. When $H < 1/2$, the record is antipersistent. This means that recognizable long-term patterns will not form, and the curve is seen to fluctuate "wildly" as trends are broken more often than for white, uncorrelated noise.

If we consider the unit shale and the unit sand to be the single event deposits that form the lithological sequence, then the process creating the sequence is one of switching between these two states. The resulting sequence may be described in terms of R/S analysis. R/S analysis of small-scale variation was done on the 1.1 m of digitized T2 core (Feder and Jøssang 1989). The record is shown in Fig. 10 as variation in pixel intensity (ξ) as a function of position z along the core. The interval length, or space lag $\tau = \Delta z = z_2 - z_1$. The cumulative function $X(z,\tau)$ is defined as

$$X(z,\tau) = \Sigma \ \{\xi(u) - \langle\xi\rangle_\tau \ \}, \text{ where } \langle\xi\rangle_\tau = 1/\tau \ \Sigma \ \xi(z)$$

This function of the dataset in Fig. 10 is plotted in Fig. 11. From this cumulative function, a range R is defined as

$$R(\tau) = \max X(z,\tau) - \min X(z,\tau) \text{ for } z_1 < z < z_2.$$

S is the standard deviation of ξ in the same interval. Log R/S versus log (interval length) is given in Fig. 18. The slope of the fitted line implies a Hurst exponent $H = 0.87$. This corresponds to a persistent distribution, as might be expected from the hierarchical nature of the sedimentary record. As mentioned earlier, persistence is also visible in Fig. 11. The value of H is in good agreement with the values obtained from the variogram analyses and the Fourier analyses of the cumulative function of intensities.

To demonstrate the value of $H = 1/2$ for a random process, Feder and Jøssang (1989) also randomized the sequence of $\xi(z)$ in Fig. 10 and then calculated R/S. The result is shown in Fig. 19. We see that the shuffling leads to $H = 0.53$, reasonably close to the expected value of 0.50.

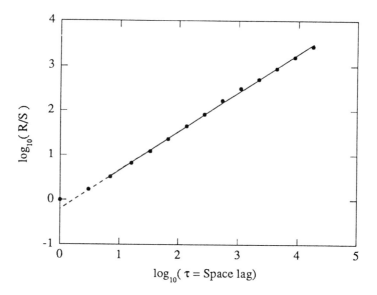

FIGURE 18. **RESCALED RANGE (R/S) ANALYSIS INTERPRETING INTENSITIES IN FIG. 10 AS STEPS IN A RANDOM WALK. THE SLOPE OF THE LINE IS H = 0.87.**

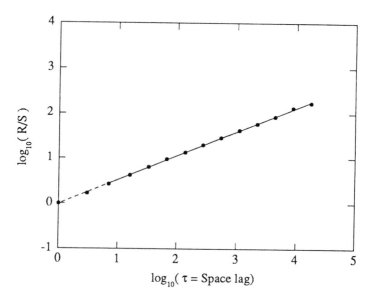

FIGURE 19. **R/S-ANALYSIS AFTER FIRST HAVING SHUFFLED THE SEQUENCE OF INTENSITIES IN FIG. 10. THE SLOPE OF THE LINE IS H = 0.53.**

Again, if we consider the unit shale and the unit sand to be the single event deposits that form the lithological sequence, then the process creating the sequence is one of switching between these two states. One might think of simulating such a process as a Markov process. Indeed, a first-order Markov chain, with probability significantly larger than 0.5 of the next state being the other than the present state, is going to yield a persistent cyclical pattern of alternating unit shales and unit sands. However, a hierarchically organised persistence cannot arise by this simple process. For a Markov process to possibly create a hierarchical pattern it has to be of a high order. That is, the probability that the next state is going to be either shale or sand is dependent on earlier states and not, or not only, the present state. This inevitably leads to a complex probability structure which commonly has no intuitive meaning. Because of this, Markov models in general have limited relevance to real sequences of sediments.

LENGTH DISTRIBUTION OF BLACK OR WHITE REGIONS

A natural question to ask when looking at the dark and light regions in Fig. 9, would be: What are the length distributions of shale and sand regions ? Fig. 20 is an intensity profile of the picture shown in Fig. 9. If we increase all intensities ξ to 1 if $\xi > \xi_\tau$ and decrease all intensities lower than the threshold value ξ_τ to zero, a record of black and white segments of varying length will appear. Given a suitable value of xt the black segments may be taken to represent shale and white intervals to represent sand (in this case xt = 128, based on the distribution in Fig. 20. In Fig. 21 is shown the cumulative distribution $N(L>\lambda)$. $N(L>\lambda)$ is the number of segments larger than λ. $N(L>\lambda)$ has been plotted against $\log_{10}\lambda$ for three cases; the black segments, the white segments, and the sum of a black plus the following white region. We see that both the black and the white segments have a region of 1.5 decade where $N(L>\lambda) = - B \log_{10}\lambda + A$. This corresponds to a distribution of line-segments $N(\lambda) \sim 1/\lambda$ that is similar to the 1/f noise of the pixel intensity variation, and which is typical of many random fractal processes. The fraction of the slopes of the curve for black line-segments relative to the slope of the white line-segments is 1.2. The fraction of the number of black pixels to the number of white pixels is roughly 1.3, which is close to the fraction of slopes. This may therefore be the reason for the difference.

This result was somewhat surprising, as an attempt to box-count the same sequence gave an inconclusive result. Box-counting, however, is sensitive to the actual spatial distribution of the pattern. The original variation in density at each scale had become fairly uniform for large chunks of "core" due to the loss of information when setting a threshold and applying it to the fairly low-contrast data set. This was equivalent to a certain randomization at small scales. The box-

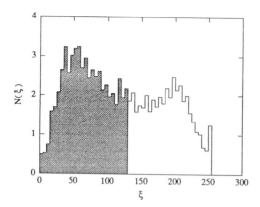

FIGURE 20. HISTOGRAM OF THE NUMBER N(ξ) (IN PERCENT) OF PIXELS HA-VING AN INTENSITY ξ. MOST SHALY REGIONS HAVE ξ < 100, AND MOST SANDY REGIONS ARE IN THE RANGE OF ξ > 160.

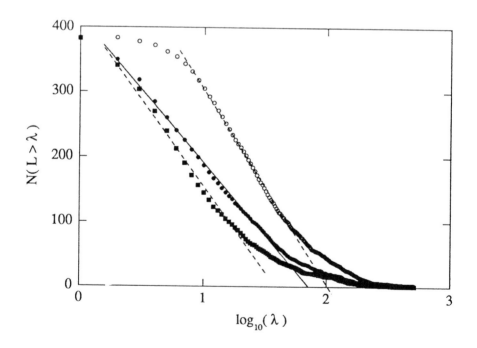

FIGURE 21. THE CUMULATIVE DISTRIBUTION OF LENGTHS λ OF THE BLACK (•) AND WHITE (■) REGIONS IN FIG. 9, AND THE COMBINATION OF A BLACK FOL-LOWED BY A WHITE REGION (o).

counting indicated this, more than recording what was left of the original hierarchy. A remnant of the hierarchy, though, is seen as variation in lengths of segments along the trace. As the cumulative distribution is recording a net, or summation statistic, it is clearly less sensitive (but not insensitive) to randomization. Due to this robustness of the cumulative distribution, it would still be able to pick up the remaining hierarchical character of the record for the scale range where most original information is preserved. Note, however, that the curves begin to taper off when λ becomes small. This is possibly due to lumping, which reduces the number of intervals at small length scales.

The curve for the combination of a black and the following white region rises more steeply than either of the other two. Simply adding the curves of the others will yield a curve with roughly the same rise as the combination curve. But why does it suddenly taper off much more strongly ? Apparently, at a certain lower length, the combination will reach a point where there are steadily fewer intervals that are shorter. This point it must reach much sooner than either of its always shorter components, unless these are strongly correlated in length. Accordingly, they continue to rise linearly until they "see" the limit of resolution.

CONCLUSION

The shale structure in T2 consists of line segments embedded in a topological dimension of 1, which means that its fractal dimension must be smaller than one, in fact it is $D_s = 0.71$.

The shale and sand structure as a whole is recorded by the pixel intensity variation. The analyses showed that the record of sand and shale forms a fractional Gaussian noise with a fractal dimension of $D = 1.8$. Since all the information in the curve was originally confined to a topological dimension of one, the curve is a kind of projection of the record in a plane. In other words, the shale structure itself has a fractal dimension of $D - 1 = 0.8$, which is close to the D_s value.

The cumulative function of intensities is a fractional Brownian motion with a Hurst exponent of 0.87 and a fractal dimension of $D = 2 - H = 1.13$.

DISCUSSION

A THREE-PARAMETER HETEROGENEITY INDEX

The scale-invariant character of the core seems a natural outset for describing the heterogenity observed. Three parameters are needed to account for scale invariance in a one-dimensional sample: D, SF, and core length. The shale factor, SF, is the constant term of Fig. 5 normalized for differences in sample size, cf. the definition in the section on "Results". In more general terms, SF is a measure of the amplitude of the shale structure (Mandelbrot 1982, Voss 1988). Since D cannot be larger than one for a topologically one-dimensional distribution, D varies between 0 and 1, as does SF. For two of the end-points of their span we have:

D = 0, SF = 0 :   Clean sand (in the vicinity of the point, which itself is
                  formally undefined; see below)

D = 1, SF = 1 :   Massive shale

In order to assess the range of the fractal dimension possible in strings or cores of natural heterolithic sequences, the number of units at various scales was calculated for a number of combinations of D and SF. The visual impression of the outcomes was assessed graphically, assuming a power law thickness distribution with median equal to two at each scale and a fractal range from one mm to 10 m. The result is sketched in Figure 22, which indicates a possible envelope of natural variation within the span of the two parameters.

Notice that higher values of D than one are not valid if the string is to exist for any length of core down to 1 m. Consider a simplified case, in which all thicknesses are equal to the base of a scale (1 mm, 1 cm, 10 cm, 1m, etc.), thus looking at minimum thicknesses, and ignoring any thickness variation. Setting D = 1 and SF = 1 means that at least one 1 m long shaly interval will exist, which can only be visualized for cores of 1 m length or more. But in that case the total length of 1 mm thick shales is also 1 m, so we are looking at a massive shale. This is the case irrespective of SF or length of core. A value of D larger than one would mean that the total length of intervals at the 1 mm scale exceed the total core length. Setting SF = 0 means that there are no 1 m thick (or thicker) shaly intervals in the core, and that the fractal range goes up to 1 m only. For instance, D = 1 and SF = 0 means that the constant term of the straight line equation is also 0. The curve then has to go through origo, at which point there exists one 1 m long shaly interval in the core. At the same time the value of SF means that there are no intervals of that length in the sequence. This contradic-

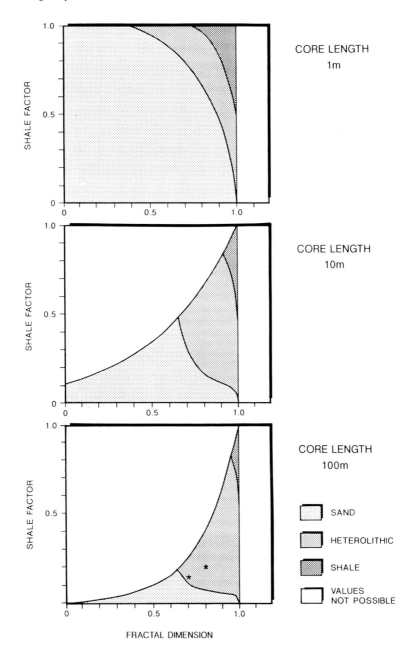

FIGURE 22.   SPAN OF D AND SF FOR VARIOUS CORE LENGTHS.   THE POSI-TIONS CORRESPONDING TO T1 AND T2 ARE MARKED BY ASTERISKS.

tion imply that the line defined by SF = 0 is never actually reached. Looking at the surroundings of SF = 0, D = 1, we find that SF = 0.10, D = 1, for instance, is a massive shale for the minimum core length. But already for a core length of 10 m this combination yields a very sandy, heterolithic sequence. So D = 1 is a pure shale only for the minimum core length. As for the case D = 0, this refers to a sequence where the density is a function of SF only, but invariably very low even for SF = 1. Also it refers to a situation where no hierarchical scaling structure exists in the sequence.

As can be seen from Fig. 22, net/gross and reservoir quality increase when D moves toward 0 from one. Thus net/gross varies with D, as well as the shale factor. Having a higher D, T1 should actually have a slightly lower net/gross than T2 if the shale factor had been the same. As mentioned earlier, SF by itself does not span the full range of shaliness from clean sand to shale. In order to cover the natural range of variation, SF and D must be used in conjunction. For a given SF > 0, increasing D means that the density of the shaly units increases, and the sandy units get more and more shaly, but without formation of new shaly units at the largest scale through coalescence.

That net/gross varies with D as well as SF means that when the sand/shale structure is fractal, net/gross will be scale-dependent within a wide range of scales. The smaller the scale chosen, the higher the net/gross will be. Commonly a particular scale is (arbitrarily) chosen, and particular cut-off values agreed upon by convention. Clearly a natural constraint is related to increasing capillary pressure as sand layers get very thin. But even so we are still faced with the question: What is the relevant net/gross scale for oil production ? or gas production ? This question indicates the inadequacy of the net/gross concept when it comes to catching up the natural complexity of heterogeneous formations.

An important potential of a fractal type of heterogeneity is the possibility of assessing its impact on permeability at various scales. The shale factor measures the magnitude of the impact for the whole interval. D measures the effect of the shaly intervals for any given shale factor. When D is close to 0, the shaly intervals are relatively open, and may contribute significantly to recovery. When D and SF get close to 1, most of the shaly intervals are relatively tight and will mainly act as barriers to vertical flow.

How the suggested three-parameter classification scheme relates to reservoir performance remains to be seen. But the three parameters definitely characterize the studied heterogeneous sequences unambiguously by their most fundamental properties: hierarchical scaling, and shale content. It seems reasonable to think that it will ge generally valid for heterogeneous formations. A classification by non-linear properties would be related to the physical processes of depositional systems and their critical self-organization, and so most likely will be able to distinguish depositional environments from each other.

A MULTIFRACTAL HETEROGENEITY DISTRIBUTION ?

The fractal dimension is related to the nested hierarchical distribution of the shaly units. Is this vertical distribution reflecting a three-dimensional fractal architecture of heterogeneity ?

There is good reason to think so. Not only does a set of vertical strings through the reservoir show a fractal distribution of shale, but modern analogies to the Tilje depositional system show a fractal distribution of depositional environments at the horizontal plane of the surface, as well as in vertical cores drilled from the surface to depths of 30-40 m. Two such analogies are presently being studied, and an example of the information they furnish is given in Figure 23.

None of this, however, will presently allow us to estimate the value of D for two or three dimensions in the Tilje reservoir. It is intuitively clear that the one-dimensional information from the cores is not sufficient for such a purpose, at least not at the outset. A range of possible geometries in three dimensions may produce the observed strings. This multifractal problem is currently under study.

Permeability often fluctuates over several orders of magnitude in heterolithic reservoirs. When measured at the core plug level , permeability records small-scale heterogeneity that repeats itself at larger scales. In this way, permeability also forms a nested hierarchy of values. How permeability distribution is affected by a fractal reservoir architecture is currently being studied.

CONCLUSIONS

1. The studied Tilje formation is a heterolithic, clastic sequence of marginally marine, tidal deposits.

2. This unit forms heterogeneous reservoirs which are extremely complex and difficult to describe quantitatively. Yet Tilje is a simplified case in the sense that only two lithologies occur; sand and shale. The sequence forms a nested hierarchy of shaly and sandy units.

3. The number of distinct, shaly intervals was counted at various scales and found to have self-similar properties in the range of 0.1 mm to 1 m. This was confirmed by various statistical analyses done on a segment of the core. The heterogeneity as seen in cores (one-dimensional samples) is well described by two parameters, a fractal dimension (D) and the shale factor (SF).

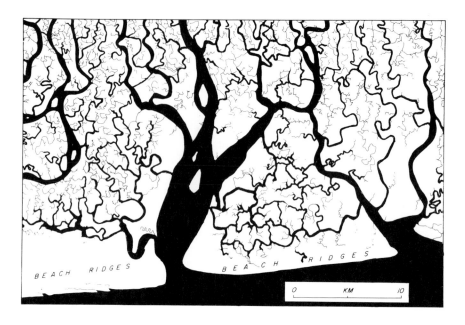

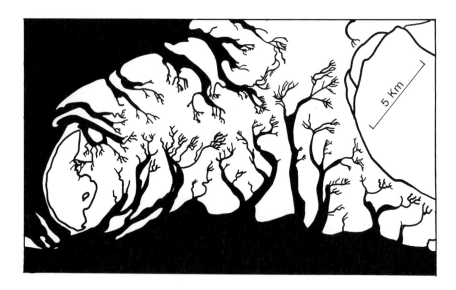

FIGURE 23. TIDAL CHANNEL SYSTEMS DRAWN FROM AIR PHOTOGRAPHS.
NOTE THE APPARENT FRACTAL PATTERN FORMED BY THE CHANNELS.
ABOVE: INTERTIDAL MANGROVE BELT OF THE NIGER DELTA (ALLEN 1965).
BELOW: GERMAN NORTH SEA NON-VEGETATED TIDAL FLAT
(REINECK AND SINGH 1975).

4. An upper sandy unit, T1, has D= 0.80 and SF= 0.13. A lower shaly unit, T2, has $D = 0.71$ and $SF = 0.20$. The value of D is probably reflecting the characteristics of the depositional system, whereas SF is a measure of shaliness in fractally heterogeneous reservoirs.

5. A series of statistical analyses confirmed that the sedimentological record is scale-invariant and can be represented as fractional Gaussian noise.

6. In three dimensions, the shale/sand architecture is a multifractal. Fractal organization is indicated by the fact that the main sand-depositing elements (tidal channels) of recent analogies show a distinctly fractal distribution at the surface.

ACKNOWLEDGMENTS

This paper benefited greatly from the contributions by Prof. Jens Feder and Torstein Jøssang at the University of Oslo, who allowed parts of their report on the statistical analyses to be included. I thank them and Benoit Mandelbrot for their support at various stages of the work, and also Einar Hinrichsen and Finn Boger who did most of the work leading up to the statistical analyses.

Dr. Thomas Hewett and Prof. Larry Lake carefully reviewed the manuscript and suggested important improvements. The paper has benefited to a large extent from their constructive criticism.

The contribution by Feder and Jøssang was made possible through a grant from the Norwegian Petroleum Directorate.

REFERENCES

Allen, J.R.L. 1965: Late Quarternary Niger delta, and adjacent areas: Sedimentary environments and lithofacies. - AAPG.Bulletin 49: 547-600

Chang, J. and Yortsos, Y.C. 1988: Pressure transient analysis of fractal reservoirs. - SPE Paper 18170: 631-643. Presented at 63rd Annual Tech. Conf. SPE, Houston

Feder, J. 1988: Fractals. - Plenum Press, New York. 283p

Feder, J. and Jøssang, T. 1989: The Distribution of Shale and Sand: Analyzing core data from Haltenbanken. - Unpubl.report. 17 p.

Gauch, H.G. 1982: Multivariate Analysis in Community Ecology. - Cambridge University Press. 298p

Hewett, T.A. 1986: Fractal distributions of reservoir heterogeneity and their influence on fluid transport. - SPE Paper 15386. 16p . Presented at 61st Annual Tech.Conf. SPE, New Orleans

Mandelbrot, B.B. 1982: The Fractal Geometry of Nature. - W.H. Freeman, New York. 468p

Mandelbrot, B.B. and Van Ness, J.W. 1968: Fractional Brownian motions, fractional noises and applications. - SIAM Review 10: 422-437

Mandelbrot, B.B. and Wallis, J.R. 1968: Noah, Joseph, and operational hydrology. - Water Resources Research 4: 909-918

Mandelbrot, B.B. and Wallis, J.R. 1969: Some long-run properties of geophysical records. - Water Resources Research 5: 321-340

Reineck, H-E. and Singh, I.B. 1975: Depositional Sedimentary Environments. - Springer-Verlag, Berlin. 439p.

Thomas, A. 1987: Fractal structure in architectural features of fracture fields in rocks. - Comptes Rendes Acad.Sci.Paris, t.304, Series 2, no.4:181-6

Thompson, A.H., Katz, A.J. and Krohn, C.E. 1987: The microgeometry and transport properties of sedimentary rock. - Advances in Physics 36: 625-694

Voss, R.F. 1988: Fractals in nature: From characterization to simulation. - in Peitgen, H-O and D. Saupe (eds.): The Science of Fractal images. Springer-Verlag, New York, pp. 21-70

ALGORITHMS FOR GENERATING AND ANALYSING
SAND–BODY DISTRIBUTIONS

S.H.Begg
J.K.Williams

BP Research
Sunbury Research Centre
Sunbury, England

ABSTRACT

Detailed descriptions are given of fast algorithms for
generating distributions of sand-bodies embedded in shale
and for analysing their connectivity at given well spacings.
The sands are modelled as simple geometrical objects whose
centres are distributed according to a scheme that
incorporates specified overlap rules along with a random
component. This allows us to model features that affect
connectivity and to generate many models that are all
consistent with known data. The technique can thus be used
to quantify uncertainty in heterogeneous, mature reservoirs,
or at the appraisal stage when there are few data. Specific
generated distributions can be chosen as the basis for
reservoir simulation studies. An emphasis is placed on
practical details and on highlighting and solving some of
the more subtle problems with this type of modelling.

1. INTRODUCTION

This paper describes fast algorithms for modelling
distributions of isolated and/or overlapping sand-bodies
that are embedded in shale. It is aimed primarily at aiding
reservoir management decisions that are based on an
assessment of sand-body connectivity in situations of
considerable uncertainty. Such situations could arise at
the reservoir appraisal stage when there are few data, or in
more mature reservoirs that are sufficiently heterogeneous
to prevent reliable correlations, if any, between wells.

RESERVOIR CHARACTERIZATION II

Because of this element of uncertainty we need to consider the many possible ways, consistent with the known data, in which sand-bodies could be distributed throughout the reservoir. This in turn requires us to be able to generate and analyse rapidly the critical features of sand-body distributions so far as connectivity is concerned.

This work differs from others in three respects. The first is that we incorporate facilities that allow us to reproduce typical ways in which sand-bodies are observed to overlap and that therefore control connectivity between them. The second is a strong emphasis on developing algorithms that are computationally efficient. Third, we describe the algorithms in detail, and in so doing highlight some problems that more simple techniques fail to cover. All of the algorithms presented here are applicable in both two and three dimensions.

It should be stressed that we are not attempting to model any particular depositional environment or to model the process by which sand-bodies were deposited, e.g. [1]. The major requirement for reservoir development purposes is to be able to describe, in a quantitative manner, how the reservoir is divided into areas of sand and areas of shale.

The approach we take is that of the Boolean or "bombing" model [2] where sand-bodies are described as simple geometrical entities whose centres are distributed in space according to given rules. In order to allow for uncertainty a random component is introduced so that we can create many sand-body distributions which all obey the known data and the general rules. Thus, although the method is statistical, the sand-body centres are not uncorrelated (except in the case where no overlap rules are given). In neither case is the distribution of sand itself uncorrelated. This is because sand-bodies are modelled by continuous objects rather than by specifying the probability of any given point being sand. Therefore, since sand-bodies tend to be large, if one point in the reservoir is sand it is likely that surrounding points will also be sand.

In the following two sections we concentrate on describing the basic algorithms for the generation process and connectivity analysis respectively. We leave illustrations of their capabilities and discussion of timing and practical applications to section 4. For brevity we will often shorten sand-bodies to sand.

2. SAND-BODY GENERATION

We implement an iterative scheme for filling a model of a section of a reservoir to a specified level with sands of given dimensions. For reasons (more fully explained later) of computational efficiency and obedience to the input data, we use the scheme shown in Figure 2.1. This consists of estimating the number of sands needed, choosing their dimensions, positioning them, and finally checking the actual proportion of the model which has been filled. If the latter is within a specified tolerance, then we finish off by calculating the statistics of the generated distribution for comparison with the input. If the proportion filled is less than that required, we generate more sands, if greater we remove some.

It is important to remember that we are attempting to generate spatial distributions of sands which characterise how they are thought to be finally distributed within a reservoir and are not trying to mimic the process of how they were laid down. The details of each part of the scheme are described in the following sections, along with the parameters used to define the model and to control the speed and convergence of the generation process.

2.1 Model definition

Consider a cuboidal section of a reservoir of length L, width W, thickness T, defined along the principal axes of the model, x, y, z, respectively. We wish to fill this with sands until they occupy a proportion, p, of the total volume, V. The geometry of a sand-body is approximated by a cuboid of length l, width w, thickness t and its axes are aligned with the principal axes of the model. These dimensions may be constant or drawn from observed or hypothetical probability density functions (pdf's) and correlations between them may be imposed. For intersecting sands the dimensions refer to those of the original sand-bodies; thus thicknesses will not be those that are observed in a vertical well – see Figure 2.2. The same applies to lengths or widths observed in a horizontal well. However, an interesting spin-off of the technique used to calculate the number of sands needed to fill the model (see 2.2 below) allows us to back out an average sand-body thickness from the values observed in a well.

The final parameters needed to describe the system are those which define the overlap relationships between sand-bodies. This is an important feature of the geology to try to model as it may have a dominant effect on the

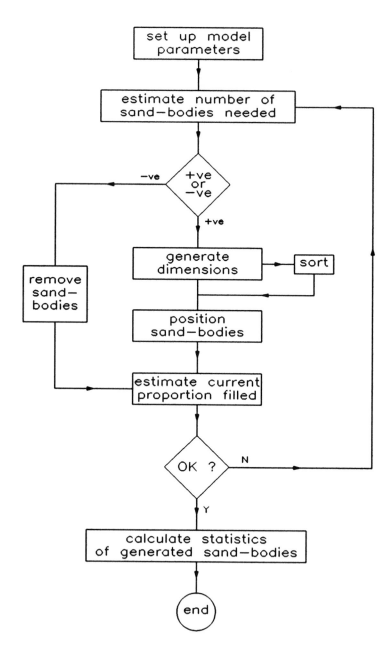

FIGURE 2.1. Sand-body generation scheme.

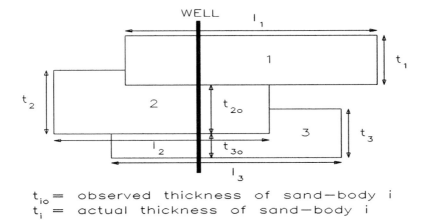

t_{io} = observed thickness of sand—body i
t_i = actual thickness of sand—body i

FIGURE 2.2. Sand-body thicknesses observed in a well

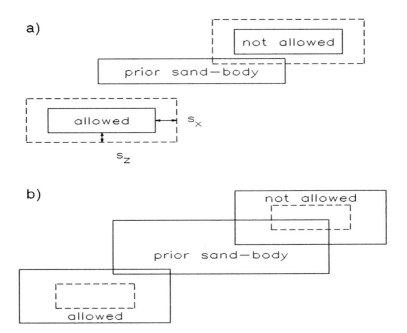

FIGURE 2.3. Constrained overlaps: a) separated, b) partial.

connectivity and conductivity of the resulting system. Three
types of relationship are allowed here. The first is where
the bodies are independent and can therefore overlap by any
amount. The second is where they are forced to be separated
by minimum distances s_x, s_y, s_z in the principal directions
- see Figure 2.3a. If s_x, s_y, s_z are negative, then the
bodies are allowed to overlap by only these amounts - see
Figure 2.3b. The third type of relationship is where there
is a preferred direction of overlap if sand centres come
within a specified range of each other. It will be shown
later (section 4.2) how these relationships can be used to
model common types of sand-body overlap. The parameters
needed to define this last case are given in section 2.4.2.

2.2 Estimation of number of sand-bodies needed

As will be shown later, the calculation of the
proportion of the model that is filled at any time during
the generation process is computationally demanding. It is
therefore desirable to be able to make a good estimate of
the number of new sands needed and then check the fill
fraction once these have been generated.

If the sands are isolated from each other then the
number needed to fill a proportion, p, of the volume, V, is
simply

$$n_e = pV/\tilde{v} \tag{1}$$

where $\tilde{v}$ is the average volume of a sand and is calculated
from the pdf's used to define the dimensions.

The calculation for independent, overlapping sands is
more difficult. If σ_v^2 is the variance of the sand volumes
(again derived from the pdf's) then it is shown in Appendix
1 that a close approximation to the number of sands needed
is

$$n_e = \frac{V \ln(1-p)}{(\tilde{v}^2/2V + \sigma_v^2/2V - \tilde{v})} \tag{2}$$

By comparing equations (1) and (2) we can write a single
equation for the number of sands

$$n_e = pV/\alpha\tilde{v} \tag{3}$$

where

$$\alpha = \frac{p(\tilde{v}/2V + \sigma_v^2/2\tilde{v}V - 1)}{\ln(1-p)} = \alpha_i \qquad : \text{independent}$$

$$\alpha = 1 \qquad\qquad\qquad\qquad\qquad\qquad : \text{separated}$$
(4)

Thus, for the independent case, $\alpha\tilde{v}$ can be thought of as the "effective" average volume of non-overlapping sands.

Now, because we are using an iterative process, we wish to be able to calculate the number of sands still required (or to be removed) once we have generated our initial estimate and checked the volume filled. This is clearly much quicker than simply adding (or removing) one at a time until we reach the desired proportion filled. If p_c is the current proportion filled, then the number required is simply given by

$$n_e = (p-p_c)V/\alpha\tilde{v}$$
(5)

It is worth noting that, at this stage, all of the above formulae apply to sand and model volumes of any shape. Also, for two- or one-dimensional models the number of sands needed is calculated by substituting respectively areas or lengths for volumes.

Finally, we consider sands that are allowed to overlap by only a limited extent. In this case α must lie somewhere between the two limits given by equations (4). It is very difficult to solve for n_e in this case. We have chosen to interpolate α linearly between these two limits using the average overlap volume (for the given constraints) between two sands. In Appendix 2 we show that this average overlap volume is given by

$$\tilde{v}_0 = \frac{l^2w^2t^2-(1-s_x)^2(w-s_y)^2(t-s_z)^2}{8[lwt-(1-s_x)(w-s_y)(t-s_z)]}$$
(6)

Using $o_r=\tilde{v}_0/\tilde{v}$ to define the ratio of this average overlap volume to the average sand volume, then $o_r=0.125$ (0.25 in 2-D) when $s_x=-\tilde{l}$, $s_y=-\tilde{w}$, and $s_z=-\tilde{t}$ (equivalent to the independent case), and $o_r=0$ when $s_x=s_y=s_z=0$ (equivalent to the isolated case). We require $\alpha=\alpha_i$ at $o_r=0.125(0.25)$ and $\alpha=1$ at $o_r=0$. A linear variation of α with o_r is assumed,

$$\alpha=1-4o_r(1-\alpha_i) \text{ in 2-D} \qquad \alpha=1-8o_r(1-\alpha_i) \text{ in 3-D}$$
(7)

This expression for α is substituted in equation (5) to calculate the number of sands needed. Note that it is worthwhile increasing n_e by up to 5% since it is easier to

remove some excess bodies than to generate and fit some more
if there are too few. The reason for this will be explained
in section 2.4.1.

2.2.1 Boundary effects

We now take account of how the model boundaries affect
the number of sands needed. At each bounding plane of the
model there will be, on average, only half the number of
sands as there are in the interior - see Figure 2.4 for the
two-dimensional case. This occurs because we have only
modelled sands whose centres lie within these planes.
Consequently there will be a depleted zone around the
boundary. This problem can be overcome by generating the
sands within a larger region given by

$$-l_m/2 \text{ to } L+l_m/2, \quad -w_m/2 \text{ to } W+w_m/2, \quad -t_m/2 \text{ to } T+t_m/2$$

where the subscript m indicates the maximum value of the
sand dimensions. This has the effect of pushing the
depleted zone to outside the actual boundaries and requires
us to use the larger volume

$$V' = (L+l_m)(W+w_m)(T+t_m) \tag{8}$$

in the expressions for the number of boxes needed.

2.2.2 Relationship between observed and deposited sand-body thickness.

The above arguments can also be applied in one-dimension
to back out an average sand thickness from the thicknesses
observed in a vertical well (or lengths for a horizontal
well). Consider a well through a number of intersecting
sands - see Figure 2.2. The average effective (or observed)
sand thickness, $\bar{t}_o$, is clearly not the same as the average
deposited (or generated) thickness, $\bar{t}$. Using the one-
dimensional version of equations (3) and (4), we have $\bar{t}_o =$
$\alpha\bar{t}$ and if $\bar{t}/T$ is small then the second-order terms in the
derivation of equation (2) - see Appendix 1 - can be
neglected to give

$$\bar{t} = \ln(1-p)\bar{t}_o/p \tag{9}$$

In a depositional environment where $\bar{t}$ represents the true
(non-eroded) average channel belt thickness and $\bar{t}_o$ the
observed thickness, then equation (9) suggests the
relationship between these two quantities is not a constant

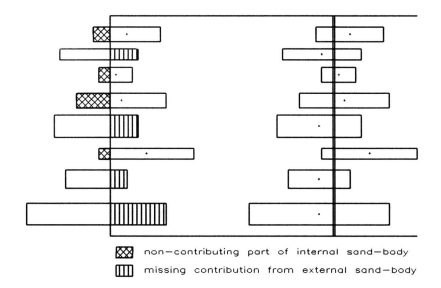

FIGURE 2.4. Sand-bodies at model boundaries

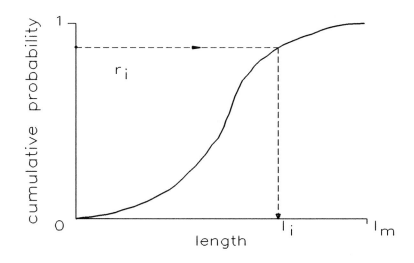

FIGURE 2.5. Selection of sand-body dimension from cdf.

but depends on the amount of sand. It would be very
interesting to check this argument using outcrop data.

2.3 Assigning dimensions to the sand-bodies

The next step in the scheme is to select dimensions for
each of the n_e sands. This is done before they are
positioned in order that, for the cases of constrained
overlap or separation, the statistics of the final
distribution closely match the input statistics. The
alternative, of "dimensioning" and positioning a body before
dealing with the next one [2] can lead to the situation
where it is impossible to position a large sand in the later
stages of the generation process. Such a sand would then
have to be rejected, biasing the final distribution towards
smaller ones. This problem can be largely overcome by
selecting the dimensions of all sands first. Then, before
positioning them, they can be sorted so that the biggest
will be laid down first. This does not guarantee that every
sand can be placed, but it removes many conflicts and has
the added advantage that it speeds up the process because
fewer attempts are necessary to find an acceptable position.

The technique used to select dimensions from the
relevant pdf's is the usual one of generating a random
number (from a uniform distribution between 0 and 1) and
using this to read off the value of the dimension concerned
from the related cumulative distribution function – see
Figure 2.5. Over a large number of selections this will
produce a set of dimensions whose pdf is the same as the one
from which they were derived. If any correlations between
the three dimensions have been specified, then clearly the
independent dimension(s) should be chosen first and used to
generate the correlated one(s).

2.4 Positioning the sand-bodies.

The method of positioning a sand when no overlap rules
are specified is to choose a point at random within the
model and centre the sand on it. The point is chosen by
generating three random numbers from a uniform 0-1
distribution and scaling these by the length, width and
thickness of the model. Recall that in order to remove
boundary effects we generate centres within a larger volume
that surrounds the model and then clip any parts of sands
that lie outside the model itself. The dimensions of the
larger volume must exceed the model by half the maximum of

the appropriate sand dimension. That is, the centres (x_c, y_c, z_c) should be generated between the limits:

$$x_c:(-l_m/2, L+l_m/2), \quad y_c:(-w_m/2, W+w_m/2), \quad z_c:(-t_m/2, T+t_m/2)$$

This is also the reason for using the larger volume, given by equation (8), in calculating the number of sands needed to fill the model. The centre of the i'th sand is thus chosen by

$$\begin{aligned}
x_{ci} &= -l_m/2 + r_{1i}(L+l_m) \\
y_{ci} &= -w_m/2 + r_{2i}(W+w_m) \\
z_{ci} &= -t_m/2 + r_{3i}(T+t_m)
\end{aligned} \tag{10}$$

where r_{1i}, r_{2i}, r_{3i}, are the three random numbers. Rather than store the sand centre and dimensions we store the co-ordinates of its faces, as this increases the speed of making overlap checks if the sands are not independent:

$$\begin{aligned}
x_{1i} &= x_{ci} - l_i/2 & x_{2i} &= x_{ci} + l_i/2 \\
y_{1i} &= y_{ci} - w_i/2 & y_{2i} &= y_{ci} + w_i/2 \\
z_{1i} &= z_{ci} - t_i/2 & z_{2i} &= z_{ci} + t_i/2
\end{aligned} \tag{11}$$

If the sands are independent then nothing further need be done before positioning the next one.

2.4.1 Separated or partially overlapping sand-bodies

If overlap criteria (extents or preferred direction) have been specified then we must check the position of the current sand against all previously positioned ones. This is the most time-consuming part of the procedure. One could calculate the actual separation (or overlap) at each comparison and then test whether or not it meets the specified criteria. However, it is more efficient to first expand (or contract) the current sand and then just check whether or not there is any overlap at all. Thus we set up temporary test co-ordinates

$$\begin{aligned}
x_{1it} &= x_{1i} - s_x & x_{2it} &= x_{2i} + s_x \\
y_{1it} &= y_{1i} - s_y & y_{2it} &= y_{2i} + s_y \\
z_{1it} &= z_{1i} - s_z & z_{2it} &= z_{2i} + s_z
\end{aligned} \tag{12}$$

If an overlap with a previous body is found we stop the checking and calculate a new centre. This process is repeated until a satisfactory centre is found, or the given maximum number of attempts is reached. In the latter case we can either disobey the overlap rules or choose a smaller sand and disobey the size statistics.

One problem that we encountered with this method is that
there was a tendency to generate too little sand within the
model. This was found to be because, in the latter part of
the generation process, the smaller sands were
preferentially placed at the model boundaries where the
constraints are less severe - i.e. where there is potential
for overlap only with sands inside the model. Also, because
the density of centres was greater nearer the edges, a
greater proportion of their volumes fell outside the model.
The problem was overcome by first estimating the number of
sands that would be expected to intersect each of the model
faces (using the appropriate 2-D versions of equation (2))
and generating them first. This gives the correct size and
density distribution at the boundaries, and later sands are
forced back into the interior.

In order to speed up the check against the previously
positioned bodies, we set up a linked-list cache structure
whose bin size is that of the maximum sand-body dimensions
(plus any separation that has been specified). This cuts
down the number of sands which we have to check against by
allowing us to pull out the indices of only those which lie
in the vicinity (+/- one bin) of the current sand. Once a
satisfactory position is found for a body it is inserted
into the list.

Note that it is much easier, and therefore quicker, to
place the sands if they are first sorted into order of
decreasing size. Thus the largest bodies are positioned
first, resulting in increased speed and fewer rule
conflicts. (The most efficient sorting algorithm we have
found is Quickersort [3], which is slightly faster than
$Nlog_2N$.) Speed can also be increased by reducing the
allowed number of positioning attempts, but this must be
traded-off against obeying the overlap rules or the
statistics. We have found a limit of 10,000 attempts to
cause few or no conflicts for many models.

2.4.2 Preferred direction of overlap.

A simple method of incorporating a preferred direction
of overlap is as follows. We first define the preferred
direction by rotating a vector (initially lying along the x
axis) through a given angle, θ_y, around the y axis in an
anti-clockwise sense. This is followed by another anti-
clockwise rotation through an angle θ_z, around the z axis -
see Figure 2.6. We also specify the size of a zone to be
placed around a sand centre ($\pm s_{cx}$, $\pm s_{cy}$, $\pm s_{cz}$) which is
divided into acceptance and rejection regions. The
acceptance region is defined by two angles about the

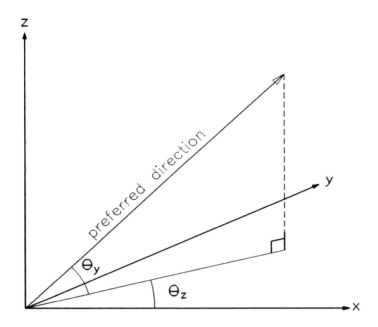

FIGURE 2.6. Definition of preferred overlap direction.

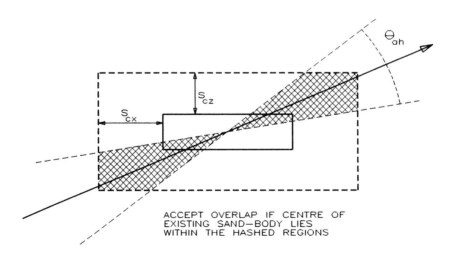

ACCEPT OVERLAP IF CENTRE OF
EXISTING SAND—BODY LIES
WITHIN THE HASHED REGIONS

FIGURE 2.7.   Constraints for preferred overlaps.

preferred direction vector; one is in the vertical plane, θ_{av}, and one in the horizontal plane, θ_{ah}. This gives an acceptance region which has the form of two truncated pyramids whose apexes are at the sand centre and whose axes lie along the preferred direction. This is illustrated for the 2-D case in Figure 2.7. The use of an ellipsoidal zone of influence and conic acceptance regions would be more elegant, but requires more computations.

We follow a procedure similar to that described in the previous section. Once a trial position has been chosen for a sand, it is tested against each of the previously positioned bodies. If the centre of an existing body comes within the specified zone then we check whether or not it lies within the acceptance region. (This is done most efficiently by comparing tangents of the relevant angles). If the existing centre does lie within the acceptance region then the trial position for the current sand is still acceptable and we move on to check against the next sand in the list, and so on. If the existing centre does not lie in the acceptance region then we generate a new centre for the current sand and start the checking process again. This procedure is repeated until an acceptable position is found or the allowed number of postioning attempts is reached.

2.5 Calculating the fraction filled.

The final part of the iterative scheme is to calculate the current proportion of the model that is filled with sand. To do this exactly, which would entail calculating complex overlap volumes, is very time consuming. Several alternative methods that will produce a close approximation are available.

The most common is a Monte Carlo integration procedure where a number of points are chosen at random within the model and tested to see if they lie within a sand or not. An estimate of the proportion filled is then the ratio of the number of points that lie within a sand to those that do not. This is very time consuming as 10,000 trials are needed to achieve an accuracy of +/-0.01, and each point must be tested against all of the sand-bodies.

A more efficient method is to set up a finely gridded representation of the model using a three-dimensional integer array initialised to zero. Each sand is then defined in terms of integer grid coordinates, and the sites of the grid that it occupies have their values set equal to one, irrespective of whether or not this was done for a previous sand. We then count the number of occupied blocks

to get the amount of sand. This technique avoids the need
to check for overlaps and has the advantage that each sand
need be dealt with only once. However, one runs into
resolution problems, particularly in 3-D, if the sands are
small compared to the model volume. This is because the
fine grid must have a cell size that is small enough to
resolve the smallest sand-body.

Our preferred technique, from both a speed and accuracy
point of view, is to average the proportions observed in
hypothetical wells. These are spaced on a rectangular grid
whose unit-cell dimensions are equal to the mean sand-body
dimensions. In order to increase the speed of identifying
which sands are penetrated by a well, we use the same
linked-list cache that was used to speed up the overlap
checks. (In the case where the sands are positioned
independently, i.e overlaps ignored, we insert them into the
cache immediately after they have been positioned.)

After calculating the current proportion we compare it
with our target value. If they are not the same, to within
a specified tolerance, we go back and calculate the number
of sands still needed. If this is positive then we generate
and position that number of sands. If negative we remove
the number. In the latter case we must be careful when
separation criteria have been specified, as the sands will
be sorted by size. Here the smallest bodies will be at the
end of the list and to remove only these would both bias the
statistics and not decrease the volume sufficiently. We
must therefore remove them at a regular interval throughout
the list.

2.6 Conditioning to observed data.

To condition the generated distributions to well data
all that need be done is to first generate dimensions for
the n_w sands observed in the wells. Then, rather than
centering the sands on the well location, we place them such
that the well penetrates a random fraction along their
lengths and widths. That is $x_1 = x_w - r_{1j} l_j$, $x_2 = x_w + (1 - r_{1j}) w_j$
etc. for the $j = 1, n_w$ well-bore sands. The only constraint
that needs to be imposed is a check to ensure that the edges
do not overlap any other wells. Having placed these sands
we then calculate the current proportion filled and proceed
as before, flagging the first n_w in the list as not to be
removed if we over-shoot the required proportion.

3. CONNECTIVITY ANALYSIS

Having generated a distribution of sands, we now want to examine how they are connected to each other or to any given well pattern. A scheme for doing this is outlined in Figure 3.1. Note that we may also wish to investigate the uncertainty in connectivity that arises from the fact that so far we have only generated one of the many possible sand distributions that are consistent with the data. As the identification of clusters is also a time-consuming process, it is most efficient to calculate connectivity for a range of well spacings on each realisation rather than to generate a lot of realisations for one spacing before doing the next.

To calculate the volume of sand connected to the wells, it is most efficient to first calculate the volume of each cluster of sands and then determine whether or not the clusters are connected to wells. In order for a cluster to be counted as connected to the wells, we introduce a "well connection criterion" which is simply the number of wells by which a cluster must be penetrated. Strictly, a cluster is connected to the well system if it is intersected by just one well. However, there are occasions when it may be necessary to specify a larger number – for example, when considering some form of injection or when the dimensions of clusters are significantly larger than the drainage radius of a well for natural depletion. In the former case we must distinguish between injectors and producers and check that a cluster is penetrated by the required number of each.

3.1 Cluster identification.

The first step in calculating cluster volumes is to identify which sands overlap to form the clusters. We start by setting up a linear array that we will use to hold an integer value which identifies the index of the cluster to which a sand belongs. Initially we set the cluster index of a sand to be the same as the sand index. The basic idea is to take one sand at a time (the current sand) and then loop over the remainder looking for overlaps. When an overlapping body is found we record its cluster index and then search the list of sands for all those with the same index. Each time we find one we reset its index to be that of the current sand. We then continue to look for any further overlaps with the current sand before taking the next one and so on. The overlap checks are performed as described in section 2.4.1.

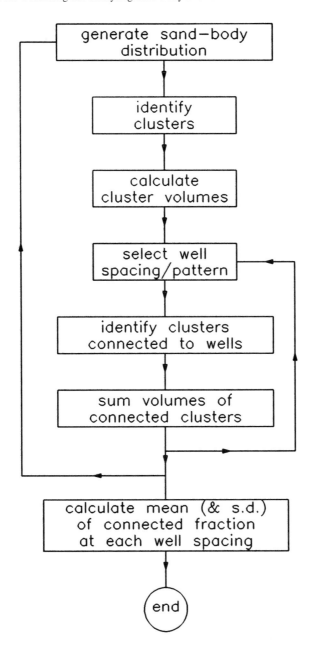

FIGURE 3.1. Connectivity analysis scheme.

Unfortunately this process does not produce cluster numbers that run conveniently from 1 to n_c (n_c=number of clusters). We therefore re-label the final list so that they do, thus making it easier to perform later manipulation and to calculate summary statistics of the clusters.

3.2 Calculation of cluster volumes.

The volume occupied by a cluster of sands is calculated in the same way as the total volume of sand (see section 2.5) with the exception of clusters consisting of just one or two sands - whose volumes can easily be calculated analytically, and therefore more efficiently. If a cluster is sufficiently compact to allow resolution of all the sands it contains, then the fine-grid method offers a quicker alternative than the "well" technique.

3.3 Calculation of connectivity.

We first calculate the number of wells to which each cluster is connected. This is done by taking each well of the current well pattern and determining which sands it intersects. When an intersecting sand is found we use its cluster index to point to a location in an array of counters that has been set up to hold the number of wells penetrating each cluster. The value of the counter at the location is then incremented by one. We also flag that that cluster is connected to the well so that we do not increment the counter again if another sand in the same cluster also intersects the well - as it is likely to do.

After this has been done for each well in the current pattern, we sweep through the array and sum the volumes of those clusters that are intersected by at least the number of wells specified in the well-connection criterion. Finally we calculate what fraction of the total sand present that this volume is.

The above procedure is then repeated for each well spacing or pattern so that we build up a table of connectivity as a function of well spacing. If more than one sand realisation is used then the whole process is repeated, and we calculate the mean and standard deviation of connectivity at each spacing.

4. RESULTS

The above algorithms have been coded up for both two-
and three-dimensional models. In this section we aim to show
that the algorithms work and to illustrate their
flexibility. An investigation of the correlation between
connectivity and types of sand distribution could be
performed using these techniques but is beyond the scope of
this work.

4.1 Estimating number of bodies needed

The use of equations (4), (5), (7) and (8) is
illustrated in Figure 4.1, which shows the estimated number
of sands versus the number actually needed to come within 1%
of the target value of p. It was found that usually no
iterations were necessary in 2-D and only one was needed in
3-D. These results were obtained for a range of p values
between 0.01 and 0.99 in both 2-D and 3-D with independently
positioned bodies. Figure 4.2 shows 2-D and 3-D results
respectively for a fixed p of 0.4 but varying the overlap
constraints (s_x, s_y, s_z) between the mean sand dimensions and
zero - totally isolated sands.

4.2 Sand-body distribution examples

Figures 4.3 to 4.7 illustrate the variety of overlap
relationships that can be modelled. Two-dimensional
examples with small numbers of sands are used because it is
difficult to see the relationships in 3-D plots or where the
sands are small. Figure 4.3 shows independent sands for
p=0.5 and Figure 4.4 shows the effect of restricting
overlaps to 25% of the mean sand dimensions whilst keeping p
fixed - the centres have moved apart and the number needed
has decreased. Using this technique we can model the kinds
of relationships that result from avulsed fluvial sands
which avoid previously deposited ones until they no longer
form topographic highs. We can go even further by enforcing
strict separation, Figure 4.5, which might result in
environments where there is periodic deposition of sand in a
background of continuous mud deposition. Figures 4.6 and
4.7 show how multi-storey and multi-lateral relationships
can be modelled using preferred alignment directions of 90
and zero degrees respectively. Typical cluster sizes and
spacings can be altered by changing the range of influence.

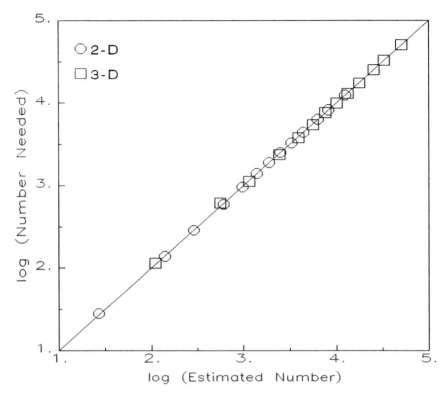

FIGURE 4.1. Comparison of estimated and actual number of sand-bodies needed to get to within 1% of the target fill proportion - independent sand-bodies.

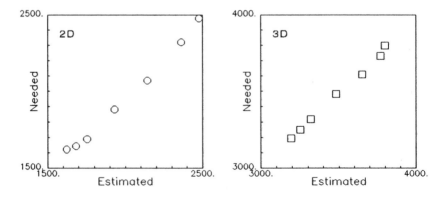

FIGURE 4.2. Comparison of estimated and actual number of sand-bodies needed to get to within 1% of the target fill proportion - variable overlap constraints from zero (isolated) to complete (independent).

FIGURE 4.3. Independently positioned sand-bodies (p = 0.3).

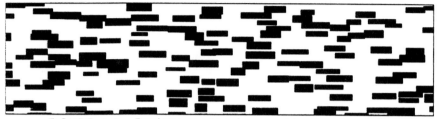

FIGURE 4.4. Partially overlapping sand-bodies (p = 0.3).

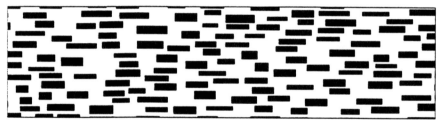

FIGURE 4.5. Separated sand-bodies (p = 0.3).

FIGURE 4.6. Multi-storey stacking of sand-bodies (p = 0.3).

FIGURE 4.7. Multi-lateral stacking of sand-bodies.

4.3 Connectivity examples

In Figure 4.8 we use a grey-scale to indicate successful
identification of the clusters. Those that are connected to
wells at a spacing of L/10 and with well connection criteria
(WCC) of 1 and 2 are shown in Figure 4.9.

The variation of connectivity with well spacing over 100
realisations of a large 3-D model (p=0.2) is shown in Figure
4.10. Here we have assumed that for the WCC=2 case we are
dealing with natural depletion and thus all wells are
producers. Consequently a cluster will be connected if it
is penetrated by any two wells. If we were considering a
flooding process then a cluster would have to be intersected
by a producer-injector pair, thus decreasing the connected
faction, particularly at longer well spacings.

Figure 4.11 shows the variation of mean connectivity at
different well-spacings for values of p above and below the
percolation thresholds of 0.28 and 0.67 [4] in 3-D and 2-D
respectively.

4.4 Typical run times

Run times are a function not only of the proportion of
the model to be filled (i.e. the number of sands) but of the
dimensionality, overlap criteria and the convergence control
parameters. It is impractical to illustrate all possible
combinations so we provide times for some typical examples
and indicate the factors most likely to speed up or slow
down the procedures. All are in seconds of CPU on a VAX
8800. Times for the most intensive parts of the generation
process are shown in Table I and are for the first iteration
where more than one was necessary.

It can be seen from Table I that timings for
independently positioned bodies are approximately linear
with N. However, introducing overlap constraints increases
the positioning time. This can become large when generating
isolated sands at proportions near the percolation
threshold. In this case the run times are found also to be
a function of the variance in the dimensions. Note that the
times for calculating the filled proportion drop rapidly as
soon as the sands become isolated, allowing us just to sum
their volumes. All of the above times are for constant
numbers of hypothetical wells used to calculate the fill
proportions. Increasing these numbers, or specifying a more
severe tolerance, will increase times still further.

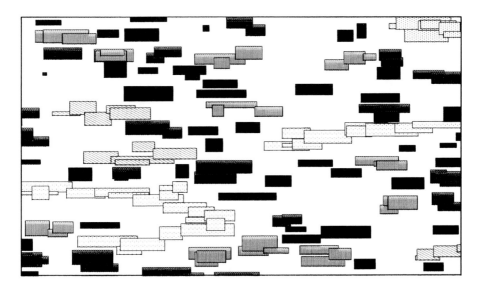

FIGURE 4.8. Cluster identification - lightness increases with increasing number of sands in a cluster.

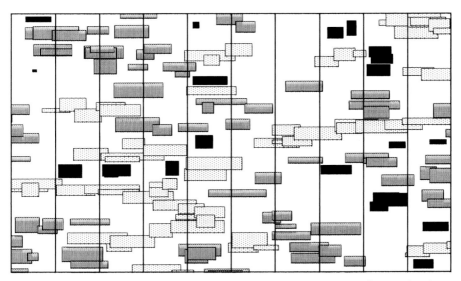

FIGURE 4.9. Sand-bodies connected at well connection criterion of one (dark grey) and two (light grey). Sands not connected to any wells are in black.

Table I - VAX 8800 CPU times (sec) for sand-body generation.

Case	Percent Overlap	Gen.	Sort	Pos.	Fill	Total	No. of sands	p
	–	0.01	–	0.00	0.26	0.3	285	0.1
	–	0.05	–	0.01	0.34	0.4	965	0.3
I2D	–	0.10	–	0.03	0.39	0.5	1875	0.5
	–	0.18	–	0.05	0.56	0.8	3256	0.7
	–	0.32	–	0.10	0.84	1.3	6228	0.9
	–	0.67	–	0.19	1.36	2.2	12456	0.99
	–	0.09	–	0.04	0.81	0.9	1122	0.1
	–	0.41	–	0.13	2.59	3.1	3798	0.3
	–	0.63	–	0.30	5.80	6.7	7381	0.5
I3D	–	1.10	–	0.52	11.51	13.1	12821	0.7
	–	2.06	–	1.03	26.68	29.8	24520	0.9
	–	4.07	–	2.03	61.29	67.4	49040	0.99
	100	0.11	0.06	0.26	1.83	2.2	2478	0.6
	80	0.12	0.06	0.24	1.70	2.1	2362	0.6
	60	0.11	0.05	0.23	1.53	1.9	2142	0.6
C2D	40	0.10	0.05	0.27	1.32	1.7	1929	0.6
	20	0.09	0.05	0.42	1.19	1.8	1754	0.6
	0	0.08	0.04	1.65	0.01	1.8	1622	0.6
	-10	0.09	0.03	28.89	0.01	29.0	1622	0.6
	100	0.31	0.12	0.90	3.04	4.4	3798	0.3
	80	0.29	0.11	0.93	2.96	4.3	3770	0.3
	60	0.27	0.09	0.90	2.71	4.0	3652	0.3
C3D	40	0.26	0.08	0.96	2.53	3.8	3482	0.3
	20	0.23	0.09	1.30	2.32	3.9	3319	0.3
	0	0.21	0.08	3.38	0.03	3.7	3194	0.3
	-10	0.21	0.08	12.10	0.03	12.4	3194	0.3

I2D/I3D = independent 2-D & 3-D
C2D/C3D = constrained 2-D & 3-D
Percent Overlap = % of mean dimensions used for (s_x, s_y, s_z)

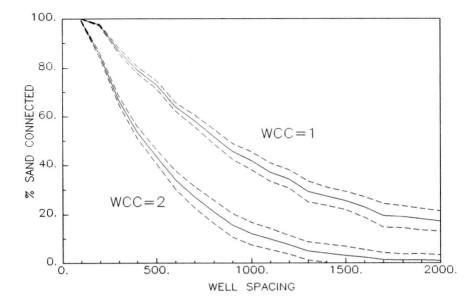

FIGURE 4.10Variation of connectivity with well spacing
and Well Connection Criterion (WCC).

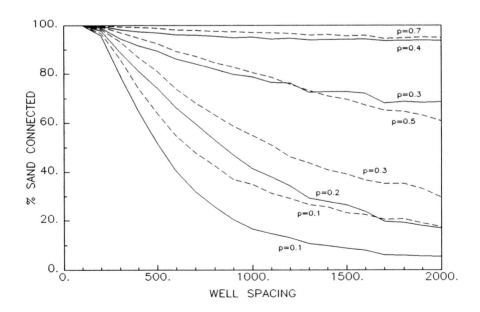

FIGURE 4.11Variation of average connectivity with well
spacing and proportion of model filled (p). Solid
lines are for 3-D and dashed lines are for 2-D.

Times for the connectivity analysis are similarly not just a function of the number of sands but also depend on the number of clusters and their range of sizes. This is illustrated in Table II, which gives times for a range of p values using independently positioned sands. All values are averages over 50 realisations.

Table II - VAX 8800 CPU times (sec) for connectivity
            calculations. First set are for 2-D and
            second set are for 3-D

p	N	N_c	N/C & (s.d.)	CPU TIME CLUS.	VOLS.	CON.
0.1	149	121	1.24 (1.36)	.02	.03	.07
0.3	507	234	2.17 (2.99)	.17	.28	.20
0.5	982	185	5.33 (10.97)	.77	.72	.55
0.7	1704	51	34.22(200.36)	2.41	.75	.59
0.1	1529	984	1.55 (1.97)	1.63	2.88	5.27
0.2	3237	1127	2.87 (6.11)	8.37	10.00	10.83
0.3	5177	720	7.20(114.38)	23.83	9.56	17.93
0.4	7419	358	20.76(360.39)	53.07	5.53	25.45

N_c =number of clusters; N/C & s.d = mean and standard deviation of the number of sands per cluster

6. CONCLUSIONS

We have developed algorithms for generating two- and three-dimensional spatial distributions of sand embedded in shales. Although we do not attempt to model the process of deposition, we have incorporated facilities which allow us to exert considerable control over the relationship between neighbouring sands. These controls enable us to model typical spatial distributions resulting from:

i) Preferred stacking directions (e.g. multi-lateral and multi-storey).

ii) Isolation due to periodic sand deposition within a background of more continuous clay/mud deposition.

iii) Partial overlap due to dependency of one sand on the position of previously deposited ones.

Additional algorithms are described for analysing a distribution of sands in terms of cluster statistics and calculating the fraction of sand connected to any given well pattern or spacing.

An emphasis is placed on the efficiency of the
algorithms so that we can quantify the uncertainty in our
models by generating hundreds of distributions containing
thousands of sands. For example, it took only 30 seconds of
VAX8800 CPU time to generate a 3-D model containing nearly
25,000 sands. Timings increase approximately linearly with
the number of sand-bodies as long as any overlap constraints
are not severe.

7. ACKNOWLEDGEMENTS

The authors would like to thank the British Petroleum
Company plc for permission to publish this paper. We also
thank Dr. P. King for help in both reviewing the manuscript
and discussions on technical issues.

8. REFERENCES

1. ALLEN, J.R.L. "Studies in fluviatile sedimentation: an
    exploratory quantitative model for the architecture
    of avulsion-controlled alluvial suites", Sedimentary
    Geology, v21 (no2), 1978, pp129-147

2. HALDORSEN, H.H. and MacDONALD, C.J. "Stochastic modeling
    of underground reservoir facies (SMURF)", SPE 16751,
    presented at the 62nd SPE Annual Technical
    Conference, Dallas, 27-30 Sept. 1987

3. SCOWEN, R.S. "Algorithm 271: Quickersort", Collected
    algorithms of the ACM"

4. KING, P.R. "Connectivity and conductivity of over-
    lapping sand-bodies", In: Proc. of the 2nd
    International Conference on North Sea Oil and Gas
    Reservoirs, Graham & Trotman, to be published in 1989

APPENDIX 1 — Estimation of the number of overlapping sand–bodies needed to fill the model to the desired level.

We want to fill a volume V to a proportion p with sand–bodies whose centres are independent and have volumes v_i, where $i = 1, n_e$. We wish to estimate n_e. Taking a point, P, at random inside the model, the desired proportion is just the probability that P lies within a sand, i.e.

$$p = \text{Prob } (P \text{ lies in a sand})$$
$$= 1 - \text{Prob } (P \text{ lies outside all sands})$$
$$= 1 - \text{Prob } (P \text{ lies outside sand 1, sand 2,} \ldots \text{sand } n_e)$$

Now the probability that P lies outside sand i is just $(V - v_i)/V = (1 - v_i/V)$. Therefore

$$p = 1 - \prod_{i=1}^{n_e} (1 - v_i/V)$$

Rearranging and taking logs

$$\ln(1 - p) = \ln \left[\prod_{i=1}^{n_e} (1 - v_i/V) \right]$$

$$= \sum_{i=1}^{n_e} [\ln(1 - v_i/V)]$$

and expanding

$$\ln(1 - p) = \sum_{i=1}^{n_e} \left(-\frac{v_i}{V} + \frac{v_i^2}{2V^2} - \frac{v_i^3}{3V^3} \ldots \right)$$

$$= \frac{-1}{V} \sum_{i=1}^{n_e} v_i + \frac{1}{2V^2} \sum_{i=1}^{n_e} v_i^2 + \text{ higher order terms}$$

Now, neglecting third order terms and above, and recalling the mean and variance of the sand–body volumes are given by

$$\bar{v} = \frac{1}{n_e} \sum_{i=1}^{n_e} v_i \qquad\qquad \sigma_v^2 = \frac{1}{n_e} \sum_{i=1}^{n_e} (v_i - \bar{v})^2$$

we get

$$\ln(1 - p) = \frac{-n_e \bar{v}}{V} + \frac{n_e \bar{v}^2}{2V^2} + \frac{n_e \sigma_v^2}{2V^2}$$

giving

$$n_e = \frac{V \ln(1 - p)}{\left(\frac{\bar{v}^2}{2V} + \frac{\sigma_v^2}{2V} - \bar{v}\right)}$$

The above is entirely general in that nothing has been said about the shape of the v_i or V. It can also be applied in two or one dimension(s) by substituting areas or lengths respectively for the volumes.

APPENDIX 2 — Calculation of average overlap area of two partially overlapping sand–bodies.

Take two equal boxes of size l by t. Keep box 1 fixed and allow box 2 to sweep through all possible locations which give a permitted overlap; this restricts the centre of box 2 to an area,

$$A^* = 4lt - (2l - 2S_x)(2t - 2S_z).$$

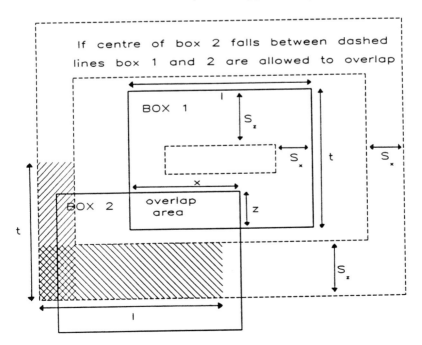

By symmetry, the average overlap area $< A >$ can be calculated by considering one–quarter of the problem and integrating the contributions from sweeping through the hatched areas in the figure.

$$< A > = \frac{4}{A^*} \int_0^{S_x} \int_0^t xz\, dz\, dx \quad + \quad \frac{4}{A^*} \int_{S_x}^l \int_0^{S_x} xz\, dz\, dx$$

$$= \frac{S_x^2 t^2 + S_z^2 l^2 - S_x^2 S_z^2}{4lt - (2l - 2S_x)(2t - 2S_z)}$$

$$= \frac{l^2 t^2 - (l^2 - S_x^2)(t^2 - S_z^2)}{4[lt - (l - S_x)(t - S_z)]}$$

The equivalent 3–D average volume of overlap can be shown to be

$$< V > \quad = \quad \frac{l^2 w^2 t^2 - (l^2 - S_x^2)(w^2 - S_y^2)(t^2 - S_z^2)}{8[lwt - (l - S_x)(w - S_y)(t - S_z)]}$$

Session 6
Posters

A smattering of information from all scales

DETERMINING OPTIMUM ESTIMATION METHODS FOR INTERPOLATION AND EXTRAPOLATION OF RESERVOIR PROPERTIES: A CASE STUDY

Allen C. Brummert
Susan E. Pool
Mark E. Portman
John S. Hancock
James R. Ammer

U.S. Department of Energy
Morgantown Energy Technology Center
Morgantown, West Virginia

EXTENDED ABSTRACT

Reservoir modeling requires estimates of reservoir properties that are interpolated and extrapolated from known
points to uncharacterized areas. Results from sophisticated
numerical reservoir models are meaningless if the results are
based on inaccurate estimates of reservoir properties. The
accuracy of four estimation methods to predict key oil and
gas reservoir properties was investigated. Similar studies
have been conducted for ore reserve estimation (Rendu, 1979;
and Sandefur and Grant, 1980), but little work has been done
to determine the best methods for estimating oil and gas
reservoir properties.

Four estimation methods were investigated.

1. Simple averaging is a mathematical method in which the
   arithmetic mean is calculated from the summation of all
   known values. The mean value is used for all unknown
   points. Averaging does not assume that a spatial relationship exists among data points.

2. Fifth-degree bicubic spline is a mathematical method
   of trend surface analysis in which a fifth-degree polynomial is used to determine the values at all unknown
   points. Fifth-degree bicubic spline assumes that a
   spatial relationship exists among the data points.

3. Inverse weighted distance squared is a mathematical
   method in which weights are assigned, based on the
   inverse of the square of the distance, from the known
   points to the unknown point. The known values are
   multiplied by their assigned weight, and these products
   are summed to determine the value of the unknown point.
   Like fifth-degree bicubic spline, inverse weighted distance squared assumes that a spatial relationship exists
   among data points.

4. Kriging is a geostatistical method used to find a set of
   weights that minimizes the estimation variance according
   to the geometry of the field or deposit. Weights are
   assigned according to their proximity to the value being
   estimated: near samples are assigned higher weights than
   distant samples. Kriging provides unbiased estimates and
   minimum variance estimates from which the associated
   error can be determined. Typically, before the data are
   kriged, histograms are generated to determine if bias
   exists in the data set. Variograms are then generated
   from the required normal or lognormal histogram distributions to determine the spatial relationship of the data
   values.

The four estimation methods were compared using actual
core data from a 42-well oil reservoir. Based on core and
geophysical well log data, the reservoir was divided into six
layers: five producing layers and one nonproducing layer.
Horizontal permeability, vertical-horizontal permeability
ratio, thickness, and porosity were examined for each of the

five producing layers. Horizontal permeability and thickness
were examined for the one nonproducing layer. In addition,
top-of-structure was examined for the first producing layer.

The estimation methods were compared by removing the
value of the reservoir property at a well, estimating the
value of the reservoir property at that well, and calculating
the mean percentage error between the estimated and actual
values. The mean percentage error was calculated for the
particular reservoir property at all wells, and then the
results were summed. This procedure was carried out for all
wells using the four estimation methods. The total mean percentage errors for each method were compared (Figure 1) and
the method with the lowest total error was assumed to be the
most accurate method. The above procedure is often referred
to as the "jackknife" or "boomerang" technique (Sandefur and
Grant, 1980). Contour maps comparing kriging (Figure 2a) and
fifth-degree bicubic spline (Figure 2b) show the diversity of
horizontal permeability values that are assigned to unknown
areas of the third producing layer using these methods.

This study revealed two, potentially case-specific
conclusions:

1. The error in the prediction of some reservoir properties
    was always less in all layers using one particular esti-
    mation method, i.e., kriging was the best estimation
    method for all porosity data sets, fifth-degree bicubic
    spline was the best for all thickness data sets, and
    averaging was the best for all vertical-horizontal
    permeability ratio data sets (Figure 1). Other point
    estimation, reservoir case-studies should be performed
    for verification.

    The estimation method with the least error varied for
    reservoir properties within each layer, i.e., for the
    first producing layer, inverse weighted distance squared
    was the best method for the horizontal permeability data
    set, averaging was the best for the vertical-horizontal
    permeability ratio data set, fifth-degree bicubic spline
    was the best for the thickness data set, and kriging was
    the best for the porosity data set (Figure 1).

2. Kriging was usually the best method for those variables
    that showed the required normal or lognormal histogram
    distribution and that fitted a stable variogram model.

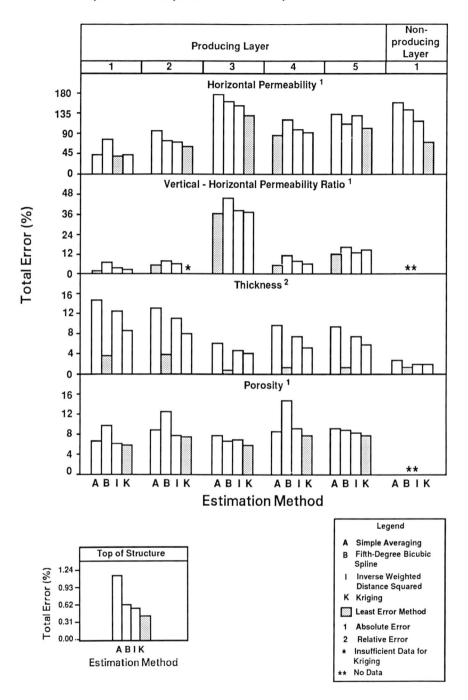

FIGURE 1. Comparison of error of estimation methods

FIGURE 2. (a). Contour map for Producing Layer 3 horizontal permeability using kriging. (b). Contour map for Producing Layer 3 horizontal permeability using fifth-degree bicubic spline. (White dots represent well locations.)

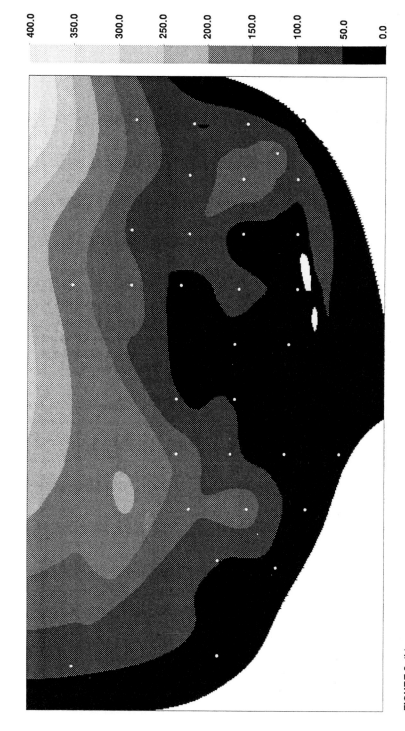

400.0
350.0
300.0
250.0
200.0
150.0
100.0
50.0
0.0

FIGURE 2. (b).

In the thickness data sets and the vertical-horizontal permeability data sets, these requirements were not completely fulfilled. However, to test the applicability of the requirements of kriging to oil and gas reservoirs, kriging was performed.

The thickness data sets did not show the required normal or lognormal histogram distribution. Zero thickness values distorted the histogram distribution, indicating a possible bimodal distribution. A normal histogram distribution was assumed in order to proceed with the kriging process. As suspected, kriging did not prove to be the best method for the thickness data sets. However, the initial steps of kriging (developing histograms and variograms) gave insight into which method may be best, i.e., fifth-degree bicubic spline may be best for bimodal histogram distributions, as was true for the thickness data sets (Figure 1).

The vertical-horizontal permeability ratio data sets did not fit a stable variogram model. The data sets contained only 13 to 14 data points from which only a limited, linear variogram model could be generated. A variogram model could not be generated for the second producing layer. Averaging was found to be the best method for the vertical-horizontal permeability ratio data sets (Figure 1). A previous study (Portman, et al., 1987) also showed averaging to be the best method for a small data set of 14 porosity values.

After the optimum estimation method was determined for each reservoir property of each layer, values were assigned to the reservoir grid for an enhanced oil recovery modeling study.

REFERENCES

Portman, M. E., H. R. Pratt, and J. S. Hancock (1987). A Statistical Comparison of Four Point-Estimation Methods. DOE-METC/EG&G Contract No. DE-AC21-85MC21353, WBS No. 9GAJ-8X. September 23, 1987.

Rendu J-M. M. (1979). Kriging, Logarithmic Kriging, and Conditional Expectation: Comparison of Theory with Actual
Results. Appl. of Comput. and Oper. Res. in the Miner.
Ind., 16th, Tucson, Arizona, October 17-19, 1979. Publ.
by Soc. of Min. Eng., AIME, New York, New York, 199-212.

Sandefur, R. I., and D. C. Grant (1980). Applying Geostatis-
ics to Roll Front Uranium in Wyoming. Engineering and
Mining Journal 181(2):90-96.

EFFECTIVE RELATIVE PERMEABILITY FOR A 1-DIMENSIONAL HETEROGENEOUS RESERVOIR

Magnar Dale

Rogaland Research Institute/Rogaland Regional College
Stavanger, Norway

I. INTRODUCTION

We study incompressible Buckley-Leverett type displacement in a heterogeneous 1D reservoir model, where relative permeability curves, absolute permeability and porosity vary on a small scale (cross bedded sequences, finely laminated material). Our main result states that effective displacement behaviour in this model can be described by a homogeneous model with a single pair of relative permeability curves. These effective relative permeability curves are dependent on the rock characteristics of the heterogeneous medium only, and explicit formulas for calculating them are presented.

II. DESCRIPTION OF THE HETEROGENEOUS MODEL

We consider a heterogeneous 1D reservoir model composed of a large number of randomly distributed homogeneous parts of relative length 1. Each part is picked from a fixed set of N homogeneous rock types, each type described by its

φ_i - porosity

k_i - absolute permeability

k_{rwi}, k_{roi} - relative permeability curves

p_i - probability of occurence

for $i = 1, .., N$. The irreducible saturation values are allowed to vary between the curves. Fig.1 shows a typical realization of such a reservoir, with $N = 2$, $1 = 0.05$, $p_1 = p_2 = 0.5$.

[1]Supported by a British-Norwegian research program financed by Den Norske Stats Oljeselskap a.s. (STATOIL).

652

We consider the common relative length 1 of the homogeneous parts as a correlation length for the parameter variation. Or, to describe _periodic_ distributions of the homogeneous parts, we let 1 denote the relative period length, and let p_i denote the relative length of each part with respect to the period length.

III. STATEMENT OF MAIN RESULT

We study standard Buckley-Leverett type displacement in the heterogeneous medium described above. Thus, initially the model is saturated with oil and water with water everywhere at its irreducible saturation. At time t = 0 one starts injection of water into the medium, the injection Darcy velocity q_T and outlet pressure p_L remaining constant in time. The response functions of interest are the outlet fractional flow of water $f_w(t)$ and the pressure distribution along the model p(x,t). In particular, we consider the response functions as the correlation length 1 goes to zero. More precisely, fixing model length L, injection velocity q_T, outlet pressure p_L and fluid viscosity ratio M, we denote by $f_{wn}(t)$ and $p_n(x,t)$ the pointwise arithmetic average of $f_w(t)$ and p(x,t) over all realizations of the medium with l = l/n, n = 1,2,3,..... In the periodic case n denotes the number of periods contained in the heterogeneous sample.

On the other hand, consider displacement in the homogeneous model of length L, whose constant parameters $\bar{k}_{rw}$, $\bar{k}_{ro}$, $\bar{k}$ and $\bar{\varphi}$ are determined from the given heterogeneous medium by

$$\bar{\varphi} = \sum_{i=1}^{N} p_i \varphi_i \dots\dots\dots\dots\dots\dots(1)$$

$$\frac{1}{\bar{k}} = \sum_{i=1}^{N} \frac{p_i}{k_i} \dots\dots\dots\dots\dots\dots(2)$$

$$\frac{1}{\bar{k}\,\bar{k}_{rw}(\bar{S})} = \sum_{i=1}^{N} \frac{p_i}{k_i k_{rwi}(S_i)} \dots\dots\dots\dots(3)$$

$$\frac{1}{\bar{k}\,\bar{k}_{ro}(\bar{S})} = \sum_{i=1}^{N} \frac{p_i}{k_i k_{roi}(S_i)} \dots\dots\dots\dots(4)$$

Here the saturation S_i for the i-th homogeneous part is to vary in the interval $[S_{wci}, 1-S_{ori}]$. The allowable N-tuples of saturation values $(S_1, \ldots, S_N)$ are uniquely determined by

$$\frac{k_{ro1}(S_1)}{k_{rw1}(S_1)} = \ldots \ldots = \frac{k_{roN}(S_N)}{k_{rwN}(S_N)} \ldots \ldots \ldots \ldots (5)$$

and $\bar{S}$ is the arithmetic average

$$\bar{S} = \sum_{i=1}^{N} p_i \varphi_i S_i \ldots \ldots \ldots \ldots \ldots (6)$$

We denote by $\bar{f}_w(t)$ and $\bar{p}(x,t)$ the response functions from this homogeneous medium.

Now our main result is the following: The homogeneous medium described above is the homogeneization of the given heterogeneous medium in the sense that for any fixed values of L, q_T, p_L and M, we have

$$\lim_{n \to \infty} f_{wn}(t) = \bar{f}_w(t) \ldots \ldots \ldots \ldots \ldots (7)$$

$$\lim_{n \to \infty} p_n(x,t) = \bar{p}(x,t) \ldots \ldots \ldots \ldots \ldots (8)$$

Thus the constant parameters (1),(2),(3),(4) of this homogeneous medium may be called effective parameters for the given heterogeneous medium.

The interpretation of the the effective relative curves is easy: They are identical to the relative permeability curves obtained by steady state measurement on a core composed of the N different rock types, each type represented once with relative length pi. - The proof of (8) and (9) consists in studying the convergence, as n ---> ∞ in the heterogeneous media, of the saturation- and pressure distribution functions for displacement in the media.

IV. SIMULATION EXPERIMENTS; EXAMPLE

To estimate the error introduced by describing displacement in the heterogeneous medium by effective parameters, simulation experiments have been performed.

A typical result for a periodic distribution of N = 2 rock types and water/oil viscosity ratio equal to 0.4, is a relative error of about 1 % both in fractional flow and differential pressure for a sample containing about 20 periods.

Fig.2 shows the rock relative permeability curves and the effective curves for a medium containing N = 3 distinct rock types, whose porosities and absolute permeabilities are .2, .3 and .35, and 20, 300 and 1000 mD, respectively. We also assume that rock type 2 occur with twice the probability of type 1 and type 3.

V. REFERENCES

(1)   Ekrann,S., and Dale,M., Averaging of relative permeability in heterogeneous reservoirs. To be presented at the European Conference on the Mathematics of Oil Recovery, Robinson College, Cambridge University, 1989.

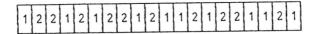

Fig 1. Random distribution of N = 2 rock types

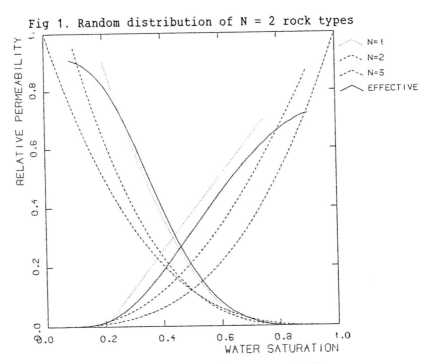

Fig.2. Effective relative permeability

POSITRON EMISSION TOMOGRAPHY (PET) IN DYNAMIC COREFLOODING STUDIES

Oyvind Dugstad
Tor Bjornstad

Institute for Energy Technology
Kjeller, Norway

Jan R. Lien

University of Bergen
Bergen, Norway

I. INTRODUCTION

Measurements of phase-distribution inside a core plug during a flooding experiment is of great importance to increase the knowledge of multiphase flow in porous media. Measuring techniques such as computer assisted X-ray tomography (CAT) and various gamma tracer or transmitting techniqus are in use. Problems are lack of spatial resolution or difficulties in applying the method when experiments are carried out under reservoir conditions. The PET can be a valuable tool which in some respect has advantages compared to other techniques. It can be used to study the movement of specific compounds which are labelled with positron emitting nuclide. The PET will be less restrictive to core-holder material than the CAT. The detectors used in the PET can be given a geometry which makes it possible to obtain 3-D infomation about the tracer distribution. The objective of this study was to examine the usefulness of the state-of-the-art PET technique for core flooding studies.

II. METHOD

PET is based on the use of positron-emitting nuclides as tracers for the fluids to be studied. The emitted positron

RESERVOIR CHARACTERIZATION II

will be slowed down in the surrounding medium. At the end of
its short range it will annihilate with an electron and
result in the emission of two photons of energy 511 keV in
diametrically opposite directions. The principle of the PET
is to detect both photons simultaneously. A coincidence dis-
criminator is needed to record only those photons which constitute a coincident pair.

The PET equipment used in this work was developed at the
University of Birmingham /1/. The instrument consists of two
position specific wire-grid detectors each of size 30x60 cm
(Fig. 1). With this detector configuration it is possible to
obtain 3-D information about the tracer distribution between
the two detectors. Each coincident event is represented by
their x-y coordinates in the two detectors. A connecting line
is computed in the space between them. The emission must have
originated somewhere on this line. Several recorded events
will eventually constitute a 3-D array of intensity distributions represented by voxels of a finite size.

Success of the technique as a tool for core flood
studies depends upon the possibilities to apply positron-
emitting tracers which are specific for the phase under
study. The half-life of the radionuclide should be sufficiently long to avoid too large activity decrease during the
experiment. Several tracer candidates are possible. In the
present experiments we have used $^{22}Na^+$ as a tracer for the
brine and ^{68}Ga-HDEHP (di-ethyl-hexyl-ortho-phosphoric acid)
as an oil tracer.

III. RESULTS AND DISCUSSION

The relatively high-energetic electromagnetic radiation
makes it possible to study flow behaviour under reservoir
conditions in ordinary steel core holder. The achievable spatial resolution has been checked with different thicknesses
of the steel core holder. A static distribution of the brine
phase as measured with ^{22}Na in a cross section of the core
plug is given in Fig. 2. The resolution deteriorates with
increasing wall thickness of the core holder, but the change
is not dramatic.

Dynamic experiments have been carried out both by
tracing the brine and the oil phases. Time-dependent phase
distributions have been obtained. However, a good resolution
in the experiment depends on the time needed to accumulate a
sufficiently good picture. We have worked with activity
levels close to the limit of what is feasible with the present PET equipment (dead-time loss in the associated electronics was the limiting factor). A typical accumulation time
for one good-quality picture was 5 min.

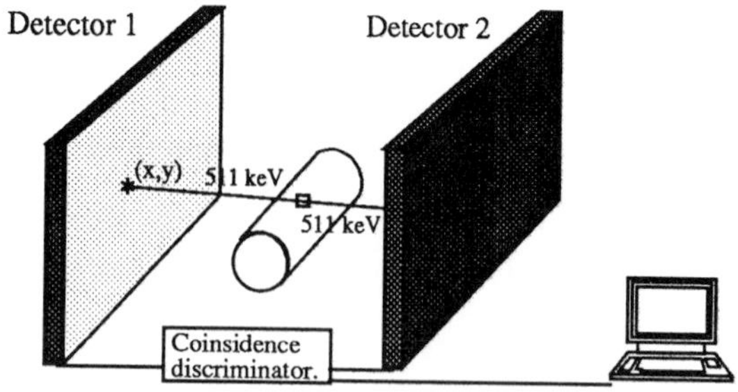

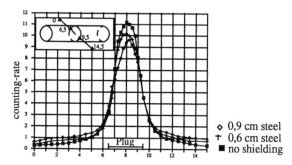

Fig. 1. The principle of the PET.

Fig. 2. The figure shows the concentration distribution along a line through the center of the plug.

V.   CONCLUSION

Positron emission tomography has shown to be a possible technique with large development potential for the study of spatial fluid saturation of various fluids during a dynamic core-flooding experiment even at reservoir conditions. With the present system, there is, however, a need for improvement of the counting electronics in order to be able to apply higher-activity sources (shorter accumulation times). One needs also to improve the data handling routines for a faster and more correct reconstruction of the spatial radiotracer distribution. Extension of the whole system by increasing the number of detectors in order to cover a larger solid angle would probably result both in a shorter accumulation time and facilitate the activity distribution reconstruction.

Reference
M.R. Hawkesworth, M.A.O'Dwyer and J. Walker, Nuc.Inst. and Meth. in Phy.Res. A 253 (1986) 145-157.

A GAS RESERVOIR MODEL FOR THE DEVONIAN SHALE OF THE APPALACHIAN BASIN[1]

Terence Hamilton-Smith
Mary Passaretti
Patrick H. Lowry
R. Michael Peterson[2]

K&A Energy Consultants, Inc.
Tulsa, Oklahoma

A reservoir model has been developed for the gas accumulation in the Devonian shale of the central Appalachian Basin. As part of the Gas Research Institute's Eastern Devonian Gas Shale Research Program, 148 feet of whole core was taken from the Lower Huron Shale of the Comprehensive Study Well No. 2 (CSW2) in central West Virginia. The very thinly bedded shale and siltstone sequence was described in detail, on the scale of the depositional events, with about 4000 distinct beds identified. Four lithotypes were identified, including two siltstones and two shales. The two siltstone types have a variety of sedimentary structures, including graded bedding and load casts, generally indicative of turbidite sedimentation. Markov analysis (Figure 1) of the vertical sequence of lithotypes showed a strong association of the two siltstone types and the green–gray shale into fining- upwards sequences, suggesting the depositional interpretation of the green–gray shale as the uppermost

---

[1]Supported by the Gas Research Insititute, Chicago, Illinois.

[2]Present address: Terra Vac, Belle Meade, New Jersey.

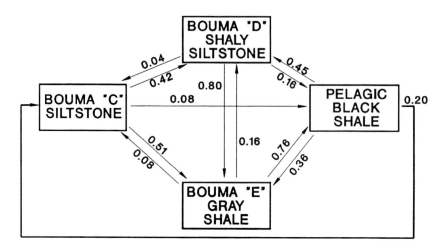

Fig. 1.   Lithotypes and Markov transition probabilities, CSW2 core.

Bouma E part of a distal turbidite. The second shale lithotype is dark gray in color and is not strongly associated with the turbidite siltstones. The result is a simple depositional model (Figure 2) of the Devonian shale section in central West Virginia. Pelagic sedimentation of dark gray to black shale in an anaerobic basin was interrupted by distal turbidite deposition of thin silt-stones and green-gray shale. Comparison of the

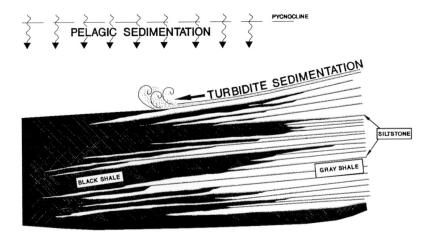

Fig. 2.   Devonian shale depositional model, central West Virginia.

depositional model with available organic geochemical data
suggests a close correlation of kerogen facies and lithotypes.

A fracture analysis was made on the core using both
geological description and dye penetrant photography. Four
types of fractures were found: petal fractures, bedding
plane partings, slickensides, and irregular mineralized
fractures. Petal fractures and bedding plane partings are
attributed to core handling and to the coring operation.

Slickensides are sub-parallel to bedding but may cut
depositional surfaces in detail, and most commonly occur
near lithological boundaries. Forty-two occurrences were
noted, with half of the total occurrences in two specific
zones from eight to ten feet thick. The slickensides are
interpreted to be associated with the imbricate thrust fault
system mapped in the Devonian shale section in the CSW2
area, part of a regional detachment front produced in the
Alleghenian orogeny of Permian age.

Irregular mineralized fractures are vertical to horizontal in orientation, and locally form networks. They are
fully to partially filled with a variety of minerals,
predominantly calcite, but also including dolomite, gypsum,
pyrite, and clay. They occur in all lithologies, but mainly
in siltstones. Eighteen occurrences were noted, with over
half of the occurrences in one specific zone ten feet thick.
There was no apparent association of this fracture type with
the slickensides. They are interpreted to represent an
independent and older fracture system, resulting from
reactivation of Precambrian basement faults prior to
Alleghenian thrust faulting.

An anomaly in the CSW2 temperature log indicates gas
entry into the well from a zone 24 feet thick in the lower
part of the cored interval. The simultaneous intersection
of the two fracture systems with relatively thickly bedded
siltstones appears to be required for effective reservoir
development. Detailed comparison of the cored sequence with
the Formation Microscanner, Gamma-ray, and Compensated
Formation Density logs will indicate to what degree these
logs can usefully identify components of the reservoir
model.

HETEROGENEITY AND EFFECTIVE PERMEABILITY OF POROUS ROCKS: EXPERIMENTAL AND NUMERICAL INVESTIGATION

C. G. Jacquin
A. Henriette
D. Guerillot

Institut Francais Du Petrole
Rueil-Malmaison, France

P. M. Alder

CNRS
Meudon, France

1. MATERIALS AND EXPERIMENTAL PROCEDURE

Two large blocks (15x15x50 cm, one limestone and one sandstone) were examined. The experimental procedure is diagramed in Fig. 1. At each step, a (initial block), b (3 intermediate blocks) and c (300 plugs), the permeabilities along the Z axis for all the elements were measured.

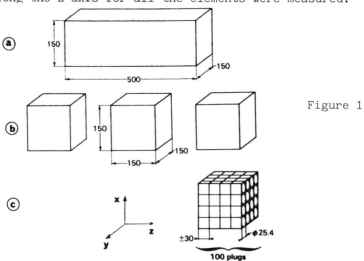

Figure 1

2. EXPERIMENTAL RESULTS

The probability density functions of log K are shown in Fig. 2. It can be seen that these histograms cannot be considered as gaussians. Permeabilities are normally assumed to be log normally distributed. Obviously, this is not true in the case studied here.

The autocorrelation coefficients of the local K were calculated. They indicate short-range correlations (for 2 to 3 elementary plugs).

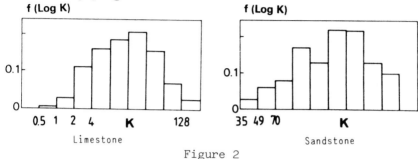

Figure 2

3. NUMERICAL ANALYSIS AND DISCUSSION

3.1. Analysis

Effective overall permeabilities were numerically derived from the elementary ones. An explicit time scheme was used. The set of difference equations describing the system has been replaced by a time-dependent set that converges to the steady solution.

3.2. Results of the numerical calculation

The numerical K_{zcal} and experimental K_{zmes} permeabilities are given in Table 1. It can be seen that the prediction is better than 20% for the intermediate blocks, and better than 10% for the initial blocks. Hence, it can be stated that the two sets of results are consistent and that they confirm each other.

3.3. Results obtained by simple averaging procedures

Table 1 gives the results obtained by:
- arithmetic averaging K_{ar}
- harmonic averaging K_{har}
- two procedures combining the preceding: K_{mix1}, K_{mix2}
- geometric averaging of K_{mix1} and $K_{mix2} = K_e$

The averaged values K_{mix1} and K_{mix2} are respectively the upper bounds and lower bounds of the numerical K_{zcal} values. The geometric average of these bounds gives very good estimate (K_e) of these numerical K_{zcal} values (Guérillot et al., 1989).

Table 1

	Block	$\bar{K}_{Z,mes}$	$\bar{K}_{z,cal}$	$\bar{K}_{ar}$	$\bar{K}_{har}$	$\bar{K}_{mix,1}$	$\bar{K}_{mix,2}$	$\bar{K}_e$
	1st intermediate	270	309	311	273	311	296	303
Sandstone	2nd intermediate	220	243	250	193	244	233	238
	3rd intermediate	150	119	129	99	120	115	117
	global	210	190	230	159	192	179	185
	1st intermediate	48	54	66	20	64	41	51
Limestone	2nd intermediate	36	35	46	12	39	29	34
	3rd intermediate	30	34	52	11	45	22	31
	global	35	38	55	13	47	22	32

3.4. Incomplete date

We examined the influence of incomplete sets of elementary data on the effective calculated permeability of the block. According to two different rules, 24 or 75 of the 300 elementary values are conserved. It can be seen (Table 2) that the prediction remains relatively good even when 92% of the initial data are omitted (Henriette et al., 1988).

Table 2

	300 data		Rule 1 78 data		Rule 2 24 data	
	$\bar{K}_{mes}$	$\bar{K}_{cal}$	$\bar{K}'_{ar}$	$\bar{K}_{cal}$	$\bar{K}'_{ar}$	$\bar{K}_{cal}$
Sandstone	210	227	234	189	230	192
Limestone	35	30	36	28	58	41

REFERENCES

1. Guérillot, D., Rudkiewicz, J.L., Ravenne, Ch., Renard, G. and Galli, A., (1989): Proceedings of the Fifth European Symposium on Improved Oil Recovery Budapest, Hungary, April 25-29.

2. Henriette, A., Jacquin, C.G. and Adler, P.M., "The Effective Permeability of Heterogeneous Porous Media". Physico-Chemical Hydrodynamics, Vol. II, No 1, pp. 63-80, 1989.

CHARACTERIZATION OF DELTA FRONT SANDSTONES FROM A FLUVIAL-DOMINATED DELTA SYSTEM

Philip Lowry
Arne Raheim

Institute for Energy Technology
Kjeller, Norway

I. INTRODUCTION

Once the decision has been made to develop a newly discovered oil or gas field, the production geology and reservoir engineering departments of the operating company are charged with the task of establishing the most efficient plan to exploit the reserves. The ability to accurately define the reservoir architecture is the key to a successful development project (Finley and Tyler, 1986). Definition of the reservoir architecture involves two primary tasks (Fig.1); the first task is to establish the geometry of the sandstone bodies which make up the reservoirs; and the second is to define reservoir heterogeneities or potential permeability barriers within the reservoir. The geologist and engineer must utilize whatever tools are available to help them build a realistic reservoir model as early in the development phase as possible. While 3-D seismic surveys are now becoming routine in certain parts of the world prior to development of a field, there are areas where these may be of limited value and so other methods are required to help in this task of establishing an accurate description of the reservoir.

One such tool that has been increasingly used, especially in the North Sea, is the stochastic reservoir simulator. The stochastic reservoir simulator programs generate a probabilistic realization of the reservoir based on transition matrices of sedimentary sequences observed in wells and input data on sandstone body and shale dimensions. While the accuracy, efficiency, and complexity of reservoir simulators has increased in recent years,

[1]Funded by the SPOR program, Norwegian Petroleum Directorate, Stavanger, Norway

665

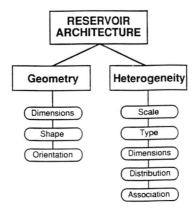

Fig. 1. Definition of reservoir architecture by establishing the geometry of
the reservoir and the character of the heterogeneities.

fundamental geologic input data to these numerical models remains relatively
limited.

In light of this limited availability of data, a project began in 1988 at the
Institute for Energy Technology to collect quantitative geologic information
on the geometry of specific sandstone bodies and the distribution of
heterogeneities within potential reservoir facies. The project has focussed on
sandstone bodies that formed in the delta-front sub-environment of fluvial-
dominated deltas. The reasons for this were two-fold; first, deltaic sandstone
bodies constitute an economically significant reservoir type and as many of
these mature reservoirs become depleted and enhanced recovery profiles are
contemplated there will be an increasing need to incorporate quantitative
geologic data into the reservoir description. The second reason for choosing
the fluvial-dominated system is based on the premise that it is potentially
more heterogeneous than wave-dominated delta systems.

A parallel outcrop study is also underway (Mjøs and Walderhaug, 1988)
at the Rogaland Research Institute in Stavanger, Norway. The focus of that
study concerns the geometry of delta-plain sandstone bodies. The goal of
both studies is to develop a small database which will provide input statistics
for a stochastic modelling program (GEOPROBE), that was developed
jointly by Scientific Software Incorporated and the Institute for Energy
Technology as part of an earlier SPOR project (Myler and Haynes, 1988).

II. DATA COLLECTION

In the initial phase of data collection the emphasis has been to focus on
the dimensions of three key reservoir sandstone bodies in the delta front sub-
environment. These are the distributary channel; the distributary mouth bar;
and the distal bar sandstones. The dimensions of the sandstone bodies of
interest are; maximum length, maximum width and maximum thickness.

Data were obtained from literature surveys, where available, and from an ongoing outcrop study. Two localities considered to be classic exposures of fluvial-dominated deltaic sequences were chosen as initial field sites. The two areas which were chosen were: a) in Central Utah: The Ferron sandstone (Turonian) member of the Mancos Shale, and b) Carboniferous deltaics of the Pochahontas Basin in eastern Kentucky (Lowry and Råheim, 1988). Although some data from Kentucky have been included here, the bulk of the field data are from the Ferron sandstone. These two areas were chosen on the basis of previous work on the regional setting and stratigraphic framework; but, most importantly because they provide analogs for important North Sea deltaic reservoirs

The Ferron sandstone, which crops out along a 70 km belt (10 km wide) (Fig. 2) has been extensively studied over the years (Hale, 1972; Cotter, 1975; Ryer, 1981; Thompson et al, 1986.) and is believed to represent a fluvial-dominated lobate delta system.

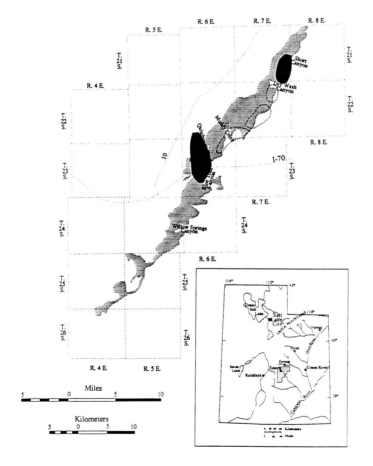

Fig. 2 Location map of the study area.

Ryer (1982) has presented an excellent stratigraphic framework of the Ferron and has recognized at least five major regressive events, which he has numbered 1 through 5. Preliminary investigations within the number 2 interval along the outcrop belt suggests that up to 5 or 6 individual sandstone bodies occur (Fig. 2). Two of the most accessible sandstone bodies within the interval have been mapped and interpreted as distributary mouth bars that are up to 7-8 km in length (along the longest axis) and 3-4 km wide.

Measurements were made on the dimensions of channel sandstones where they were observed during the field work. In the Ferron sandstone, most of the channels observed were relatively homogeneous, sandstone-filled distributary channels. In the Kentucky outcrops, channels ranged from small scale (10 m wide), sandstone filled lenticular forms; to large scale (1 km wide) laterally accreting, heterogeneous channel-fill complexes.

III. SANDSTONE BODY DIMENSIONS

The stochastic modelling program GEOPROBE requires sandstone body dimension data as primary input parameters. Presented below are examples of the type of input data that can be used for the three key sandstone bodies in the delta front sub-environment.

A. Distributary Channels

Figure 3 is a display of the channel dimension data separated into the two major channel types, distributary channels and crevasse splay channels.

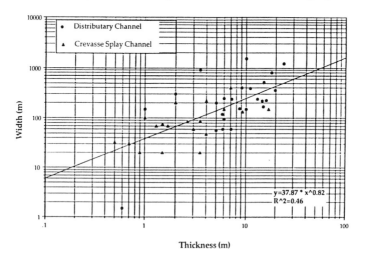

Fig. 3. Width versus thickness graph for distributary and crevasse splay channels

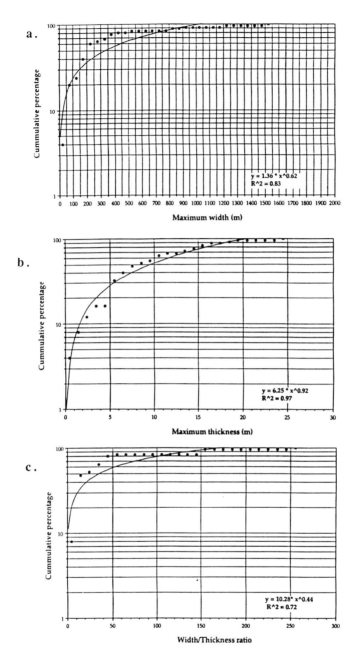

Fig. 4. Exceedance graph for distributary channel sandstone bodies
a). Channel width.   b). Channel thickness.   c). Channel width-
thickness ratio.

The best fit line represents a logarithmic fit to the data. No attempt has been made to further subdivide these data into smaller categories such as the degree of sinuosity, the nature of the fill or the type of delta system. In many cases this type of information was not available and since the data-set is still relatively small, further subdivision is not justified. This plot shows that given a thickness of 10 m, one can expect an individual channel width to be on the order of 250 m. Note that the thickness measurement represents the maximum thickness of the channel sandstone.

Figure 4 is a graph of the exceedance probability for simple channel geometric attributes. Figure 4a depicts exceedance values for the maximum distributary channel width. The graph shows 90% of the channels are less than 850 m wide; 50% are less than 325 m wide; and 20% are less than 75 m wide. Figure 4b is a display of the exceedance data for distributary channel thickness. Ninety percent of the distributary channels shown in this graph are less than 18 m thick; 50% of the channels are less than 9 m thick; and 20% are less than 3.5 m thick. Figure 4c shows a width-thickness exceedance plot for distributary channels. Ninety percent of the channels have a width to thickness ratio of less than 130:1 and 50% have width/thickness ratios less than 40:1, 20% of the channels have ratios less than 10:1.

B. Distributary Mouth Bars

Table 1 is a summary of the descriptive statistics for the distributary mouth bar sandstone body data-set.

	Mean	Std. Dev.	Minimum	Maximum	Range	Count
Width (km)	3.477	1.921	1.5	8	6.5	9
Thickness (m)	18.589	12.86	1.2	42	40.8	10
Length (km)	6.318	2.409	2.4	9.6	7.2	8
Length/Width ratio	2.106	.505	1.5	3	1.5	7
Width/Thickness ratio	367.686	378.789	83.33	1333.33	1250	9
Length/Thickness ratio	622.906	592.55	205.71	2000	1794.29	8

Table 1. Descriptive statistics for the distributary mouth bar sandstone bodies.

Figure 5 depicts the relationships between distributary mouth bar width, thickness and length. The data are displayed in a log-log format and a best fit line has been calculated using a logarithmic curve fitting routine. In figure 5a with a correlation co-efficient of 0.74 the graph reveals that given a maximum thickness of the distributary mouth bar sandstone of 30 m, the corresponding length should be approximately 8 km. Figure 5b is a thickness versus width plot with a significantly lower correlation co-efficient (0.36). This graph predicts that a 30 m thick distributary mouth bar sandstone should be 4 km in width. Figure 5c is a graph of width versus length of distributary

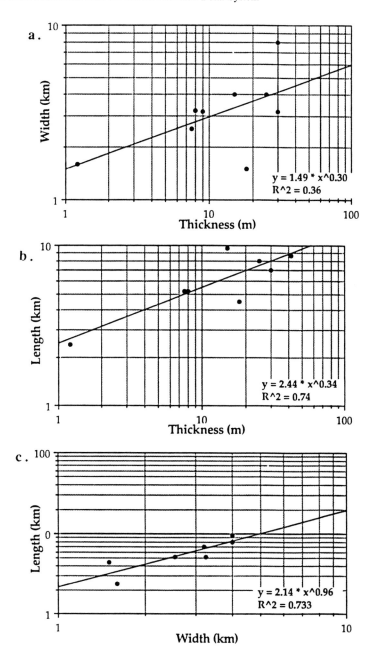

Fig. 5. Relationships between width, thickness and length of distributary
    mouth bar sandstones.
      a). Thickness versus width. b). Thickness versus length.   c) Width
        versus length.

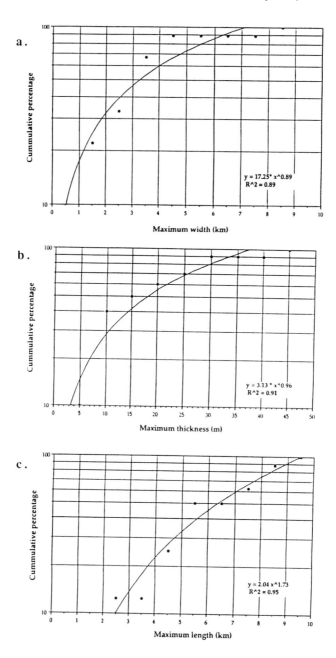

Fig. 6. Exceedance graph for distributary mouth bar sandstone bodies
a).Width.   b). Thickness.   c). Length.

mouth bar sandstones. In this example the correlation co-efficient is higher than the previous graph (0.73) and shows that a 4 km wide distributary mouth bar sandstone will be approximately 8 km long.

Figure 6 depicts the variation in dimensions of the distributary mouth bars. In figure 6a, the graph shows that 90% of the distributary mouth bar sandstones (within this data set) will be less than 30 m thick; 50% of the distributary mouth bar sandstones will be less than 18 m thick; and 20% will be less than 7 m thick.. Figure 6b shows exceedance values for distributary mouth bar length. In this graph, for example, 90% of the distributary mouth bar sandstones are less than 9 km in length; 50% are less than 6.5 km long; and 20% are less than 3.5 km long. Figure 6c is an exceedance plot of the width of the distributary mouth bar sandstones and shows that 90% are less than 6.5 km wide; 50% are less than 3.25 km wide; and 20% are less than 1.2 km wide.

C. Distal Bar sandstones

In detailed lithofacies mapping of individual distributary mouth bar sandstones it became apparent that the underlying distal bar sandstones showed significant spatial differences in interval thickness, individual thickness and degree of amalgamation. These observations may provide important information to enable one to determine relative position of a well within the sandstone body. Here, however we have not attempted to distinguish different regions of the sandstone body and present a summary of all the distal bar sandstone thickness data within the number two regressive interval. Figure 7 shows the exceedance curve for the distal bar sandstone thickness. In the example shown, 90% of the distal bar sandstones are less than 0.55 m thick and 70% are less than 0.10 m thick.

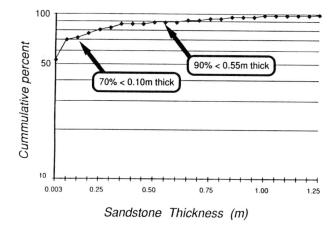

Fig. 7. Graph of the exceedance percentage of distal bar sandstones in the number 2 delta interval.

IV. DISCUSSION

What we have presented here are the results from a very small data-set. These results are obviously a first approximation only. The graphs clearly show a wide degree of scatter and therefore considerable uncertainty. However, as this type of data-set is added to, one can begin to differentiate the data into progressively more detailed categories such as, for example, tectonic setting, deltaic regime, channel sinuosity and channel fill type. It is only with a significantly larger database that one can begin to discriminate between local depositional conditions and genetically inherent attributes of a sandstone body.

While the geologic literature can provide some of the necessary quantitative data, it is clear that continued collection of field data at the reservoir scale in areas of high quality three dimensional outcrops will be required. It is the next generation of facies models which will define the three dimensional variability of lithofacies within individual sandstone bodies that will enable the quantification of heterogeneities within the reservoir. With an understanding of the distribution of, and type of heterogeneities, more reliable fluid-flow models of the reservoir will be developed. In addition to improving our understanding of the hydrodynamics of a sandstone body, the detailed lithofacies mapping will also aid in the interpretation of relative positioning within a specific sandstone body.

Clearly the use of mini-permeameters and core plug data from outcrops will also allow one to assess any relative variations in porosity and permeability within the sandstone body. When coupled to the three dimensional internal facies model it will further improve our ability to describe the flow units within a reservoir sandstone. However, it is essential that these data be calibrated to porosity and permeability data from equivalent depositional environments in the subsurface.

Another important source of three dimensional data can be obtained from shallow drilling or vibro-coring projects in modern environments. However, care should be taken in the interpretation of modern sand body geometry when predicting preserved sandstone body geometry, since clearly the two can differ.

The task of establishing a quantitative geologic database for improved reservoir description is daunting. This task should require collaboration between companies and other interested groups to avoid duplication of effort and to promote the rapid and efficient development of a statistically significant database.

V. CONCLUSIONS

1. There is need to provide quantitative geologic to accurately define the architecture of sandstone reservoirs. This will involve two phases, the first phase concerns the collection of data on the gross geometry of sandstone bodies, and the second phase will concern the collection of data describing the distribution of heterogeneities within sandstone bodies.

2. Based on the data collected so far for delta front sandstones the following results can be reported:

a). Ninety percent of the distributary channels are less than 850 m wide
Fifty percent of the distributary channels are less than 325 m wide
Twenty percent of the distributary channels are less than 75 m wide

b). Ninety percent of the distributary channels are less than 18 m thick
Fifty percent of the distributary channels are less than 9 m thick
Twenty percent of the distributary channels are less than 3.5 m thick

c). Ninety percent of the distributary channels have width/thickness ratios less than 130:1
Fifty percent of the distributary channels have width/thickness ratios less than 40:1
Twenty percent of the distributary channels have width/thickness ratios less than 10:1

d). Ninety percent of the distributary mouth bar sandstones are less than 30 m thick.
Fifty percent of the distributary mouth bar sandstones are less than 18 m thick.
Twenty percent of the distributary mouth bar sandstones are less than 7 m thick.

e). Ninety percent of the distributary mouth bar sandstones are less than 9 km in length (strike).
Fifty percent of the distributary mouth bar sandstones are less than 6.5 km in length (strike).
Twenty percent of the distributary mouth bar sandstones are less than 3.5 km in length (strike).

f). Ninety percent of the distributary mouth bar sandstones are less than 6.5 km in width (dip).
Fifty percent of the distributary mouth bar sandstones are less than 3.25 km in width (dip).
Twenty percent of the distributary mouth bar sandstones are less than 1.2 km in width (dip).

g). Ninety percent of distal bar sandstones are less than 0.55 m thick
Seventy percent of distal bar sandstones are less than 0.10 m thick.

VI. REFERENCES

Cotter, E. 1975, Deltaic deposits in the Upper Cretaceous Ferron Sandstone, Utah, *in*, M. L. S. Broussard, ed., *Deltas, models for exploration*: Houston Geol. Soc., p. 471-484.

Finley, R. J. and Tyler, N., 1986, Geological characterization of sandstone reservoirs, in L.W. Lake and H.B. Carroll, ed, Reservoir Characterization, Academic Press, New York, p. 1-38.

Hale, L. A., 1972, Depositional history of the Ferron formation, central Utah, in , Plateau-Basin and range transition zone, central Utah, Utah Geol.Assoc. Pub. 2, p. 29-40.

Lowry, P. and Råheim, A., 1988, "Establishment of a Geologic database: Part 1, Geometry and lithologic heterogeneities in deltaic sandstone bodies from Central Utah and Eastern Kentucky", *SPOR Summaries of Projects September 1988*, Norwegian Petroleum Directorate, Stavanger.

Mjøs, R., and Walderhaug, O. 1988, Crevasse splay sandsteiner i Ravenscar Gruppen, Yorkshire, Rogalands Research Institute, Stavanger, SPOR Report 1/88

Myler, N. and Haynes, P., 1988, "Reservoir description of the Statfjord sands using PERMOPROBE and GEOPROBE", IFE Report KR/F-88/035, 30p

Ryer, T. A., 1981, Deltaic coals of Ferron sandstone member of the Mancos shale: Predictive model for Cretaceous coal-bearing strata, Am. Assoc. Petroleum Geologists Bull., v. 65, p. 2323-2340.

Ryer, T. A., 1982, Cross-section of the Ferron Sandstone Member of the Mancos Shale in the Emery Coal field, Emery and Sevier Counties., U.S. Geol. Survey Map, MF-1357

Thompson, S. L., Ossian, C. R. and Scott, A. J., 1986, Lithofacies, inferred processes, and log response characteristics of shelf and shoreface sequences, Ferron Sandstone, central Utah: *in* T. F. Moslow and E. G. Rhodes, eds., *Modern and Ancient Shelf Clastics: A Core Workshop*, Soc. Econ. Paleontologists Mineralogists Core Workshop No. 9, p. 325-361.

CONSTRUCTING A FLOW-UNIT MODEL
FOR FLUVIAL SANDSTONES
IN THE PEORIA FIELD, COLORADO

Mark A. Chapin
David F. Mayer

Colorado School of Mines
Golden, Colorado

I. INTRODUCTION

By identifying and mapping the distribution of facies and bounding surfaces of three coalesced levels of meanderbelts within the Lower Cretaceous J Sandstone in the Peoria field, Colorado, and associating characteristic reservoir properties with these facies, fluid-flow units and barriers were distinguished and mapped. Within a meanderbelt, trough cross-stratified and clean ripple-laminated point bar sandstones acted as flow units, and clay-rich abandonment fills acted as barriers to flow. Hydraulic communication between vertically-stacked meanderbelts depended on the muddy sandstones with abundant clay inclusions above the basal scour surface. The model constructed explained the different levels of oil-water contacts originally observed and the production-injection behavior of the field.

RESERVOIR CHARACTERIZATION II
677

II. PROCEDURE FOR CONSTRUCTING THE
   ROCK-PROPERTY MODEL

The interaction of primary depositional units, tectonic features, and diagenesis controls the flow of fluids in a reservoir. When the distribution of sedimentary facies has the greatest affect on fluid flow, detailed sedimentology and facies analysis can be used to define the spatial distribution of porous and permeable rock units. The following procedure has been used to construct a rock-property model for the Peoria field.

A. Identify the Reservoir Lithofacies and Bounding Surfaces

Grain sizes, clay content, and physical and biological sedimentary structures were identified in cores from the Peoria field. These descriptions were used to group small-scale lithologic types into facies based on common associations of these features. Five facies were distinguished from the core descriptions.

Crevasse splay and crevasse channel sandstones were combined together based on similar rock-property characteristics and well log responses. Mudstones from coastal plain marshes, lakes, and brackish bays were grouped as fine-grained vertical accretion deposits. Three facies were used to describe the point bar sandstones which are the main productive units in the Peoria field. The first of these facies consists of cross-stratified, massive, and planar-laminated sandstones and represents the basal channel fill of a point bar. The ripple laminated sandstones with occasional mudstone drapes representing the lower energy deposits of the middle to upper point bar were considered together. The top of the point bar was considered as the third point bar facies and consisted of rippled sandstone, containing wavy or flaser beds, abundant mud drapes, convolute bedding, and burrows.

The two facies representing the lower and middle point bar and the crevasse splay sandstones were found to be the reservoir lithofacies. Potentially important bounding surfaces included basal scour surfaces and lateral accretion surfaces.

B. Determine Rock-Property Patterns within the Lithofacies
   and Across the Bounding Surfaces

Flow units and barriers were distinguished by permeability obtained from core plugs. Permeability distributions were determined for each facies. The middle and lower point bar facies were found to be the major reservoir facies and the facies which contained the top of the point bar deposits was found to act as a barrier to flow. The crevasse splay and verticle accretion facies showed very irregular permeability distributions reflecting the variability within these units. Small-scale permeability variations within the reservoir facies were considered to be negligible because relatively little verticle/horizontal anisotropy was measured.

Mecury injection capillary pressure was used to distinguish characteristic pore types for each facies. Capillary pressure curves measured on cores taken from mud breakup layers and along basal channel scours showed that these features act as barriers to flow. Relative permeability curves appeared to correlate with the reservoir facies.

C. Determine the Spatial Distribution of Facies and Meanderbelts

This step consisted of two parts, first, recognizing the arrangement of lithofacies and bounding surfaces internal to discrete, mappable sandbodies and, second, determining the geometries and interconnectedness among the discrete, mappable sandbodies. Discrete, mappable bodies within the Cretaceous J Sandstone in the Peoria field were several levels of meanderbelt sandstones composed internally of coalesced point bars and abandonment fills. Three levels of meanderbelt sandbodies with varying degrees of interconnectedness were distinguished in the Peoria field.

Constructing a rock-property model from these data required mapping internally consistent rock-property units and their connections. The lower and middle point bar facies were mapped separately in each meanderbelt. The connectivity between vertically-overlapping meanderbelts varied depending on the abundance of mudstone inclusions overlying basal scour surfaces.

III. VERIFICATION OF THE MODEL

The initial production of wells completed in the reservoir facies and injection behavior and response during waterflooding were generally consistent with the geologic and flow-unit models. However, uncertainty still exists in the interpretations since bounding surfaces and lateral accretion surfaces cannot be precisely identified and their influence on fluid flow accurately determined and because of uncertainty in production and injection data. Two elevations of oil-water contacts along the downdip western flank of the field were explained by the concentration of abandonment fill facies and lack of hydraulic communication between coalesced channelbelts in the center of the field.

RESERVOIR CHARACTERS OF
THE UNDERSALT TIGHT GAS-BEARING FORMATION AT WENLIU AREA
DONGPU DEPRESSION,CHINA

MENGHUI LIU
CHENGLIN ZHAO
ZAIXING JIANG

Department of Petroleum Exploration
University of Petroleum,China   Dongying
Shandong,PRC

I. ON SANDSTONES AS RESERVOIRS

The 4th member of Shahejie formation of Eogene,the study member,has a great burial,generally at 2700-4000m.And it has been subjected to intense and complicated diagenetic modifications.

Primary porosity in the sandstone reduced gradually,even to a compressible degree.

Cementation is also very common.Clastic rocks have been tightly consolidated from the repeatitive intergranular infilling and precipitation of many kinds of chemically-genetic minerals.These minerals include mainly quartz,carbonate minerals, clay minerals,zeolites, feldspar,pyrite and salt,etc.

But it should be noticed that there were still some factors and processes which helped form the reservoir space.

The early stage of cementation has detained compaction.The content of carbonate cements in sandstone and siltstone is usually 10-20%.Grains,   suported by calcite and dolomite,even at the depth up to 4000m do not usually touch each other.

Secondary porosity is the main reservoir space.Secondary porosity account for about 70-80% of the total volume of the reservoir porosity.

Secondary porosity was produced by the dissolution of carbonate cements, calcite and ferric calcite. Gypsum and halite were also liable to have been dissolved, but they are not so

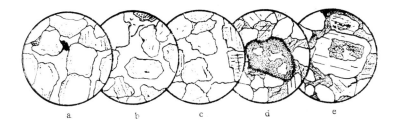

Fig. 1. Secondary porosity in sandstones: a. elongate pore, W61-57; b.and c. enlarged pore,W22-29; d. intragranular pore; e. intracrystal pore filled with calcite.

abundant as the former. Other dissoluble components,eg.carbonate sand, ooid and carbonate mud, tended to help porosity. In the terrigenous components, feldspar dissolution is most common (Fig.1).

Mud matrix has played a role. The existence of abundant mud has stopped chemical precipitation and preserved the porosity. Statistics has shown that if the total content of mud exceeds 10%, the rock is highly porous and permeable; while when the content of mud is lower than 10%, both porosity and permeability get poor.

II.   ON MUDSTONE AS RESERVOIR

Observations showed that there were a lot of cracks and caves in mudstone. They were formed by secondary dissolution and interlayer contraction.The dissolved compositions include clay minerals,halite,gypsum and carbonate minerals. Contractive cracks are common, they sometimes occur by groups parallel to strata and connect to other pores and cracks (Fig.2).

III.   SUMMARY

Owing to a deep burial, compaction and cementation were dominant to dissolution. But in certain conditions (geotemperature,pore water chemistry and pressure-dissolution),porosity and permeability can be improved to be effective by dissolution of some compositions.

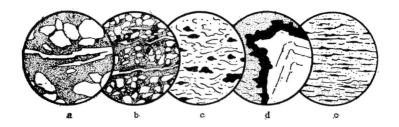

Fig. 2.   Contractive and dissolved cracks in mudstone in thin sections.

    Precipitation, dissolution and replacement have repeated for many times not only on dissoluble carbonate and sulfate minerals, but also on indissoluble clay minerals, feldspar and even quartz.

REFERENCES

    Wescott,W.A. (1983). AAPG Bulletin.67,6.

DARCY'S AND CAPILLARY-PRESSURE LAWS DERIVED FROM SPACE AVERAGING FOR IMMISCIBLE TWO-PHASE FLOW IN POROUS MEDIA

D. Pavone

Institut Francais Du Petrole
France

I. INTRODUCTION

Porous media are characterized by several length scales. The microscopic pore scale is the most fundamental one for flow. Nevertheless, measurements are made on a length scale of hundreds of pores (millimetric scale), and observations need a larger scale, extending from centimeters to decimeters.

The microscopic level must be investigated by using idealized and well-known mathematical structures, such as bundles of nonintersecting tubes [1] or two-dimensional networks [2]. Statistical approaches may involve the percolation theory [3] or fractal geometry [4].

Space averaging is different. It was introduced to give a theoretical justification of the Darcy and capillary pressure equations for two-phase flow. Space averaging does not required geometrical assumptions, but flow parameters, such as permeability, cannot be derived from space averaging calculation. Space averaging leads to macroscopic equations but fails to predict constitutive parameters, which must be evaluated empirically. Attempts made to justify these laws can be divided in two groups. The first uses volume averaging over a representative elementary volume [5], the second assumes that porous media are periodic and that space averaging is done over a period [6].

II. MACROSCOPIC EQUATIONS

Details about the research described here can be found in [7,5]. Starting from mass, momentum and energy balance equations, which are proved on the microscopic scale, equations on a macroscopic scale are derived.

Firstly, microscopic values for one phase are extended to the whole medium. Values are set at zero everywhere but in their own phase. Microscopic equations can be written everywhere if derivatives are taken in the sense of distributions. Secondly, space averaging of the microscopic equations is performed over a representative elementary volume. Following Marle [5], if (f) is a locally defined function (called a microscopic function), an averaging technique $(F = f * m)$ that involves a weight function m is introduced. F is called the macroscopic function. If m is of class C^∞, F is of class C^∞ too, which is very convenient.

RESERVOIR CHARACTERIZATION II

683

Macroscopic functions are defined in such a way that macroscopic equations look very much like microscopic ones. Thirdly, to solve the problem of closure, the Thermodynamics of Irreversible Processes (TIP) is used to produce phenomenological equations.

III.   PHASE a MASS BALANCE EQUATION

By starting from the fundamental mass balance equation for fluid a and by averaging this equation, the following macroscopic mass balance equation can be derived:

$$\frac{\partial (\Phi S_a R_a)}{\partial t} + \vec{\nabla} \cdot \left(R_a \vec{U}_a \right) = 0 \tag{1}$$

($\Phi = 1 - \chi_s * m$) is the porosity. ($\chi_i = 1$ in phase i and zero elsewhere). The saturation (S_a) is defined by ($\Phi S_a = \chi_a * m$). The macroscopic density (R_a) is defined by ($\Phi S_a R_a = \rho_a * m$), where ρ_a is the microscopic density. The mean filtration velocity ($\vec{U}_a$) is defined by ($R_a \vec{U}_a = (\rho_a \vec{v}_a) * m$), where $\vec{v}_a$ is the local velocity.

IV.   DARCY'S AND CAPILLARY PRESSURE LAWS

The TIP is used to solve the problem of closure as follows. The entropy source is presented in a bilinear form. If all terms of the form are zero at the equilibrium state, the TIP concludes that linear phenomenological equations exist between these terms. This leads to the conclusion that there are three rank two tensors (K_{aa}, K_{ab} and K_{aq}) such that:

$$\vec{U}_a = K_{aa} \left[-\vec{\nabla} (P_a) + R_a \vec{g} \right] + K_{ab} \left[-\vec{\nabla} (P_b) + R_b \vec{g} \right] + K_{aq} \vec{\nabla} (1/T) \tag{2}$$

Where P_i is the pressure in phase i, $\vec{g}$ is gravity and T is temperature. Eq. (2) is a Darcy-like equation for two-phase flow in porous media, with coupling between the two phases and the temperature gradient. The TIP says that tensors (K_{ii}) depend on all variables except on their derivatives. Of course, we can introduce the permeability tensor K, the relative permeability k_{ra} and the viscosity μ_a ($K_{aa} = K k_{ra}/\mu_a$), but it is not possible to say: i) that K and k_{ra} do not depend on μ_a and ii) that k_{ra} depends only on the saturation.

From the TIP, we can ascertain that there is a parameter ($\pi \geq 0$) such that:

$$P_a - P_b = \gamma_{ab} \cos\theta \frac{\partial A_{as}}{\partial (\Phi S_a)} + \gamma_{ab} \frac{\partial A_{ab}}{\partial (\Phi S_a)} + \pi \frac{\partial (\Phi S_a)}{\partial t} \tag{3}$$

γ_{ab} is the interfacial tension, θ is the constant wettability angle, A_{as} (A_{ab}) is the interfacial area between phase a and the solid (phase b). t is time. Eq. (3) is a capillary pressure equation. The first two terms look very much like the Young-Dupré microscopic equation terms; the third is a dynamic term. Capillary pressure appears to be twofold, i.e. i) a static capillary pressure taking into account the interface areas created beetwen the liquid/liquid and liquid/solid phases, as well as ii) a dynamic capillary pressure. Note that we do not obtain the standard equation $P_a - P_b = P_c(S_a)$.

Capillary pressure laws are almost identical on the microscopic scale (capillary) and on the macroscopic scale (porous media). Nevertheless, the example given

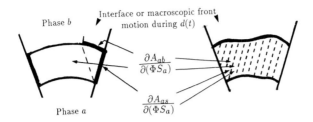

Fig. 1: Location of the interfacial areas involved in the capillary pressure
in a conical capillary and in a conical sample.

in Fig.1 shows that the locations of the areas involved in the capillary pressure law are different. The Laplace law, which is proved in a capillary, is certainly not valid in porous media, such as for the Chuoke equation [8].

V. SUMMARY AND CONCLUSIONS

This paper shows how the space averaging of microscopic equations, valid at the pore level, is used: i) to define macroscopic variables, and ii) to find equations that link these variables. In addition to balance equations, two phenomenological laws are derived from the Thermodynamics of Irreversible Processes. One is a Darcy-like equation; the other one is a capillary-pressure equation.

The Darcy-like equation is the well known "generalized Darcy's law" plus couplings between the flow of both phases and the temperature. The static part of the capillary pressure law looks very much like the microscopic Young-Dupré law. The dynamic part of capillary pressure is proportional to the time derivative of the saturation.

The static capillary pressure is compared in a capillary and in a porous medium. Even with identical laws, capillary pressure is different in these two cases.

REFERENCES

[1] Dullien, F.A.L. (1975). A.I.Ch.E. J., 21,299.

[2] Mohanty, K.K., Davis, H.T. and Scriven, L.E. (1980).
Proc. 55th Fall Technical Conf. of the SPE, SPE 9406.

[3] Lenormand, R and Zarcone, C. (1985). Phys. Rev. Letters. V. 54, N. 20.
2226-2229.

[4] Jacquin, C.G. and Adler, P.M. (1987). T.I.P.M. 2-6, 571-596.

[5] Marle, C.M. (1982). Int. J. Engng. Sci., 20,5:643-662.

[6] Auriault, J.L. and Sanchez-Palencia E. (1986). J. of Theor. and Appl.
Mech., no special, pp 141-156.

[7] Pavone, D. (1989). Rev. Inst. Franc. Petr., 44, 1, 29-41.

[8] Chuoke, R.L., van Meurs, P. and van der Poel, C. (1959). Trans. AIME,
216, 188-194.

GEOLOGICAL KNOWLEDGE IN THE GEOSTATISTICAL MODELING OF RESERVOIR PROPERTIES

J. L. Rudkiewicz
R. Eschard

Institut Francais du Petrole
Rueil-Malmaison, France

and Group Heresi

Institut Francais du Petrole
and Ecole des Mines de Paris/
Centre de Geostatistique

1. THE STOCHASTIC MODEL HERESIM

The stochastic model **HERESIM** (an acronym for Heterogeneous Reservoir Simulation) has been developed to extrapolate high resolution data known at the wells from log and core analysis onto the whole reservoir.

In most silici-clastic reservoirs, either fluvial, deltaic, tidal or even turbiditic, heterogeneity is mainly related to sharp differences between lithofacies. A realistic stochastic simulation of such reservoirs must be consistent with a geological scheme, based on depositional concepts. Thus HERESIM generates a regular high resolution grid of lithofacies, coded as discrete values, prior to any lithology linked properties, such as porosity and/or permeability. The cell's size is suited to the size of the heterogeneity to be modeled : in our examples, we used 0.5 m vertically and 2-4 m horizontally. Our poster precisely describes the input parameters used to generate the lithology grid, and shows how they can be linked to geological knowledge.

The reservoir lithology is modeled with a geostatistical formalism, and represented as a discrete stochastic function (1)(2). Let us just recall that a conditioned multigaussian function is simulated, then truncated into lithofacies at threshold points.

A reservoir is usually divided in several significantly different geological units, limited by continuous surfaces (unconformity ...), which may represent time-lines. Each unit is deposited in an homogeneous sedimentary environment, that HERESIM models with a unique set of parameters :
1. lithology at each well
2. structure of correlation lines
3. global ratio of each lithology
4. vertical and horizontal correlation length.

2. GEOLOGICAL AND STATISTICAL PARAMETERS

The *lithology at each well* is usually deduced from log and core analysis. We illustrate this step with combined analysis of cores and gamma-ray logs, based on field data from the Middle Jurassic Ravenscar Group (Yorkshire, Great-Britain).

The *correlation lines* inside each geological unit and their relations to the limits are important geological features. They can be either conformable or unconformable and represent various depositional styles : tabular, erosion, truncation ...

Usually statistical parameters are deduced from the analysis of large data sets. At the appraisal stage of a field, only few wells have been drilled, which may not provide enough data to fulfil the correct statistical inference of parameters.

The *global ratio of each lithology* is more or less an extension of the net/gross ratio. It gives the proportion of each lithofacies along every correlation line. It can be computed from only few well data, once the correlation lines have been defined. We show how proportions along correlation lines are used. The shape of the computed proportion curves may indicate the sedimentary environment. They depend on the external geometry and the sedimentary nature of the sandstone bodies (braided or meandering channel, crevasse splay or littoral barrier).

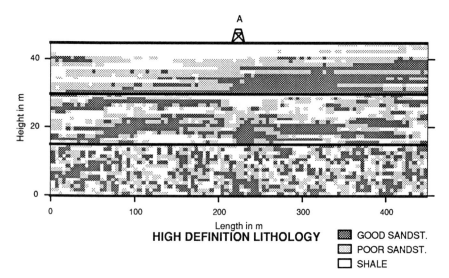

Figure 1: Stochastic simulation of a reservoir divided into three tabular sub-units. From bottom to top, the horizontal correlation length are respectively 10, 100 and 1000 m, all other parameters are unchanged.

The *vertical and horizontal correlation length* or *ranges* are used to compute the correlation between two points as a function of the distance that separates them. This function is called the variogram. At the appraisal stage, there are ususally too few data to infere a correct horizontal variogram. Then, geological assumptions play a major role. With different variogram ranges, either disconnected or highly correlated bodies may be simulated, once the lithology ratios are known (Fig. 1).

REFERENCES

1. Matheron, G., Beucher, H., de Fouquet, C., Galli, A., Guerillot, D., Ravenne, C. (1987). SPE 16753, Dallas.

2. Rudkiewicz, J.L., Guerillot, D., Galli, A., and *Group Heresi*. (1989). to be published in "North Sea Oil and Gas Reservoirs: an International Conference". Graham and Trotman Ltd. London.

A CROSSPLOT TECHNIQUE FOR DISCRIMINATION OF VARIOUS SANDSTONE FACIES IN BARRIER ISLAND SANDSTONE DEPOSITS

Bijon Sharma
Susan Jackson
IIT Research Institute
National Institute for Petroleum and Energy Production Research
Bartlesville, OK 74005

ABSTRACT

A crossplot technique has been developed which discriminates between important barrier and nonbarrier sandstone facies in a barrier island sandstone in Bell Creek (MT) field based on interpretation of density, resistivity, and gamma ray logs.

A distinct range of rock and reservoir properties, such as grain size, clay content and water saturation, occurs in barrier and nonbarrier facies due to differing depositional processes. For example, clay-rich lower shoreface and lagoonal sandstones tend to have higher water saturations compared to sandstones from higher productive foreshore and shoreface facies. Move productive facies usually also tend to be better sorted due to wave-dominated deposition and exhibit much higher porosities and permeabilities. In this technique, the formation resistivity, R_t, determined from induction or any other deep penetration logs is plotted against the corresponding porosity (ϕ) values. Porosity

Work performed for the U. S. Department of Energy under cooperative agreement DE-FC22-83FE602149.

689

can be determined from sonic, density or neutron logs, but in sandstone reservoirs such as Bell Creek field, which is locally very clayey, best porosity estimates can be obtained from interpretations of density logs.

In Figure 1 density log derived porosity values have been sequentially plotted against corresponding deep resistivity values for well W-4 in Unit 'A' of Bell Creek field. Major facies boundaries in well W-4, determined from core descriptions and comparison with neighboring wells, can be identified by distinct linear patterns on the plot. Alluvial, valley-filled sediments, high-productive barrier island sandstones, and lower productive lower shoreface sandstones exhibit distinct patterns on the crossplot. For finer facies distinction and verification of results from the ϕ, R_t crossplot, plots of gamma ray (GR) versus porosity (ϕ) have proved very useful. The porosity vs. gamma ray crossplot for well W-7 (Figure 2) illustrates the resolution obtained by ϕ, GR crossplot in separating different facies.

Twenty-six crossplots of wells in Units 'A', 'B', and 'C' of Bell Creek field were generated, and from the facies identified by the crossplot a dip-oriented stratigraphic section was constructed (Fig. 3). The product of average porosity values for discrete zones and the thicknesses of the zones indicates storage capacities of various zones in a facies. The stratigraphic section depicts the spatial distribuiton of the different rock units with distinct petrophysical properties.

This crossplot technique may be used to construct facies models in barrier island deposit. These models can be used in planning extension drilling, waterfloods, or enhanced oil recovery projects.

For better predictions of productivity of sandstones in a barrier island depositional environment, the distribution of clay content, porosity, permeability, and oil saturation can be studied in each facies individually before the total effect from all facies is evaluated at a particular location.

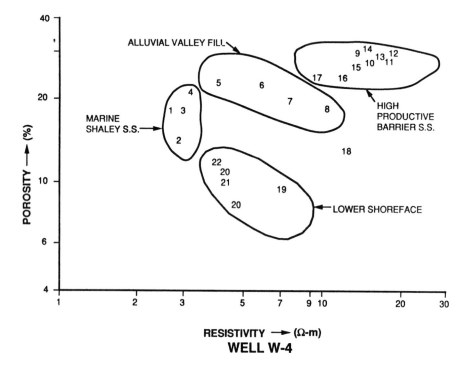

Figure 1. Depth sequential crossplot of resistivity and
porosity to distinguish various sandstone facies
in well W-4.

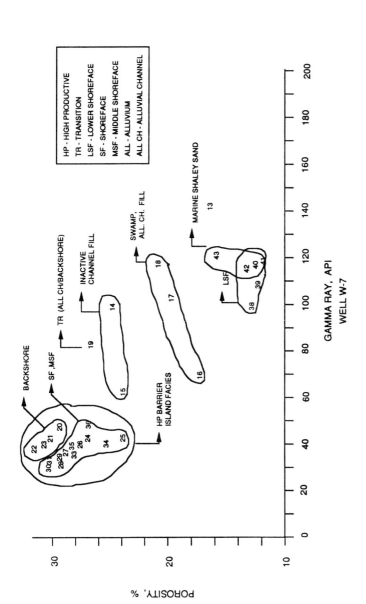

Figure 2. Depth sequential crossplot of gamma ray and porosity to distinguish various sandstone facies in well W-7.

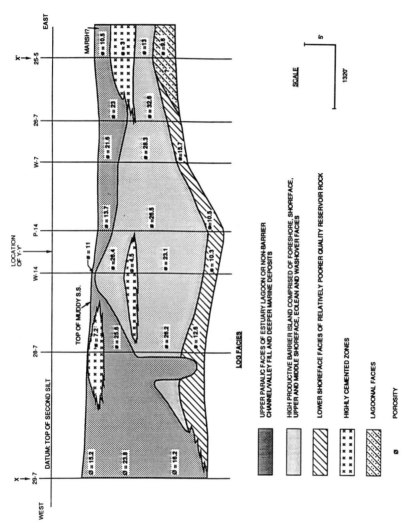

Figure 3. Dip-oriented stratigraphic section across barrier island deposit at Bell Creek obtained from facies analysis.

DRILLING FLUID DESIGN BASED ON RESERVOIR CHARACTERIZATION

Michael P. Stephens

M-I Drilling Fluids Company
Houston, Texas

I. INTRODUCTION

The interactions between drilling fluid and the hydrocarbon reservoir can be key factors controlling the productivity of an oil or gas well. An inappropriate drilling fluid composition can adversely affect porosity, permeability, and wettability of the reservoir in the near-wellbore region. To design a drilling fluid to maximize production, the reservoir should be characterized with respect to: 1) reservoir fluid content and composition; 2) pore geometry and size; and, 3) mineral content and distribution. Drilling fluid design to maximize productivity incorporates controlling both the chemistry of the liquid phases and the sizes of particles suspended in the fluid. The liquid phases of the drilling fluid should be chosen to prevent destabilization of reservoir mineral material as well as to prevent precipitation of solids or formation of emulsions when reservoir fluids and drilling fluid filtrates are mixed. The solid phases of the drilling fluid should be chosen to yield an appropriate size distribution to prevent the invasion of particles into the pore network of the reservoir. Using reservoir characterization, drilling fluids can be formulated to minimize the risk of damaging production and to maximize the rate of production and recovery of hydrocarbons.

II. RESERVOIR FLUID CONTENT AND COMPOSITION

Oil, gas, and brine comprise the reservoir fluid content. Drilling fluid filtrates have the potential to interact with certain components of each phase of reservoir fluid.

Gas frequently contains carbon dioxide. In reservoirs with high carbon dioxide content, the interaction with calcium in the drilling fluid filtrate can precipitate calcium carbonate scale in the reservoir porosity. In addition, highly alkaline mud filtrates can shift carbonate chemical equilibria to increase concentration of carbonate ion in the

RESERVOIR CHARACTERIZATION II
Copyright © 1991 by Academic Press, Inc.
All rights of reproduction in any form reserved.

water phase. This increase in dissolved carbonate can also result in the precipitation of calcium or magnesium carbonate scale in the reservoir pore space decreasing permeability in the near wellbore region.

Many of the younger, less mature crude oils contain significant quantities of organic acids. The organic acid components of the oil can react with highly alkaline mud filtrate to create surface active compounds that can alter reservoir wettability and decrease interfacial tension between oil and connate brine. The decrease in interfacial tension can allow an emulsion to form in the reservoir pore space blocking the permeability in the near wellbore region. A reservoir wettability reversal from preferentially water wet to preferentially oil wet decreases relative permeability to oil in the near wellbore region.

Connate formation brines contain dissolved ions that vary from one reservoir to another. Individual ion components of the connate brine can precipitate a variety of scales when mixed with drilling fluid filtrates. Unless specifically tailored to exclude certain ions, drilling fluid filtrates are likely to contain sodium, potassium, calcium, magnesium, hydroxide, carbonate, sulfate and chloride ions in appreciable quantities. Common components of connate formation waters that lead to precipitation of scale include calcium, magnesium, sulfate, and carbonate.

The drilling fluid formulations available to avoid formation damage caused by incompatibility between reservoir fluids and drilling fluid filtrates include low filtrate loss drilling fluids, oil based drilling fluids, low alkalinity drilling fluids, and chemically controlled drilling fluid and filtrate compositions.

III. PORE GEOMETRY AND SIZE

The invasion of drilling fluid particles into the reservoir pore space has long been recognized as a cause of formation damage. The size and geometry of reservoir pore spaces plays an important role in determining whether drilling fluid particles will damage the permeability in the near wellbore region. Conventional drilling fluids are designed to deposit a filter cake at the face of the well and prevent drilling fluid particles from invading the pore spaces of the reservoir. The drilling fluid must contain a distribution of particle sizes appropriate to the size and geometry of the reservoir pore openings for proper filter cake development.

Typical drilling fluids contain a particle size range (1 - 50 micrometers) that will bridge and properly form a filter cake on sandstone with pores diameters up to about 120 to 150 micrometers. Very coarse sand or conglomerate reservoir zones may be damaged by invasion of drilling fluid particulate material because the pore openings are larger and the drilling fluid does not bridge. Reservoirs with a significant amount of fracture permeability may also be subject to significant damage from drilling fluid particulate invasion depending upon the fracture width. Limestone reservoirs exhibit a variety of porosity patterns that may inhibit proper filter cake development. If the filter cake does not properly develop and protect the reservoir from invasion of mud particles then the permeability in the near wellbore region can be severely damaged.

There are a variety of drilling fluid technologies that can be applied to minimize formation damage from drilling fluid particle invasion of the reservoir. For zones with larger pore openings, a number of size graded materials known collectively as lost circulation additives can be incorporated into the drilling fluid to bridge the larger pore openings. Drilling fluids based on sized salt, with water soluble particles, have been applied to coarse zones. Acid soluble drilling fluid formulations have been used when reservoir acidization is contemplated. Clear brine drilling fluids without particulate additives have been successfully used in reservoirs with fracture permeability.

IV. MINERAL CONTENT AND DISTRIBUTION

The interaction between clay mineral components of the reservoir and drilling fluid filtrate may cause a reduction in permeability of the near wellbore region. While some clay containing reservoirs are unaffected by drilling fluid filtrate, both clay fines migration and clay swelling can affect reservoir permeability.

Pore lining authigenic clay is probably most affected by aqueous drilling fluid filtrate. Quantities of authigenic clay lining the pore openings as low as 2 percent of the total mineral content can be mobilized by drilling fluid filtrate. When this clay becomes mobilized it can lodge in pore constrictions and severely damage permeability. Mobilization of pore lining clay is promoted by low salinity, high alkalinity, and the presence of clay dispersant in the filtrate.

Clay masses present in oil and gas reservoirs consist of clasts of clay material and mineralogically unstable grains that have diagenetically altered to clay. These clay masses are also able to reduce permeability when they interact with aqueous drilling fluid filtrate. Both swelling of smectite components of these clays and dispersion of clay masses can contribute to permeability damage. Swelling and dispersion of these clay masses depends on mineralogical cementation of the clay masses. Clay masses that are cemented with silica or carbonate are much less reactive to filtrate than uncemented clay masses. Swelling and dispersion of clay masses are promoted by low salinity, high alkalinity and the presence of clay dispersant in the filtrate.

Drilling fluid formulations that are used to minimize swelling and dispersion of clays in the reservoir include low filtrate loss fluids, non dispersed polymer based fluids, potassium based fluids, high salinity drilling fluids, low alkalinity fluids, and oil based drilling fluids.

V. SUMMARY AND CONCLUSION

Basic elements of reservoir description are important for proper drilling fluid design. Characterization of reservoir fluids, porosity, and mineral content can be used to tailor drilling fluid composition and properties to maximize hydrocarbon production.

GEOSTATISTICAL CHARACTERIZATION OF RESERVOIRS FOR OPTIMAL FIELD DEVELOPMENT

Mojtaba Taheri
Marcel Chin-A-Lien
Eduardo Rodriguez

INTEVEP
Venezuela, South America

I. INTRODUCTION ·

Many problems encountered in field development are consequences of reservoir heterogeneities caused by variations in depositional architecture and diagenesis, and complexities inherent in the interpretation of multicomponent data assemblages of different nature, scale, order of magnitude and realibility.

To overcome these problems and to understand the combined influence of geological, petrophysical and production parameters on reservoir behavior, we propose to use the concept of the multivariate statistics to analyse such multicomponent information simultaneously, taking into account the effect of their combined interaction.

These methods require treatment of each parameter and thereafter normalization of factors through a particular process such as factor analysis, principal component analysis, etc. Then the behavior of each reservoir parameter is studied by establishing its spread, spatial distribution, background values and by correlating their empirical distribution with the different depositional environments.

The quantitative models elaborated this way serve as a framework for Cluster Analysis and Characteristic Analysis, which permit to establish the relative importance of subsurface parameters, and enable to characterize the quality and potential of reservoirs for optimal field development.

These methods were applied in the Arecuna Field, located in the Orinoco Oil Belt in eastern Venezuela. The most important oil accumulations in this area are found in the basal sands of the Oficina Formation, which were deposited during the second transgressive-regressive Miocene cycle. Based on the sedimentological framework and facies relationships, it has been interpreted that the productive sands were deposited within a fluvio-deltaic setting, and were derived from rocks of the Guayana Shield in the south.

II. RESERVOIR QUALITY AND POTENTIAL

In Arecuna field, more than 45 different characteristics or factors in 25 productive horizons were analized. From the large data set of parameters, 7 were selected as being the most indicative and important, and that contained sufficient data for statistical evaluation. From the 25 oil bearing horizons studied, the S3-4 interval is chosen to illustrate the methodology used in this study.

In this interval 4 main genetic facies are recognized: distributary channel fill and associated levee, delta front and coastal barrier, creavasse splay, and flood plain. The most prolific oil saturated sands are found in a vast elongated east-west belt, in north-central Arecuna, where delta front and channel facies interfinger. The porosity, permeability and shale content are genetically controled by the associated facies with the best values clustered in the delta front and central channel facies.

Cluster Analysis is used to correlate the reservoirs based on assemblages of different subsurface properties, which permitted to delineate and visualize 4 different groups of sand qualities. These groups are denominated in decreasing order of quality as A, B, C and D respectively (Table I).

TABLE I. Sand quality characteristics, Arecuna field

	QUALITY			
PARAMETER	A	B	C	D
Net sand (ft)	> 40	40-25	25-15	< 15
Net oil sand (ft)	> 35	35-20	20-10	< 10
Porosity (%)	> 29	29-25	25-22	< 22
Oil saturation (%)	> 85	85-70	70-60	< 60
Permeability (md)	> 2000	2000-1000	1000-500	< 500
Resistivity (ohm/m)	> 50	50-25	25-15	< 15
Shale content (%)	< 5	5-15	15-25	> 25

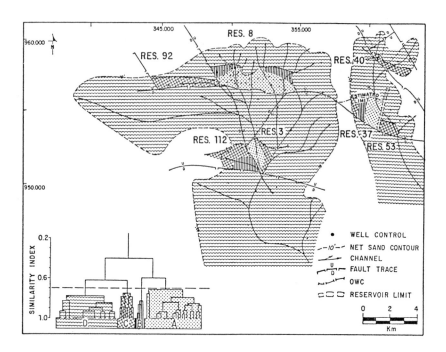

Fig. 1.   Reservoir quality, S3-4 interval, Arecuna field

The results are presented in a reservoir quality map (Fig.
1), which in addition indicates the areal sand development,
fault traces and oil-water contacts. As can be seen, changes
in quality from zone to zone are generally transitional and
only in the presence of a fault or an oil-water contact is
the variation in quality abrupt.

The second multivariate statistics (Characteristic
Analysis) is used to compare and rank the reservoirs, based
on combination of seven sand properties, and also to identify
the relative contribution of each parameter on reservoir
potential. This analysis indicates that porosity is the most
important and determinant factor, while the shale content has
the least significant effect on reservoir productivity.

III.   CONCLUSIONS

The quantitative approach developed in this study,
allowed to classify and map the sand qualities and to determine the degree of influence or contribution of subsurface
parameters on reservoir potential in Arecuna field. This
methodology can be applied as well to other areas for the
optimum selection and development of prospective zones.

SANDBODY GEOMETRY IN FLUVIAL SYSTEMS - AN EXAMPLE FROM THE JURASSIC OF YORKSHIRE, ENGLAND

Olav Walderhaug
Rune Mjos

Rogland Research Institute
Stavanger, Norway

The geometric characterization of subsurface reservoirs can be improved by outcrop studies of analogs. The Middle Jurassic Ravenscar Group of Yorkshire, England, contains possible analogs of the fluvial sediments within the Brent Group of the North Sea.

The fluvial sediments in Yorkshire dominantly contain two types of sandbodies; fluvial channel sandstones and crevasse splay sandstones. Fluvial channel sandstones of the Saltwick Formation have widths of 100 - 700 m and thicknesses of 4 - 19 m (Fig. 1). The channel sandstones in the Saltwick Formation dominantly formed by vertical accretion, and the thicker sandstones consist of several stories. The ratio between width and thickness is between 15 and 40 for 85 % of these channel sandstones (Fig. 1). Fluvial channel sandstones in the Scalby Formation dominantly have widths of around 100 m and less and thicknesses between 4 m and 12 m (Fig. 1). The channel sandstones in the Scalby Formation may be subdivided into: 1. ribbon sandstones dominated by vertical accretion, thicknesses of 4 - 12 m and width/thickness ratios below 15, 2: laterally accreted channel sandstones with thicknesses of 4 - 6 m and width/thickness ratios of 20 - 25, and 3. channel sandstone complexes built up of several ribbon sandstones and with a total width of several hundred m and width thickness ratios of around 35.

Crevasse splay sandstones form the Saltwick, Scalby and Cloughton Formations have thicknesses of up to 2.5 m, widths up to 2.5 km and maximum lengths of 3 - 4 km (Fig. 2). Ratios between width and thickness and length and thickness are 100

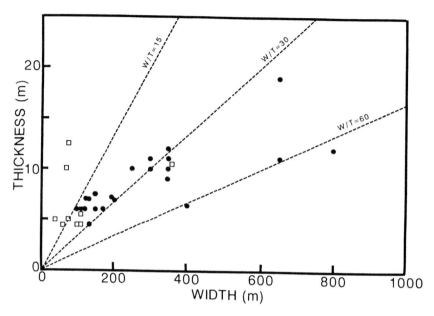

Fig. 1. Thickness plotted against width for fluvial
channel sandstones. Open squares represent measurements from
the Scalby Formation, measurements from the Saltwick
Formation shown by dots.

- 2000 and 500 - 2000 respectively. Thickness of individual
crevasse splay lobes remains relatively constant throughout
most of the lobe. The Cloughton Formation also contains composite crevasse splay sandstones (subdeltas) with thickness
up to 6 m and lateral extent greater than 6.5 km. The composite crevasse splay units are built up of up to 1.5 m thick
shoestring sandstones with widths of around 0.5 km and rooted
tops.

Crevasse splay sandstones in the Ravenscar Group are probably in most cases in contact with fluvial channel sandstones deposited in the fluvial channels that fed the crevasse
systems, but crevasse splay sandstones were also observed to
be in contact with fine-grained channel-fill deposits. Crevasse splay sandstones are commonly cut by younger fluvial
channel sandstones.

Petrographic study of 50 thin-sections from the studied
sandstones indicates that porosities in channel sandstones
are on average around 5 % higher than in crevasse splay sandstones. This is mainly due to greater amounts of quartz overgrowths in the crevasse splay sandstones. Increased quartz

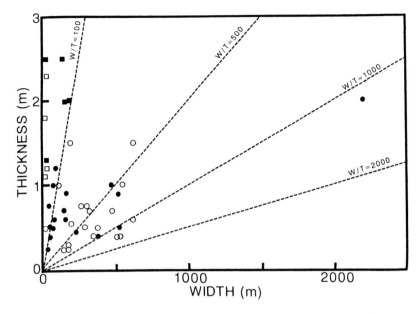

Fig. 2. Thickness plotted against width for sections through crevasse splay sandstones from the Saltwick, Cloughton and Scalby Formations. Squares represent sections through crevasse splay channels, circles represent sections through crevasse splay lobes. Open symbols represent cases where full width could not be measured due to termination of exposures.

cementation in the crevasse splay sandstones may be due to their greater content of clay lamina which evolved into sty-lolites during burial and to diffusion of dissolved quartz from enclosing siltstones into the thin crevasse splay sand-stones.

The determined width/thickness ratios and dimensions of fluvial channel sandstones and crevasse splay sandstones in-creases our ability to make geometric predictions outwards from wells, increases the reliability of correlations between wells, and also provides constraints on the dimensions and shapes of sandbodies. Incorporation of this type of informa-tion into a reservoir model will reduce the uncertainty in the model and contribute to more reliable predictions of reservoir performance. The relevance of the presented data and data from other outcrop studies for a given subsurface reservoir must, however, be carefully evaluated in each specific case.

Analyzing Permeability Anisotropy with a Minipermeameter

ABSTRACT

Gordon R. Young
Manmath N. Panda
Larry W. Lake

Department of Petroleum Engineering
The University of Texas at Austin
Austin, Texas 78712

Inasmuch as potential gradients and velocity vectors are not necessarily co-linear during flow through permeable media, permeability must be a tensorial property. The complete tensorial character is generally unacknowledged, usually because of ensuing mathematical complexities; consequently, the impact of anisotropy on fluid flow is unclear. Moreover, laboratory measurement of anisotropy is usually so time-consuming and expensive that only scalar estimates are extracted. This poster outlines a new procedure for estimating permeability anisotropy with a minipermeameter.

The minipermeameter is a rapid and non-destructive means of measuring gas permeability using relatively simple apparatus and an interpretation procedure which is based on a general form of Darcy's law. An isotropic estimate of permeability follows from the application of the theoretical development given in Goggin et al.[1] This work converted the basic gas flow rate and tip pressure measurements of the minipermeameter into permeability through the use of a shape factor which depends only on .the tip and sample geometry. The isotropic procedure also allows for corrections for gas slippage and high-velocity flow effects.

The anisotropy procedure consists of using an elliptical tip which has different shape factor values which now depends on the permeability ratio (k_x/k_y) and also on whether the tip is oriented along the major or the minor axes of the permeability tensor. Further details are in Young.[2] We demonstrate the basics of the procedure (neglecting non-Darcy corrections) and illustrate typical results on four core plug samples of varying degrees of visual cross-bedding. The permeability ratio varies between 1.3 and 3.5. Our analysis indicates that cross-bedding can be an unreliable indicator of anisotropy.

[1]Goggin, David J., Richard Thrasher and Larry W. Lake, "A Theoretical and Experimental Analysis of Minipermeameter Response Including Gas Slippage and High Velocity Flow Effects," In Situ, 12, nos. 1 and 2,1988, pp. 79-116.

[2]Young, Gordon R., "Determining Permeability Anisotropy from a Core Plug Using a Minipermeameter, M.S. Thesis, May 1989.

RESERVOIR CHARACTERIZATION II
Copyright © 1991 by Academic Press, Inc.
All rights of reproduction in any form reserved.

Index